STUDIES IN LOGIC

AND

THE FOUNDATIONS OF MATHEMATICS

VOLUME 135

ELSEVIER
AMSTERDAM • LAUSANNE • NEW YORK • OXFORD • SHANNON • TOKYO

LOGICAL FRAMEWORKS FOR TRUTH AND ABSTRACTION

An Axiomatic Study

LOGICAL FRAMEWORKS FOR TRUTH AND ABSTRACTION

An Axiomatic Study

Andrea CANTINI
Department of Philosophy
University of Florence
Florence, Italy

N·H

1996

ELSEVIER
AMSTERDAM • LAUSANNE • NEW YORK • OXFORD • SHANNON • TOKYO

ELSEVIER SCIENCE B.V.
Sara Burgerhartstraat 25
P.O. Box 211, 1000 AE Amsterdam, The Netherlands

ISBN: 0 444 82306 9

This book is printed on acid-free paper.

Transferred to digital printing 2006.

PREFACE

This book is concerned with logical systems, which are usually termed *type-free* or *self-referential* and emerge from the traditional discussion on logical and semantical paradoxes. We will consider non-set-theoretic frameworks, where forms of *type-free abstraction* and *self-referential truth* can consistently live together with an underlying theory of combinatory logic.

However, this is *not* a book on paradoxes; nor we aim at a grand logic à la Frege-Russell, inspired by a foundational program. We shall rather investigate type-free systems, in order to show that:

(i) there are rich theories of self-application, involving both operations and truth, which can serve as foundations for property theory and formal semantics;

(ii) these theories give new outlooks on classical topics, such as inductive definitions and predicative mathematics;

(iii) they are promising as far as applications are concerned.

This way of looking is justified by the history of the antinomies in our century. In spite of isolated foundational and philosophical traditions, the research arising from paradoxes has become progressively closer to the mainstream of mathematical logic and it has received substantial impulse during the last twenty years: a number of significant developments, techniques and results have been cropping up through the work of several logicians (see below for our main debts). Therefore a major aim of this book is to attempt a unifying view of relevant research in the field, by dwelling on connections with well-established logical knowledge and on applicable theories and concepts.

However, the present work is far from being comprehensive. We do not treat illative combinatory logic (with the exception of a system of Ch.VI, investigated by Flagg and Myhill 1987), nor we deal with the Barwise-Etchemendy approach to self-reference via non-well-founded sets. Another significant direction, which is only touched upon in two sections of chapter XIII, is the systematic development of the general theory of semi-inductive definitions (in the sense of Herzberger, Gupta and others).

The project started some years ago, when Prof. A. S. Troelstra kindly suggested an English translation of the author's monograph (Cantini 1983a) about theories of partial operations and classifications in the sense of Feferman (1974). The attempted translation soon shifted towards a thorough expanded revision of the old text, and eventually gave rise to an entirely new set of notes at the end of 1988. After a stop of almost two years, these notes were taken up again, fully rewritten and reorganized. The manuscript was submitted to the editor for final refereeing in October 1993.

The content and the results of the present version are disjoint from the 1983 monograph; they partly overlap with the 1988 notes, except for a different choice of primitive notions and for the addition of Ch.VI, parts of Ch.XIII and the epilogue. Ch.VIII offers a development of topics, contained in the author's paper "Levels of Truth" (to appear in the *Notre Dame Journal of Formal Logic*, 1995): we thank the Editors for granting the permission of using parts of that paper in Ch.VIII of this book.

Acknowledgments. The present work owes a great deal to the writings of several logicians, and even if I tried hard to make a complete list of my debts in the text and in the reference list, I am sure that there are omissions: I apologize for them.

As to the proper content of the book, pertaining to type-free abstraction and self-referential truth, I would like to underline my intellectual debt with the following papers (listed in alphabetical order): Aczel (1980), Feferman (1974), (1984), (1991), Fitch (1948), (1967), Friedman and Sheard (1987), Kripke (1975), Myhill (1984), Scott (1975).

Prof. W. Buchholz offered an invaluable help in correcting errors of any kind and in proposing technical improvements. I owe a special thank to him, also because the topics I dealt with were not touching his main research interests.

I am grateful to Prof. S. Feferman and to Prof. G. Jäger for keeping me informed over the years about their own research on type-free systems and proof theory, and for important advice. Jäger's Ph.D. student, T. Strahm made useful critical comments on the first chapter.

Dr. R. Giuntini and Dr. P. Minari undertook the final proof-reading of chapters I-VIII and XII-XIV; I warmly thank them for a host of useful remarks and corrections.

I am deeply indebted to Dr. A. P. Tonarelli for proof-reading the remaining chapters and for eagle-eyed assistance in the unrewarding task of preparing the final manuscript.

Of course, I must stress that I am fully responsible for all errors, to be found in the whole work.

I am grateful to the Alexander-von-Humboldt Stiftung (Germany) for granting me a "Wiederaufnahme" of a research fellowship at the Ludwig-Maximilians-Universität München in Sommer Semester 1991, when the present work was at a difficult stage.

Partial support to the present project was granted by the Italian National Research Council (CNR) and the Italian Ministry for University, Scientific Research and Technology (MURST).

Last but not least, this work is dedicated to my children Giulia and Francesco.

Firenze, April 1995

CONTENTS

Preface		V
Contents		IX
Introduction		1
PART A: COMBINATORS AND TRUTH		11
I	Introducing operations	13
	1. The basic language	14
	2. Operations I: general facts	15
	3. Operations II: elementary recursion theory	18
	4A. The Church-Rosser theorem	22
	4B. Term models	26
	5. The graph model	28
	6. An effective version of the extensional model D_∞	34
	Appendix	39
II	Extending operations with reflective truth	43
	7. Extending combinatory algebras with truth	45
	8. The theory of operations and reflective truth: simple consequences	51
	9A. Type-free abstraction, predicates and classes	55
	9B. Operations on predicates and classes	59
	10A. The fixed point theorem for predicates	63
	10B. Applications to semantics and recursion theory	68
	11. Non-extensionality	73
	Appendix I: a property theoretic definition of the fixed point operator for predicates	76
	Appendix II: on the explicit abstraction theorem	77
	Appendix III: independence of truth predicates from the encoding of logical operators	80

PART B: TRUTH AND RECURSION THEORY 83

III Inductive models and definability theory 85
12. Inductive models and the induction theorem 86
13. The envelope of an inductive model 88
14. The uniform ordinal comparison theorem for inductive models 91
15. Applications of the uniform ordinal comparison theorem 97

IV Type-free abstraction with approximation operator 103
16. Approximating properties by classes 104
17. The approximation theorem for extensional operations and the fixed point theorem for monotone operations 109
18. Topology displayed: basic definitions 113
19. The representation theorem for explicitly CL-continuous operators 117
Appendix: alternative proofs 122

V Type-free abstraction, choice and sets 125
20. Choice principles and the distinction between operations and functions 126
21. Admissible hulls: elementary facts 131
22. A model of admissible set theory 137
23. The boundedness theorem 144

PART C: SELECTED TOPICS 149

VI Levels of implication and intensional logical equivalence 151
24. Myhill's levels of implication 152
25. Formal deducibility based on levels of implication and its proof-theoretic strength 158
26. Introducing an intensional equivalence relation 162
27. The infinitary reduction relation $\Rightarrow$ 165
28. The Church-Rosser theorem for $\Rightarrow$ 169
29. A model of type-free logic based on intensional equivalence 174

VII On the global structure of models for reflective truth 177
30. The lattice of fixed point models for the neutral minimal theory 179
31. The sublattice of intrinsic fixed point models and the cardinality theorem 186

32. Variations on the encoding technique: non-modularity and other oddities 192
33. A model for an impredicative extension of reflective truth 198
34. On Kripke's classification of self-referential sentences 203
35. On the consistency of coinduction principles 207
Appendix: a variant to the basic operator Γ and the restriction axiom 209

PART D: LEVELS OF TRUTH AND PROOF THEORY 213

VIII Levels of reflective truth 215
36. A language and axioms for reflective truth with levels 218
37. Simple consequences 220
38. Universes and the Weyl extended iteration principle 225
39. A recursion-theoretic model 230
40. Levels of truth and predicatively reducible subsystems of second-order arithmetic 238
41. Consistency of a reducibility principle for classes 244
42. Levels of truth and impredicative subsystems of second-order arithmetic 248
Appendix: on projectibility and stronger reflection 253

IX Levels of truth and predicative well-orderings 257
43. On well-orderings 258
44. Ramified hierarchies 261
45. Predicative well-orderings I 269
46. Predicative well-orderings II 277

X Reducing reflective truth with levels to finitely iterated reflective truth 285
47. A sequent calculus STLR for a theory of reflective truth with levels 286
48. Basic properties of STLR 291
49A. Elimination of the full level induction schema 293
49B. Elimination of unbounded level quantifiers 297
50. The infinitary sequent calculus IT_n^∞ of n-iterated reflective truth 303
51. Embedding $STLR_n$ into IT_n^∞ 305

XI Proof-theoretic investigation of finitely iterated reflective truth 311
52. The ramified system RS_n 312
53. Cut elimination 316
54. Some derivable sequents of RS_n 320

55. Embedding IT_n^∞ into RS_n 324
56. The upper bound theorem for IT_n^∞ 327
57. Upper bound theorems for TLR and its subsystems 329
58. Conclusion: the conservation theorems 335
Appendix: primitive recursive cut elimination for RS_n 338

PART E: ALTERNATIVE VIEWS 349

XII Non-reductive systems for type-free abstraction and truth 351
59. The core system VF^- and transfinite induction 352
60. Supervaluation models of VF^- 357
61. An abstract sequent calculus and truth 358
62. Cut elimination and related properties 364
63. A provability interpretation and the upper bound theorem 369
64. Reconciling supervaluation models with provability interpretation 375

XIII The variety of non-reductive approaches 379
65. An inconsistency 380
66. On a truth theory of Friedman and Sheard 383
67. Fitch's models 386
68. Introducing semi-inductive definitions 390
69. Semi-inductive models for reflective truth 394

XIV Epilogue: applications and perspectives 401
70A. A logical theory of constructions: informal motivations 402
70B. A logical theory of constructions: basic syntax 403
71. Axioms for the computation relations 407
72. Extending the logical theory of constructions with higher reflection 411
73. Proof-theoretic reduction 416
74. Perspectives: related work in Artificial Intelligence and Theoretical Linguistics 419
75. Sense and denotation as algorithm and value: subsuming theories of reflective truth under abstract recursion theory 422

Bibliography 425

Index 441

List of Symbols 453

INTRODUCTION

"There never were set-theoretic paradoxes, but the property-theoretic paradoxes are still unresolved" (K. Gödel, as reported by J. Myhill 1984).

"... the theory of types brings in a new idea for the solution of the paradoxes, especially suited to their intensional form. It consists in blaming the paradoxes not on the axioms that every propositional function defines a concept or a class, but on the assumption that every concept gives a meaningful proposition, if asserted for any arbitrary object or objects as arguments" (K. Gödel 1944)

1. **Informal ideas.** The starting point of our investigation is the idea that the notions of *predicate application* and *property* are susceptible of independent study; in particular, these intuitive notions should be kept distinct from their counterparts of *set-theoretic membership* and *set*, as it is readily seen through a brief comparison.

According to the iterative conception, a set is always a collection of mathematical entities of a given type (possibly, sets of lower rank); thus it *has its being in its members,* and equality among sets is ruled by the extensionality principle. Sets are conceived as completed totalities, generated by *language independent operations and iterations thereof.* The membership relation is a *standard mathematical relation*: this means that $a \in b$ is a well-defined proposition, whenever a and b are sets. Moreover, if we reflect upon the intuitive picture of the cumulative hierarchy, we come to know that $\in$ is well-founded and does not allow self-application.

By contrast, a property *is an abstract object, which is grounded in a concept, i.e. a function,* not in the objects which fall under it (Frege 1984, p.199); it has no a priori bound on its extension, and it usually depends on some sort of *explicit* or *implicit finite specification.* Properties satisfy the so-called *unrestricted abstraction or comprehension principle* (AP):
every condition $A(x)$ determines a property $\{x : A\}$, which applies to all and only those things of which $A(x)$ holds true.
Of course, on the face of the well-known paradoxes, AP introduces elements which are open to dispute and to multiform solutions; for instance, as Gödel's citation suggests, the predication relation – henceforth η – cannot be always meaningful, and therefore the laws of standard (classical) logic cannot be valid.

The present approach, to be developed in various forms in this book, tries to keep the regimentation for predication and abstraction at a minimum; we maintain that $\{x : A\}$ is an individual term and that η applies to statements possibly involving η itself. Thus we are looking for flexible, type-free theories of predication. More specifically, we are influenced by the tradition of illative combinatory logic in the sense of Curry and Fitch, by the work of Feferman (1975) on partial classifications and of Aczel (1980) on Frege structures. The inspiring idea is that properties and predication can be adequately explained in terms of the primitive notions of *function* and *truth.*

As to the notion of function, we cannot expect to deal with functions in set-theoretic sense. In fact properties, given in intension, may apply to anything in a given realm, without type restrictions; and the same must hold of the functions underlying the properties themselves. Thus we are driven to understand functions essentially as rules of constructions (or, in short, operations) in the sense of combinatory logic. In contrast to the set-theoretic conception, operations are prior to their graphs and have no a priori bound on their domain; in particular, *they do support non-trivial forms of self-application.* On this view, *it is natural to identify properties with object-correlates of functions, and to reduce the abstraction operation to familiar λ-abstraction*; formally, $\{x : A\}$ simply becomes a λ-term of the form $\lambda x[A]$, where $[A]$ is a term of combinatory logic, canonically representing the function defined by the condition A (of any given language).

The second point concerns the reduction of predication to a primitive notion of reflective (or self-applicable) truth. Indeed, the expression

$$y\eta\{x : A\}$$

is analyzed as: " the result of applying the function represented by $\{x : A\}$ to the argument y turns out to be true". Therefore, if we let T stand for the truth predicate, $y\eta\{x : A\}$ is defined as $T(\{x : A\}y)$ (with juxtaposition of $\{x : A\}$ and y as application), and the abstraction principle AP becomes obviously derivable from the basic law of λ-abstraction (i.e. we convert $\{x : A\}y$ to the term $[A[x := y]]$, the result of replacing x with y in $[A]$).

Of course, these preliminary considerations do not solve the main problem of specifying the basic features of the truth predicate T. Nevertheless, they direct our attention towards the study of *simple mathematical objects*, namely expansions of combinatory algebras by reasonably closed truth sets. The typical structure (essentially) consists of a pair

$$\mathcal{M}^+ = \langle \mathcal{M}, \mathcal{T} \rangle,$$

where (i) $\mathcal{M}$ is a combinatory algebra, i.e. a model of Curry's combinatory logic; (ii) $\mathcal{T}$ is a subset of M (= the domain of $\mathcal{M}$), which assigns a semantical structure to $\mathcal{M}$.

These expansions are uniformly described by means of operators Γ from the power-set of M into itself, acting as *abstract valuation schemata.* Informally, if $X \subseteq M$, $\Gamma(X)$ represents the set of "truths" we come to know by means of the semantic rules of Γ on the basis of X. A central role in this book is played by an operator Γ, which essentially embodies Kleene's three valued non-strict interpretation of logical constants. In general, if Γ is monotone and reflects a cumulative conception of knowledge, the natural candidates for $\mathcal{T}$ will be those subsets of M, that cannot be further extended with new information by means of Γ, i.e. the so-called *fixed points* of Γ, satisfying $\Gamma(X) = X$. Among these sets, a special role will be played by the minimal ones: they are technically the most interesting objects for the recursion-theoretic and proof-theoretic investigations. Conceptually, they reflect the idea that abstraction is not the mere description of an independent logical realm, but rather a process with its own logic implicit in Γ.

In order to provide a few intuitions behind the construction of the first part of the book, it may be suggestive to regard $\mathcal{M}$ as an abstract syntax, in which formal languages can be processed and defined. In particular, elements of M may be thought of as symbolic expressions, to be combined and identified according to the operations and laws of combinatory logic. M will typically include (notations for) natural numbers or any other chosen ground type, but also and most important, objects representing functions. The objects associated to computable functions can be seen as (functional) programs, implementing effective algorithms. On the other hand, still pursuing the computational analogy, properties – as representatives of (generally non-computable) propositional functions – can be considered as programs implementing a sort of *generalized algorithms.* While application of an effective algorithm to an input produces a *computation*, possibly converging to *a value,* a property $\{x : P(x)\}$ is ultimately applied to an object, in order to produce a *verification* that the object itself satisfies the given condition, according to the rules specified by the truth set $\mathcal{T}$.

We wish to conclude by raising three conceptual points. First of all, the notion of truth T is not understood as a formalized truth predicate in the usual metamathematical sense: T classifies *the objects of a combinatory algebra,* and not an *inductively defined* collection of sentences ! In this sense, T, like set-theoretic membership, *does not depend upon a specific language.* As it should be clear from the sketched schema of interpretation $\mathcal{M}^+$, *the predicate T is a primitive concept, which is prior to the specification of any formalism and is the source of the abstract notion of proposition.* The underlying philosophy is that there are certain objects in our universe $\mathcal{M}$, which carry information and can be called propositions; they can be partitioned into atomic or complex. Atomic propositions are simply grasped and reflect implicit (synthetic) knowledge, to be accepted as given. On the other side, complex propositions correspond to some sort of construction via logical operators; thus they require a(n analytic) process, in order to be understood (think of the search for verification), and they are controlled by the truth predicate T.

As a second point, we like to stress the importance of having *operations acting on classifications.* Indeed, the fact that operations and classifications live together has the consequence of a symmetry, lacking to set theory: not only we can *classify* operations, but we can *operate* on classifications. So we can treat classifications, which depend on parameters, as operations; this is generally impossible in set theory. It follows that many constructions and statements acquire an "explicit character" and greater uniformity.

A final comment is left for the choice of non-extensional basic notions. In general, even if we make use of intensional data (like definitions or enumerations), we never appeal to specific features of them, and thus we obtain results with an *intrinsic* character. Moreover, we find that the non-extensional language helps to avoid "strong logical principles" and to carry out proofs in rather weak systems (just as remarked in Kreisel 1971, p. 170); it often permits *uniform and explicit statements* of the results obtained, which do not obscure the appreciation of proper extensional aspects. On the contrary, non-extensional and extensional features are free to interact in a unified framework. As it will be clarified by the introduction of the approximation structure in chapter III and its applications in the subsequent chapters, the essential interplay of these aspects leads to rather smooth generalizations of the Myhill-Shepherdson theorem (§17), to the appreciation of extensional choice principles (§20), and to a satisfactory "internal" treatment of inductive definitions (boundedness and covering; §23).

2. **Organization and contents.** As we previously explained, the starting point of the book is the need for an independent logical approach to the notions of predicate application, property, abstraction, truth. The arrangement of the material reflects the increasing logical complexity of the truth notions that are met in the text. The different proposals, though generally motivated by model-theoretic constructions, are developed in axiomatic style. This is mainly because we wish to emphasize the connections with standard concepts of mathematical logic and deductive systems for (substantial parts of) mathematics.

Proof-theoretic considerations and conservative extension results play an important role in classifying the various systems: very loosely, we tend to stress the importance of frameworks not stronger than Peano arithmetic and to distinguish predicative from impredicative systems. We also underline that type-free systems should not be opposed to type theories; we regard the former as a sort of generalized type assignment systems, in which types are left implicit and emerge from the theory itself.

More concretely, the book is divided into five parts, which group together relatively homogeneous topics. The read thread can be described as follows. By and large, the first three parts form a sort of independent essay on a first-order theory of reflective truth over combinatory logic, whose truth axioms essentially stem from Fitch's extended basic logic (Fitch 1948) through Scott (1975) and Aczel (1977). The notion of reflective truth explicitly refers to Feferman (1991). After the general results of Part A, the theory is motivated and enriched by means of recursion-theoretic investigations (part B), by showing its unifying power and studying its semantics (part C). Parts D and E explore alternative routes. Part D deepens the intuitions underlying the systems of parts A-C by use of proof-theoretic techniques and by relativizing the concept of truth. Part E is experimental in character and scans over a variety of approaches, which are still subject of investigation.

To give the reader a closer idea of what is in the book, we shall survey the content of the single chapters. A more detailed account can be found in the introductory section to each chapter.

Part A: it offers a general introduction to the basic notions of operation and reflective truth. The basic aim is to illustrate, both axiomatically and semantically, a consistent notion of type-free logical structure, which will be fundamental in the whole book.

The opening chapter contains the necessary preliminaries on (expanded) combinatory logic, which is here taken as the core of a classical first-order theory of operations OP. There is an introduction to concrete models of OP, as they form the ground structures in the entire book.

In the second chapter, we inductively expand combinatory algebras with a notion of self-referential truth, which naturally generalizes the familiar Tarskian semantical clauses, in order to cope with a situation of partiality. The given expansions only depend on the isomorphism type of the underlying combinatory algebras. By inspection of the model-theoretic construction, we are led to a minimal axiomatic first-order system MF^-, which contains a version of the Kripke-Feferman axioms for reflective truth and yields a theory of partial and total properties (= classes henceforth), satisfying natural closure conditions. For instance, classes are provably closed under Feferman's join and elementary comprehension principles; moreover, MF^- is provably closed under inductive definitions (though not capable of showing the corresponding induction schemata). We also consider extensions of MF^- with various number-theoretic induction principles.

Part B: we show that there is a two-sided link between generalized recursion theory and languages with operations and self-referential truth. Not only inductive definitions are crucial for building up models of self-referential languages, but these languages offer smooth formulations of non-trivial definability results.

In chapter III we prove that classes (properties) in the inductive model over a given combinatory algebra $\mathcal{M}$ exactly define the hyperelementary (inductive) subsets of $\mathcal{M}$ in the sense of Moschovakis (1974). The recursion-theoretic approach suggests to extend the minimal system by simple approximation conditions on properties. The new axioms, together with MF^-, a powerful generalized induction schema GID and number-theoretic induction for classes, form an axiomatic system PW_c+GID, which is still conservative (actually proof-theoretically reducible to) over the theory of operations and hence over Peano arithmetic.

In chapter IV we show that PW_c+GID proves a number of interesting consequences (separation and reduction principles) and, above all, an analog of the Myhill-Shepherdson theorem for operations which are η-extensional (i.e. extensional with respect to the predication relation). The results can be restated in topological terms via a natural generalization of the positive information topology.

In chapter V, we produce models for admissible set theory and a boundedness theorem for inductive sets, again provably in PW_c+GID.

Part C: it is a natural complement to the previous parts. In chapter VI, the reader will find two alternative type-free logics. The first system, due to Myhill (1972, 1980), relies on a logic with levels of implication. The second system, inspired by Aczel-Feferman (1980), offers a type-free logic with a definitional equivalence relation on formulas, which is inspired by conversion in combinatory logic. Both systems are formally interpreted in the theory PW_c+GID of chapters IV-V.

Chapter VII offers a general outlook on the global structure $FIX(\mathcal{M})$ of fixed point models of NMF^-(=MF^- without a consistency axiom) over arbitrary combinatory algebras $\mathcal{M}$. We prove that $FIX(\mathcal{M})$ only depends on the isomorphism type of $\mathcal{M}$ and that the set of sentences A such that TA holds in every structure of $FIX(\mathcal{M})$, for arbitrary $\mathcal{M}$, is axiomatizable. It is shown that $FIX(\mathcal{M})$ is a very rich and intricate non-distributive complete lattice; a few applications to consistency results and to formal semantics are thereby outlined (see connection with Kripke 1975).

Part D: it focuses on proof theory and the foundations of mathematics. We investigate a type-free logic TLR, which is able to internalize to a certain extent quantification on classes and negative semantic information. The intuitive idea is that truth is the (direct) limit of local self-referential truth predicates, which are related one another by a directed pre-order of levels.

Formally, we present TLR and its variants in chapter VIII. Among its consequences, it is possible to introduce a notion of "mathematical universe" with nice closure properties and interpret non-trivial subsystems of second-order arithmetic (ranging from versions of predicative analysis, like Friedman's ATR_0, to the so-called Δ^1_2-CA).

In chapter IX we develop the prerequisites for a proof-theoretic analysis of TLR: in particular, we describe a well-ordering proof of the so-called Feferman-Schütte ordinal. Chapter X proves that the theory of truth with levels is proof-theoretically reducible to (infinitary) theories of finitely iterated self-referential truth IT^{∞}_n; on the other hand, each IT^{∞}_n is shown to be reducible to fragments of predicative analysis in chapter XI. The methods used include cut elimination for ramified systems in ω-logic and asymmetrical interpretations à la Girard.

Part E: we are concerned with logics of truth and type-free abstraction, which are based upon non-reductive, non-truth functional semantical valuation schemata. In contrast to the reductive schema underlying the semantics of chapter II, we study systems which are well-behaved with respect to logical consequence (e.g. a tautology is always classified as true; this does not work under a partial semantics à la Kleene).

Chapter XII investigates a minimal system VF endowed with a simple supervaluation monotone semantics; VF naturally justifies principles of generalized inductive definitions (in contrast to what happens with the theories of parts A-C, it yields a model of the theory of elementary inductive definitions ID_1). We also develop an alternative interpretation for VF by means of proof-theoretic infinitary methods.

Chapter XIII addresses the problem of extending the logic of truth, as codified in VF. We discuss a refinement of supervaluation methods; but the new point is the introduction of semi-inductive definitions (in the sense of Herzberger 1982) and the application of the related notion of stable truth. We also consider consistent though ω-inconsistent logics of truth, due to Friedman, Sheard and Mc Gee.

The epilogue (chapter XIV) discusses prospective applications of type-free systems, as they result from the literature. In particular, we consider a logical theory of constructions, that has been investigated in Theoretical Computer Science and is strictly linked with the theories of part D. We conclude with a short survey of applications in other fields.

3. **How to use the book.** The interdependence of the chapters is roughly indicated in the diagram below:

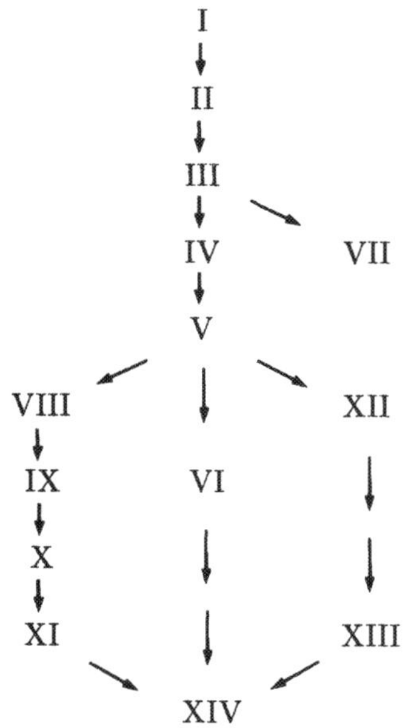

Certain parts of the book, once suitably combined, offer a non-conventional approach to:
1) generalized recursion theory and inductive definability (part A + part B);
2) predicative mathematics and subsystems of analysis (part A + part B + + part D).
If we disregard the recursion-theoretic and proof-theoretic parts, the book can serve as an introduction to:
3) formal semantics (part A + III + part C + VIII (§§36-39) + part E).
If the reader has in mind possible connections with logics for Artificial Intelligence, Theoretical Computer Science or semantics for natural languages, the suggestion 3) can be profitably modified to:
4) part A + part B + VIII (§§36-39) + part E.

Some chapters have appendices, containing additional details for proofs or suggestions for alternative routes: they can be always skipped without prejudice of understanding the later developments.

4. **Prerequisites.** The text is intended for readers who are familiar with the topics usually covered in an advanced undergraduate or basic graduate logic course. Thus we assume acquaintance with the elements of first-order logic

and model theory, recursion theory, set theory and proof theory, as they are developed in good standard textbooks, or in the corresponding chapters of the *Handbook of Mathematical Logic* (Barwise 1977).
In particular, it is useful to have a preliminary knowledge of the basic facts of hyperarithmetic and inductive definability (see Aczel 1977a, Moschovakis 1974). For the proof theory of Chapters VIII-XI, a previous exposure to sequent calculi and ω-logic would be helpful (e.g. see Schwichtenberg 1977 or the textbooks of Takeuti 1975, Schütte 1977, Girard 1987, Pohlers 1989). The simple topological notions of Ch. IV can be obtained from any standard textbook in general topology. Ch. VII presupposes a few elementary facts about partially ordered sets and lattices, usually met in logic courses (consider the classical reference of Birkhoff 1967). In Ch. VIII we hinge upon some advanced results of admissible set theory, to be found in Barwise (1975), Hinman (1977); however, the basic definitions and results are briefly recalled there.

5. **General notations and conventions.** A number of notations are adopted in the whole text. We summarize them below.

5.1. *Internal and bibliographical references.* The book is structured in five parts from A to E; each part is subdivided into chapters; the chapters are organized in sections, which are numbered in progressive order. Within each section, each specific item (subsection, definition, remark, axiom, rule, theorem, lemma or corollary) is usually assigned a pair "$m.n$" of numbers: "$m.n$" refers to the nth item of the mth section. Sometimes, for finer classifications and reference, we allow the use of three (and exceptionally four) numbers (e.g. 37.4.1 locates the first sub-item of the 4th-item of section 37). In some cases, we specify the class, which the referred item belongs to (e.g. we may speak of theorem 3.2 or definition 34.5).
References to publications are given by means of the author's name followed by the year of publication, possibly followed by a letter in the case of more publications by the same author in the same year.

5.2. *Definitions.* $:=$ is used as the definition symbol (definiendum on the left of $:=$, definiens on the right), while $\equiv$ stands for literal identity, unless it is specified otherwise.

5.3. *Variables and substitution.* We adopt the standard notions of free and bound variable; FV(E) is the set of free variables of the expression E. $\mathrm{E}[x := t]$ denotes the substitution of x with t in E. $\mathrm{E}(\mathrm{E}')$ means that E' possibly occurs as a subexpression of E.

5.4. *Logical Symbols.* As usual we use $\forall$, $\exists$, $\neg$, $\rightarrow$, $\wedge$, $\vee$, $\leftrightarrow$. For bracketing, we adopt the usual conventions; $\forall$, $\exists$, $\neg$ bind stronger than the other symbols, while $\wedge, \vee$ bind more than $\rightarrow$ and $\leftrightarrow$. To enhance

readability, dots may be used instead of brackets as separating symbols. $A \wedge B. \rightarrow C$, $A \rightarrow . B \vee C$, $\exists x.A$ stand for $(A \wedge B) \rightarrow C$, $A \rightarrow (B \vee C)$ $\exists xA$ (respectively); $\lambda x.ts$ shortens $\lambda x(ts)$, etc... Sometimes, we make use of bounded quantifiers as abbreviations: if $R := \eta, \in$, $\forall xRa.A$, $\exists xRa.A$ shorten $\forall x(xRa \rightarrow A)$, $\exists x(xRa \wedge A)$. If bounded quantifiers are iterated, we write: $(\forall xRa)(\forall yRb)(\ldots)$, $(\forall xRa)(\exists yRb)(\ldots)$, or even $\forall xRa.\forall yRb.(\ldots)$, $\forall xRa.\exists yRb.(\ldots)$, for the proper $\forall x(xRa \rightarrow \forall y(yRb\ldots))$, $\forall x \exists y(xRa \wedge yRb\ldots)$ (respectively).
We shorten logical equivalence on the metalevel (i.e. "if and only if") by the standard "iff" . "$\exists!x$" stands for "there is exactly one x". Sometimes, we adopt $\Rightarrow$ as implication on the metalevel.

5.5. *Logical Complexity.* The logical complexity of any given formula A is the number of distinct occurrences of logical symbols in A.

5.6. *Set-theoretic symbols.* We use the standard notations:
$\in$, $\notin$ (negation of $\in$), ω (the set of natural numbers, but also the first infinite ordinal), $\emptyset$, $\cup, \cap$, $\mathcal{P}(X)$ (power set of X), $X-Y$, $\subset, \subseteq, \supset, \supseteq$, $\times$ (Cartesian product), $f: X \rightarrow Y$ (to be read as "f is a function from X to Y"), c_Z (characteristic function for the set Z). If $k, m \in \omega$,

$$[k,m] := \{i \in \omega : k \leq i \leq m\};\ (k,m] := \{i \in \omega : k < i \leq m\};$$

$$(k,m) := \{i \in \omega : k < i < m\};\ [k,m) := \{i \in \omega : k \leq i < m\}.$$

$\{x : \ldots\}$ is the set of objects satisfying the condition $(\ldots)$; $\{a_1, \ldots, a_n\}$ is the set containing exactly the elements $a_1, \ldots, a_n$. $\langle \ldots \rangle$ denotes set-theoretic n-tuple operation, unless otherwise specified.
We warn the reader that set-theoretic symbols will be sometimes adopted as abbreviations for corresponding non-extensional operations on properties and predicates. But possible ambiguities will be spared by the context. The arithmetical symbols are the standard ones.

5.7. *Provability and standard Tarskian semantics.*
$\mathcal{M} \models A$ stands for "A holds in the structure $\mathcal{M}$".
$S \vdash A$ means that A is derivable from S by means of classical logic (unless otherwise specified).

5.8. *Inductive proofs.* We often carry out proofs by induction (either in the metatheory or within axiomatic theories). As a rule, we adopt the acronym IH as a shortening for "induction hypothesis".

PART A

COMBINATORS AND TRUTH

“ Μήπω μέγ' εἴπης: ἔτι γάρ, ὦ μακάριε, ἔστι, καὶ ταῦτά γε τῶν ἀποριῶν ἡ μεγίστη καὶ πρώτη. περὶ γὰρ αὐτὴν αὐτοῦ τὴν ἀρχὴν οὖσα τυγχάνει”
(Plato, Soph.238a)

CHAPTER 1

INTRODUCING OPERATIONS

§1. The basic language
§2. Operations I: general facts
§3. Operations II: elementary recursion theory
§4A. The Church-Rosser theorem
§4B. Term models
§5. The graph model
§6. An effective version of the extensional model D_∞
Appendix

This chapter contains an elementary introduction to combinatory logic. The topic is highly developed, but the chapter has quite a limited aim: that of yielding all the necessary prerequisites and making the book self-contained.
According to the informal ideas outlined in the general introduction, we aim at investigating an axiomatic notion of abstract logical system, whose ground structure (*the abstract syntax*) is a combinatory algebra, extended with suitable built-in operations and with a primitive notion N of natural number. The choice of N is largely a matter of convenience and tradition; the basic constructions do not depend on the initial stock of built-in predicates and operations.

The central aim of this chapter is to clarify what we understand by ground structure. We begin in the axiomatic style and we describe a formal system OP for a type-free theory of operations; we then outline three basic semantic constructions. We underline that the basic constructions can be carried out in OP itself.

After the description of the formal language (§1), we define OP and we discuss its general features (§2; closure under β-conversion, fixed point theorem, relation with λ-calculus), while §3 reviews some basic facts on recursion theory. We then present the term models of OP, which are based on the fundamental Church-Rosser theorem (§4). In §5 we give an elementary description of the Plotkin-Scott graph model $P\omega$, together with its recursive submodel RE and Engeler's construction. Finally, following an elegant procedure, due to Scott (1976, 1980), we show how to isolate an extensional submodel D_∞ of RE.

§1. The basic language

We describe an axiomatic theory of operations OP, which is a first-order extension of pure combinatory logic by simple number-theoretic notions. OP is proof-theoretically equivalent to PA, the elementary system of Peano arithmetic, and it will constitute the basis of all systems to be investigated in this book.

The basic language $\mathfrak{L}$ contains:

(i) countably many individual variables $x_1, x_2, x_3, \ldots$;

(ii) the logical constants $\neg$, $\wedge$, $\forall$;

(iii) the individual constants K (constant function combinator), S (composition combinator), SUC (successor), $PRED$ (predecessor), $PAIR$ (ordered pair operation), $LEFT$ (left projection), $RIGHT$ (right projection), $\bar{0}$ (zero), D (definition by cases on numbers);

(iv) the binary function symbol Ap (application operation) and the predicate symbols N (natural numbers), T (truth), $=$ (equality).

Terms are inductively defined from variables and constants via application of Ap. We use x, y, z, u, v, w, f, g as metavariables; while t, t', t'', s, s', r, r', etc., are metavariables for terms. We write (ts) instead of $Ap(t,s)$, and outer brackets are usually omitted, while the missing ones are restored by associating to the left; for instance, xyz stands for $((xy)z)$.

We adopt familiar shorthands for special terms: $t+1 := SUCt$ (= the successor of t); $\langle t,s\rangle := PAIRts$ (= the ordered pair composed by t and s); $(t)_1 := LEFTt$ (= the left projection of t) and $(t)_2 := RIGHTt$ (= the right projection of t).

Formulas are inductively generated by means of the logical operations from atomic formulas (atoms, in short) of the form $t = s$, Nt, Tt.

A, B, C are syntactical variables for formulas of $\mathfrak{L}$. As to the syntactical notions of free and bound variable, substitution, etc., we follow the standard conventions and terminology (Shoenfield 1967). In particular, if E is an expression (term or formula), $E(x)$ means that x may occur free in E, while $E[x := t]$ stands for the result of substituting t for the free occurrences of x (provided t is substitutable for x in E). FV(E) is the set of free variables of the expression E; $x \in \mathrm{FV}(E)$ means that "x occurs free in E".

The remaining logical symbols are defined classically:

$$\exists xA := \neg\forall x\neg A;\ A\vee B := \neg(\neg A\wedge\neg B);\ A\to B := \neg A\vee B\ ;$$

$$A\leftrightarrow B := (A\to B)\wedge(B\to A).$$

We stick to the usual convention that $\neg$ and quantifiers bind more than the remaining connectives, while $\wedge$, $\vee$ bind more than $\to$ and $\leftrightarrow$; sometimes dots are used in place of parentheses (see §5 of the introduction).

As usual, a numeral is any term obtained from the constant zero by means of a finite number of successor applications; if $n \in \omega$ ($\omega =$ the set of natural numbers), $\overline{n}$ stands for the n-th numeral, i.e. the term built up from $\overline{0}$ with n applications of SUC.

We now recall the standard definition of λ-abstraction in combinatory logic.

1.1. DEFINITION. If t is an arbitrary term of $\mathfrak{L}$, $\lambda x.t$ is introduced by induction on the notion of $\mathfrak{L}$-term:

(i) $\lambda x.x := SKK$;
(ii) $\lambda x.t := Kt$ if $x \notin \mathrm{FV}(t)$;
(iii) $\lambda x.(ts) := S(\lambda x.t)(\lambda x.s)$, if $x \in \mathrm{FV}(ts)$

Of course $\lambda x.t$ has exactly the same free variables as t, minus x. Coding of n-tuples can be obviously defined by iteration of $\langle\ ,\ \rangle$.

1.2. We inductively put : $\langle t\rangle := t$ and $\langle t_1, \ldots, t_{k+1}\rangle := \langle\langle t_1, \ldots, t_k\rangle, t_{k+1}\rangle$. If $1 \leq i \leq k$, we define projections for k-tuples:

$$(t)_1^k := LEFT^{(k-1)}t \; ; \; \text{ if } i > 1, \; (t)_i^k := RIGHT(LEFT^{(k-i)}t),$$

where $LEFT^{(0)}t := t$ and $LEFT^{(k+1)}t := LEFT(LEFT^{(k)}t)$. If the upper index is clear, e. g. if $t = \langle t_1, \ldots, t_k\rangle$, we write $(t)_i$ for $(t)_i^k$.

§2. Operations I : general facts

OP, our basic theory of operations, is based on PC, classical predicate calculus with equality in Hilbert style presentation (the equality axioms being given by the formula

$$\forall x \forall y \forall z(x = x \wedge (x = y \rightarrow y = x) \wedge (x = y \wedge y = z \rightarrow x = z\ )) \wedge$$
$$\wedge \forall x \forall y \forall z(x = y \rightarrow (zx = zy \wedge xz = yz) \wedge (Nx \rightarrow Ny) \wedge (Tx \rightarrow Ty)).$$

The non-logical axioms of OP are listed in 2.1 below.

2.1. *Axioms for operations and natural numbers*

COMB $\forall x \forall y \forall z(Kxy = x \wedge Sxyz = xz(yz))$;

PAIR $\forall x \forall y((\langle x, y\rangle)_1 = x \wedge (\langle x, y\rangle)_2 = y)$;

NAT.1 $N\overline{0} \wedge \forall x(Nx \rightarrow (N(x+1) \wedge \neg(x+1) = \overline{0} \wedge PRED(x+1) = x))$;

NAT.2 $\forall x \forall y \forall u \forall v(Nx \wedge Ny \wedge \neg x = y \rightarrow Dxxuv = u \wedge Dxyuv = v)$;

NIND $A(\overline{0}) \wedge \forall x(A(x) \rightarrow A(x+1)) \rightarrow \forall x(Nx \rightarrow A(x))$ (A arbitrary).

Clearly OP states that the universe is a combinatory algebra, which is suitably extended with natural numbers. It is here important to mention a natural extension of OP, which includes the extensionality axiom for operations:

Ext_{op} $\qquad \forall x(zx = yx) \rightarrow z = y.$

CONVENTION. If S is any system containing NIND, S^{-} = S minus NIND. Similarly, if S^{-} is any system not containing NIND, S = S^{-} plus NIND.

We now recall the basic elementary facts underlying the pure theory of combinators and λ-calculus. Of course, the results below are quite standard and are quickly surveyed for the reader's sake. To this aim, we first restrict our attention to the induction free fragments OP^{-} and $\mathrm{OP}^{-}+\mathrm{Ext}_{op}$ of OP. λ-abstraction, as defined in 1.1, commutes with substitution and satisfies α- and β-conversion. It also satisfies η- and ξ-conversion, in presence of extensionality; projections of k-tuples, as defined in 1.2, verify the due equations.

2.2. PROPOSITION

1. *The following formulas are provable in* OP^{-}, *for arbitrary* t, s:

 (i) *α-conversion* $\qquad \lambda x.t = \lambda y.t[x := y]$;

 (ii) *β-conversion:* $\qquad (\lambda x.t)s = t[x := s]$;

 (iii) $\qquad (\lambda y.t)[x := s] \; = \; \lambda y.(t[x := s])$.

Proviso : *in* (i) $y \notin \mathrm{FV}(t)$; *in* (iii) $y \notin FV(s) \cup \{x\}$.

2. *The following formulas are provable in* $\mathrm{OP}^{-} + \mathrm{Ext}_{op}$, *for arbitrary* t, s:

 (i) *ξ-conversion:* $\qquad \forall x(t = s) \rightarrow \lambda x.t = \lambda x.s$;

 (ii) *η-conversion:* $\qquad \lambda x.tx = t \; (x \notin FV(t))$.

3. *For each* k, $i \in \omega$ *such that* $1 \leq i \leq k$, OP^{-} *proves:*

$$(\langle x_1, \ldots, x_k \rangle)_i = x_i.$$

PROOF. 1. (i)-(iii) follow by straightforward induction on the build up of t.
2. (i)-(ii): by β-conversion and Ext_{op}.
3: argue by induction on k with the axiom *PAIR*. □

In (i) and (iii) = can be replaced by literal identity $\equiv$. Closed term are sometimes called *combinators.* In combinatory logic every operation has a fixed point; indeed, a stronger result holds:

2.3. THEOREM (*Fixed-point*). *We can find a combinator* FP *such that* $\mathrm{OP}^{-} \vdash \forall f(f(FPf) = FPf)$.

PROOF: define $FP := \lambda f.(\lambda x.f(xx))(\lambda x.f(xx))$. Then we have by β-conversion:

$$FPf = (\lambda x.f(xx))(\lambda x.f(xx)) = f((\lambda x.f(xx))(\lambda x.f(xx))) = f(FPf). \quad \square$$

FP is Curry's *paradoxical combinator* Y; for other fixed point combinators, see Barendregt (1984), pp.131-32.

We now introduce, following Scott (1980a) and Meyer (1982), certain combinators, which characterize (the first-order theory of) λ-calculus as a fragment of OP^-.

2.4. DEFINITION

(i) $I := \lambda x.x$ and $1 := S(KI)$;
(ii) $1_0 := I$ and $1_{k+1} := S(K1_k)$.

2.5. LEMMA (*provable in* $\mathrm{OP}^- + \mathrm{Ext}_{op}$)

MS.1 $\forall x(yx = zx) \rightarrow 1y = 1z$;
MS.2 $1_2K = K$;
MS.3 $1_3S = S$;
MS.4 $1 = I$.

The proof is a routine application of extensionality and the axioms for K and S. The reason for these esoteric equations is that MS.1-MS.3 axiomatize ξ-conversion over OP^- and yield all together the full strength of Ext_{op}.

2.6. Let $\mathrm{OP}\lambda^- := \mathrm{OP}^- +$ MS.1-MS.3 .

2.7. PROPOSITION

(i) $\mathrm{OP}\lambda^-$ *proves the schema of* ξ-*conversion* (2.2.2 (i)).
(ii) $\mathrm{OP}\lambda^- + (1 = I)$ *proves* Ext_{op}.

PROOF. (ii) is immediate by MS.1. As to (i), assume $\forall x(t = s)$; then by β-conversion

$$\forall x((\lambda x.t)x = (\lambda x.s)x),$$

which yields with MS.1

$$1(\lambda x.t) = 1(\lambda x.s). \tag{1}$$

But MS.2-MS.3 imply $Kx = 1(Kx)$ and $Sxy = 1(Sxy)$; for instance, we have:

$$Sxy = S(K1_2)Sxy = ((K1_2)x(Sx))y =$$

$$= (1_2(Sx))y = S(K1)(Sx)y = (K1)y(Sxy) = 1(Sxy).$$

Hence by definition of λ, we have $\lambda x.t = 1(\lambda x.t)$ and the conclusion follows by (1). □

Of course, we can assume λ-abstraction as primitive, and define: $\mathcal{L}(\lambda) := (\mathcal{L} \dot{\cup} \{\lambda\}) - \{K, S\}$. In the new language $\mathcal{L}(\lambda)$, terms are inductively generated by the new clause: if t is a term, $\lambda x.t$ is a term and $\mathrm{FV}(\lambda x.t) = \mathrm{FV}(t) - \{x\}(= x$ is bound in $\lambda x.t)$. We also identify terms differing only by the names of their bound variables (e.g. $\lambda x.x$ is identified with $\lambda y.y$).

2.8. DEFINITION. Λ is the first-order theory in the language $\mathcal{L}(\lambda)$, which is obtained from OP by omitting COMB and by assuming the schemata of β- and ξ-conversion (in β-conversion it is understood that s is substitutable for x in t).

2.9. THEOREM. *We can define maps* $C : \mathcal{L}(\lambda) \mapsto \mathcal{L}$ *and* $L : \mathcal{L} \mapsto \mathcal{L}(\lambda)$ *such that, if* $\mathfrak{T} = \mathrm{OP}\lambda^-$ $(\mathrm{OP}^- + \mathrm{Ext}_{op})$, $\mathfrak{S} = \Lambda^-$ $(\Lambda^- + \mathrm{Ext}_{op})$, *then for every* $\mathcal{L}$*-formula* A *and every* $\mathcal{L}(\lambda)$*-formula* B,

(i) $\mathfrak{T} \vdash A \leftrightarrow C(L(A))$ *and* $\mathfrak{S} \vdash B \leftrightarrow L(C(B))$;

(ii) $\mathfrak{T} \vdash A$ *iff* $\mathfrak{S} \vdash L(A)$; $\mathfrak{S} \vdash B$ *iff* $\mathfrak{T} \vdash C(B)$.

(The exponent $^-$ means that the number-theoretic induction axioms are omitted).

As to the verification of 2.9, $C(B)$ is the $\mathcal{L}$-formula which results by replacing in A each term of the form $\lambda x.t$ with the corresponding term, built up from K, S and defined in §1; $L(A)$ is the $\mathcal{L}(\lambda)$-formula, which is obtained from A by replacing K, S by $\lambda x \lambda y.x$ and $\lambda x \lambda y \lambda z.xz(yz)$ in the given order (λ primitive symbol). The proof requires some work and computations (apply 2.2, 2.5, 2.7 or see Barendregt 1984).

After 2.9, we henceforth identify $\mathrm{OP}\lambda$ $(\mathrm{OP}+\mathrm{Ext}_{op})$ with Λ $(\Lambda+\mathrm{Ext}_{op})$. It is well-known that the inclusions $\mathrm{OP} \subset \mathrm{OP}\lambda \subset \mathrm{OP}+\mathrm{Ext}_{op}$ are proper, i.e. there are models of OP which refute MS.1-MS.3 and there are models of $\mathrm{OP}\lambda$ falsifying MS.4 (see §§4-5).

§3. Operations II : elementary recursion theory

In this section we review a few basic results of elementary recursion theory on ω, formalized in the setting of combinatory logic. Since this topic is thoroughly developed in the literature (see Barendregt 1984, Hindley-Seldin

1986), we only outline the main steps.

CONVENTION. In formal contexts m, n, k (also with indices or apices) are syntactical variables for bound number variables; thus we write $\forall m$, $\exists m$ instead of $\forall x(Nx \to \ldots)$, $\exists x(Nx \wedge \ldots)$. Informally n, m, k, p, q range over natural numbers in contexts of the form $n \in \omega$, $k \in \omega$; $\overline{n}$, $\overline{m}$, $\overline{k}$, etc., denote numerals.

3.1. LEMMA. OP *proves the following sentences:*

(i) $A(\overline{0}) \wedge \forall n(A(n) \to A(n+1)) \to \forall n A(n)$, *A arbitrary;*

(ii) $\forall n(n = \overline{0} \vee \exists m(m+1 = n))$;

(iii) $\forall n \forall m(n+1 = m+1 \to n = m)$;

(iv) $\forall n(\neg n = \overline{0} \to N(PREDn) \wedge (PREDn)+1 = n)$;

(v) $\forall n(n+1 \neq n) \wedge K \neq S$.

PROOF. (i)-(ii): apply N-induction, identity logic and the consequence of the axiom NAT.1 $N\overline{0} \wedge \forall x(Nx \to N(x+1))$ to the formulas

$$A'(x) := Nx \wedge A(x) \text{ and } A(x) := (x = \overline{0} \vee \exists y(Ny \wedge y+1 = x)).$$

(iii): if $n+1 = m+1$, also $PRED(n+1) = PRED(m+1)$, whence $n = m$ by use of $\forall n(PRED(n+1) = n)$.
(iv): if $\neg n = \overline{0}$, then there exists an m such that $n = m+1$, by (ii); hence $m = PRED(m+1) = PRED(n)$ and $n = m+1 = PRED(n)+1$.
Since Nm, also $N(PREDn)$.
(v): the first conjunct follows by N-induction, NAT.1 and (iii); it also implies $\exists x \exists y(x \neq y)$; but $K = S$ yields $\forall x \forall y(x = y)$ by a well-known trick.
□

3.2. LEMMA (*Existence of recursor on natural numbers*). *We can find a closed term R_N such that, provably in* OP:

$$\forall m \forall y \forall z(R_N \overline{0} yz = y \wedge R_N(m+1)yz = zm(R_N myz)).$$

PROOF: choose $t = \lambda w \lambda x \lambda y \lambda z.Dx\overline{0}y(z(PREDx)(w(PREDx)yz))$.
Define $R_N := FPt$ (by 2.3); then apply β-conversion and definition by cases on N.□

Clearly, we can define the standard primitive recursive functions and predicates by means of R_N and then verify by NIND that the usual definitions are correct; in particular, we have a formula $x < y$ which represents the standard less-relation on numbers and satisfies its elementary properties, provably in OP.

3.3. DEFINITION

(i) $f: N \to N := \forall n.N(fn)$ ($= f$ is a number-theoretic operation);
(ii) $\forall n < m.A := \forall n(n < m \to A)$ and $\exists n < m.A := \exists n(n < m \wedge A)$.

3.4. LEMMA. *We can find a closed term μ, such that, provably in* OP:

$$gk = \overline{0} \ \wedge \forall m < k.\ gm > \overline{0} \to \mu g = k.$$

PROOF. By fixed point theorem, we find a term h such that $h = \lambda g \lambda z.D\overline{0}(gz)z(hg(z{+}1))$, and we choose $\mu := \lambda g.(hg\overline{0})$. Assume that $gk = \overline{0} \wedge \forall m < k.\, gm > \overline{0}$. Then $hgk = k$ and $hgm = hg(m+1)$, if $m < k$. By induction we verify $hgk = hg\overline{0} = k$. □

3.5. DEFINITION. A partial number-theoretic function $F: \omega \to \omega$ is *representable* in a theory $\mathfrak{T}$ (in the language $\mathfrak{L}$) iff there is a closed term f such that:

$$F(n_1, \ldots, n_k) \simeq n \text{ iff } \mathfrak{T} \vdash f\overline{n}_1 \ldots \overline{n}_k = \overline{n};$$

($n_1, \ldots, n_k$ are arbitrary natural numbers; $\simeq$ is Kleene's notation and $F(n_1, \ldots, n_k) \simeq n$ means that $F(n_1, \ldots, n_k)$ is defined and has value n).

3.6. THEOREM. *The partial recursive functions are representable in* OP.

PROOF. SUC, $K\overline{0}$, $\lambda x_1 \ldots \lambda x_n.x_i$ represent the initial functions successor, constant-zero, projections (respectively). The recursor and minimalization operators exist by 3.3-3.4; the substitution operator is immediately available by λ-abstraction. □

3.6.1. REMARK. The representing combinators in 3.6 can always be chosen in normal form and such that, if $F(n_1, \ldots, n_k)$ diverges, then $f\overline{n}_1 \ldots \overline{n}_k = \Omega = (\lambda x.xx)(\lambda x.\, xx)$; cf. §4 below and Barendregt, cit. ,p.179.

By 3.6, we denote the standard primitive recursive number-theoretic predicates (e.g. the ordering relation on ω) by their customary symbols. Strictly speaking, if P is a primitive recursive predicate, Px stands for the quantifier-free formula $fx = \overline{0}$, where f is a term representing the characteristic function of P. It is also clear that OP can provably formalize the standard facts of elementary recursion theory à la Kleene. In the following, we shall adopt the bracket notation $\{a\}(x) \simeq y$ without distinguishing it from its formal presentation in OP. We conclude with a few observations.

3.7. First of all, the distinction between operations and functions in the set-theoretic sense has interesting conceptual consequences. Let Church's thesis be the statement:

CT $$\forall f(f: N \to N. \to \exists n \forall m(\{n\}(m) \simeq fm)).$$

Then CT is consistent with the basic theories we consider in this book, even if full classical logic is used (see 4.11 below).

3.8. It is well-known (see Barendregt 1984, Hindley-Seldin 1986) that numerals, successor, predecessor, definition by cases on numerals and pairing are representable in the theory CL of pure combinatory logic. By CL we here understand the subsystem of OP^-, formalized in the sublanguage of $\mathcal{L}$, which only contains the function symbols Ap, K, S, the predicate $=$, variables and logical operators. The only non-logical axioms of CL are COMB and $(\neg K = S)$. Here follow the basic steps.

(a) Let $T := K$ and $F := KI$ (I is the identity combinator).
Pairing: $PAIRxy := \lambda u.uxy$; $LEFTx := xT$; $RIGHTx := xF$.
Then $\mathrm{CL} \vdash Txy = x \wedge Fxy = y \wedge (PAIRx_1x_2)_i = x_i$ $(i = 1, 2)$.

(b) Numerals: $\overline{0} := I$; $SUC\overline{n} := PAIR\overline{n}K$; $PRED\overline{n} := LEFT\overline{n}$;
$Z\overline{n} := (RIGHT\overline{n})FT$.

Then, for arbitrary $n, m \in \omega$,

$$\mathrm{CL} \vdash Z\overline{0} = T \wedge \ Z(SUC\overline{n}) = F$$

and $\mathrm{CL} \vdash \neg\ \overline{0} = SUC\overline{n} \wedge (SUC\overline{n} = SUC\overline{m} \to \overline{n} = \overline{m})$ (apply $\neg K = S$). By fixed point, choose

$$G = \lambda x\lambda y.(Zx)(Zy)((Zy)(Zx)(G(LEFTx)(LEFTy)))$$

and let $D = \lambda x\lambda y\lambda a\lambda b.(Gxy)ab$. Then CL proves that, if $\overline{n} = \overline{m}$, $G\overline{n}\,\overline{m} = T$, else $G\overline{n}\,\overline{m} = F$. Hence by the properties of T, F we are done.

Notice that, once we choose to enlarge combinatory logic by standard numerals $\overline{0}$, $SUC\overline{0}$, etc., and we assume SUC as a primitive constant, we are forced to introduce $PRED$ and D: without them, it would be impossible to define a number-theoretic recursion operator (see Curry et al. 1972, vol.II § 13.A.3, theorem 2).

3.9. *On full definition by cases.*
Let DIS be a new constant satisfying the axiom

3.9.1 $$\forall x \forall y((x = y \wedge \mathrm{DIS}xy = \overline{0}) \vee (\neg x = y \wedge \mathrm{DIS}xy = \overline{1})).$$

Then 3.9.1 is inconsistent with CL.

PROOF (folklore). Let $\mathrm{Neg}(x) = \mathrm{DIS}x1$. By 2.3 we can find an e such that $e = \mathrm{Neg}(e)$. Hence we have that $e = \overline{1}$ implies $e = \mathrm{Neg}(\overline{1}) = \overline{0}$, and $\neg e = \overline{1}$ implies $e = \mathrm{Neg}(e) = \overline{1}$. □

It follows:

3.9.2. CL plus the statement "*everything is a number*" is inconsistent.

Indeed, if we apply the above trick to D, we get an e such that $\neg Ne$.

3.9.3. There cannot exist an injective operation f from the universe into the natural numbers (define by D an operation h such that $hx = \overline{1}$, if $fx = f\overline{0}$ and $hx = \overline{0}$, if $\neg fx = f\overline{0}$; any fixed point of h leads to a contradiction).

§4A. The Church-Rosser theorem

We are going to construct term models for the non-extensional theory OP of operations. The strategy is well-known and it relies upon a fundamental result of Church and Rosser (1936). In order to ensure that the given theory of rules is consistent, we prove that computations – regardless of the various patterns we may follow – give unambiguous results. This technique regards the equality relation, as inductively generated by an asymmetric reduction relation, which splits the computation process into basic atomic steps. For a thorough treatment of the subject, we send the reader to Barendregt (1984), Hindley-Seldin (1986).

In the following we deal with the term fragment of the basic language $\mathcal{L}$ (§1), which contains individual variables and the individual constants K, S, SUC, $PRED$, $\overline{0}$, D, $PAIR$, $LEFT$, $RIGHT$. Terms are inductively generated by application from variables and individual constants.

4.1. DEFINITION OF REDUCTION. (a) The reduction relation $\geq$ is the smallest binary relation among terms, which satisfies the clauses below:

(i) $\geq$ is reflexive (R) and transitive (T);
(ii) $\geq$ is preserved by application, namely:
$t \geq s$, $t' \geq s'$ imply $tt' \geq ss'$; (A)
(iii) $Kts \geq t$; (K)
(iv) $Strs \geq ts(rs)$; (S)
(v) $LEFT(PAIRts) \geq t$ and $RIGHT(PAIRts) \geq s$; (P)
(vi) $PRED(SUC\overline{n}) \geq \overline{n}$; (SUC)
(vii) $D\overline{n}\,\overline{n}tr \geq t$; $(D.1)$
$D\overline{n}\,\overline{m}tr \geq r$, if $n \neq m$; $(D.2)$

(b) We then read $t \geq r$ as "t reduces to r" and the clauses (iii)-(vii) are called *proper reductions.*

(c) The terms which are on the left (right) of the proper reductions are called *redexes* (*contracta*).

(d) A term t is *normal*, or *in normal form* (in short $t \in$ NF), iff no subterm of t (t included) is a redex. For instance, every numeral or basic constant is normal; however, there are terms without normal form, the most typical among them being $\Omega = (\lambda x.xx)(\lambda x.xx)$.

A reduction statement $t \geq s$ can be conveniently regarded as a derivable formula of a formal system, where (R) and proper reductions play the role of axioms, while the inference rules are (T) and (A). Hence $t \geq s$ holds iff there exists a derivation in tree-form which is locally correct with respect to the axioms and the rules, and which has $t \geq s$ at its root; thus we can recursively assign a length to reductions:

4.2. DEFINITION (reduction with length). $t \geq_n s$ iff

(i) $0 = n$ and $t \geq s$ is a proper reduction or an application of (R), or

(ii) there are k, $m < n \in \omega$ and either $t \geq_k r$, $r \geq_m s$, for some r, or $t' \geq_k r'$, $t'' \geq_m r''$ and $t \equiv t'\ t''$, $s \equiv r'\ r''$, for some t', t'', r', r''.
(In the following sections of this chapter $\equiv$ is literal identity).

4.3. THE CHURCH-ROSSER PROPERTY CR
CR states that the reduction relation $\geq$ is directed (confluent): for all t, t', t'', if $t \geq t'$ and $t \geq t''$, there exists a term s such that $t' \geq s$ and $t'' \geq s$.

We verify that CR holds for a relation RED, whose transitive closure is $\geq$.

4.4. (i) $t\, RED_n\, s$ is defined by replacing everywhere $\geq$ by RED in 4.2 and by omitting the transitivity rule in 4.2 (ii);

(ii) $t\, RED\, s :=$ " $t\, RED_k\, s$, for some $k \in \omega$".

Obviously, $t \geq s$ holds iff $t \geq_k s$ holds for some k.

4.5. LEMMA. *$t \geq s$ iff there are $t_1, \ldots, t_k$ such that $t_1 \equiv t$, $t_k \equiv s$ and $t_1\, RED\, t_2, \ldots, t_{k-1}\, RED\, t_k$.*

The proof is immediate by induction on the length of the derivation. If we replace $\geq$ by RED in 4.3, we have a statement of the Church-Rosser property for RED. It is also clear that we can obtain, by simple diagram chasing:

4.6. LEMMA. *If* CR *holds for RED,* CR *holds for $\geq$.*

4.7. LEMMA (*Analysis of RED*)

(i) *If C is an arbitrary constant and $C\, RED\, t$, then $t \equiv C$.*

(ii) *If $C \equiv K, S, D$ and $Cs\, RED\, t$, then there is a t' such that $s\, RED\, t'$ and $t \equiv Ct'$.*

(iii) *If $C \equiv S, D, PAIR$ and $Ct_1t_2\, RED\, t$, there are t'_1, t'_2 such that $t_i\, RED\, t'_i$ $(= 1,\ 2)$ and $t \equiv Ct'_1t'_2$.*

(iv) *If* $C \equiv D$ *and* $Ct_1t_2t_3$ *RED* t, *there are* t_1', t_2', t_3' *such that* t_i *RED* t_i' $(i = 1, 2, 3)$ *and* $C\, t_1'\, t_2'\, t_3' \equiv t$.

PROOF. (i) $C\, RED\, t$ can hold only by (R).

(ii) Let $Cs\, RED_0\, t$: since $C \not\equiv PRED$, $LEFT$, $RIGHT$, we must have applied (R) and we are done. Let $Cs\, RED_m\, t$ be derived by (A): we have $C\, RED_n\, t'$ and $s\, RED_k\, t''$, for some n, k, t', t'' such that $m > k$, n and $t \equiv t't''$. One application of (i) yields the conclusion.

(iii) If $Ct_1t_2\, RED_0\, t$, t must have the form Ct_1t_2, because $C \not\equiv K$. If m, $k < n$ and $Ct_1\, RED_k\, s_1$, $t_2\, RED_m\, s_2$, we can find by (ii) a t_1' such that $t_1\, RED\, t_1'$ and $s_1 \equiv Ct_1'$: hence we have t_1', $t_2' \equiv s_2$, such that $t_i\, RED\, t_i'$, $(i = 1, 2)$ and $t \equiv Ct_1't_2'$.

(iv): by similar arguments resting on (iii). □

4.8. THEOREM. CR *holds for* RED.

PROOF. Assume $t\, RED_n\, s_1$, $t\, RED_m\, s_2$: we produce a term r such that $s_1\, RED\, r$, $s_2\, RED\, r$. We argue by induction on $l = n+m$.
If $l = 0$ and (R) is applied on both sides, choose $r \equiv s_1 \equiv s_2$. If there is a proper reduction, choose r as the result of the contraction. Let $l > 0$.

Case 1. One of the given derivations has length 0. By symmetry, it is not restrictive to assume $t\, RED_0\, s_1$. We analyse its derivation $\mathfrak{D}_1$.

1.1 (R) is the inference applied in $\mathfrak{D}_1$: choose $r \equiv s_2$.

1.2. (R) is not applied. We have to distinguish a few subcases.

1.2.1. Let (K) be the inference applied in $\mathfrak{D}_1$. Then $t \equiv Kt_1t_2$ and $s_1 \equiv t_1$. Since $m > 0$, the last inference of the derivation $\mathfrak{D}_2$ (of $t\, RED_m\, s_2$) must be (A): $Kt_1\, RED_k\, r_1$, $t_2\, RED_n\, r_2$, where $r_1r_2 \equiv s_2$ and k, $n < m$.
By 4.7 (ii), there is a t_1' such that $r_1 \equiv Kt_1'$ and $t_1\, RED\, t_1'$: t_1' is the right choice.

1.2.2. Let (P) be the inference applied in $\mathfrak{D}_1$: for definiteness, assume that we apply left projection. Then $t \equiv LEFT(PAIR\, t_1t_2)$; in $\mathfrak{D}_2$ we have, for some r_1, r_2,

$$LEFT\, RED\, r_1,\ (PAIRt_1t_2) RED\, r_2.$$

Hence by 4.7 (i) and (iii), $r_1 \equiv LEFT$ and we find t_1', t_2' such that t_i *RED* t_i' $(i = 1, 2)$ and $r_2 \equiv PAIR\ t_1'\ t_2'$. Choose $r \equiv t_1'$.

1.2.3 Let (S) be applied in $\mathfrak{D}_1$. Then $t \equiv St_1t_2t_3$ and $s_1 \equiv t_1t_3(t_2t_3)$; by 4.7 (iii), and noting that (A) is applied in $\mathfrak{D}_2$, we find t_1', t_2', t_3' such that $t_i\, RED\, t_i'$ $(1 \le i \le 3)$ and $s_2 \equiv St_1't_2't_3'$. Choose $r \equiv t_1't_3'(t_2't_3')$.

1.2.4 Let $(D.1)$ be the last inference of $\mathfrak{D}_1$: then $t\, RED\, t_1$ and

$t \equiv D\overline{n}\,\overline{n}\,t_1t_2$, $t_1 \equiv s_1$. By 4.7 (iv), we find t'_3, t'_4, t'_1, t'_2 such that $s_2 \equiv Dt'_3t'_4t'_1t'_2$ and $\overline{n}\,RED\,t'_3$, $\overline{n}\,RED\,t'_4$, $t_1\,RED\,t'_1$, $t_2\,RED\,t'_2$. Since $\overline{n}$ is normal, $t'_3 \equiv t'_4 \equiv \overline{n}$: hence we can choose $r \equiv t'_1$.
If $D.2$ is applied, $t \equiv D\overline{n}\,\overline{m}t_1t_2$ with $\overline{n}$ distinct from $\overline{m}$ and $s_1 \equiv t_2$. By normality of $\overline{n}$, $\overline{m}$ and 4.7 (iv), we find a t'_2 such that $t_2\,RED\,t'_2$ and $s_2\,RED\,t'_2$.

1.2.5. Assume $t \equiv PRED(SUC\overline{n})$ and $s_1 \equiv \overline{n}$. Then for some s', $s_2 \equiv PREDs'$, where $(SUC\overline{n})RED\,s'$: but $SUC\overline{n}$ is normal, hence $s' \equiv SUC\overline{n}$ and we can choose $r \equiv \overline{n}$.

Case 2. Assume n, $m > 0$: hence (A) is the last inference in both derivations. We may suppose

2.1 $t_1\,RED_k\,t'_1$, $t_2\,RED_m\,t''_1$ and $t \equiv t_1t_2$, $s_1 \equiv t'_1t''_1$, where k, $m < n$;

2.2 $t_1\,RED_p\,t'_2$, $t_2\,RED_q\,t''_2$ and $s_2 \equiv t'_2\,t''_2$, where p, $q < m$.
But $k + p$, $m + q < n+m$ and by IH, there are t', t'' such that

$$t'_1\,RED\,t',\ t'_2\,RED\,t' \text{and } t''_1\,RED\,t'',\ t''_2\,RED\,t''.$$

By (A), $s_1 \equiv t'_1t''_1\,RED\,t't''$ and $t'_2t''_2 \equiv s_2\,RED\,t't''$: hence we choose $r \equiv t't''$. □

4.8.1. REMARK. It is essential to the proof that D be restricted to numerals and, more generally, to closed normal terms (see Church's calculus of $\lambda\delta$-conversion in Church 1941, Barendregt 1984). There are counterexamples to CR in the case, where $PAIR$ satisfies surjectivity, namely $PAIR((LEFTt)(RIGHTt)) \geq t$ (see Barendregt 1984, p.403). However, this last reduction gives rise to a consistent convertibility relation by §5.

If we apply 4.8, 4.7 and the definition of normal form, we get

4.9. COROLLARY. (i) CR *holds for* $\geq$.
(ii) *If* t', $t'' \in \mathrm{NF}$ *and* $r \geq t'$, $r \geq t''$, *then* t' *and* t'' *coincide* (*uniqueness of normal form*).

4.9.1. REMARK. (a) A notion of reduction which enjoys CR, can be defined for the λ-formalism, plus ξ- and η-conversion (see §2).

(b) There exist several reduction strategies. However, for combinatory logic, there is a standard reduction procedure SR, whose main virtue is condensed in the Standardization Theorem (Curry-Feys 1958). If t has a normal form, then SR terminates and yields the normal form of t.

(c) The property of having a normal form is recursively enumerable, but not recursive, by a classical result of Church.

§4B. Term Models

We are ready to introduce the syntactical models of the theory of operations.

4.10. DEFINITION

(i) CTM := $\{t : t$ is a closed term$\}$;
TM := $\{t : t$ is an arbitrary term$\}$.

(ii) The *open term model* **TM** is the structure $\langle$TM, $*$, $=$, $\mathfrak{C}$, $N, T\rangle$, where the empty set interprets the truth predicate T and

1. $\mathfrak{C} = \{K, S, D, PAIR, LEFT, RIGHT, SUC, PRED, \overline{0}\}$;
2. $*$: $\text{TM}^2 \rightarrow \text{TM}$ is the operation of juxtaposition of terms (i.e. application);
3. $= \subseteq \text{TM}^2$ and $t = s$ holds iff $t \geq r$ and $s \geq r$, for some term r;
4. $N \subseteq \text{TM}$ and $t \in N$ iff $t \geq \overline{n}$, for some $n \in \omega$.

(iii) The *closed term model* **CTM** is the substructure of **TM**, whose support is TM.

4.11. THEOREM. *Let* $\mathcal{M}$:= **CTM** *or* **TM**. *Then* $\mathcal{M}$ *is a non-trivial model of* OP+CT, *but* $\mathcal{M}$ *is not a model of* OPλ (*for notations* 2.6, 3.7).

PROOF (sketch). $\mathcal{M}$ is non-trivial, because K, S are normal and $\mathcal{M}$ falsifies $K = S$ (by unicity of normal form). Again the Church-Rosser theorem ensures that $=$ is transitive and that $(Nt \wedge t = s \rightarrow Ns)$ holds in $\mathcal{M}$. In $\mathcal{M}$, $=$ preserves application, simply because $\geq$ is closed under (A). The axioms on the special constants and N are true by definition of $\geq$ and since the numerals of $\mathcal{M}$ are isomorphic to the standard numbers. Consider the closed term $1_2K = S(K1)K$ where $1 = S(K(SKK))$. Then $\neg K = 1_2K$ holds in $\mathcal{M}$ because the given terms are different normal terms; thus the axiom MS.2 of OPλ (see 2.5) fails in $\mathcal{M}$.
Assume that $f : N \rightarrow N$ (see 3.3) is true in $\mathcal{M}$; we define $F : \omega \rightarrow \omega$ by $F(n) = m$ iff $f\overline{n} \geq \overline{m}$. By hypothesis, F is total and its graph is trivially recursively enumerable: hence F is recursive. This – together with the fact that $\mathcal{M}$ is a model of OP and OP provably formalizes the standard results of elementary recursion theory – yields the soundness of CT. □

4.12. COROLLARY. *Let* $\mathcal{M}$:= **CTM** *or* **TM**. *If* t *and* s *are closed terms,*

(i) $\mathcal{M} \models t = s$ *iff* OP $\vdash t = s$;

(ii) $\mathcal{M} \models Nt$ *iff* OP $\vdash t = \overline{n}$, *for some numeral* $\overline{n}$.

PROOF. From right to left, apply 4.11. As to the opposite direction, it suffices to verify that $t \geq r$ implies OP $\vdash t = r$. □

The proof of CR is constructive and it can be carried out in the system PA of first-order Peano arithmetic (PA is described in the appendix). Hence we have:

4.13. THEOREM. OP *is interpretable in the system* PA *of Peano Arithmetic.*

4.13 can be sharpened, once we realize that CR is (at least) provable in PRA, the system of primitive recursive arithmetic. This remark naturally leads to a subsystem of OP, which is tailored for PRA. Let $\exists(+)$ be the smallest class of $\mathcal{L}$-formulas which is generated from atoms of the form Nt, $t = s$ by means of $\wedge, \vee$ and existential quantification; $\exists(+)$-NIND is N-induction schema restricted to $\exists(+)$-conditions. If we define

$$\mathrm{OP}_1 := \mathrm{OP}^{-} + \exists(+)\text{-NIND},$$

we can prove:

4.14. THEOREM. *If* f *is a combinator such that* $\mathrm{OP}_1 \vdash f : N \rightarrow N$, *then* f *defines a primitive recursive function.*

Of course, the proof depends upon a careful formalization of 4.11 (for more details, see Troelstra 1973, Troelstra and van Dalen 1988, Jäger and Strahm 1994, and the Appendix). 4.13-4.14 can be further strengthened by adding to OP and OP_1 some truths of the closed term model **CTM**. An important example (to be applied in Ch.VI) is the enumeration axiom:

EA $$\exists f \forall x \exists y (Ny \wedge fy = x).$$

EA holds in **CTM**, because there is a closed term which enumerates CTM (cf. Appendix).

4.15. THEOREM. *Theorem* 4.13 (4.14) *remains true if we replace* OP (OP_1) *respectively by the system* OP+EA+CT (OP_1+EA).

We do not know whether Church's thesis can be conservatively added to OP_1+EA. We also mention that there are consistency results involving continuity of extensional operations, encoding type-2-functionals (see Beeson 1985, Troelstra and van Dalen 1988).

§5. The graph model

In this section we describe the classical graph model $P\omega$, due to Plotkin and Scott, its recursive submodel RE and Engeler's D_M-models. $P\omega$, D_M and RE verify OPλ, i.e. OP plus the Meyer-Scott axioms (cf. 2.5) and hence they are models of (the extended) λ-calculus with ξ-conversion, but without full extensionality (cf. 2.6). The construction of RE can be carried out in OP and this fact yields another method to interpret OP plus the Meyer-Scott axioms into Peano arithmetic (see §2). While RE validates Church's thesis, $P\omega$ is a model of a strong choice principle AC_N on natural numbers: OP+AC_N yields a model of full second-order arithmetic.

Let us fix a few preliminaries. First of all, we let a, b, c, d, x, y, z, u range over elements of $P\omega = \{x : x \subseteq \omega\}$, where ω is the set of natural numbers. Of course, if a, $b \in P\omega$, $a = b$ stands for extensional equality; n, m, k, p, i, j range over ω. We also adopt the lambda notation informally, i.e. to name (set-theoretically defined) functions.
If T is the Kleene predicate, $W_k = \{n : \exists m T(k,n,m)\}$ = the k-th r.e. set (r.e = recursively enumerable) and $RE = \{x : x \in P\omega \wedge \exists n (W_n = x)\}$.

We also put $(n,m) = \frac{1}{2}(n+m)(n+m+1)+m$; $\lambda x \lambda y.(x,y)$ is a primitive recursive bijection of $\omega \times \omega$ onto ω. We also define a canonical enumeration of the finite subsets of ω:

$$e_n = \{n_0, \ldots, n_{p-1}\}, \text{ provided } n_0 < \ldots < n_{p-1} \text{ and } n = \sum_{i<p} 2^{n_i}, \text{ and } e_0 = \emptyset.$$

F, G, H range over the set of operators (= total functions) from $P\omega$ to $P\omega$. We set

$$F \subseteq G := F(a) \subseteq G(a), \text{ for every } a \in P\omega.$$

5.1. DEFINITION

(i) $GRAPH(F) := \{(n, m) : m \in F(e_n)\}$;

$FUN(a) := \lambda b.\{n : \exists k(e_k \subseteq b \wedge (k,n) \in a)\}$ (for $a \in P\omega$);

(ii) $a \cdot b := (FUN(a))(b)$;

(iii) F is *continuous* iff $F(a) = \cup\{F(e_k) : e_k \subseteq a\}$ (for every $a \in P\omega$);

(iv) F is *r.e.* iff $GRAPH(F)$ is r.e;

(v) F is *effectively continuous* iff F is continuous and r.e.

$P\omega$ is a topological space under the so-called positive information topology, where the basic open sets have the form $\mathcal{O}_k = \{a : a \subseteq P\omega \wedge e_k \subseteq a\}$, for some $k \in \omega$ (cf. Scott 1976 or Barendregt 1984).

5.2. LEMMA

(i) *Continuous operators are monotone, i.e.* $a \subseteq b$ *implies* $F(a) \subseteq F(b)$;
(ii) $P\omega$, RE *are closed under the application operation* $\cdot$;
(iii) $a \cdot x = \cup\{a \cdot e_k : e_k \subseteq x\}$ *and* $a \cdot x = \cup\{e_k \cdot x : e_k \subseteq a\}$;
$x \subseteq y$ *implies* $a \cdot x \subseteq a \cdot y$ *and* $x \cdot a \subseteq y \cdot a$;
(iv) $FUN(a)$ *is continuous; if* a *is r.e, so is* $FUN(a)$;
(v) RE *is closed under effectively continuous operators;*
(vi) *If* $F \subseteq G$, *then* $GRAPH(F) \subseteq GRAPH(G)$;
$a \subseteq b$ *implies* $FUN(a) \subseteq FUN(b)$.

The verification of 5.2 is straightforward by definitions of continuity, application, FUN, $GRAPH$ and the closure properties of RE.

5.3. LEMMA

(i) *If* F *is continuous,* $FUN(GRAPH(F)) = F$.
(ii) *Let* $H(a) = GRAPH(FUN(a))$: *then* H *is a closure operator* (H *is monotone and it satisfies* $a \subseteq H(a)$ *and* $H(H(a)) \subseteq H(a)$).
(iii) F *is effectively continuous iff* $F = FUN(a)$, *for some r.e set* a.

PROOF. (i) By continuity and definition of FUN and $GRAPH$, we have

$$\begin{aligned} m \in F(a) &\Leftrightarrow m \in F(e_n), \text{ for some } e_n \subseteq a \text{ (continuity)};\\ &\Leftrightarrow (n,m) \in GRAPH(F), \text{ for some } e_n \subseteq a;\\ &\Leftrightarrow m \in FUN(GRAPH(F))(a) \text{ (by 5.1 (i))}.\end{aligned}$$

(ii): by (i), 5.2 (iv)-(vi).
(iii): choose $a = GRAPH(F)$ and apply (i). □

By theorem 2.9, it is enough to interpret in $P\omega$ the elementary theory Λ of 2.8, whose language $\mathfrak{L}(\lambda)$ contains the constants $\overline{0}$, SUC, $D, PRED$, $PAIR$, $LEFT$, $RIGHT$, the binary symbol Ap, the λ-operator and N (unary predicate constant). If C is a basic constant, $[\![C]\!]$ is the interpretation of C in $P\omega$, indeed an r.e subset of ω. We freely use the improper notation (a,n), whenever $a = \{n_1,\ldots,n_k\}$ instead of (p,n), p being the canonical code of a.

5.4 *Realization of* $\mathfrak{L}(\lambda)$ *in* $P\omega$ *and RE.*

(i) $[\![\overline{0}]\!] = \{0\}$;

(ii) $[\![SUC]\!] = \{(\{n\}, n+1) : n \in \omega\}$;

(iii) $[\![PRED]\!] = \{(\{n+1\}, n) : n \in \omega\}$;

(iv) $[\![PAIR]\!] = \{(\{n\}, (0, 2n)) : n \in \omega\} \cup \{(0, (\{m\}, 2m+1)) : m \in \omega\}$;

(v) $[\![LEFT]\!] = \{(\{2k\}, k) : k \in \omega\}$;

(vi) $[\![RIGHT]\!] = \{(\{2k+1\}, k) : k \in \omega\}$;

(vii) $[\![D]\!] = \{(\{m\}, (\{n\}, (l, (k, i)))) : k, l, m, n, i \in \omega \wedge$
$\wedge [(m = n \wedge i \in e_l) \vee (\neg m = n \wedge i \in e_k)]\}$;

(viii) The unary predicate N is interpreted by $[\![N]\!] = \{\{n\} : n \in \omega\}$, while identity is interpreted by extensional equality.

5.5. *Assignments.* An assignment is a map $\rho : \omega \to P\omega$. If $P\omega$ is replaced by RE, ρ is called RE-assignment. Once ρ is given, we define, as usual, the assignment $\rho(i := a)$ such that $\rho(i := a)(i) = a$; $\rho(i := a)(j) = \rho(j)$ if $j \neq i$. If $x := x_i$, we simply write $\rho(x := a)$ instead of $\rho(i := a)$.

5.6. INDUCTIVE DEFINITION OF THE VALUE $[\![t]\!]_\rho^M$ (t arbitrary term of $\mathcal{L}(\lambda)$, ρ assignment, $M = P\omega$, RE).

(i) If C is one of the basic constants, then $[\![C]\!]_\rho^M = [\![C]\!]$;

(ii) $[\![x_i]\!]_\rho^M = \rho(i)$;

(iii) $[\![rs]\!]_\rho^M = [\![r]\!]_\rho^M \cdot [\![s]\!]_\rho^M$;

(iv) $[\![\lambda x.s]\!]_\rho^M = GRAPH(\lambda a.[\![s]\!]^M_{\rho(x := a)}) = \{(k,m) : m \in [\![s]\!]_{\rho(x := e_k)}\}$.

N.B. In (iv) λx is the syntactical operator applied to the variable x; λa is ranging over $D = P\omega, RE$.

Clearly, $[\![t]\!]_\rho^M \in P\omega$ by definition of $GRAPH$ and 5.2 (ii).

5.7 LEMMA. *For every term t and every valuation ρ,*

(i) $\lambda a.[\![t]\!]^M_{\rho(x := a)}$ *is a continuous operator;*

(ii) *if y is free for x in t and s is free for x in t,*

$[\![t]\!]^M_{\rho(x := a)} = [\![t\ [x := y]]\!]^M_{\rho(y := a)}$;

$[\![t\ [x := s]]\!]^M_\rho = [\![t]\!]^M_{\rho(x := [\![s]\!]^M_\rho)}$;

(iii) *if ρ is an RE-assignment, $[\![t]\!]^M_\rho$ is r.e and hence $\lambda a.[\![t]\!]^M_{\rho(x := a)}$ is effectively continuous; moreover $[\![t]\!]^{P\omega}_\rho = [\![t]\!]^{RE}_\rho$.*

PROOF: by induction on the definition of t. We only check that $\lambda a.[\![t]\!]^M_{\rho(x := a)}$ continuously depends on the interpretation of the variables.

First note that $\lambda a.[\![t]\!]^M_{\rho(x := a)}$ is monotone (apply 5.2 (iii)).

As to continuity, if t is atomic, the conclusion is trivial. Assume $t = rs$; pick any $m \in [\![t]\!]^M_{\rho(x := a)}$: then for some $e_k \subseteq [\![s]\!]^M_{\rho(x := a)}$, we have

$(k,m) \in [\![r]\!]^M_{\rho(x := a)}$. By IH, for every $i \in e_k$, there is a $k(i)$ such that $e_{k(i)} \subseteq a$ and $i \in [\![s]\!]^M_{\rho(x := e_{k(i)})}$. Since e_k is finite, we can find a j such that, for every $i \in e_k$, $e_{k(i)} \subseteq e_j \subseteq a$; by monotonicity, $e_k \subseteq [\![s]\!]^M_{\rho(x := e_j)}$. Again by IH, $(k,m) \in [\![r]\!]^M_{\rho(x := e_n)}$ for some n such that $e_n \subseteq a$. Choose $p = \max(j,n)$: by monotonicity

$$e_n \subseteq [\![s]\!]^M_{\rho(x := e_p)} \text{ and } (k,\, m) \in [\![r]\!]^M_{\rho(x := e_p)},$$
$$m \in [\![t]\!]^M_{\rho(x := e_p)} \text{ by definition.}$$

Assume $(k,m) \in [\![\lambda y.t]\!]^M_{\rho(x := a)}$: then by definition $m \in [\![t]\!]^M_{\rho(x := a,\, y = e_k)}$. Hence by IH, there is an n such that $e_n \subseteq a$ and $m \in [\![t]\!]^M_{\rho(x := e_n,\, y = e_k)}$, i.e. by definition $(k,m) \in [\![\lambda y.t]\!]^M_{\rho(x := e_n)}$. This concludes the verification of continuity, since the other direction follows by monotonicity. □

Once $M = RE$, $P\omega$ is fixed or clear from the context, we simply write $[\![t]\!]_\rho$ instead of $[\![t]\!]^M_\rho$. Let ρ be an assignment in M and let $\mathcal{M}$ be the structure for the language $\mathcal{L}(\lambda)$ whose domain is M: then the initial conditions

$$\mathcal{M} \models (t = s)\ [\rho] \text{ iff } [\![t]\!]_\rho = [\![s]\!]_\rho \text{ and}$$

$$\mathcal{M} \models (Nt)\ [\rho] \text{ iff } [\![t]\!]_\rho \in [\![N]\!] \text{ iff for some } k \in \omega,\ [\![t]\!]_\rho = \{k\},$$

fully determine the standard satisfaction relation for $\mathcal{L}(\lambda)$

$$\mathcal{M} \models A[\rho] \Leftrightarrow \rho \text{ satisfies } A \text{ in } \mathcal{M},$$

once we interpret T as the empty set. Then we can state:

5.8. THEOREM. $\mathcal{M}$ *(with* $M = P\omega$, RE*) is a model of* Λ *(and hence* OPλ*) plus the surjective pairing axiom* $\forall x(PAIR(LEFTx)(RIGHTx) = x)$. *Moreover* $\mathcal{M} \models \neg 1 = I$ *and hence* Ext_{op} *fails in* $\mathcal{M}$ *(cf.* 2.5*).*

PROOF. We have to verify the schemata of ξ-, α- and β-conversion (see 2.2). First observe that $=$ is preserved by application (5.2 (iii)). As to (ξ), assume $[\![t]\!]_{\rho(x := a)} \subseteq [\![s]\!]_{\rho(x := a)}$, for all $a \in \mathcal{M}$. Then, by definition of inclusion for operators, $\lambda a.[\![t]\!]_{\rho(x := a)} \subseteq \lambda a.[\![s]\!]_{\rho(x := a)}$; hence:

$$[\![\lambda x.t]\!]_\rho \subseteq [\![\lambda x.s]\!]_\rho \text{ (by 5.2 (vi) and 5.6 (iv)).}$$

(α)-conversion is essentially the first part of 5.7 (ii). As to (β)-conversion, if $y = y_i$, $\rho(i) = a$,

$$[\![(\lambda x.t)y]\!]_\rho = [\![\lambda x.t]\!]_\rho \cdot a = FUN(GRAPH\ (\lambda c.[\![t]\!]_{\rho(x := c)}))(a) =$$

$$= [\![t]\!]_{\rho(x := a)} = [\![t[x := y]]\!]_{\rho} \text{ (apply 5.3, 5.7).}$$

Hence $\mathcal{M}$ is a model of λ-calculus. By definition of $[\![N]\!]$, it is easy to check that $[\![N]\!]$ is the least subset of RE, which contains $\{0\}$ and is closed under $[\![SUC]\!]$. Hence the N-induction schema is satisfied; all the remaining axioms are verified with 5.4 (as to the surjective pairing axiom, observe that if $n \in a$, either $n = 2m$ and hence $m \in [\![LEFTa]\!]$ or $m = 2m+1$ and $m \in [\![RIGHTa]\!]$). A simple computation also shows that $1 = I$ (= MS.4 of §2) fails in RE; hence Ext_{op} is false in RE (and $P\omega$). □

$P\omega$ and RE greatly differ as to the interpretation of the space of number-theoretic operations.

5.9. THEOREM. *RE is a model of Church's thesis and of the enumeration axiom* (*see* EA, *after* 4.14).

PROOF. EA: this is a by-product of the enumeration theorem for closed $\mathcal{L}(\lambda)$-terms (cf. Scott 1976, Barendregt 1984, p.166 or Appendix), plus the following fact: to each r.e. set a, we can effectively associate a closed $\mathcal{L}(\lambda)$-term t_a, such that $[\![t_a]\!] = a$ (Scott, cit. or Beeson 1985, pp.133-134).
CT: assume $b \in RE$ and for every $a \in [\![N]\!]$, $b \cdot a \in [\![N]\!]$. Define $F(n) = m$ iff $m \in b \cdot \{n\}$. Then F is a function (as $b \cdot \{n\}$ is a singleton) and is total ; but F has an r.e. graph: hence F is recursive . □

5.9.1. REMARK. The cited result for $RE = \{[\![t]\!] : t$ closed term of $\mathcal{L}(\lambda)\}$ should be contrasted with the fact that $P\omega^0 \subset RE$ ($\subset$ strict inclusion), $P\omega^0$ being the set $\{[\![t]\!] : t$ closed term of *pure* lambda calculus$\}$. This follows from 5.7 (iii), 5.8 and the fact that that $P\omega^0$ *is not a model* of ξ-conversion 2.2 by Barendregt (1984, p.514).

5.10. THEOREM (*Extension*). *Let X be a subspace of the topological space Y. If $F : X \rightarrow P\omega$ is continuous, then there exists a continuous function $\overline{F} : Y \rightarrow P\omega$, which extends F.*

PROOF: one can define, if $y \in Y$,

$$\overline{F}(y) = \cup \{ \cap \{F(x) : x \in X \cap \mathcal{U}\} : y \in \mathcal{U}, \mathcal{U} \text{ open in } Y\}. \qquad \square$$

If we endow ω with the discrete topology and we identify $[\![N]\!]$ with ω, ω can be regarded as a subspace of $P\omega$, and hence by 5.10 and choice, we have:

5.11. COROLLARY

(i) *If $F : \omega \rightarrow P\omega$, then there is a continuous function $\overline{F} : P\omega \rightarrow P\omega$, which extends* (*i.e.* $\overline{F}(\{n\}) = F(n)$, *for* $n \in \omega$).

(ii) *If $R \subseteq \omega \times P\omega$ and the domain of R is ω, then R has a continuous choice function, i.e. there exists a continuous operator F such that for every $n \in \omega$, $R(n, F(n))$.*

5.12. DEFINITION

(i) AC_N is the schema

$$\forall n \exists y A(n,y) \to \exists f \forall n A(n, fn) \text{ (A arbitrary formula of $\mathcal{L}$ or $\mathcal{L}(\lambda)$).}$$

(ii) $\mathrm{AC}_N!$ is the schema obtained from AC_N by replacing $\exists y$ with $\exists!\, y$(= there exists a unique y).

As an immediate application of 5.11, we obtain:

5.13. THEOREM. *$P\omega$ is a model of* $\mathrm{OP}\lambda+\mathrm{AC}_N$.

It is worth noting that $\mathrm{OP}\lambda+\mathrm{AC}_N!$ yields a model of full second-order arithmetic (cf.40.2), if we interpret variables ranging over sets of numbers simply as operations from ω to $\{0,1\}$. We shall see later that the restriction to numbers in the choice schemata cannot be neglected (because of an inconsistency); however, there are models of λ-calculus where the choice schema holds, if the opening universal quantifier is bounded by a type generated from ω (one applies the models of Flagg 1989).

5.14. *Generalization: Engeler models*

In order to study computability over an arbitrary structure $\mathcal{M}$ with (non-empty) domain M, one would like to expand $\mathcal{M}$ with a set of additional objects, which represent programs or rules of constructions. Engeler (1981) produces simple generalizations of the model $P\omega$, the so-called D_M-models, which meet the above desiderata.
Below, lower case Greek letters α, β, γ stand for finite subsets of M. If $\alpha \subseteq M$ is a (finite) subset of M, we write $(\alpha \to b)$ for the ordered pair $\langle \alpha, b \rangle$. The notation is suggestive of the situation, where the elements of M are atoms of a given language and $(\alpha \to b)$ is a program clause with b as head and α as body (see Engeler 1988).

5.14.1. DEFINITION

(i) $G_0(M) = M$;

$$G_{k+1}(M) = G_k(M) \cup \{(\alpha \to b) : \alpha \subseteq G_k(M),\ \alpha \text{ finite and } b \in G_k(M)\};$$

$$G(M) = \cup \{G_n(M) : n \in \omega\}.$$

(ii) If X, $Y \subseteq G(M)$ and $\mathcal{P}(G(M))$ is the power set of $G(M)$, we define application by

$$X*Y := \{b \in G(M) : \exists \alpha \subseteq Y.(\alpha \to b) \in X\}.$$

(iii) If $F: \mathcal{P}(G(M)) \to \mathcal{P}(G(M))$ is a continuous function (i.e. it is $\subseteq$-monotone and preserves $\bigcup$), λ-abstraction is interpreted with the natural generalization of $GRAPH$:

$$\lambda_M(F) := \{\langle\beta, a\rangle : \beta \subseteq G(M) \text{ and } a \in F(\beta)\}.$$

(iv) Finally, we put

$$D_M := \langle\mathcal{P}(G(M)),\ *,\ \lambda_M\rangle.$$

It is then clear how to adapt the previous results in order to make sense of the following

5.15. THEOREM. *D_M is a λ-model of* OPλ.

5.16. REMARK. Graph models form a significant class of λ-models and they have been studied in general. We mention that D_M and $P\omega$ are not isomorphic; $P\omega$ has no non-trivial automorphism (Schellinx 1991). For extensive information on models of λ-calculus and combinatory logic, the interested reader might consult Scott (1982), Longo (1983), Koymans (1984), Barendregt (1984), Lambek and Scott (1986), Hindley and Seldin (1986), Asperti and Longo (1990).

§6. An effective version of the extensional model D_∞

Henceforth we work within the model RE of the previous section; we use capital letters X, Y, Z, P, Q for r.e. sets and we freely write $X(Y)$ for $X \cdot Y$, and $\lambda X.P(X)$ for the graph of the operator $X \mapsto P(X)$.
By 5.3 (ii), RE satisfies the laws:

(η^-): $$Q \subseteq \lambda X.Q(X)$$

(ξ^*): $$Z \subseteq Y \text{ implies } \lambda X.Z(X) \subseteq \lambda X.Y(X).$$

6.1. DEFINITION (after Scott 1976, 1980)

(i) $P \circ Q := \lambda X.P(Q(X))$;

(ii) $D_0 := \lambda X.X = I = \{(n,m) : m \in e_n\}$;

$D_{n+1} := \lambda P.D_n \circ P \circ D_n$.

Of course the operation $\circ$ is associative .

6.2. LEMMA

1. *We can find a primitive recursive function σ such that for every $n \in \omega$,*

$$W_{\sigma(n)} = D_n.$$

2. *For every n,*

(i) $D_n \subseteq D_{n+1}$;

(ii) $D_n \circ D_n = D_n$.

PROOF. 1. The definition of $\lambda X.W_a \circ X \circ W_a$ is uniform in a and hence there exists a primitive recursive function τ such that

$$W_{\tau(a)} = \lambda X.W_a \circ X \circ W_a.$$

If we let $\sigma(0)$ = index for the r.e. set I, $\sigma(n+1) = \tau(\sigma(n))$, we are done.
2. If $n = 0$, (ii) is trivial. By (η^-), $Y \subseteq \lambda X.Y(X)$; hence by (ξ^*), $D_0 = \lambda Y.Y \subseteq \lambda Y \lambda X.Y(X) = D_1$, which verifies (i).
Induction step: we assume $D_n \subseteq D_{n+1}$ and $D_n = D_n \circ D_n$. Then we have by monotonicity on the right and on the left (5.2 (iii)) plus IH,

$$D_n(P(D_n(X))) \subseteq D_{n+1}(P(D_{n+1}(X))). \tag{+}$$

$D_{n+1} \subseteq D_{n+2}$ is a consequence of (+) and (ξ^*). As to (ii), we have by associativity of $\circ$ and IH,

$$\begin{aligned}(D_{n+1} \circ D_{n+1})(P) &= D_{n+1}(D_{n+1}(P)) = D_{n+1}(D_n \circ P \circ D_n) = \\ &= D_n \circ (D_n \circ P \circ D_n) \circ D_n = (D_n \circ D_n) \circ P \circ (D_n \circ D_n) = \\ &= D_n \circ P \circ D_n = D_{n+1}(P).\end{aligned}$$

Hence $D_{n+1} \circ D_{n+1} = D_{n+1}$ by (ξ^*). □

6.3. DEFINITION $\qquad D_\infty := \cup\{D_n : n \in \omega\}.$

6.3.1. FACT. *D_∞ is r.e.*

PROOF: by 6.2.1 D_∞ is the union of a primitive recursive family of r.e. sets. □

N.B. By the proof of 5.9 and 5.9.1, there is a closed term t of $\mathcal{L}(\lambda,)$ with $[\![t]\!]^{P\omega} = D_\infty$; but no such term exists in *pure λ-calculus* by Longo (1983, p.170).

We now verify that the collection of r.e. fixed points of $FUN(D_\infty)$ is the required extensional λ-model.

6.4. MAIN LEMMA

(i) $\qquad D_\infty = D_\infty \circ D_\infty$

(ii) $\qquad D_\infty = \lambda P.D_\infty \circ P \circ D_\infty.$

PROOF. (i) $D_\infty \subseteq D_\infty \circ D_\infty$: by monotonicity of application, 6.2.2 (ii). $D_\infty \circ D_\infty \subseteq D_\infty$: if $m \in D_\infty(D_\infty(e_k))$, by continuity, there exists a j such that $e_j \subseteq D_\infty(e_k)$ and $m \in D_\infty(e_j)$. Since e_j is finite, we find a q such that $e_j \subseteq D_q(e_k)$ and by monotonicity of application $m \in D_\infty(D_q(e_k))$. By continuity of application in the first coordinate, and since $\lambda x.D_x$ is increasing, we finally get, for p big enough, by 6.2 and monotonicity of application,

$$m \in D_p(D_p(e_k)) = D_p(e_k) \subseteq D_\infty(e_k).$$

(ii) : similar. □

6.4.1. REMARK. Since $I = \lambda X.X \subseteq D_\infty$ (by 6.1 and 6.2), 6.4 (i) states that D_∞ is a *closure* operator on RE (see 5.3 (ii)). By 6.4 (i) we also have

$$\mathrm{FIX}(D_\infty) = \{X : X \in RE \wedge D_\infty(X) = X\} = \{D_\infty(X) : X \in RE\}.$$

Moreover, $I \subseteq D_\infty$ and 6.4 (i)-(ii) imply that $D_\infty(D_\infty) = D_\infty$.

6.5. DEFINITION

(i) $\mathbb{D}_\infty := \mathrm{FIX}(D_\infty)$

(ii) Given any assignment $\rho : \omega \rightarrow \mathbb{D}_\infty$, we inductively define the value $[\![t]\!]_\rho^{D_\infty}$:

$$\begin{aligned} [\![x_i]\!]_\rho^{D_\infty} &= \rho(i); \\ [\![ts]\!]_\rho^{D_\infty} &= [\![t]\!]_\rho^{D_\infty} \cdot [\![s]\!]_\rho^{D_\infty}; \\ [\![\lambda u.t]\!]_\rho^{D_\infty} &= \lambda X.D_\infty([\![t]\!]_{\rho(u := D_\infty(X))}^{D_\infty}). \end{aligned}$$

$[\![t]\!]_\rho^{D_\infty}$ is well-defined by the theorem below; we simply write $[\![t]\!]_\rho$ instead of $[\![t]\!]_\rho^{D_\infty}$ and we neglect the interpretation of the additional constants of Λ.

6.6. THEOREM. $\mathbb{D}_\infty$ *is a non trivial applicative substructure of* RE, *which can be expanded to a model of* OPλ *and extensionality for operations* Ext_{op}.

PROOF. $\mathbb{D}_\infty$ has at least two distinct elements $\mathrm{T} := \omega$ and $\perp := \emptyset$. Indeed, we have that $D_\infty(\mathrm{T}) = \mathrm{T}$ holds by remark 6.4.1. Trivially $\perp \subseteq D_\infty(\perp)$ and $D_0(\perp) \subseteq \perp$. Assume by IH $D_n(\perp) \subseteq \perp$. Then $D_n(\perp(D_n(X))) \subseteq \perp$ (IH and since $\perp(X) = \perp$); by ξ^* we get

$$D_{n+1}(\perp) = \lambda X.D_n(\perp(D_n(X))) \subseteq \lambda X.\perp = \perp.$$

Now let $D_\infty(X) = X$, $D_\infty(Y) = Y$: we then obtain, by the main lemma and hypothesis,

$$X(Y) = D_\infty(X)(Y) = (D_\infty \circ X \circ D_\infty)(Y) = D_\infty(X(D_\infty(Y))) = D_\infty(X(Y)).$$

This shows that $\mathbb{D}_\infty$ is an applicative substructure of RE; hence it satisfies left and right monotonicity of application. Moreover, for every term t, $\mathbb{D}_\infty$ is closed under the operation $\rho \mapsto [\![t]\!]_\rho$, provided $\rho : \omega \to \mathbb{D}_\infty$. This is verified by induction on t. In particular, by lemma 6.4, IH and β-conversion in RE:

$$D_\infty([\![\lambda x.t]\!]_\rho) = D_\infty \circ [\![\lambda x.t]\!]_\rho \circ D_\infty =$$
$$= \lambda Z.D_\infty((\lambda X.D_\infty([\![t]\!]_{\rho(x\,:=\,D_\infty(X))}))\,(D_\infty(Z))) =$$
$$= \lambda Z.D_\infty([\![t]\!]_{\rho(x\,:=\,D_\infty(D_\infty(Z)))}).$$

But $D_\infty \circ D_\infty = D_\infty$; hence

$D_\infty([\![\lambda x.t]\!]_\rho) = \lambda Z.D_\infty([\![t]\!]_{\rho(x\,:=\,D_\infty(Z))}) = [\![\lambda x.t]\!]_\rho$, i.e. $[\![\lambda x.t]\!]_\rho \in \mathbb{D}_\infty$.

β-conversion in $\mathbb{D}_\infty$: since RE is a model of β-conversion, we have for $[\![x]\!]_\rho = P$,

$$[\![(\lambda y.t)x]\!]_\rho = \lambda Y.D_\infty([\![t]\!]_{\rho(x\,:=\,D_\infty(Y))}) \cdot P = D_\infty([\![t]\!]_{\rho(x\,:=\,D_\infty(P))}).$$

But $[\![t]\!]_{\rho(x\,:=\,D_\infty(P))}$ and P are fixed points of D_∞, whence the conclusion.

ξ-conversion in $\mathbb{D}_\infty$ (cf. 2.2): let $\rho : \omega \to \mathbb{D}_\infty$ and assume

$$[\![t]\!]_{\rho(x\,:=\,P)} = [\![s]\!]_{\rho(x\,:=\,P)}, \text{ for every } P \in \mathbb{D}_\infty.$$

Then $[\![t]\!]_{\rho(x\,:=\,D_\infty(P))} = [\![s]\!]_{\rho(x\,:=\,D_\infty(P))}$, for every $P \in RE$ and hence by closure of $\mathbb{D}_\infty$ under $\rho \mapsto [\![-]\!]_\rho$, we have for every $P \in RE$,

$$D_\infty([\![t]\!]_{\rho(x\,:=\,D_\infty(P))}) = D_\infty([\![s]\!]_{\rho(x\,:=\,D_\infty(P))}).$$

By ξ-conversion in RE (5.8),

$$\lambda X.D_\infty([\![t]\!]_{\rho(x\,:=\,D_\infty(X))}) = \lambda X.D_\infty([\![s]\!]_{\rho(x\,:=\,D_\infty(X))}),$$

whence by 6.5 (ii) $[\![\lambda x.t]\!]_\rho = \lambda x.[\![s]\!]_\rho$.

η-conversion (see 2.2): it corresponds to the equation $[\![\lambda x.fx]\!]_\rho = [\![f]\!]_\rho$ which holds by 6.5 and 6.4 (ii). On the other hand, ξ- and η-conversion imply Ext_{op}. (α)-conversion is left to the reader. Finally, $\mathbb{D}_\infty$ can be expanded to a full model of OP by choosing the denotations of the terms given by 3.8. □

6.7. REMARK (Park's theorem). $\mathbb{D}_\infty$ and RE share an important feature: in these models the paradoxical combinator FP (see § 2.3) coincides with Tarski's fixed point operator, namely, if $X \in \mathbb{D}_\infty$ (or RE), $FP(X)$ satisfies the condition: $X(Z) \subseteq Z$ implies $FP(X) \subseteq Z$, i.e. $FP(X)$ is the $\subseteq$-least fixed point of the operator $FUN(X)$ (for a proof, see Scott 1980).

By straightforward arithmetization of the preceding model, we immediately obtain:

6.8. THEOREM. $\mathrm{OP}+\mathrm{Ext}_{op}$ *is interpretable in* PA.

However, if we wish to refine 6.8 with OP_1 and PRA in place of OP and PA respectively, it is not clear how to deal with D_∞, in presence of restricted inductions. The difficulty can be overcome by considering term models of $\mathrm{OP}+\mathrm{Ext}_{op}$.

According to the equivalence theorem of 2.9, it is enough to produce models for the system $\Lambda+\mathrm{Ext}_{op}$, i. e. the variant of OP based on λ-abstraction as primitive; of course, $\Lambda_1+\mathrm{Ext}_{op}$ is obtained from $\Lambda+\mathrm{Ext}_{op}$ by assuming the restricted induction schema $\exists(+)$-NIND of 4.13 in place of the full N-induction schema.

Let $\geq_{\beta\eta}$ be the least reflexive transitive relation which preserves application and is closed under the clauses (A), (P), (SUC), $(D.1)$, $(D.2)$ of 4.1 and

(β) $\quad (\lambda x.t)s \geq_{\beta\eta} t[x := s]$;

(η) $\quad \lambda x.tx \geq_{\beta\eta} t$ (x not free in t);

(ξ) $\quad$ from $t \geq_{\beta\eta} s$ infer $\lambda x.t \geq_{\beta\eta} \lambda x.s$.

By adapting the standard argument of Barendregt (1984), it turns out that $\geq_{\beta\eta}$ satisfies the Church-Rosser property (cf. 4.3); hence the associated conversion relation is a non-trivial congruence relation on the set $\mathrm{TM}_{\beta\eta}$ of all terms in the language of $\Lambda+\mathrm{Ext}_{op}$, *provably in primitive recursive arithmetic.*

To be definite, let $\mathbf{TM}_{\beta\eta}$ be the structure $\langle \mathrm{TM}_{\beta\eta}, *, =_{\beta\eta}, \mathfrak{C}, N_{\beta\eta}\rangle$, where $\mathrm{TM}_{\beta\eta}$ is the set of all terms in the language of Λ and

1. $\mathfrak{C} = \{D, PAIR, LEFT, RIGHT, SUC, PRED, \overline{0}\}$;

2. $* : \mathrm{TM}^2_{\beta\eta} \rightarrow \mathrm{TM}_{\beta\eta}$ is the operation of juxtaposition of terms (i.e. application);

3. $=_{\beta\eta} \subseteq \mathrm{TM}^2_{\beta\eta}$ and $t =_{\beta\eta} s$ holds iff $t \geq_{\beta\eta} r$ and $s \geq_{\beta\eta} r$, for some $r \in \mathrm{TM}_{\beta\eta}$;

4. $N_{\beta\eta} \subseteq \mathrm{TM}_{\beta\eta}$ and $t \in N_{\beta\eta}$ iff $t \geq_{\beta\eta} \overline{n}$, for some n.

Now $\mathbf{TM}_{\beta\eta}$ is a non-trivial model of $\Lambda+\mathrm{Ext}_{op}$, and satisfies CT, as in 4.11. Thus by standard arithmetization, we get:

6.9. THEOREM. OP+Ext_{op}+CT (*respectively* OP_1+Ext_{op}) *is interpretable in* PA (PRA *respectively*).

A final question involves the consistency of OP+Ext_{op}+EA+CT. At present, we can only state the following partial result:

6.10. THEOREM. OP+Ext_{op}+EA *is consistent.*

For the proof, consider the theory OP(ω), which includes OP plus an infinitary ω-rule for terms :

$$\frac{tr = sr, \text{ for each closed term } r}{t = s}.$$

Then define $\mathbf{CTM}(\omega) := \langle \text{CTM}, *, =_{\omega}, \mathfrak{C}, \text{N}_{\omega}\rangle$, where CTM is the set of $\mathcal{L}_{op}$-closed terms and

1. $\mathfrak{C} := \{K, S, D, PAIR, LEFT, RIGHT, SUC, PRED, \overline{0}\}$;
2. $* : \text{CTM}^2 \to \text{CTM}$ is the operation of juxtaposition of terms (i.e. application);
3. $=_{\omega} \subseteq \text{CTM}^2$ and $t =_{\omega} s$ holds iff $\text{OP}(\omega) \vdash t = s$;
4. $N_{\omega} \subseteq \text{CTM}$ and $t \in N_{\omega}$ iff $\text{OP}(\omega) \vdash Nt$.

By Barendregt (1984, Ch.XVII and Ch.XIX, p. 508), OP(ω) is consistent; hence $\mathbf{CTM}(\omega)$ is non-trivial and makes Ext_{op} and EA obviously true (see also Flagg-Myhill 1987).

Appendix

This appendix contains a few details about results, which were quickly summarized in §4. First of all, we deal with the provability of the Church Rosser theorem in PRA. It is fairly obvious to see that the proof of §4 works in PA; however, it is not entirely obvious that only suitably restricted instances of number-theoretic induction are needed.

A reminder on the chosen metatheory PRA is in order: the basic language is a standard first-order language containing $\overline{0}$ (zero), successor and function symbols for primitive recursive functions; terms and formulas are defined as usual. (An occurrence of) A quantifier $\exists$ ($\forall$) in a formula is bounded, if $\exists$ occurs in the context $\exists x(x < t \wedge \ldots)$ (respectively $\forall x(x < t \rightarrow \ldots)$). Formulas, which only contain bounded number quantifiers, are called bounded (or Δ_0); formulas of the form $\exists x B$ ($\forall y \exists x B$) are called Σ_1 (Π_2), provided B is bounded. PRA is the formal system, based on classical predicate calculus, which includes Peano axioms for zero and successor, defining equations for primitive recursive functions and number-theoretic induction for bounded formulas. By Σ_1-induction = Σ_1-IND, we mean the induction schema extended to Σ_1-formulas; the rule of Π_2-induction, Π_2-INDR, is the rule: if A is a Π_2-formula, infer $\forall x A$ from $A(\overline{0}), \forall x(A(x) \to A(x+1))$. PA can be identified with the extension of PRA which contains the number-theoretic induction schema for arbitrary formulas.

A.1. THEOREM (Parsons 1972). PRA, $\mathrm{PA}_1 = \mathrm{PRA}+\Sigma_1\text{-IND}$ *and* $\mathrm{PRA}+\Pi_2$-INDR *have the same* Π_2*-theorems.*

A.1 grants that the provably recursive functions of the three systems are exactly the primitive recursive functions.

We know from §4 that there exists a formal calculus $\mathcal{T}$, for deriving expressions of the form $tREDs$ (t, s terms of OP). $\mathcal{T}$ has the axioms (R), (K), (S), (P), (SUC), ($PRED$), (D.1)-(D.2) of 4.1, while the inference rule is (A); a $\mathcal{T}$-derivation is a finite sequence, whose elements are either axioms or else expressions, obtained by previous ones by application of (A). If d is a derivation, ending with $tREDs$, we put $\mathrm{Head}(d) = t$ and $\mathrm{Tail}(d) = s$. It is folklore to find: 1) a bounded formula $\mathrm{Dim}(d)$, which formally represents in PRA the metamathematical predicate "d is a derivation in the $\mathcal{T}$-calculus"; 2) terms representing the functions Head and Tail. Moreover, $tREDs$ is $\mathcal{T}$-derivable iff $tREDs$ holds according to 4.4 (ii).

It is easy to see that (modulo encoding of tuples) the statement of the crucial inversion lemma 4.7 has the form $\forall x R(x, t(x))$, where $t(x)$ is a term of PRA, actually built-up by inspecting the proof, and R is a bounded formula. As a consequence of bounded induction, we get:

A.2. LEMMA. PRA *proves the formalization of* 4.7.

The Church-Rosser theorem $\mathrm{CR}(RED)$ for RED has the form of a bounded condition

$$(\forall d < x)(\forall d' < x)(x = d+d' \wedge \mathrm{Dim}(d) \wedge \mathrm{Dim}(d') \wedge \mathrm{Head}(d) = \mathrm{Head}(d') \to$$

$$\to (\mathrm{Dim}(\mathrm{CR}_1(d, d')) \wedge \mathrm{Dim}(\mathrm{CR}_2(d, d')) \wedge \mathrm{Head}(\mathrm{CR}_1(d, d')) = \mathrm{Tail}(d) \wedge$$

$$\wedge \mathrm{Head}(\mathrm{CR}_2(d, d')) = \mathrm{Tail}(d') \wedge \mathrm{Tail}(\mathrm{CR}_1(d, d')) = \mathrm{Tail}(\mathrm{CR}_2(d, d'))),$$

where $\mathrm{CR}_1(x, y)$, $\mathrm{CR}_2(x, y)$ are primitive recursive terms which can be explicitly extracted from the proof of 4.8. If we apply bounded induction on x, A.2 and we mimic the content of 4.8, we get

A.3. LEMMA. $\mathrm{PRA} \vdash \mathrm{CR}(RED)$.

Now we need CR for the transitive closure $\mathrm{TC}(RED)$ of RED, provably in PRA; hence we extend $\mathcal{T}$ to the system $\mathcal{T}^*$ with the transitivity rule (T) and we have an obvious notion of $\mathcal{T}^*$-derivation and a corresponding bounded formula Dim*, which represents it in PRA. Let:

$$R_0(d,a,b) := \mathrm{Dim}(d) \wedge \mathrm{Head}(d) = a \wedge \mathrm{Tail}(d) = b;$$

$$R^*(d,a,b) := \mathrm{Dim}^*(d) \wedge \mathrm{Head}(d) = a \wedge \mathrm{Tail}(d) = b.$$

As R_0,R^* are bounded, the following formula is Π_2:

$$C(n) := \forall d \forall d' \forall a \forall b \forall c (R_0(d,a,b) \wedge R^*(d',a,c) \wedge \mathrm{lh}(d') = n \rightarrow$$
$$\rightarrow \exists r \exists r' \exists x (R^*(r,b,x) \wedge R^*(r',c,x));$$

(here $\mathrm{lh}(x)$ is the primitive recursive term which computes the length of a finite sequence).

By A.3, we can show PRA $\vdash C(\overline{0})$ and PRA+$C(n) \vdash \mathrm{C}(n+1)$; hence a first application of Π_2-INDR together with Parsons's theorem yields:

A.4. LEMMA. PRA $\vdash \forall n C(n)$.

Again Π_2-INDR applied to the condition

$$B(n) := \forall d \forall d' \forall a \forall b \forall c (\mathrm{lh}(d') = n \wedge \mathrm{R}^*(d',a,b) \wedge R^*(d,a,c) \rightarrow$$
$$\rightarrow \exists r \exists r' \exists x (R^*(r,b,x) \wedge R^*(r',c,x)),$$

together with A.1, implies that PRA proves $\forall n B(n)$, whence:

A.5. THEOREM. PRA *proves the Church-Rosser property for the transitive closure of RED.*

Now let:

$\mathrm{CT}(x) :=$ "x is the code of a closed term of $\mathcal{L}_0$";

$\mathrm{NUM}(x) :=$ "x is the code of a numeral of $\mathcal{L}_0$";

$\mathrm{CONV}(x,y) := \exists z \exists d \exists d' (R^*(d,x,z) \wedge R^*(d',y,z));$

$\mathrm{NAT}(x) := \exists z \exists d (R^*(d,x,z) \wedge \mathrm{NUM}(z)).$

Clearly $\mathrm{CT}(x)$, $\mathrm{NUM}(x)$ are bounded while $\mathrm{CONV}(x,y)$ and $\mathrm{NAT}(x)$ are Σ_1. If $A \in \mathcal{L}$, A_{cm} is the $\mathcal{L}_0$-formula which is obtained by replacing:
1) each atom of the form Nt, $t = s$ respectively by $\mathrm{NAT}(t)$, $\mathrm{CONV}(t,s)$;
2) each quantifier $\forall x$, $\exists y$ respectively by $\forall x(\mathrm{CT}(x) \rightarrow \ldots)$, $\exists x(\mathrm{CT}(x) \wedge \ldots)$.
Now we give a more explicit statement of 4.15:

A.6. THEOREM
(i) *If* OP+EA $\vdash A(x)$, *then* PA $\vdash \mathrm{CT}(x) \rightarrow A_{cm}(x)$;
(ii) OP_1+EA $\vdash A(x)$ *implies* $\mathrm{PA}_1 \vdash \mathrm{CT}(x) \rightarrow A_{cm}(x)$.

As to the proof of A.6 (ii), we apply A.5 and we remark that each instance of $\exists(+)$-N-induction is sent into a suitable version of Σ_1-induction. As to EA, it suffices to see that its verification in the closed term model requires only Σ_1-induction (at most). The claim is made apparent by the following informal argument. First, fix a primitive recursive bijection J^* from $\omega \times \omega$ onto $\omega - \{0,\ldots,8\}$ such that $\mathrm{J}^*(n,m) > n$, m (we can choose the modified Cantor pairing function

$\mathrm{J}^*(n,m) := \frac{(n+m)^2+3n+m+18}{2}$). Then primitive recursively define a Gödel numbering GD of closed terms as follows:

$$\mathrm{GD}(\overline{0}) = 0;\ \mathrm{GD}(SUC) = 1;\ \mathrm{GD}(PRED) = 2;\ \mathrm{GD}(D) = 3;$$

$$\mathrm{GD}(PAIR) = 4;\ \mathrm{GD}(LEFT) = 5;\ \mathrm{GD}(RIGHT) = 6;\ \mathrm{GD}(K) = 7;$$

$$\mathrm{GD}(S) = 8;\ \mathrm{GD}(Ap(t,s)) = \mathrm{J}^*(\mathrm{GD}(t), \mathrm{GD}(s)).$$

The fixed point theorem for operations ensures the existence of a closed term E such that

A.7. If $0 \leq n \leq 8$, then:

$$E\overline{n} = \text{the unique constant C such that } \mathrm{GD(C)} = n\text{; else:}$$

$$E\overline{n} = (E(\overline{n})_0)(E(\overline{n})_1);$$

($(\overline{n})_0$,$(\overline{n})_1$ are the terms representing in OP^- the projections of J^*). A straightforward induction on closed terms yields that $E\,\overline{n} = t$, where $\overline{n}$ is the numeral representing the value of $\mathrm{GD}(t)$. Clearly, Parsons's theorem (ii) entails 4.14.

CHAPTER 2

EXTENDING OPERATIONS WITH REFLECTIVE TRUTH

§ 7. Extending combinatory algebras with truth
§ 8. The theory of operations and reflective truth: simple consequences
§ 9A. Type-free abstraction, predicates and classes
§ 9B. Operations on predicates and classes
§ 10A. The fixed point theorem for predicates
§ 10B. Applications to semantics and recursion theory
§ 11. Non-extensionality
Appendix I
Appendix II
Appendix III

We introduce an axiomatic framework MF^- (=Minimal Framework without number-theoretic induction) and we derive a set of simple, but significant consequences of MF^-. The minimal fixed points of a natural monotone operator over arbitrary combinatory algebras yield set-theoretic models of MF^-. This kind of models (in short, *inductive models*) are generated by means of natural elementary semantic clauses.

The informal intuition is probably due to Curry and Fitch, and it freely takes inspiration from the ideas of illative combinatory logic, later reinterpreted by Aczel with the notion of *Frege structure*. The main intuition can be summarized as follows: MF^- describes an *abstract logical system*, i.e. a pair given by an *abstract syntax* and a *semantics*. More explicitly, we can imagine a non-empty set U of objects (if you like, *terms*), which is endowed with a two-fold structure. The syntax establishes the rules of combining elements of U; application is the basic combination mode and U is an applicative structure with strong closure conditions and self-referential abilities (indeed a combinatory algebra). Furthermore, the syntax identifies two objects of U, whenever they are computationally equivalent in a precise sense, specified by a conversion relation.

It must be stressed that these features are quite general and that they can be reasonably specialized, as soon as we specify U with additional constraints (for instance, we can always assume a set of primitive number-theoretic operations, if we are interested in foundational applications).

On the other hand, the semantic structure comes on the scene, as soon as we *assert* equalities and *classify* elements of U, e.g. we state that an object truly enjoys a certain property, or that an algorithm yields well-defined values for arguments of a certain type. At this stage, we content ourselves by choosing the simplest alternative, i.e. a truth predicate T, naturally extending the standard Tarskian truth conditions. The basic T-clauses reflect the idea of a *reductionist semantics*: truth is assigned to certain basic syntactic objects, (representing) atomic propositions, and it propagates to more complex entities by means of appropriate reductive clauses for logical operations. In agreement with the reductive spirit, atomic propositions *do not refer to the truth predicate* T and their semantic value only depends on the combinatory structure. Nevertheless, T *strictly extends* the limits of Tarskian semantics. T itself becomes a *propositional constructor* and it directly applies to expressions explicitly using T, like $T[\neg T\mathrm{t}]$ ($[A]$ stands for the term representing the sentence A). Of course, such "higher order" expressions cannot be regarded as atomic, and they have a definite truth value, only if they can ultimately be reduced to well-defined atoms (cf. Kripke 1975).

There is, however, a price to pay, as one might expect from Tarski's theorem: T cannot be consistent and complete at the same time. In spite of this limitation, a logical system in the previous informal sense yields a reasonable environment for an extended logic: predication, abstraction and the notions of proposition and predicate (total or partial) can be easily introduced by means of T and the combinatory structure, and the resulting theory of abstraction has non-trivial aspects.

In details, § 7 shows how to expand any given model of the theory of operations with a reductive notion of self-referential truth, which satisfies natural axioms. These axioms give rise to the basic axiomatic system MF^{-}, whose consequences are first discussed in § 8. In §§ 9A-B we define the predicate abstraction operation $\{x : A\}$ via λ-abstraction, and we introduce a consistent reformulation of the type-free comprehension principle AP, together with a few closure conditions on total properties (= classes). It turns out that AP can be finitely axiomatized by means of four primitive predicates and eight generating operations. In §§ 10A-B we exploit a kind of second recursion theorem for predicates, which yields fixed point solutions to a class of significant conditions (positive operators, definable in the language of MF^{-}). It follows that the system $\mathrm{MF} := \mathrm{MF}^{-}$plus the number-theoretic induction schema, is proof-theoretically stronger than first-order arithmetic PA, but still predicatively reducible in the sense of Feferman. We then apply the fixed point technique to the formalization of semantics and we obtain analogues of recursion-theoretic results, due to Rice and Myhill. This last point naturally hints at possible connections with

Generalized Recursion Theory, to be pursued in parts B and C. The final section §11 shows that extensionality for classes and properties is violated in MF^-.

§7. Extending combinatory algebras with truth

We canonically associate to each OP^--model an interpretation of the truth predicate, which satisfies natural closure conditions. To this aim, we first introduce canonical terms representing the "logical" functions, which are defined by $\mathcal{L}$-formulas.

7.1. DEFINITION. (a) We choose:

$$ID := \lambda x\lambda y.\langle\overline{1},\langle x,y\rangle\rangle;\quad TR := \lambda x.\langle\overline{2},x\rangle;\quad NAT := \lambda x.\langle\overline{3},x\rangle;$$

$$NEG := \lambda x.\langle\overline{4},x\rangle;\quad AND := \lambda x\lambda y.\langle\overline{5},\langle x,y\rangle\rangle;\quad ALL := \lambda x.\langle\overline{6},x\rangle.$$

(b) We then define the map $A \mapsto [A]$ by induction on the notion of $\mathcal{L}$-formula:

(i) $[t = s] = (IDt)s;\quad [Ns] = NATs;\quad [Tt] = TRt;$

(ii) $[\neg A] = NEG[A];\quad [A \wedge B] = AND[A][B];\quad [\forall xA] = ALL(\lambda x[A]).$

7.1.1. FACT

(i) *If* $L_1,\ L_2 \in LOG_1 = \{NAT,\ NEG,\ TR,\ ALL\}$, *then* $\mathrm{OP}^- \vdash L_1x = L_2y \to L_1 = L_2\ \wedge\ x = y;$

(ii) *if* $G_1,\ G_2 \in LOG_2 = \{ID, AND\}$, *then* $\mathrm{OP}^- \vdash G_1xy = G_2x'y' \to\ G_1 = G_2 \wedge\ x = x' \wedge\ y = y';$

(iii) *if* $L_1 \in LOG_1,\ L_2 \in LOG_2$, *then* $\mathrm{OP}^- \vdash \neg\ L_1x = L_2yz;$

(iv) *if* $L_1,\ L_2$ *are distinct elements of* $LOG_1 \cup LOG_2$, *then* $\mathrm{OP}^- \vdash \neg\ L_1 = L_2.$

(Verification by pairing axioms and β-conversion).

We stress that $[A]$ and A have the same free variables in common. It would be possible to trivialize TR to $\lambda x.x$; but we stick to the present choice, since it better suits to the generalizations of chapter VIII.

7.2. *Further notations and terminology*

(i) We henceforth write TA as a shorthand for $T[A]$.

(ii) To increase readability, we keep using $\neg$, $\wedge$, $\forall$, etc., and infix notation, instead of the terms NEG, AND, ALL, etc.
Thus $t \wedge s$, $\forall f, \forall x.t$, $\neg t$ stand for the terms $(ANDt)s$, $ALLf$, $ALL(\lambda x.t)$,

$NEGt$ (in the given order); we also adopt the obvious shorthands $\neg\neg t$, $t \vee s$, $t \rightarrow s$, in place of $\neg(\neg t)$, $\neg(\neg t \wedge \neg s)$, $(\neg t \vee s)$ (respectively). As to the existential operator, we define:

$$\exists(f) := \neg(\forall(\lambda x.\neg(fx))) \text{ and } \exists x.t := \exists(\lambda x.t).$$

(iii) $\mathcal{L}_{op}$ is the (operational) fragment of $\mathcal{L}$, which omits the predicate T. The atoms of $\mathcal{L}_{op}$, i.e. Nt, $t = s$ are called *elementary atoms*, e-atoms for short. Atoms of the form Tt of the full language are called *T-atoms*. $\mathcal{L}$ and $\mathcal{L}_{op}$ clearly have the same terms.

We now fix a model $\mathcal{M}$ of OP^{-} (i.e. OP without number-theoretic induction) with domain M.

7.3. DEFINITION

(i) $\mathcal{L}_{op}(\mathcal{M})$, $\mathcal{L}(\mathcal{M})$ are the languages $\mathcal{L}_{op}$, $\mathcal{L}$ (respectively), expanded with distinct individual constants, for each element of M. If t is a closed term of the expanded languages, $\mathcal{M}(t)$ denotes the (unique) value of t in M. For the sake of simplicity, we shall use a, b, c, d, $e, \ldots$, both for the elements of M and the corresponding constants (we identify $\mathcal{M}(a)$ with a).

(ii) If P is a unary predicate (possibly T itself), $\mathcal{L}_{op}(P)$ is the language $\mathcal{L}_{op}$ expanded with P; so, Pt is a new atomic formula ($=$ atom) of $\mathcal{L}_{op}(P)$.

(iii) Let S be any subset of M: $\langle \mathcal{M}, S\rangle$ is the realization of $\mathcal{L}_{op}(P)$, which interprets P by the set S: if t is an arbitrary closed term of $\mathcal{L}_{op}(\mathcal{M})$, $\langle \mathcal{M}, S\rangle \models Pt$ iff $\mathcal{M}(t) \in S$. If $P = T$, then $\mathcal{L} = \mathcal{L}_{op}(T)$ and $\langle \mathcal{M}, S\rangle$ is the realization of $\mathcal{L}$, which interprets T by the set S.

(iv) $PFOR(x)$ is the $\mathcal{L}_{op}$-formula:

$$\exists y \exists z(x = [y = z] \vee x = [Ny] \vee x = [Ty] \vee x = (\neg y) \vee x = (y \wedge z) \vee x = \forall y);$$

$$M\text{-}PFOR = \{a \in M : \mathcal{M} \models PFOR(a)\}.$$

If $a \in M\text{-}PFOR$, we say that a is (the code of) a *pseudo-formula* (p-form, for short).

(v) We define:

$$\begin{aligned}
\Gamma(x, P) := \exists u \exists v((&x = (\neg u) \wedge \neg PFOR(u)) \vee (x = (\neg\neg u) \wedge Pu) \vee \\
&\vee (x = [u = v] \wedge u = v) \vee (x = [\neg(u = v)] \wedge \neg(u = v)) \vee \\
&\vee (x = [Nu] \wedge Nu) \vee (x = [\neg Nu] \wedge \neg Nu) \vee \\
&\vee (x = [\neg Tu] \wedge P(\neg u)) \vee (x = [Tu] \wedge Pu) \vee \\
&\vee (x = (u \wedge v) \wedge Pu \wedge Pv) \vee (x = [\neg(u \wedge v)] \wedge (P(\neg u) \vee P(\neg v))) \vee \\
&\vee (x = (\forall u) \wedge \forall y P(uy)) \vee (x = \neg(\forall u) \wedge \exists y P(\neg(uy))));
\end{aligned}$$

If $S \subseteq M$, we put:

$$\Gamma(S) := \{a \in M : \langle \mathcal{M}, S \rangle \models \Gamma(a, P)\}.$$

(vi) A subset S of M is *consistent* (*complete*) iff for every $a \in M$, either $a \notin S$ or $(\neg a) \notin S$ ($a \in S$ or $(\neg a) \in S$).

(vii) Put $\mathcal{M}(N) := \{\mathcal{M}(t) : t \text{ numeral}\}$; then $\mathcal{M}(SUC)\hat{}$ ($\mathcal{M}(PRED)\hat{}$) is the unary function $\mathcal{M}(N) \mapsto \mathcal{M}(N)$, represented by SUC ($PRED$) in $\mathcal{M}$.

$\mathcal{M}$ is an *ω-model* iff the structure $\langle \mathcal{M}(N), \mathcal{M}(\overline{0}), \mathcal{M}(PRED)\hat{}, \mathcal{M}(SUC)\hat{} \rangle$ is isomorphic with $\langle \omega, 0, pred, suc \rangle$ (= set of natural numbers with zero, predecessor and successor).

(viii) $S \subseteq M$ is *Γ-dense* (*Γ-closed*) iff $S \subseteq \Gamma(S)$ ($\Gamma(S) \subseteq S$).

Once $\mathcal{M}$ is a fixed combinatory algebra of domain M and b, $c \in M$, we shall write bc, $\neg b$, $\forall b$, $b \wedge c$, instead of the proper $\mathcal{M}(Ap(b,c))$, $\mathcal{M}(NEGb)$, $\mathcal{M}(ALLb)$, $\mathcal{M}(ANDbc)$ (in the given order); we also let $id(b,c)$ ($tr(b)$) stand for $\mathcal{M}([b = c])$ (respectively $\mathcal{M}([Tb])$).

7.3.1. REMARK. $\Gamma(x, S)$ formalizes the clauses of the intended semantic schema, to be used for interpreting the truth predicate T. As to $PFOR(x)$, it defines the range of application of T: we stress that *p-forms are not inductively defined entities* (like sentences), but only objects of the ground algebra, possibly representing semantical information.

7.4. LEMMA

(i) *If* $S \subseteq M$, $\Gamma(S) \subseteq M\text{-}PFOR$.

(ii) $\Gamma : \mathcal{P}(M) \to \mathcal{P}(M)$ *is monotone*: $S \subseteq S'$ *implies* $\Gamma(S) \subseteq \Gamma(S')$.

(iii) *Assume* a, b, $f \in M$:
if $a \notin M\text{-}PFOR$, $(\neg a) \in \Gamma(S)$;
$\mathcal{M}([A]) \in \Gamma(S)$ *iff* A *holds in* $\mathcal{M}$ (A *closed e-atom or a negated e-atom*);
$(a \wedge b) \in \Gamma(S)$ *iff* $a \in S$ *and* $b \in S$;
$(\neg(a \wedge b)) \in \Gamma(S)$ *iff* $(\neg a) \in S$ *or* $(\neg b) \in S$;
$(\forall f) \in \Gamma(S)$ *iff* $(fa) \in S$, *for all* $a \in M$;
$(\neg(\forall f)) \in \Gamma(S)$ *iff* $(\neg(fc)) \in S$, *for some* $c \in M$;
$(\neg tr(a)) \in \Gamma(S)$ *iff* $(\neg a) \in S$;
$(tr(a)) \in \Gamma(S)$ *iff* $a \in S$;
$(\neg\neg a) \in \Gamma(S)$ *iff* $a \in S$.

(iv) *If* S *is consistent and* Γ*-dense* (*complete and* Γ*-closed*), *then* $\Gamma(S)$ *is consistent and* Γ*-dense* (*complete and* Γ*-closed*).

PROOF. (i): trivial.

(ii): $\Gamma(x,P)$ is positive in P, i.e no negated atom Pt occurs in $\Gamma(x,P)$.
(iii): by 7.1, 7.3, 7.1.1 and the pairing axioms.
(iv): we repeatedly apply (iii). Let S be consistent and Γ-dense; we claim:

$$a \notin \Gamma(S) \text{ or } (\neg a) \notin \Gamma(S). \tag{$*$}$$

Case 1. $a \notin M\text{-}PFOR$: then $a \notin \Gamma(S)$ follows by (i) above.

Case 2. $a = \mathcal{M}([A])$ with $A = Nb$, $(b = c)$ or $a = (\forall b)$, $tr(b)$, $(b \wedge c)$. Then $(*)$ is a consequence of the consistency of S and (iii). For instance, if $a = (b \wedge c) \in \Gamma(S)$, then $b \in S$ and $c \in S$. Were $(\neg a) \in \Gamma(S)$, we ought to have $(\neg b) \in S$ or $(\neg c) \in S$: either alternative contradicts the consistency of S.

Case 3. $a = (\neg b)$. Assume by contradiction $(\neg b) \in \Gamma(S)$, $(\neg\neg b) \in \Gamma(S)$. Then $b \in S$ and by Γ-density $b \in \Gamma(S)$, which also implies $b \in M\text{-}PFOR$.

3.1. $b = \neg c$: then we have $(\neg\neg c) \in \Gamma(S)$ and $(\neg c) \in S$, whence $c \in S$ and $(\neg c) \in S$, against the consistency of S.

3.2. $b = (c \wedge d)$, $(\forall c)$, $tr(c)$, $\mathcal{M}([A])$ with A e-atom or negated e-atom. By Γ-density, we are led to case 2.

Note that if S is Γ-dense, so is $\Gamma(S)$ by Γ-monotonicity. If S is complete and Γ-closed, the argument is similar. □

7.5. DEFINITION. $FIX(\Gamma, \mathcal{M}) := \{S \subseteq M : \Gamma(S) = S\}$.
$FIX(\Gamma, \mathcal{M})$ is the set of fixed points of Γ over $\mathcal{M}$. In chapter VII we shall investigate the global structure of the fixed points of Γ; however, in the following we concentrate upon $\mathbb{O}(\mathcal{M}) :=$ the $\subseteq$-least fixed point, which is generated from below by transfinite iteration of Γ.

7.6. DEFINITION (by recursion on ordinals).

(i) $\mathbb{O}(\mathcal{M}, 0) = \emptyset$;
(ii) $\mathbb{O}(\mathcal{M}, \alpha+1) = \Gamma(\mathbb{O}(\mathcal{M}, \alpha))$;
(iii) $\mathbb{O}(\mathcal{M}, \lambda) = \bigcup_{\alpha<\lambda} \mathbb{O}(\mathcal{M}, \alpha)$ (λ limit).

7.7. LEMMA

(i) *$\alpha < \beta$ implies $\mathbb{O}(\mathcal{M}, \alpha) \subseteq \mathbb{O}(\mathcal{M}, \beta)$;*
(ii) *$\mathbb{O}(\mathcal{M}, \alpha) \subseteq M\text{-}PFOR$ and $\mathbb{O}(\mathcal{M}, \alpha)$ is consistent and Γ-dense, for each α.*

PROOF. (i). We verify by induction on β:

$$\text{for each } \alpha < \beta,\ \mathbb{O}(\mathcal{M}, \alpha) \subseteq \mathbb{O}(\mathcal{M}, \beta). \tag{$*$}$$

If $\beta = 0$ or β is a limit, the claim $(*)$ is trivial. Assume that $(*)$ holds and let $\alpha < \beta+1$. It suffices to check $\mathbb{O}(\mathcal{M}, \beta) \subseteq \mathbb{O}(\mathcal{M}, \beta+1)$. If $\beta = 0$, we are done; if $\beta = \delta+1$, we have $\mathbb{O}(\mathcal{M}, \delta) \subseteq \mathbb{O}(\mathcal{M}, \delta+1)$ by IH, which implies $\mathbb{O}(\mathcal{M}, \beta) \subseteq \mathbb{O}(\mathcal{M}, \beta+1)$ by Γ-monotonicity. If β is a limit and $\gamma < \beta$, then

by IH and Γ-monotonicity $\mathbb{O}(\mathcal{M},\gamma) \subseteq \mathbb{O}(\mathcal{M},\gamma+1) \subseteq \mathbb{O}(\mathcal{M},\beta+1)$. Hence $\mathbb{O}(\mathcal{M},\beta) \subseteq \mathbb{O}(\mathcal{M},\beta+1)$.
(ii) $\mathbb{O}(\mathcal{M},\alpha)$ is Γ-dense by (i) above and hence $\mathbb{O}(\mathcal{M},\alpha) \subseteq M\text{-}PFOR$ by 7.4 (i). On the other hand, $\mathbb{O}(\mathcal{M},\alpha)$ is consistent by induction on α, using (i) and lemma 7.4 (iv). □

The $\subseteq$-chain $\{\mathbb{O}(\mathcal{M},\alpha) \colon \alpha \in ON\}$ cannot be strictly $\subseteq$-increasing by the well-known Cantor's theorem: hence, there exists an ordinal $\delta < \mathrm{card}(\mathcal{M})^+$ ($^+ :=$ successor operation on cardinals), such that $\mathbb{O}(\mathcal{M},\delta) = \mathbb{O}(\mathcal{M},\delta+1)$.

7.8. DEFINITION. We set:

$$\mathbb{O}(\mathcal{M}) := \mathbb{O}(\mathcal{M},\delta)$$

where $\delta =$ the least α such that $\mathbb{O}(\mathcal{M},\alpha) = \mathbb{O}(\mathcal{M},\alpha+1)$.

Then $\mathbb{O}(\mathcal{M}) = \bigcup\{\mathbb{O}(\mathcal{M},\alpha) \colon \alpha \in ON\}$.

7.9. PROPOSITION. *Let $\mathcal{M}$ be a model of* OP^- (OP). *$\mathbb{O}(\mathcal{M})$ is consistent and is the $\subseteq$-least fixed point of* Γ:

$$\Gamma(\mathbb{O}(\mathcal{M})) \subseteq \mathbb{O}(\mathcal{M}) \ (\Gamma\text{-}closure); \qquad (*)$$

$$if\ \Gamma(S) \subseteq S,\ then\ \mathbb{O}(\mathcal{M}) \subseteq S \ (\Gamma\text{-}induction). \qquad (**)$$

PROOF. The consistency follows by lemma 7.7, while (*) holds by choice of δ in 7.8. As to (**), simply prove $\mathbb{O}(\mathcal{M},\alpha) \subseteq S$ by transfinite induction on α, applying Γ-closure of S and Γ-monotonicity. □

7.10. THEOREM. (i) *If $\mathcal{M} \models \mathrm{OP}^-$, the structure $\langle \mathcal{M}, \mathbb{O}(\mathcal{M})\rangle$ satisfies the universal closures of the following $\mathfrak{L}$-formulas:*

$T.1$	$TA \leftrightarrow A$, *if* $A = (x=y)$, Nx, $(\neg x = y)$, $\neg Nx$;		
$T.2.1$	$TTx \leftrightarrow Tx$;	$T.2.2$	$T\neg Tx \leftrightarrow T\neg x$;
$T.3$	$T\neg\neg x \leftrightarrow Tx$;		
$T.4.1$	$T(x \wedge y) \leftrightarrow Tx \wedge Ty$;	$T.4.2$	$T\neg(x \wedge y) \leftrightarrow T\neg x \vee T\neg y$;
$T.5.1$	$T(\forall f) \leftrightarrow \forall x T(fx)$;	$T.5.2$	$T\neg(\forall f) \leftrightarrow \exists x T\neg(fx)$;
$T.6$	$\neg(Tx \wedge T\neg x)$ (= CONS);		
RES	$Tx \rightarrow PFOR(x)$;		$\neg PFOR(x) \rightarrow T\neg x$.

(ii) *If $\mathcal{M}$ is an ω-model (i.e the denotation of N is isomorphic with the standard set of natural numbers), $\langle \mathcal{M}, \mathbb{O}(\mathcal{M})\rangle$ satisfies the N-induction schema for arbitrary formulas of $\mathfrak{L}$.*

PROOF. Part (ii) is trivial. As to part (i), CONS is true in $\mathbb{O}(\mathcal{M})$ by 7.9, while $Tx \rightarrow PFOR(x)$ holds by 7.7 (ii). The other axioms are immediate consequences of 7.4 (iii) and the fact that $\mathbb{O}(\mathcal{M})$ is Γ-closed and Γ-dense. □

7.10.1. REMARK. Let $\mathbb{O}(\mathcal{M})^d := \{a : a \in M \text{ and } (\neg a) \notin \mathbb{O}(\mathcal{M})\}$. The reader can easily check that $\mathbb{O}(\mathcal{M})^d \models T.1\text{-}T.5$ + RES + COMP $:= \forall x(Tx \vee T\neg x)$; $\mathbb{O}(\mathcal{M})^d$ is the $\subseteq$-largest fixed point of Γ over $\mathcal{M}$ (see Ch.VII).

7.11. DEFINITION. (i) The theory MF^- (= *minimal framework for self-referential truth and abstraction*) is the finite extension of OP^- by means of axioms $T.1$-$T.6$ of 7.10.

(ii) MF is MF^- plus the schema of N-induction for arbitrary $\mathcal{L}$-formulas:

$$\mathrm{NIND} := A(\overline{0}) \wedge \forall x(A(x) \rightarrow A(x+1)) \rightarrow \forall x(Nx \rightarrow A(x)).$$

NB. RES is omitted (unless we explicitly mention it).

In the sequel, we mostly deal with MF^- or with subsystems of MF, which contain restricted forms of number-theoretic induction. The restriction axioms RES will play a marginal role in our investigation; also, the second restriction axiom is certainly a matter of convention (for alternatives, see Ch.VII). However, RES is needed for a full characterization of models of theories of reflective truth in Ch.VII, as it can be guessed from the following:

7.12. PROPOSITION (*Alternative axiomatization of* MF^-+RES) *Let the fixed point axiom for truth* FPT *be the sentence*

$$\forall x(Tx \leftrightarrow \Gamma(x, T)),$$

where $\Gamma(x,T)$ *is obtained from the formula* $\Gamma(x,P)$ *of* 7.3 (v) *by replacing every subformula of the form* Pt *with* Tt . *Then we have*:

$$\mathrm{MF}^- + \mathrm{RES} \subseteq \mathrm{OP}^- + \mathrm{FPT} + \mathrm{CONS} \textit{ and } \mathrm{MF}^- + \mathrm{RES} \vdash \mathrm{FPT}.$$

The verification makes use of the independence properties of 7.1.1; we underline that by 7.12 MF^-+RES is a genuine *fixed point theory* in the sense of Feferman (1982), and it axiomatizes the property of being an arbitrary fixed point of Γ (see Ch.VII).

§8. The theory of operations and reflective truth: simple consequences

In this section we start working axiomatically within the system MF$^-$ without number-theoretic induction; since we are interested in general properties of truth and propositions, the number-theoretic axioms are not needed. Towards the end of the section, we sketch a version of MF$^-$, where the consistency axiom CONS is replaced by its dual, i.e. completeness.

First of all, we must distinguish between $T\neg t$, which can be read as "t is internally false", from $\neg Tt$; so we define a notion of internal falsehood F:

8.1. $$Fx := T\neg x.$$

8.2. PROPOSITION. *The following formulas are provable in* MF$^-$ *without consistency*:

(i) $Tx \leftrightarrow FFx$; $\qquad Fx \leftrightarrow TFx \leftrightarrow FTx$;

(ii) $T(x \vee y) \leftrightarrow Tx \vee Ty$; $\qquad F(x \vee y) \leftrightarrow Fx \wedge Fy$;

(iii) $T\exists(f) \leftrightarrow \exists x T(fx)$; $\qquad F\exists(f) \leftrightarrow \forall x F(fx)$.

Closure under cut: MF$^-$ *proves*

(iv) $T(x \rightarrow y) \rightarrow (Tx \rightarrow Ty)$.

PROOF. (i)-(ii): apply T.3 and T.2, T.4. As to (iii), recall the definition of $\exists(f)$ and apply β-conversion 2.2 (ii) and T.5. The statement (iv) is a consequence of (ii) and consistency. □

8.3. DEFINITION

(i) $Prop(x) := Tx \vee T\neg x$ = "x is a proposition".

(ii) $Propfun_n(f) := \forall x_1 \ldots \forall x_n Prop(fx_1 \ldots x_n)$ = "f is a n-ary propositional function" ($n \geq 1$; if $n = 1$, we simply omit the index).

Clearly we have in pure logic :

8.3.1. $$Propfun_{k+1}(f) \leftrightarrow \forall x Propfun_k(fx) \quad (k \geq 1);$$

hence we can restrict our attention to unary propositional functions.

We now investigate the closure properties of $Prop$ under standard logical operations and the behaviour of T, whenever T is restricted to $Prop$. Abbreviation: $Prop(A) := Prop([A])$.

8.4. LEMMA. MF$^-$ *proves*:

(i) $Prop(A)$, whenever $A = (\neg)Nx$, $(\neg)x = y$;

$Tx \rightarrow Prop(x)$;

(ii) $Prop(x) \leftrightarrow Prop(Tx) \leftrightarrow T(Prop(x))$;

(iii) $Prop(x) \leftrightarrow Prop(Prop(x))$;

(iv) $Prop(x) \leftrightarrow Prop(\neg x)$;

(v) $Prop(x) \wedge Prop(y) \rightarrow Prop(x \wedge y) \wedge Prop(x \vee y)$;

(vi) $Prop(x) \wedge (Tx \rightarrow Prop(y)) \rightarrow Prop(x \rightarrow y)$;

(vii) $Propfun(f) \rightarrow Prop(\forall(f)) \wedge Prop(\exists(f))$;

(viii) $Prop(x{*}y) \rightarrow Prop(x) \vee Prop(y)$ $(* = \vee, \wedge, \rightarrow)$;

(ix) $Prop(\mathrm{Q}(f)) \rightarrow \exists x Prop(fx)$ $(\mathrm{Q} = \exists, \forall)$;

(x) $\neg F(Prop(x))$.

PROOF: straightforward application of 8.2 and T-axioms. As to the final point, if $FProp(x)$ is assumed, we have $FTx \wedge FFx$ (8.2 (ii)), whence $Fx \wedge Tx$ (by T.2.1), against consistency T.6. □

8.4.1. REMARK. 8.4 (x) implies $T(\forall x(Prop(x) \rightarrow fx)) \leftrightarrow T(\forall f)$; this means that internal truth (i.e. truth with respect to T) disregards quantification on propositions; thus, there is no hope to produce propositions by means of $\forall x(Prop(x) \rightarrow \ldots)$, except for trivial cases.

We stress that the internal truth predicate is partial and that 8.4 (viii)-(ix) cannot be improved by replacing $\vee$, $\exists$ in the right member of the implications with $\wedge$, $\forall$ respectively; for instance, there are disjunctive propositions with a member which is not itself a proposition. Therefore the behaviour of logical operators on $Prop$ is *non-strict*. By 8.4 (x), the notion of proposition is essentially *external and positive*; we cannot come to know that p is *not* a proposition by adopting the semantical schema embodied by T.

8.5. PROPOSITION. MF^- *proves:*

(i) $\exists x(\neg Prop(x)) \wedge \neg Propfun(\lambda x.[Prop(x)])$;

(ii) $\exists x \exists y(Prop(x \vee y) \wedge \neg(Prop(x) \wedge Prop(y)))$;

(iii) $\exists f(Prop(\exists f) \wedge \neg \forall x Prop(fx))$.

PROOF. (i) We consider the fixed point L of $\lambda x.[Fx]$, i.e. $L = [FL]$ (apply 2.2). Then $Prop(L)$ implies both TL and FL (by 8.2 (i)), against consistency. If $\lambda x.[Prop(x)]$ were a propositional function, we could conclude by 8.4 (x) $\forall x.TProp(x)$, which contradicts the previous result.
(ii): choose $y = [\overline{0} = \overline{0}]$ and $x = [L]$.

(iii): choose $f = \lambda y.[Fy]$. $\Box$

On the other hand, T is consistent and complete on propositions and satisfies the standard Tarski conditions; indeed, the essential content of 8.2 and 8.4 can be summarized as follows:

8.6. THEOREM. MF$^-$ *proves:*

(i) $Prop(A) \wedge (TA \leftrightarrow A)$, *whenever* $A = (\neg)\ x = y,\ (\neg)Nx$;

(ii) $Prop(x) \rightarrow Prop(\neg x) \wedge (T\neg x \leftrightarrow \neg Tx)$;

(iii) $Prop(x) \wedge (Tx \rightarrow Prop(y)) \rightarrow Prop(x \rightarrow y) \wedge (T(x \rightarrow y) \leftrightarrow (Tx \rightarrow Ty))$;

(iv) $Propfun(f) \rightarrow Prop(\forall f) \wedge (T(\forall f) \leftrightarrow \forall x T(fx))$;

(v) $Prop(x) \rightarrow Prop(Tx) \wedge (T(Tx) \leftrightarrow Tx)$.

8.6.1. REMARK. (i) 8.6 shows that MF$^-$ essentially contains the (classical) theory of Frege structures (see Aczel 1977, 1980).
(ii) The Curry paradox (Curry 1942). We cannot consistently add to MF$^-$ a strengthened introduction axiom for implication, which omits the hypothesis $Prop(x)$ in 8.6 (iii): MF$^-$ + (*) is inconsistent, where (*) is the statement

$$(Tx \rightarrow Prop(y)) \rightarrow ((Tx \rightarrow Ty) \rightarrow T(x \rightarrow y)). \qquad (*)$$

Indeed, we can find c such that $c = c \rightarrow y$ ($c = FP(\lambda x.[x \rightarrow y])$, see 2.3) and clearly $Tc \rightarrow Ty$ (by 8.2 (ii), consistency and $=$-logic). If we assume (*), we can infer $T(c \rightarrow y)$, i.e. Tc, whence Ty by 8.2 (iv): contradiction (choose $y = [\overline{0} = \overline{1}]$).

As to the general Tarski schema $TA \leftrightarrow A$, it can be justified "from left to right" and also for positive conditions.

8.7. DEFINITION. (i) A formula B is T-free if T does not occur in B.

(ii) The collection T-Pos (T-Neg) of T-positive (T-negative) formulas is inductively generated by the following clauses:

1. each e-atom is both T-positive and T-negative; each atom of the form Ts ($\neg Ts$) is T-positive (T-negative);
2. if B is T-positive (T-negative), $\neg B$ is T-negative (T-positive);
3. if B, C are T-positive (T-negative), then so is $B \wedge C$;
4. if B is T-positive (T-negative), then so is $\forall x B$.

8.8. THEOREM

(i) *The soundness schema*: MF$^-$ *proves* $(TA \to A)$, *for arbitrary* A;
(ii) *if* A *is* T*-positive* (T*-negative*), MF$^-$ *minus consistency proves*:

$$A \to TA \ (\neg A \to FA, \ \textit{respectively});$$

(iii) *if* A *is* T*-free*, MF$^-$ *minus consistency proves* $TA \vee T\neg A$.

PROOF. (i): by induction on A and by considering the form of B whenever $A = \neg B$. If A is an atom, we apply $T.1$ and $T.2.1$, while, if A is a conjunctive or universally quantified formula, we use IH, $T.4.1$, $T.5.1$ plus β-conversion. If $A = \neg Tt$, we apply $T.2.2$ and consistency; in the remaining cases, we make use of $T.4.2$, $T.5.2$ coupled with IH.
(ii): by simultaneous induction on the definition of T-Pos and T-Neg.
(iii): by (ii) and tertium non datur. □

We conclude with a simple, but useful duality property, whose semantic content will be made clear in Ch.VII (§30).

8.9. DEFINITION
1. $\hat{}$ is the (unique) map of the basic language $\mathcal{L}$ into itself such that
(i) $\hat{}$ is the identity map on e-atoms and $(Tt)\hat{} = \neg Ft$;
(ii) $\hat{}$ commutes with the logical operations:

$$(A \wedge B)\hat{} = A\hat{} \wedge B\hat{}, \ (\forall x A)\hat{} = \forall x(A\hat{}), \ (\neg A)\hat{} = \neg(A\hat{}).$$

2. Put COMP $:= \forall x(Tx \vee T\neg x)$ (Completeness);

NMF (= the neutral MF) is MF minus CONS, where CONS = $T.6$;

MF$\hat{}$:= NMF + COMP.

As usual, NMF$^-$ is NMF without N–induction.

3. $x \Rightarrow y := (Tx \to Ty) \wedge (Fx \to Fy)$ and $x \Leftrightarrow y := (x \Rightarrow y) \wedge (y \Rightarrow x)$.

Then CONS $\leftrightarrow$ (COMP)$\hat{}$ and COMP $\leftrightarrow$ (CONS)$\hat{}$, provably in NMF (use axiom T.3); more generally, we can easily check :

8.10. LEMMA. NMF *proves*:
(i) $A\hat{}\hat{} \leftrightarrow A$;
(ii) $(x \Rightarrow y)\hat{} \leftrightarrow (y \Rightarrow x)$; $(x \Leftrightarrow y)\hat{} \leftrightarrow (x \Leftrightarrow y)$.

8.11. THEOREM (*Self-duality of* NMF)
For every A, NMF $\vdash A$ *iff* NMF $\vdash A\hat{}$. *The same holds if the restriction axioms* RES *are added to* NMF.

PROOF. By the previous lemma, it is enough to check the theorem from left to right. The verification runs by induction on the length of the formal proof of A in NMF; the induction step is immediate by definition of $\hat{}$ and IH. If A is an axiom, either it is self-dual (i.e. equivalent to its $\hat{}$-transform, like NIND, T.3) or it can be proved by the axiom lying on the same line in the statement 7.10 (e.g. $(T.2.1)\hat{}$ requires T.2.2). □

8.12. COROLLARY. *For every* A, $\mathrm{MF} \vdash A$ *iff* $\mathrm{MF}\hat{} \vdash A\hat{}$. MF *and* MF$\hat{}$ *have the same* T*-free theorems and hence they are equiconsistent.*

NMF is a possible axiomatic counterpart of the four-valued approach to semantics (Belnap 1977, Woodruff 1984, Visser 1984), according to which self-reference leads to *underdefined* (neither true nor false), as well as to *overdefined* (both true and false) sentences. For a general account of the NMF-models, we send the reader to Ch.VII.

§9A. Type-free abstraction, predicates and classes

We will show that MF^- supports a reasonable theory of type-free abstraction. To this aim, we observe that internal truth yields a well-behaved notion of *general predicate application* (in short *predication*), and that the underlying combinatory structure grants a systematic notation for *partial* predicates defined by *abstraction.* Furthermore, if we identify total predicates with propositional functions, we obtain a rich domain, satisfying natural closure conditions for abstraction. Henceforth, we shall adopt Feferman's terminology by using the shorter term *class* instead of *propositional function.* Of course, as we already know from the previous section, there exists a stumbling block in any theory of abstraction, based on such an identification: the notion of propositional function (or class) is itself non-total and this is an essential limitation for deriving impredicative fragments of second-order logic. On the other hand, the limitation is not surprising, in view of the reductive, predicativistic interpretation, which is suggested by the $\subseteq$-minimal model of §7.

9.1. DEFINITION

(i) $(x_1 \ldots x_n)\eta y := T(yx_1 \ldots x_n)$; $(x_1 \ldots x_n)\overline{\eta} y = F(yx_1 \ldots x_n)$;

(ii) $\{x_1 \ldots x_n : A\} := \lambda x_1 \ldots \lambda x_n[A]$("the n-ary predicate defined by A");

(iii) $Cl(y) := \forall x(x\eta y \vee x\overline{\eta} y)$ ("y is a class");

$CL := \{x : Cl(x)\}$.

Note that $Cl(y) = Propfun(y)$. We also recall (see 8.10):

$$x \Leftrightarrow y := (Tx \leftrightarrow Ty) \wedge (Fx \leftrightarrow Fy) \text{ and } A \Leftrightarrow B := [A] \Leftrightarrow [B].$$

9.1.1. REMARK. (i) The definition does not ensure the injectivity condition

$$[x\eta y] = [u\eta v] \rightarrow x = u \wedge y = v \qquad (*)$$

If $(*)$ is needed, choose $PD := \lambda xy.[x = x \wedge y = y \wedge yx]$ and define $x\eta y := T(PDxy)$, $x\overline{\eta}y := T\neg(PDxy)$. Then $(*)$ is met and we can prove in NMF^- the formula:

$$T(PDxy) \leftrightarrow T(yx) \wedge T\neg(PDxy) \leftrightarrow T\neg(yx).$$

(ii) Of course, η, $\overline{\eta}$, $\{\ \}$ might be accepted as primitive symbols of $\mathcal{L}$ and the definitions of 9.1 (i) would become axioms. A similar choice might be advisable in applications, or if one wishes to avoid combinatory logic (see appendix I and Ch.XIV).

9.2. PROPOSITION. (i) *The Abstraction principle* AP: *for every formula* A, NMF^- (*i.e* MF^- *minus consistency*) *proves*:

$$\forall u_1 \ldots \forall u_n((u_1 \ldots u_n)\eta\{x_1 \ldots x_n : A\} \Leftrightarrow A[x_1 := u_1, \ldots, x_n := u_n]);$$

(ii) NMF^- *proves*:

$$((x_1 \ldots x_n)\eta y \leftrightarrow T[(x_1 \ldots x_n)\eta y]) \wedge ((x_1 \ldots x_n)\overline{\eta} y \leftrightarrow F[(x_1 \ldots x_n)\eta y]);$$

(iii) $\mathrm{NMF}^- \vdash (u\eta\{x : A\} \leftrightarrow TA[x := u]) \wedge (u\overline{\eta}\{x : A\} \leftrightarrow FA[x := u])$;

(iv) $\mathrm{NMF}^- \vdash T[Cl(x)] \leftrightarrow Cl(x) \leftrightarrow x\eta CL$;

(v) $\mathrm{MF}^- \vdash Cl(x) \leftrightarrow (\neg(y\eta x) \leftrightarrow y\overline{\eta}x)$.

PROOF. (i) Assume $n = 1$: then $T(u\eta\{x : A\}) \leftrightarrow T(\lambda x.[A])u \leftrightarrow TA[x := u]$ (by T.2.1, β-conversion and $[A][x := u] = [A[x := u]]$). A similar argument works for F (we need T.2.2.). (ii)-(v): left to the reader. □

By the Russell paradox, there exist predicates, which are not classes, and the notion of class does not determine a class. More generally:

9.3. PROPOSITION. *Let* $r = \{x : \neg x\eta x\}$. *Then*:

$$\mathrm{MF}^- \vdash \neg\exists x(Cl(x) \wedge \forall u(u\eta x \leftrightarrow u\eta r)) \wedge \neg\exists y(Cl(y) \wedge \forall u(u\eta y \leftrightarrow Cl(u))).$$

PROOF. Let x be a class such that $\forall u(u\eta x \leftrightarrow u\eta r)$. Then by AP we have: $x\eta x \leftrightarrow x\eta r \leftrightarrow x\overline{\eta}x$, whence by consistency $\neg(T(x\eta x) \vee F(x\eta x))$, i.e. $\neg Cl(x)$. If y is a class, which exactly contains all classes, $b = \{x : x\eta y \wedge \neg x\eta x\}$ is a class and $b\eta y$: hence $b\eta b \leftrightarrow (b\eta y \wedge b\overline{\eta}b) \leftrightarrow b\overline{\eta}b$, i.e. b is not a class: contradiction ! □

n-ary predication can be reduced to unary one by adding parameters:

9.4. LEMMA (*Parametrization*). NMF^- *proves*:

$$\forall x_1 \ldots \forall x_{n+m}((x_1 \ldots x_{n+m})\eta\{u_1 \ldots u_{n+m} : A\} \Leftrightarrow$$
$$\Leftrightarrow (x_{n+1} \ldots x_{n+m})\eta\{u_{n+1} \ldots u_{n+m} : A[u_1 := x_1, \ldots, u_n := x_n]\}.$$

We now consider some useful approximations to the naive abstraction principle, i.e. to versions of AP where $\Leftrightarrow$ is replaced by the standard biconditional.

9.5. DEFINITION

(i) A formula B is *elementary* in the list $x_1, \ldots, x_n$ iff B is built up from e-atoms, negated e-atoms, T-atoms of the form $t\eta x_i$ and their negations $\neg t\eta x_i$ ($1 \le i \le n$), by means of $\wedge$, $\vee$, $\forall y$, $\exists y$ ($y \notin \{x_1, \ldots, x_n\}$);

(ii) a formula B is *quasi-elementary* in $x_1, \ldots, x_n$ iff B is built up from e-atoms and negated e-atoms, arbitrary T-atoms, and negated T-atoms of the form $\neg t\eta x_i$ ($1 \le i \le n$), by means of $\wedge$, $\vee$ and $\forall y$, $\exists y$ ($y \notin \{x_1, \ldots, x_n\}$).

9.5.1. REMARK. (i) B is T-positive iff B is (up to logical equivalence) quasi-elementary in the empty list of variables.
(ii) If B is elementary, then B is trivially quasi-elementary; moreover, the negation of an elementary formula is always elementary (up to logical equivalence).

The notion of *elementary condition* for type-free languages is adapted from Feferman (1975). We say that B is (quasi-) elementary *tout court* iff B is (quasi-) elementary in some list $x_1, \ldots, x_n$. With the notions of 9.5, we obtain a useful generalization of 8.8 (i)-(ii):

9.6. LEMMA

(i) *Let A be quasi-elementary in $x_1, \ldots, x_n$. Then:*

$$\mathrm{MF}^- \vdash Cl(x_1) \wedge \ldots \wedge Cl(x_n) \to (A \leftrightarrow TA).$$

(ii) *If A is elementary in $x_1, \ldots, x_n$,*

$$\mathrm{MF}^- \vdash Cl(x_1) \wedge \ldots \wedge Cl(x_n) \to Prop(A).$$

PROOF. (i) By 8.8 and induction on A, using the hypothesis on $x_1, \ldots, x_n$, whenever $A = \neg t\eta x_i$. (ii): by (i), classical logic and 9.5.1 (ii). □

Lemma 9.6 and the abstraction principle 9.2 immediately imply :

9.7. COROLLARY. (i) *If $A(v, x_1, \ldots, x_n)$ is quasi-elementary in $x_1, \ldots, x_n$,*

$$\mathrm{MF}^- \vdash Cl(x_1) \wedge \ldots \wedge Cl(x_n) \to \forall y(y\eta\{v : A(v, x_1, \ldots, x_n)\} \leftrightarrow A(y, x_1, \ldots, x_n)).$$

(ii) *If $A(v, x_1, \ldots, x_n)$ is elementary in $x_1, \ldots, x_n$,*

$$\mathrm{MF}^- \vdash Cl(x_1) \wedge \ldots \wedge Cl(x_n) \to Cl(\{v : A(v, x_1, \ldots, x_n)\}).$$

Corollary 9.7 yields the so-called *elementary comprehension schema,* in short EC (Feferman 1975). It may be asked whether classes are closed under a strengthened schema, where "elementary" is replaced by some reasonable notion of "second-order condition", e.g. the formula A admits a standard interpretation in second-order logic. In Ch.VII, we shall prove that a second-order impredicative comprehension schema is consistent with MF, but there are models (e. g. the inductive model of §7), which falsify it.

9.8. DEFINITION

(i) $\forall x\eta t.A := \forall x(x\eta t \to A)$; $\quad \exists x\eta t.A := \exists x(x\eta t \wedge A)$;

(ii) "f is a family of classes indexed by a " $:= \forall x\eta a.Cl(fx)$;

(iii) $\{\langle x_1,\ldots,x_n\rangle : A(x_1,\ldots,x_n)\} :=$

$:= \{x : x = \langle (x)_1,\ldots,(x)_n\rangle \wedge A((x)_1,\ldots,(x)_n)\}$ (for $(x)_i$, see §1);

(iv) $\Sigma(a,f) := \{\langle x,y\rangle : x\eta a \wedge y\eta(fx)\}$ (= generalized direct sum or join);

(vi) $\Pi(a,f) := \{g : \forall x\eta a.\,(gx)\eta(fx)\}$ (= generalized product).

(iii) is justified in NMF$^-$ with pairing axioms:

9.8.1. $\quad \langle a_1,\ldots,a_n\rangle\eta\{\langle x_1,\ldots,x_n\rangle : A(x_1,\ldots,x_n)\} \Leftrightarrow A[x_1 := a_1,\ldots,x_n := a_n]$.

9.9. PROPOSITION (*The Join Principle* J). *CL is closed under generalized sums over families of classes, indexed by classes. Formally,* MF$^-$ *proves:*

(i) $\forall u(u\eta\Sigma(b,f) \leftrightarrow \exists x\exists y(u = \langle x,y\rangle \wedge x\eta b \wedge y\eta(fx)))$;

(ii) $\forall x\eta b.Cl(fx) \wedge Cl(b) \to Cl(\Sigma(b,f))$.

PROOF. (i) is an immediate application of 9.7 (i) and 9.5.1.
(ii): let f be a family of classes indexed by the class b and let:

$$A(u) := \exists x\exists y(u = \langle x,y\rangle \wedge x\eta b \wedge y\eta(fx)).$$

By T.1, $Cl(b)$, 8.2 (ii)-(iii), T.4.2, 9.2 (v) we get:

$$\begin{aligned} FA(u) &\leftrightarrow \forall x\forall y F(u = \langle x,y\rangle \wedge x\eta b \wedge y\eta(fx)) \\ &\leftrightarrow \forall x\forall y(u \neq \langle x,y\rangle \vee F(x\eta b) \vee F(y\eta(fx))) \\ &\leftrightarrow \forall x\forall y(u \neq \langle x,y\rangle \vee \neg(x\eta b) \vee y\bar{\eta}(fx)) \\ &\leftrightarrow \forall x\forall y(u = \langle x,y\rangle \wedge x\eta b \to y\bar{\eta}(fx)) \\ &\leftrightarrow \forall x\forall y(u = \langle x,y\rangle \wedge x\eta b \to \neg(y\eta(fx))) \\ &\leftrightarrow \neg A(u). \end{aligned}$$

Together with (i) and 9.2 (i), this yields:

$$\neg(u\eta\Sigma(b,f)) \leftrightarrow \neg A(u) \leftrightarrow FA(u) \leftrightarrow u\overline{\eta}\Sigma(b,f). \square$$

9.9.1. REMARK. 9.9 proves the so-called join principle J; 9.9 and 9.7 show that MF contains Feferman's system EM+J for explicit mathematics (Feferman 1975, 1979; Beeson 1985; cf. appendix II).

As an exercise, the reader may verify the dual principle for Π:

9.10. PROPOSITION (*Closure under generalized products*). MF^- *proves*:

(i) $Cl(b) \rightarrow \forall g(g\eta\Pi(b,f) \leftrightarrow \forall x\eta b.\,(gx)\eta(fx))$;

(ii) $Cl(b) \wedge \forall x\eta b.Cl(fx) \rightarrow Cl(\Pi(b,f))$.

Similar arguments prove that *Prop*, the notion of (internal) proposition, is closed under infinitary conjunctions and disjunctions in the following sense: if f is a family of propositions indexed by any class c (for instance $c = \{x : Nx\}$), there exist propositions $\bigwedge\{fx : x\eta c\}$, $\bigvee\{fx : x\eta c\}$, satisfying:

$$T(\bigwedge\{fx : x\eta c\}) \leftrightarrow \forall x\eta c.T(fx); \qquad T(\bigvee\{fx : x\eta c\}) \leftrightarrow \exists x\eta c.T(fx).$$

§9B. Operations on predicates and classes

We extend the standard operations of the algebra of (extensional) classes and relations to the general domain of partial properties. In particular, each definable predicate can be generated starting from four primitive predicates by means of eight predicate operations.

9.11. DEFINITION

1. Initial Predicates:

$$\mathbb{APP} := \{\langle x,y,z\rangle : x = yz\};$$

$$\mathbb{ID} := \{\langle x,y\rangle : x = y\};$$

$$\mathbb{N} := \{x : Nx\};$$

$$\mathbb{T} := \{x : Tx\}.$$

2. Basic Operations:

Singleton	$\{a\} := \{x : x = a\}$;
Complement	$-a := \{x : \neg x\eta a\}$;
Intersection	$a \cap b := \{x : x\eta a \wedge x\eta b\}$;
Domain	$dom(a) := \{x : \exists y.\langle x,y\rangle\eta a\}$.

3. Combinatorial operations:

Expansion	$Exp(a) := \{\langle x,y\rangle : y\eta a\};$
Converse	$Conv(a) := \{\langle x,y\rangle : \langle y,x\rangle \eta a\};$
Cycle	$Cyc(a) := \{\langle x,y,z\rangle : \langle z,x,y\rangle \eta a\};$
Transpose	$Trans(a) := \{\langle x,y,z\rangle : \langle x,z,y\rangle \eta a\};$

(x, y, z, a, b denote distinct variables; remind that $\langle x,y,z\rangle = \langle\langle x,y\rangle ,z\rangle$).

4. We say that CL is closed under a given n-ary operation H, provably in a theory $\mathfrak{T}$, if $Cl(a_1) \wedge \ldots \wedge Cl(a_n) \rightarrow Cl(H(a_1,\ldots,a_n))$ is provable in $\mathfrak{T}$;

5. $a \equiv b := \forall x(x\eta a \Leftrightarrow x\eta b);$

6. $a \subseteq b := \forall x(x\eta a \rightarrow x\eta b)$ and $a =_e b := a \subseteq b \wedge b \subseteq a$.
{Clearly, if a and b are classes, then $a =_e b \leftrightarrow a \equiv b$)}.

7. EXPL is the collection of $\mathfrak{L}_{op}$-terms, which is inductively generated by the clauses: $\mathbb{APP}$, $\mathbb{ID}$, $\mathbb{N}$, $\mathbb{T} \in$ EXPL; if t is a variable or a constant of $\mathfrak{L}_{op}$, $\{t\} \in$ EXPL; if $t \in$ EXPL and $s \in$ EXPL, then $dom(t)$, $t \cap s$, $-t$, $Conv(t)$, $Exp(t)$, $Cyc(t)$, $Trans(t)$ are elements of EXPL.

If $t \in$ EXPL, we say that t is *an explicit predicate of* $\mathfrak{L}$. The subcollection ELP of *elementary predicates* is inductively generated as EXPL, except that we omit $\mathbb{T}$ from the initial clauses of EXPL, and we add the condition: if t is a variable, $Exp(t) \in$ ELP.

A trivial application of elementary comprehension 9.7 (ii) yields:

9.12. LEMMA. *CL contains $\mathbb{APP}$, $\mathbb{ID}$, $\mathbb{N}$ and is closed under the operations of* 9.11.2-9.11.3, *provably in* MF$^-$.

9.13. THEOREM (*Explicit abstraction*)

(i) *For every formula A of $\mathfrak{L}$, for every n, we can effectively define a term* $\tau_n(A) \in$ EXPL *with* $\mathrm{FV}(\tau_n(A)) = \mathrm{FV}(A) - \{x_1,\ldots,x_n\}$, *such that, provably in* MF$^-$:

$$\tau_n(A) \equiv \{\langle x_1,\ldots,x_n\rangle : A\}. \qquad (\bigstar)$$

(ii) *Assume that A is elementary in u (where u may be a finite list of variables): then the term $\tau_n(A)$ can be chosen in* ELP.

PROOF. (i) The argument parallels the well-known class theorem for Gödel-Bernays set theory. We proceed by induction on the built-up of A. The inductive step is easily handled by means of complement, intersection and domain; therefore we only need to find, for each n, a predicate $\tau_n(A)$, satisfying (★) above, whenever $A = Nt$, Tt, $t_1 = t_2$. On the other hand, $t_1 = t_2$ is equivalent (mod $\Leftrightarrow$) to $\exists y(y = t_1 \wedge y = t_2)$. Now we can find, uniformly in r, n, i with $1 \leq i \leq n$, a term $\sigma_n^i(r)$ such that:

$$\sigma_n^i(r) \equiv \{\langle x_1, \ldots, x_n\rangle : x_i = r\}.$$

Hence:

$$\{\langle x_1, \ldots, x_n\rangle : t_1 = t_2\} \equiv dom(\sigma_{n+1}^{n+1}(t_1) \cap \sigma_{n+1}^{n+1}(t_2)).$$

The definition of $\sigma_n^i(r)$ can be reduced to the construction of elementary predicates $\mathbb{APP}_n(i,j,k)$ and $\mathbb{ID}_n(i,j)$, where $1 \leq i$, $j \leq n$ such that:

$$\mathbb{APP}_n(i,j,k) \equiv \{\langle x_1, \ldots, x_n\rangle : x_i = x_j x_k\}$$
$$\mathbb{ID}_n(i,j) \equiv \{\langle x_1, \ldots, x_n\rangle : x_i = x_j\}. \qquad (1)$$

In turn, (1) is verified by induction on n with the help of $\mathbb{APP}$, $\mathbb{ID}$ and the combinatorial functions (details are in appendix II). A similar argument works if $A = Nt$ or Tt where, of course, we must use $\mathbb{N}$ and $\mathbb{T}$.

(ii) It is enough to deal with the atomic case $A(x_1, u) = t\eta u$, where x_1, u are distinct variables, possibly free in t. But there is an elementary predicate $\sigma_2^2(t) \equiv \{\langle x_1, x_2\rangle : x_2 = t\}$; if we choose $ext(u,t) = dom(\sigma_2^2(t) \cap Exp(u))$, $ext(u,t)$ is elementary and $ext(u,t) \equiv \{x_1 : t\eta u\}$. □

By application of elementary comprehension and closure under generalized sums and products, we easily have:

9.14. PROPOSITION. *CL is closed, provably in* MF$^-$, *under the operations defined in the following list.*

1. *Pair:* $\{a,b\} := \{x : x = a \vee x = b\}$;
2. *Direct Sum:* $a \oplus b := \{\langle x,y\rangle : (x\eta a \wedge y = \overline{0}) \vee (x\eta b \wedge y = \overline{1})\}$;
3. *Cartesian Product:* $a \otimes b := \{\langle x,y\rangle : x\eta a \wedge y\eta b\}$;
4. *Exponentiation:* $[a \to b] := \{f : \forall x\eta a.(fx)\eta b\}$;
5. *Universe and empty class:*

$$V := \{x : x = x\}\ ,\ \emptyset := \{x : \neg x = x\};$$

6. *Generalized union and intersection:*

$$\bigcup_{x\eta b} fx := \{u : \exists x\eta b.\, u\eta(fx)\},\ \bigcap_{x\eta b} fx := \{u : \forall x\eta b. u\eta(fx)\},$$

provided b is a class and $f : b \to CL$.

9.14.1. REMARK. In analogy with the generalized closure condition of *Prop* under implication, it holds, provably in MF$^-$:

$$Cl(a) \wedge (\exists x(x\eta a) \rightarrow Cl(b)) \rightarrow Cl([a \rightarrow b]) \wedge \forall f(f\eta[a \rightarrow b] \leftrightarrow \forall x\eta a.\,(fx)\eta b).$$

We end up the section by considering a natural question: is there any reasonable counterpart at the level of predication of the set-theoretic power set operation? Let us try two obvious alternatives:

$$P(a) := \{x : x \subseteq a\}; \quad P^{+}(a) := \{x : x\eta CL \wedge x \subseteq a\}.$$

Unfortunately, the former is unmanageable, while CL is not closed under the operation P^{+}:

9.15. PROPOSITION

(i) $\mathrm{MF}^{-} \vdash \neg\forall y \forall x(x \subseteq y \rightarrow x\eta P(y))$;

(ii) $\mathrm{MF}^{-} \vdash \forall x(x\eta P^{+}(y) \leftrightarrow Cl(x) \wedge x \subseteq y)$;

(iii) $\mathrm{MF}^{-} \vdash \neg\forall x(Cl(x) \rightarrow Cl(P^{+}(x)))$.

PROOF.(i): choose $x = CL$ and apply 9.3.
(ii) is obvious. As to (iii), choose $y = V$ (= universal class) and verify that $Cl(P^{+}(V))$ implies $Cl(CL)$: contradiction ! □

Clearly, both operations $\lambda x.P(x)$ and $\lambda x.P^{+}(x)$ are $\subseteq$-monotone (e.g. $a \subseteq b$ implies $P(a) \subseteq P(b)$, etc.); however, in the present framework we can also define an intensional power set operation:

9.16. DEFINITION (Weak power set). $Pw(b) := \{x : \exists y(x = y \cap b)\}$.

Then by elementary comprehension 9.7 we have, provably in MF^{-}:

9.17. $$Cl(Pw(b)) \wedge \forall u(u \subseteq b \rightarrow \exists v(v\eta Pw(b) \wedge v =_{e} u)) \wedge \wedge \forall u(u\eta Pw(b) \rightarrow u \subseteq b)),$$

where $=_{e}$ is the extensional equality of 9.11.6.

The interest of $Pw(b)$ is limited by the fact that, even if b is a class, we cannot predict whether any given subproperty of b is itself total. A weakening of the power set operation along constructivistic patterns, looks more promising and it leads to consider the decidable subclasses of a given class x. If c is a class, a decidable subclass of c has the form $\{x : x\eta c \wedge fx = \overline{0}\}$, for some $f : c \rightarrow 2$ (here $2 = \{\overline{0}, \overline{1}\}$).

9.18. DEFINITION. $Pd(a) := \{y : \exists f \exists y(y = \{x : x\eta a \wedge fx = \overline{0}\} \wedge f\eta[a \rightarrow 2]\}$.

Then we have:

9.19. PROPOSITION. MF^{-} *proves*:

$$Cl(c) \rightarrow (Cl(Pd(c)) \wedge \forall b(b\eta Pd(c) \rightarrow Cl(b) \wedge b \subseteq c)).$$

We underline that $Pd(c)$, for certain c's, can be very large: it is consistent to assume that $Pd(\mathbb{N})$ is a model of second-order comprehension (cf. 5.13).

§ 10A. The fixed point theorem for predicates

We know from the previous sections that the logical universe described by MF^-, though not well behaved under impredicative quantification over classes and propositions, is closed under natural infinitary constructive operations, which go beyond the limits of elementary logic. We now consider the problem of inductive definitions; we shall see that MF^- can prove the existence of solutions to a number of recursive conditions, but in general such solutions cannot be shown to define total predicates, or to be extremal (i.e. minimal or maximal). The main tool is the property-theoretic analogue of the second recursion theorem. Indeed, if we combine the fixed point theorem for operations and the abstraction schema, we get the simple and fundamental:

10.1. THEOREM (*Fixed point for predicates*)

(i) *Let $A(x,y,v)$ be a formula with the free variables shown only. We can find – uniformly in A – a term $IxyA(x,y,v)$, whose only free variable is v, such that* MF^- *proves:*

$$IxyA(x,y,v) = \{x : A(x, IxyA(x,y,v), v)\} \wedge$$
$$\wedge \forall u(u\eta IxyA(x,y,v) \Leftrightarrow A(u, IxyA(x,y,v), v)).$$

(ii) *If $A(x,y,v)$ is quasi-elementary in v (a fortiori if A is T-positive, see* 8.7), *and we assume that v is a class, $\Leftrightarrow$ can be replaced by $\leftrightarrow$.*

PROOF. Choose:

$$IxyA(x,y,v) := FP(\lambda y.\{x : A(x,y,v)\}).$$

Hence $u\eta IxyA(x,y,v) \Leftrightarrow A(u, IxyA(x,y,v), v)$ by 2.3 and 9.2. The second part follows with 9.7. □

10.1.1. NOTATION. 10.1 introduces a new variable binding operator I. If x, y are clear from the context, we shorten $IxyA(x,y,v)$ to $I(A,v)$; we also let $I(A,v) := I(A)$ if the dependence on the variable v is not explicitly needed. Of course, the definition of I makes sense in the general case where v is a finite (possibly empty) list of variables, apart from the bound x and y.

10.2. *Examples*

(i) *The notion IS of "iterative set" (or "hereditary class").*

Let $ITS(x,y) := Cl(x) \wedge x \subseteq y$ and choose $IS(x) := x\eta I(ITS)$. Then we have, by 10.1(ii) and exploiting the classhood of x:

10.2.1 $$\mathrm{MF}^- \vdash IS(x) \leftrightarrow Cl(x) \wedge \forall z(z\eta x \rightarrow IS(z)).$$

In chapter V we shall prove that IS (actually, a generalization) yields a well-behaved set-theoretic universe.

(ii) *The intensional notion of (finite) type over* $\mathbb{N}$.
Let $TP(x,y)$ be the formula:

$$[(x = \mathbb{N}) \vee \exists c \exists b(x = [b \rightarrow c] \wedge b\eta y \wedge c\eta y) \vee \exists b \exists c(x = b \otimes c \wedge b\eta y \wedge c\eta y)].$$

If $\mathrm{TYPE}_f := I(TP)$, we get, provably in MF^-, that TYPE_f contains $\mathbb{N}$ and is closed under Cartesian product and exponentiation.

(iii) *The constructive second number class* $\mathcal{O}$.
Let:

$$Ord(x,y) := x\eta\mathbb{N} \wedge [(x = \overline{1}) \vee \exists k(x = 2^k \wedge k\eta y) \vee$$
$$\vee \exists k \forall n \exists m(\{k\}(n) \simeq m \wedge m\eta y \wedge x = 3^k)]$$

(for the notations 2^k, 3^k, recall the convention 3.6.1). If $\mathcal{O} := I(Ord)$, MF^- proves the fixed point axiom

10.2.2. $$\forall m(m\eta\mathcal{O} \leftrightarrow Ord(m, \mathcal{O})).$$

By inspecting the examples above, we are naturally led to lift to the present framework the notion of (positive) operator, which is familiar from the standard theory of inductive definitions (Moschovakis 1974).

10.3. DEFINITION

(i) A formula $A(v)$ is *operative* in v iff either v does not occur in A or A belongs to the least class of formulas, which is inductively generated by means of $\wedge, \vee$, $\forall y$, $\exists y$ (y distinct from v) from formulas of the form Nt, $\neg Nt$, $t = s$, $\neg t = s$, $t\eta v$, Tr, $\neg Tr$, provided v does not occur in t, r, s.

(ii) If A is a formula, which is operative in v and T-positive, the term $\lambda v.\{x_1 \ldots x_n : A(x_1, \ldots, x_n, v)\}$ is called an *operator*, n being the arity of the operator; A may contain free variables $\notin \{x_1, \ldots, x_n, v\}$.

(iii) An operator $\lambda v.\{x_1 \ldots x_n : A(x_1, \ldots, x_n, v)\}$ is *elementary* if every T-atom, which occurs in A, has the form $t\eta v$.

(iv) An operator $\lambda v.\{x_1, \ldots, x_n : A(x_1, \ldots, x_n, v)\}$ is *existential* iff no universal quantifier occurs in A, except possibly for universal bounded number quantifiers of the form $\forall m < n$ (see 3.1).

10.3.1. CONVENTION. We henceforth identify any given operator with the formula defining it; thus we simply call $A(x_1, \ldots, x_n, v)$ an operator in v *tout court*, and $x_1, \ldots, x_n$ show its arity. In the following we only consider operators of arity 1 (or 2 at most; this is not restrictive by pairing).

The idea is that any given operator defines a monotone operation (with respect to $\subseteq$ of 9.11), transforming relations (represented by) v into relations (represented by) $\{x_1 \ldots x_n : A(x_1, \ldots, x_n, v)\}$. The notion of *existential operator* is suggested by the formula involved in the definition of $TYPE_f$, while $Ord(x, y)$ is elementary and $ITS(x, y)$ is not.

In set theory (or in a suitable fragment of second-order logic), the Knaster-Tarski theorem grants the existence of fixed points; here we have, by induction on the build-up of the given operator and by 10.1 (ii):

10.4. PROPOSITION. (i) *If $A(v)$ is operative in v, then in pure logic*

$$A(v) \wedge v \subseteq u \rightarrow A(u).$$

Hence if $A(x, v)$ is an operator, $\lambda v.\{x : A(x, v)\}$ is $\subseteq$-monotone, provably in MF^-:

$$a \subseteq b \rightarrow \{x : A(x, a)\} \subseteq \{x : A(x, b)\}.$$

(ii) *For every operator $A(x, v)$, $I(A) := IxvA(x, v)$ is a fixed point, provably in* MF^-:

$$\forall u(u\eta I(A) \leftrightarrow A(u, I(A))).$$

10.4.1. REMARK. We warn the reader that the class of T-positive formulas, which are operative in the variable v, is not closed under substitution: e.g. $A(x, v) = x\eta v$ is an operator, but $A(v, v) = v\eta v$ is not.

Proposition 10.4 raises two natural problems. We first wonder whether *there exists a more mathematical characterization of operators, which does not refer to the syntax*: the question will be answered in the positive in Ch. IV, where we shall prove that operators coincide with *extensional* (*or $\subseteq$-monotone*) *operations* and have a natural topological interpretation. A second question concerns the apparent weakness of 10.4 (ii): it only offers implicit solutions to the given condition and states nothing about the *minimality* (*or the maximality*) of the solution $I(A)$, which is essential to argue by (generalized) induction on the given $I(A)$. In the next chapter, we shall see that such limitation is essential: MF^- is consistent with different hypotheses on $I(A)$ and hence MF^- is formally unable to distinguish among them.

However, proposition 10.4 (ii) already establishes a simple link with standard subsystems of second-order arithmetic (= analysis, in short).

Indeed, let us consider the standard language $\mathcal{L}^{\frown}$ of PA (= first-order arithmetic), expanded by a new unary predicate symbol P; an *arithmetical operator* is simply a formula $A(x, P)$ of $\mathcal{L}^{\frown}$, which is positive in P (i.e. A is logically equivalent to a formula, built up from atoms of the form Pt, $t = s$, $\neg t = s$ by means of $\vee$, $\wedge$, $\exists$, $\forall$), and has the free variable shown only. $\mathcal{L}(\mathrm{ID}_1)$, the language of the elementary theory of inductive definitions, is obtained from the language of PA, by adding a distinct unary predicate symbol I_A, for each arithmetical operator $A(x, P)$.

10.5. DEFINITION. $\widehat{\mathrm{ID}}_1$ is the first-order theory in the language $\mathcal{L}(\mathrm{ID}_1)$, which contains Peano arithmetic PA (with the induction schema, extended to all formulas of $\mathcal{L}(\mathrm{ID}_1)$) and the fixed point axiom:

$FP(A)$ $\qquad\qquad$ $\forall x(I_A x \leftrightarrow A(x, I_A))$,

for each arithmetical operator A (of course, $A(x, I_A)$ results from $A(x, P)$ by replacing each subformula of the form Pt with $I_A t$).

It is clear that every arithmetical operator becomes an elementary operator in the sense of 10.3 above, once we replace "Px" with "$x\eta v$" (we can assume that v is a fresh variable); thus, if we choose $\mathbb{N}$ as the range of individual variables and we interpret "$I_A x$" by "$x\eta I(A)$", we readily obtain:

10.6. PROPOSITION. $\widehat{\mathrm{ID}}_1$ *is interpretable in* MF.

Since $\widehat{\mathrm{ID}}_1$ has the same arithmetical theorems as the subsystem Σ^1_1-AC or, equivalently, Predicative Analysis of levels $< \epsilon_0$ (cf. Aczel 1977, Feferman 1982 and Ch.VIII), we also have a lower bound on the arithmetical content of MF (which turns out to be sharp by the proof theory of chapters IX-XI).

At this stage of the investigation, it becomes essential to calibrate the strength of the number-theoretic induction available; so we explicitly introduce two finitely axiomatized subsystems of MF, which restrict NIND to arbitrary properties and to classes respectively.

10.7. DEFINITION (Subsystems of MF with restricted number-theoretic induction).

(i) Set $Clos(y) := (\overline{0}\eta y) \wedge \forall x(x\eta y \rightarrow (x+1)\eta y)$. Then:

Property N-induction P-NIND: $\forall y(Clos(y) \rightarrow \mathbb{N} \subseteq y)$;

Class N-induction CL-NIND: $\forall y(Clos(y) \wedge Cl(y) \rightarrow \mathbb{N} \subseteq y)$;

(for $\subseteq$ see 9.11.6).

(ii) $\mathrm{MF}_p := \mathrm{MF}^- + \text{P-NIND}$; $\quad \mathrm{MF}_c := \mathrm{MF}^- + \text{CL-NIND}$.

Later, we shall prove that the inclusions $\mathrm{MF}_c \subseteq \mathrm{MF}_p \subseteq \mathrm{MF}$ are proper; while MF_c is proof-theoretically equivalent to OP, MF_p already proves the consistency of OP. If we restrict our attention to existential operators, then we can give a standard inductive definition of minimal fixed points.

10.8. LEMMA. *Let $A(x,u)$ be an existential operator. Then we can find a term $\lambda x.I_A^x$ such that, provably in* MF:

(i) $\forall n(I_A^0 = \{x \colon \neg x = x\} \wedge I_A^{n+1} = \{x : A(x, I_A^n)\})$;

(ii) $\forall n \forall p(n < p \to I_A^n \subseteq I_A^p)$;

(iii) $\forall x(A(x, I_A) \to \exists k A(x, I_A^k))$.

PROOF. (i)-(ii): we apply recursion on $\mathbb{N}$ (see 3.2), N-induction and proposition 10.4 (i). As to (iii), proceed by outer induction on the build up of A; here it is necessary that A is existential and we apply the true arithmetical schema:

$$\forall n(n < m \to \exists k B) \to \exists j \forall n(n < m \to \exists k(k < j \wedge B)),$$

which is provable by induction on m. □

Now choose:

$$I_A := \{x : \exists n(x \eta I_A^n)\}.$$

10.9. THEOREM. MF *proves (for B arbitrary, A existential operator)*:

(i) $\forall x(A(x, I_A) \to x \eta I_A)$;

(ii) $\forall x(A(x, B) \to B(x)) \to \forall x(x \eta I_A \to B(x))$.

PROOF. (i): by 10.8 (iii), (i).
(ii): if we assume the premise of (ii), $\forall n \forall x(x \eta I_A^n \to B(x))$ is derivable by means of N-induction and monotonicity. □

10.9.1. REMARK. If the existential operator defined by $A(x,u)$ is elementary in u, then by property N-induction, elementary comprehension 9.6 and 9.14, we have:

$$\mathrm{MF}_p \vdash \forall n Cl(I_A^n) \wedge Cl(I_A). \tag{1}$$

Under the same hypothesis on A, (1) implies that 10.8 (ii)-(iii) are provable in MF_p. Hence, 10.9 (i) together with

$$\forall x(A(x, u) \to x \eta u) \to \forall x(x \eta I_A \to x \eta u),$$

is already derivable in MF_p. (1) also holds if every T-atom occurring in $A(x,u)$ – except those of the form $t\eta u$ – is of the form $s\eta r$ where r can be proven to be a class. Hence 10.9 (i) and the special case of 10.9 (ii) with $B(x) = x\eta u$ are derivable in MF_p.

§ 10B. Applications to semantics and recursion theory

The fixed point theorem 10.1 turns out to be a significant tool in general: in most applications, one only needs the existence of a solution and not its minimality/maximality. We illustrate the theme by showing the existence of a partial satisfaction predicate for $\mathfrak{L}$ and then by proving two abstract versions of well-known results, due to Rice and Myhill.

First of all, OP is obviously sufficient (see § 2) to carry out a primitive recursive arithmetization of the $\mathfrak{L}$-syntax; so, we can fix an effective Gödel numbering $\ulcorner\ \urcorner$ and we let $\ulcorner E \urcorner$ stand for the (canonical term of $\mathfrak{L}$ representing the) Gödel number (in short *gn*) of the expression E. For later applications, it is also convenient to define the satisfaction predicate over Gödel numbers of arbitrary terms, possibly encoding formulas via $[\ldots]$. To this aim, we say that t is a formula-term if $t = [A]$, for some formula of $\mathfrak{L}$.

10.10. DEFINITION

(i) If f is any term, $f(\,{}^{i}_{x}\,)$ is the operation defined by

$$f(\,{}^{i}_{x}\,)i = x,\ f(\,{}^{i}_{x}\,)n = fn, \text{ if } \neg n = i.$$

Here we suppose that n, k, i range over $\mathbb{N}$. Clearly $f({}^{i}_{x})$ is well-defined, uniformly in i, x (apply definition by cases on $\mathbb{N}$; see 2.1, NAT.2).

(ii) By 3.6, we may assume that there are formulas and terms in the T-free part $\mathfrak{L}_{op}$ of the language $\mathfrak{L}$, defining the following notions:

$Ter(x) :=$ x is the *gn* of a term;
$For(x) :=$ x is the *gn* of a formula of $\mathfrak{L}$;
$Fort(x)\!:=$ x is the *gn* of a formula-term $[A]$;
$ftr(x) :=$ the *gn* of $[A]$ if x is the *gn* of A (i.e. $ftr(\ulcorner A \urcorner) = \ulcorner [A] \urcorner$);
$vr(x) :=$ the *gn* of the x-th variable;
$tr(x) :=$ the *gn* of $[Tt]$, if x is the *gn* of t;
$id(x,y) :=$ the *gn* of $[t = s]$ if $x = \ulcorner t \urcorner$, $y = \ulcorner s \urcorner$;
$nat(x) :=$ the *gn* of $[Nt]$ if $x = \ulcorner t \urcorner$;
$neg(x) :=$ the *gn* of $[\neg A]$, if x is the *gn* of $[A]$;
$and(x,y) :=$ the *gn* of $[A \wedge B]$ (x *gn* of $[A]$, y *gn* of $[B]$);
$all(x,y) :=$ the *gn* of $[\forall x_i A]$ ($x = \ulcorner x_i \urcorner$, y *gn* of $[A]$);

$app(x, y) :=$ the gn of (ts) (x gn of t, y gn of s).

10.11. LEMMA. *There exists a term Val such that, provably in* OP^-,

$$Val(\ulcorner t \urcorner, f) = t[x_0 := f\bar{0}, \ldots, x_n := f\bar{n}],$$

for every term t with free variables in the list $x_0, \ldots, x_n$.
In particular, OP^- *proves*

$$Val(\ulcorner [A] \urcorner, f) = [A[x_0 := f\bar{0}, \ldots, x_n := f\bar{n}]],$$

for every $\mathfrak{L}$*-formula A with free variables in the list* $x_0, \ldots, x_n$.

PROOF: we define an operation Val such that:

$Val\ (\ulcorner v_i \urcorner, f) = f\bar{i}$;
$Val\ (\ulcorner c \urcorner, f) = c$ if c is a constant;
$Val\ (\ulcorner ts \urcorner, f) = Val(\ulcorner t \urcorner, f)Val(\ulcorner s \urcorner, f)$.

Val is well-defined by means of the operation D and fixed point theorem 2.3 (it is essential that the operation $\ulcorner\ \urcorner$ is N-valued and that the language has a finite number of constants). □

In order to introduce the satisfaction predicate, let $S(x, v)$ be the formula saying that x has the form $\langle m, f \rangle$ and

1. $m = id(n, k)$ and $Val(n, f) = Val(k, f)$, or
2. $m = nat(n)$ and $N(Val(n, f))$, or
3. $m = tr(n)$ and $T(Val(n, f))$, or
4. $m = neg(n)$ and $\neg(\langle n, f \rangle \eta v)$, or
5. $m = and(n, k)$ and $(\langle n, f \rangle \eta v \wedge \langle k, f \rangle \eta v)$, or
6. $m = all(vr(i), n)$ and $\forall x(\langle n, f(^i_x) \rangle \eta v)$.

Then, if we apply the fixed point for properties (10.1) and we define

$$Sat(x, y) := \langle x, y \rangle \eta I(S),$$

we have:

10.12. THEOREM. (i) *For every* $\mathfrak{L}$*-formula A,*

$$\mathrm{MF}_c \vdash Sat(\ulcorner [A] \urcorner, f) \Leftrightarrow A[x_0 := f\bar{0}, \ldots, x_n := f\bar{n}];$$

($x_0, \ldots, x_n$ *contain all the free variables of A*).

If $V(x) := Sat(x, K\bar{0})$, *we have for every* $\mathfrak{L}$*-sentence C*:

$$\mathrm{MF}_c \vdash V(\ulcorner [C] \urcorner) \Leftrightarrow C.$$

(ii) *Moreover Sat satisfies, provably in* MF_c, *the adequacy conditions:*

1. $Sat(neg(n), f) \Leftrightarrow \neg Sat(n, f)$;

2. $Sat(and(n,m),f) \leftrightarrow Sat(n,f) \wedge Sat(m,f)$;
3. $Sat(all(vr(n),m),f) \leftrightarrow \forall u.Sat(m,f(\,^n_u\,))$;
4. $Sat(tr(n),f) \leftrightarrow T(Val(n,f))$;
5. $Sat(nat(n),f) \leftrightarrow N(Val(n,f))$;
6. $Sat(id(n,m),f) \leftrightarrow Val(n,f) = Val(m,f)$.

(iii) $\dot{\mathrm{MF}}_c \vdash \neg V(\ulcorner[L]\urcorner) \wedge \neg V(\ulcorner[\neg L]\urcorner)$, *for some $\mathfrak{L}$-sentence L.*

PROOF: the definition of Sat is already available. Condition (i) follows by metamathematical induction on A, while (ii) is a consequence of the definition of Sat and (iii) is Tarski's theorem. □

Of course, by the first clause of 10.12 (ii), V cannot be an adequate truth definition; however, if A is a sentence of OP, $\mathrm{MF}_c \vdash V(\ulcorner[A]\urcorner) \leftrightarrow A$ and we obtain:

10.13. COROLLARY. MF_p *proves the consistency of* OP *and hence is proof-theoretically stronger than Peano Arithmetic.*

PROOF. Let $Prov_{op}(x)$ be the formalization of "x is the gn of a formula provable in OP". Then we apply P-NIND with $X = \{x : Nx \wedge V(ftr(x))\}$ and we verify the formalized soundness theorem:

$$\forall n(Prov_{op}(n) \rightarrow n\eta X). \qquad \square$$

The argument does not work in the subsystem with class N-induction only, because the predicate involved is strictly partial.

After a glimpse into formal semantics, we illustrate the fixed point technique by lifting two well-known propositions of Recursion Theory to the present setting. Indeed, we can easily formalize a general form of Rice's theorem (see Rogers 1967, Fitting 1981), which is a useful tool to test the "non-classhood" of certain properties.

10.14. DEFINITION. We recall that $x =_e y := \forall u(u\eta x \leftrightarrow u\eta y)$ (extensional equality modulo η);

(i) b is *extensional* iff, for every x, y, if $x =_e y$ and $y\eta b$, then $x\eta b$.
(ii) $b \subset c := (b \subseteq c) \wedge \neg(c \subseteq b)$;
b is *proper* if $b \subset V = \{x : x = x\}$.

For instance, $\{u : \exists y(y\eta u)\}$ is extensional; but $CL = \{x : Cl(x)\}$ is not extensional (provably in MF^-). In fact, recall that $\mathbb{N} = \{x : Nx\}$ is a class and clearly, if r is the Russell property of 9.3, $b = \{x : Nx \vee r\eta r\} =_e \mathbb{N}$. Were b a class, we should have $\forall u((Nu \vee r\eta r) \vee (\neg Nu \wedge r\bar{\eta} r))$. Since $r\bar{\eta} r$, $r\eta r$ are provably false, we should have $\forall u.Nu$, against remark 3.9.2.

10.15. THEOREM (*"Rice generalized"; provable in* MF^-). *If a is non-empty, extensional and proper, then there exists no class c such that $c =_e a$.*

PROOF: by contradiction, assume that d is a class such that $d =_e a$ and choose b, c such that $b\eta a$ and not $c\eta a$. Then by 10.1 we can find a property I such that $I = \{x: (x\eta b \wedge \neg I\eta d) \vee (x\eta c \wedge I\eta d)\}$; so, by classhood of d,

$$\forall x(x\eta I \leftrightarrow (x\eta b \wedge \neg I\eta d) \vee (x\eta c \wedge I\eta d)). \tag{1}$$

If $I\eta a$, then $I\eta d$ and by (1) and consistency $\forall x(x\eta I \leftrightarrow x\eta c \wedge I\eta d)$, i.e. $I =_e c$, whence by extensionality, $c\eta a$: contradiction. If $\neg I\eta a$, then $\neg I\eta d$ and $I =_e b$, whence $I\eta a$: contradiction! □

10.15.1. APPLICATION
$\{x: \exists y(y\eta x)\}$, $\{x: c \subseteq x\}$ (c is a class), cannot be made equiextensional to classes. The same holds for the property of being a finite property. Indeed, let $f: k \mapsto b$ (k in $\mathbb{N}$) mean that f is a bijection between $\{x: Nx \wedge x < k\}$ and b, and assume that $\mathrm{FIN} = \{b: \exists k \exists f(f: k \mapsto b)\}$ is a class. Then b is an element of FIN iff, for some k and f, $f: k \mapsto b$. Clearly, if $c =_e b$ holds, also $f: k \mapsto c$, i.e. $c\eta\mathrm{FIN}$. Hence, since FIN is non-empty, we must have $V \subseteq \mathrm{FIN}$, which is absurd ($\mathbb{N}$ is not in FIN provably in MF_c).

We outline an extension of Myhill's theorem about recursive equivalence of creative sets (Rogers 1967). Being a creative property, again implies non-classhood. We closely follow the proof, given by Fitting (1981).

10.16. DEFINITION. Let $a \cup b := \{x: x\eta a \vee x\eta b\}$. Then:
(i) b is *creative* iff $\exists f \forall a(a \cap b =_e \emptyset \rightarrow \neg(fa)\eta(b \cup a))$.
(ii) b is *m-reducible* to c (in short $b \leq_m c$) iff $\exists f \forall x(x\eta b \leftrightarrow (fx)\eta c)$.

10.16.1. Example: $b = \{x: x\eta x\}$ is creative. Indeed, choose $f = \lambda x.x$. Then, if $a \cap b =_e \emptyset$, we get $\neg a\eta(b \cup a)$.

10.17. THEOREM (MF^-). *b is creative iff every c is m-reducible to b.*

The proof is a consequence of two lemmata.

10.17.1. LEMMA. *If b is creative and $b \leq_m c$, then c is creative.*

PROOF. Assume, for some f,

$$\forall x(x\eta b \leftrightarrow (fx)\eta c). \tag{1}$$

There exists an operation g, such that

$$a \cap b =_e \emptyset \rightarrow \neg(ga)\eta(a \cup b). \tag{2}$$

Choose $h = \lambda x.f(g(tx))$, where $tx = \{u: (fu)\eta x\}$. The claim is that

$$a \cap c =_e \emptyset \rightarrow \neg((ha)\eta(a \cup c)). \tag{3}$$

First note that $a \cap c =_e \emptyset$ implies $(ta) \cap b =_e \emptyset$ (by (1)). Hence by (2), $g(ta)$ does not belong to $(ta) \cup b$, which implies by definition $\neg(ha)\eta(a \cup c)$. □

10.17.2. LEMMA

We can define a term f such that, if y is in b, then $fy =_e \{g(fy)\}$ and if y is not in b, then $fy =_e \emptyset$.

PROOF. Define $hey = \{x : y\eta b \wedge x = gey\}$. By fixed point, choose f such that $hfy = fy$: hence

$$\forall x(x\eta fy \leftrightarrow y\eta b \wedge x = gfy). \qquad \square$$

PROOF of 10.17. Assume that b is creative; for some g, we have

$$c \cap b =_e \emptyset \rightarrow \neg(gc)\eta(c \cup b). \tag{1}$$

It follows:

$$c =_e \emptyset \rightarrow \neg(gc)\eta b; \tag{2.1}$$

$$c =_e \{x\} \wedge \neg x\eta b \rightarrow \neg gc = x. \tag{2.2}$$

By 10.17.2, we find an operation f such that:

$$u\eta c \rightarrow fu =_e \{g(fu)\}; \tag{3}$$

$$\neg u\eta c \rightarrow fu =_e \emptyset.$$

It is immediate to check that c is m-reducible to b via $h = \lambda x.g(fx)$ using (2)-(3).

In the other direction, assume that every property is m-reducible to b: then $\{x : x\eta x\} \leq_m b$. Therefore b is creative by 10.16.1 (example) and 10.17.1. □

10.18. COROLLARY (*provable in* MF^-)

No creative property can be equiextensional to a class.

PROOF: pick any property c not equiextensional to a class (such c exists by 9.3 or 10.15.1). Were b creative and equiextensional to some class b', we should have that c is extensionally equivalent to $c' = \{x : (fx)\eta b'\}$, for some f. But c' is a class: contradiction. □

Moreover, by Rice's theorem, the property of being creative does not define a class.

§11. Non-extensionality

The informal intuition behind the theory MF^- and its extensions obviously suggests an interpretation of properties and classes, which is radically opposed to a world of extensional entities. We show that extensionality fails, already in a very weak fragment of MF^-. It is impossible, in general, to identify arbitrary empty properties.

11.1 THEOREM. (Gilmore 1974)

(i) *We can produce terms t, s such that $(t =_e s \wedge \neg t = s)$ is provable in intuitionistic logic with equality plus the schema $\forall u(u\eta\{x: A\} \leftrightarrow A(u))$, where A is T-positive.*

(ii) *We can show in MF^- that t and s are classes.*

PROOF. Define:

$$\kappa := \{x: x\eta x\}$$
$$fy := \{x: x = \kappa \ \wedge \ y\eta x\}$$
$$a := \{u: fu = fu \cap \emptyset\},$$

where, as usual, $\emptyset := \{x: \neg x = x\}$. We first verify

$$fa \cap \emptyset =_e fa. \tag{1}$$

Trivially, $u\eta fa \cap \emptyset$ implies $u\eta fa$. Let $u\eta fa$; then $u = \kappa \wedge a\eta u$, i.e. $a\eta\kappa$, whence $a\eta a$, which yields $fa \cap \emptyset = fa$, i.e. $u\eta fa \cap \emptyset$, by $=$-logic.
We now prove:

$$\neg fa \cap \emptyset = fa. \tag{2}$$

If $fa \cap \emptyset = fa$, then $a\eta a$, i.e. $a\eta\kappa$, whence $\kappa\eta fa$ and by (1) $\kappa\eta\emptyset$: contradiction !
(1)-(2) complete the proof of (i).
Ad(ii). $fa \cap \emptyset$ is a class, because we have $\forall u(u\overline{\eta} fa \cap \emptyset)$. As to the classhood of fa, if $\neg u = \kappa$, we trivially obtain:

$$(u = \kappa \wedge a\eta u) \vee \neg u = \kappa \vee a\,\overline{\eta}\,u,$$

i.e. by T.1 $u\eta fa \vee u\overline{\eta} fa$. Assume $u = \kappa$; by (2), we have $a\overline{\eta} a$, i.e. $a\overline{\eta} u$, whence $u\overline{\eta} fa \vee u\eta fa$. It follows that $(u\eta fa \vee u\overline{\eta} fa)$ holds for arbitrary u, i.e. $Cl(fa)$. □

Gilmore's tricky empty classes are constructed via a detour through properties, which are not classes (like κ; see 10.16.1, 10.18). However, no essential use is made of the combinatory structure underlying the semantic structure. Gordeev found a quicker refutation of extensionality for classes; in his proof, the self-referential aspect is entirely absorbed by the fixed point for operations, and the argument has the advantage of being intrinsic (no detour through properties) and acceptable in a fully constructive context. The informal idea is to consider the fixed point of the property "to be equal to the empty property":

11.2. *Gordeev's Paradox.* MF^- proves:

$$\exists a(Cl(a) \wedge a = \{x : x = \emptyset \wedge x = a\} \wedge a =_e \emptyset \wedge \neg a = \emptyset).$$

PROOF. By 10.1 we can find an f such that $f = \{u : u = \emptyset \wedge u = f\}$.
By axiom $T.1$, f and $\emptyset$ are classes. Moreover, $f =_e \emptyset$: trivially $\emptyset \subseteq f$ and $u\eta f$ implies $u = \emptyset$ and $u = f$, i.e. $u\eta\emptyset$ and $f \subseteq \emptyset$. Now $f = \emptyset$ yields $f\eta\emptyset$, which is a contradiction. □

11.3. REMARK

(i) 11.1-11.2 imply that there are very simple non-extensional operations on classes ; here we say that $g : CL \rightarrow CL$ is extensional if $a =_e b$ implies $ga =_e gb$ for every a, $b \in CL$.
For instance, if we choose $gx = \{x\} = \{u : u = x\}$ and f as in the proof of 11.2, we have $f =_e \emptyset$, but $\neg \{f\} =_e \{\emptyset\}$: hence g is non-extensional.

(ii) 11.2 can be generalized; indeed MF^- proves

$$\forall a(Cl(a) \rightarrow \exists b(Cl(b) \wedge \neg a = b \wedge a =_e b)).$$

Hint (by Minari): choose $b = \{x : (a = b \wedge \neg a\eta a) \vee (\neg a = b \wedge x\eta a)\}$.

11.1-11.3 do not completely settle the situation: for instance, it remains to see whether it is consistent to assume:

$$a \equiv b \rightarrow a = b, \qquad (*)$$

where $a \equiv b := a =_e b \wedge \forall x\ (x\overline{\eta} a \leftrightarrow x\overline{\eta} b)$ and a, b are not classes.
A straightforward modification of Gordeev's argument shows that $(*)$ leads to inconsistencies. Choose $E = \{x : \neg x = x \vee r\eta r\}$, where $r = \{x : \neg x\eta x\}$ and $G = \{x : (x = E \wedge\ x = G) \vee r\eta r\}$. It is easy to verify that, in MF^-:

$$\neg Cl(E) \wedge E =_e \emptyset \wedge \neg Cl(G) \wedge E \equiv G \wedge \neg E = G.$$

We don't know whether the non-uniform version of abstraction $\exists y(\forall u(u\eta y \Leftrightarrow A(u)))$ is consistent with extensionality for properties. Of course, it is possible to maintain extensionality for properties, as soon as a specific equality relation for properties is adopted. This suggests to extend the language with a new sort of variables, ranging over properties in extension, in order to distinguish *properties as data types* (or partial functions) from *properties as formal constructs* (or names); the interested reader can consult Jäger (1987), Marzetta (1993).

11.4. *On formalizing mathematical arguments*

Beginning with the work of Bishop (1967), it has been demonstrated that non-extensionality of the basic notions does not affect the formalization of mathematical practice. Since we always work within specific mathematical structures, we are simply required to make explicit the equality relation, which is appropriate to the structures in question. In other words, mathematical practice has to cope with different kinds of "equalities", which are simply congruence relations and have to be explicitly given: think of the introduction of "equality" in the different number systems or when a basic equality on a given domain U of urelements is lifted to the type structure, built upon U. Moreover, equality on a given structure may depend on the presentation of the structure itself; for instance, consider equality on Cauchy reals *versus* equality on Dedekind reals.

The reader versed in foundational applications will find the philosophy of constructive and explicit mathematics appropriate here (see Bishop 1967, Feferman 1979, 1985, Beeson 1985). Accordingly, one has to consider not classes (or properties), but objects of the form $\mathcal{W} = \langle W, =_w \rangle$, where $=_w$ is an equivalence relation on W. This also means that the notion of operation itself has to be specialized. If $\mathcal{W} = \langle W, =_w \rangle$ and $\mathcal{U} = \langle U, =_u \rangle$ are structures (of the same similarity type), a function $f : \mathcal{W} \to \mathcal{U}$ is an operation $f : W \to U$, which "preserves equality":

$$x,\ y \text{ are in } W \text{ and } x =_w y \Rightarrow fx =_u fy.$$

Incidentally, we recall that, according to the views of Poincaré (1913, p.133), a class is well-defined *not only if it is predicatively defined*, but also if it is possible to produce *a predicative definition of the equality* on elements of that class.

As usual in the constructive practice, the power set operation $\mathcal{P}(\mathcal{W})$ can be replaced by the function space $2^{\mathcal{W}}$ (see §9). Of course, the classical step from a structure $\mathcal{W}$ to its quotient $\mathcal{W}/E$, modulo a given congruence relation, cannot be carried out: if two elements x, y of W are E-equivalent, it is not true in general that the corresponding equivalence classes $[x]_E$, $[y]_E$ are equal (in the sense of ground equality). At any rate, one can use equivalence classes as "contextually eliminable symbols"; one should also not forget that, in axiomatic set theory, the definition of $[x]$ requires some trickery.

To conclude, let us recall that the present framework and its extension look promising for the treatment of categorial concepts; some hints may be found in Feferman (1977a) (although the theory is somewhat different), and interesting developments are due to Gilmore (1990).

Appendix I: a property theoretic definition of the fixed point operator for predicates

In §10A we discussed a fixed point theorem for predicates which heavily relies on the full definitional strength of combinatory logic. It may be of interest to observe that the essence of the construction can be recovered under much weaker conditions.

1. DEFINITION. Let $A(x,y)$ be a formula with the free variables shown. We define:

$$D(A) := \{\langle x, f\rangle : A(x, \{u : \langle u, f\rangle \eta f\})\};$$

$$V(A) := \{x : \langle x, D(A)\rangle \eta D(A)\}.$$

The definition of $V(A)$ is essentially in Visser (1989, pp. 695-96), though in semantical context (η being replaced by a satisfaction predicate).

The verification of the following lemma is straightforward:

2. LEMMA (*"Second diagonalization"*).

(i) *If $A(x,y)$ is an arbitrary $\mathfrak{L}$-formula with the free variables shown only, then we can find a term $V(A)$, not containing the paradoxical combinator FP, such that* MF$^-$ *proves:*

$$\forall x(x\eta V(A) \Leftrightarrow A(x, V(A))).$$

(ii) *If A is T-positive, $\Leftrightarrow$ can be replaced by $\leftrightarrow$.*

We underline that $V(A)$ can be used in place of $I(A)$ in all the relevant applications of this book. This is important when *we wish to weaken the operational basis of the theory and to avoid full combinatory logic.* In fact, the lemma suggests the consideration of a *pure property theory* PT, based on the following language:

(i) three binary predicates $=$, η, $\bar{\eta}$ and a unary predicate T;

(ii) three function symbols for pairing function $\langle -, - \rangle$ and left and right projection $(\ )_1$, $(\ )_2$;

(iii) the operators [ ] and { }.

Terms and formulas of PT are defined by simultaneous induction: variables are terms; if t, s are terms, $t = s$, $t\eta s$, $t\bar{\eta}s$, Tt are formulas; if A is a formula, $[A]$ and $\{x : A\}$ are terms (the free variables of $[A]$ are exactly the free variables of A; the free variables of $\{x : A\}$ are exactly the free variables of A minus x); if A and B are formulas, $\neg A$, $A \wedge B$, $\forall x A$ are formulas; if t, s are terms, then $\langle t, s\rangle$ and $(t)_1$, $(t)_2$ are terms. We again use TA as an abbreviation for $T[A]$.

3. DEFINITION. PT is the theory based on classical first-order logic with identity, which contains the following axioms and schemata:

AP: $\forall x(x\eta\{u\colon A\} \leftrightarrow TA[u := x]) \wedge \forall x(x\overline{\eta}\{u\colon A\} \leftrightarrow T\neg A[u := x])$;

$T.1.$ $TA \leftrightarrow A$, if $A = (x = y)$, Nx, $\neg(x = y)$, $\neg Nx$, $x\eta y$, $x\overline{\eta}y$;

$T.2.1.$ $TTA \leftrightarrow TA$; $T.2.2.$ $T\neg TA \leftrightarrow T\neg A$;

$T.3.$ $T\neg\neg A \leftrightarrow TA$;

$T.4.1.$ $T(A \wedge B) \leftrightarrow TA \wedge TB$; $T.4.2.$ $T\neg(A \wedge B) \leftrightarrow T\neg A \vee T\neg B$;

$T.5.1.$ $T\forall xA \leftrightarrow \forall xTA$; $T.5.2.$ $T\neg\forall xA \leftrightarrow \exists xT\neg A$;

CONS $\neg(TA \wedge T\neg A)$;

SYM $(x\eta y \leftrightarrow T\neg(x\overline{\eta}y)) \wedge (x\overline{\eta}y \leftrightarrow T\neg(x\eta y))$;

PAIR $\forall x\forall y((\langle x, y\rangle)_1 = x \wedge (\langle x, y\rangle)_2 = y)$.

Clearly PT can be interpreted into MF^- and considered as a fragment of MF^-. By inspection, $D(A)$ and $V(A)$ can be regarded as PT-terms and hence we have:

4. PROPOSITION. *The second diagonalization lemma holds for* PT.

Of course, we can expand PT with the notion of natural number, some sort of number-theoretic induction schema and axioms for successor, zero, plus, times and additional number-theoretic operations. The resulting theory would have a non-trivial mathematical content; yet, in comparison with MF^- and its extension, PT has models in which $=$ is decidable and is compatible with the axiom that everything is a number. Thus it could be a basis for applicative refinements of the theory of abstraction (cf. Ch. XIV).

Appendix II: on the explicit abstraction theorem

We give details for the proof of the explicit abstraction theorem 9.13; then we conclude with a few comments about the applications of explicit abstraction and related problems.

1. LEMMA (*Combinatorial operators*)
We can define ELP-terms $Prod(a,b)$, V, V^n, Sep, Sep^-, $RE(n,k,a)$ (*for* $1 \leq n, k$), $LE(n,k,a)$, (*for* $1 \leq n$, $1 \leq k \leq 3$), $Ins(k,i,a)$ (*for* $1 \leq k$, $i \in \{2,3\}$), *which satisfy, provably in* MF^-:

1.1. $\forall u(u\eta Prod(a,b) \leftrightarrow \exists x \exists y(u = \langle x, y\rangle \wedge x\eta a \wedge y\eta b));$

$\forall u(u\eta V^n \leftrightarrow \exists x_1 \ldots \exists x_n(u = \langle x_1, \ldots, x_n\rangle)).$

1.2 $\forall u(u\eta Sep(a) \leftrightarrow \exists x \exists y \exists z(u = \langle x, y, z\rangle \wedge \langle x, \langle y, z\rangle\rangle\eta a));$

$\forall u(u\eta Sep^{-}(a) \leftrightarrow \exists x \exists y \exists z(u = \langle x, \langle y, z\rangle\rangle \wedge \langle x, y, z\rangle\eta a)).$

1.3. *Expansion to the right:*

$$\forall u(u\eta RE(n,k,a) \leftrightarrow$$
$$\leftrightarrow \exists x_1 \ldots \exists x_n \exists y_1 \ldots \exists y_k(u = \langle x_1, \ldots, x_n, y_1, \ldots, y_k\rangle \wedge \langle x_1, \ldots, x_n\rangle\eta a)).$$

1.4. *Expansion to the left:*

$$\forall u(u\eta LE(n,k,a) \leftrightarrow$$
$$\leftrightarrow \exists x_1 \ldots \exists x_n \exists y_1 \ldots \exists y_k(u = \langle y_1, \ldots, y_k, x_1, \ldots, x_n\rangle \wedge \langle x_1, \ldots, x_n\rangle\eta a)).$$

1.5. *k-Insertion:*

$$\forall u(u\eta Ins(k,3,a) \leftrightarrow \exists x \exists x_1 \ldots \exists x_k \exists y \exists z(u = \langle x, x_1, \ldots, x_k, y, z\rangle \wedge \langle x, y, z\rangle\eta a));$$
$$\forall u(u\eta Ins(k,2,a) \leftrightarrow \exists x \exists x_1 \ldots \exists x_k \exists y(u = \langle x, x_1, \ldots, x_k, y\rangle \wedge \langle x, y\rangle\eta a)).$$

PROOF. 1.1. Choose:
$Prod(a,b) = Conv(Exp(a)) \cap Exp(b)$; $V = dom(\mathbb{ID})$, $V^{k+1} = Prod(V^k, V)$.

1.2. Choose $Sep(a) = Cyc^2(Conv(a))$ and $Sep^{-}(a) = Conv(Cyc(a))$, where $Cyc^2(t) = Cyc(Cyc(t))$.

1.3: by iteration of $RE(n,1,a) = Conv(Exp(a \cap V^n))$.

1.4: we inductively define

$$F(1,a) = Exp(a);\ F(2,a) = Sep(Exp(a));$$
$$F(n{+}3,a) = Sep(F(n{+}2, Sep^{-}(a))).$$

Then we verify by induction on n:

$$u\eta F(n,a) \leftrightarrow \exists v \exists x_1 \ldots \exists x_n(u = \langle v, x_1, \ldots, x_n\rangle \wedge \langle x_1, \ldots, x_n\rangle\eta a).$$

Finally choose $LE(n,k,a) = V^{k+n} \cap F(n,a)$.

1.5: let

$G(a) = Sep(Cyc^2(Sep(Exp(Cyc(a)))))$. Then we recursively define:

$$Ins(1,3,a) = G(a);\ Ins(k{+}1,3,a) = G(Ins(k,3,a)).$$

The case of 2-insertion is similar. □

2. LEMMA. *We can find ELP-predicates* $\mathbb{APP}_n(i,j,k)$, $\mathbb{ID}_n(i,j)$, $\mathbb{N}_n(i)$, $\mathbb{T}_n(i)$, *such that:*

2.1 $\mathbb{APP}_n(i,j,k) \equiv \{\langle x_1,\dots,x_n\rangle : x_i = x_j x_k\}$;

2.2 $\mathbb{ID}_n(i,j) \equiv \{\langle x_1,\dots,x_n\rangle : x_i = x_j\}$;

2.3 $\mathbb{N}_n(i) \equiv \{\langle x_1,\dots,x_n\rangle : Nx_i\}$;

2.4 $\mathbb{T}_n(i) \equiv \{\langle x_1,\dots,x_n\rangle : Tx_i\}$;
(all the variables are distinct and $1 \le i,\ j,\ k \le n$*; in* 2.1, $3 \le n$ *and in* 2.2, $2 \le n$*).*

PROOF. 2.1: we proceed by induction on $n \ge 3$. If $n = 3$, $\mathbb{APP}_n(i,j,k)$ is obviously obtained from $\mathbb{APP}$ by permutation and hence *Cyc* and *Trans* suffice. If $n > 3$, let u, v, w denote distinct variables in the set $\{x_i,\ x_j,\ x_k\}$.

Case a): u, v, w already occur in the list $x_1,\dots,x_{n-1}$; then we apply the induction hypothesis and expansion to the right.

Case b): $u = x_n$, but $v,\ w \ne \{x_{n-1}\}$. It is enough to observe that IH applies to $Trans\{\langle x_1,\dots,x_n\rangle$: $x_i = x_j x_k\}$.

Case c): $v = x_n$, $w = x_{n-1}$. Apply left expansion and possibly insertion to $\mathbb{APP}$, whenever $u \ne x_{n-2}$.

The proof of 2.2 – 2.4 are similar. □

3. LEMMA. *Let* r *be a term of* $\mathcal{L}_{op}$. *For every* n, i, *such that* $1 \le i \le n$, *and* x_i *not occurring in* r, *we can effectively find an elementary term* $\sigma_n^i(r)$ *with the free variables of* r *except* $x_1,\dots,x_n$, *such that, provably in* MF$^-$:

$$\{\langle x_1,\dots,x_n\rangle : x_i = r\} \equiv \sigma_n^i(r).$$

PROOF. Induction on r. If r is a constant or a variable $v \notin \{x_1,\dots,x_n\}$, we apply right and left expansions to the singleton $\{r\}$. If r is a variable $v \in \{x_1,\dots,x_n\}$, by assumption $v = x_k$ with $k \ne i$, and hence lemma 2.2 applies. If $r = r_1 r_2$, we use IH, *dom*, $\cap$ and lemma 2.1. □

Remarks and problems

(a) The collection ELP (see 9.11) of elementary terms is well behaved with respect to substitution (if $t(y)$ is elementary, $z \notin \{y\}$, z free for y in t, then $t[y := z]$ is elementary), and renaming of bound variables; this should be contrasted with the collection of elementary formulas. By the theorem 9.13 (ii) we can avoid the syntactical notion of elementary formula in defining $\{x : A\}$, if A is elementary in y.

(b) Add a new unary predicate $ECl(x) :=$ "x is an explicit class" to the language of MF$^-$ with the axioms: 1) ECl contains $\mathbb{ID}$, $\mathbb{APP}$, $\mathbb{N}$; 2) ECl is closed under the combinatorial and basic operators of 9.11; 3) ECl is closed under join; 4) ECl is the $\subseteq$-least such property (i.e. the schema saying that any "externally definable" property with the same closure conditions of ECl, contains ECl). Note that 4) immediately implies $\forall x(ECl(x) \rightarrow Cl(x))$. Call the resulting extension of MF^- (MF_c, etc.) EMF^- (EMF_c, etc.). Show that all these extensions are consistent with the construction principle CP:

$$\forall x(Cl(x) \rightarrow \exists y(ECl(y) \wedge x =_e y)).$$

CP can be regarded as a sort of abstract Suslin-Kleene theorem (Moschovakis 1974). Prove that EMF_c+CP is a conservative extension of MF_c and hence of OP.

(c) Consider the language $\mathcal{L}_{op}$, expanded with the new unary predicate Cl, the binary relation $\in$, and $\mathbb{APP}$, $\mathbb{ID}$, $\mathbb{N}$, singleton $\{-\}, -, \cap$, dom, $Trans$, $Conv$, Cyc, Exp and Σ (all regarded as primitive individual constants). Let $\mathcal{S}$ be the first-order extension of OP in the modified language, which includes, besides the number-theoretic induction schema, finitely many axioms, stating that: 1) Cl contains $\mathbb{APP}$, $\mathbb{ID}$, $\mathbb{N}$; 2) Cl is closed under singleton, complement, domain, intersection, transpose, cycle, converse, expansion and join, and it satisfies the intended equivalences for each of the given operations (e.g. for Exp we require:

$$Cl(a) \rightarrow Cl(Exp(a)) \wedge \forall x(x \in Exp(a) \leftrightarrow \exists y \exists z(\langle y, z\rangle = x \wedge z\eta a)).$$

Then $\mathcal{S}$ is a possible reformulation of Feferman's system for explicit mathematics $EM_0 + J +$ "every operation is total" (Feferman 1979), in a language, which avoids the use of countably many constants.

Appendix III: independence of truth predicates from the encoding of the logical operators

The notion of truth we introduced and axiomatized in §7, clearly involves the use of the special terms

$$ID := \lambda xy.\langle\overline{1}, \langle x, y\rangle\rangle;\ \ TR := \lambda x.\langle\overline{2}, x\rangle;\ \ NAT := \lambda x.\langle\overline{3}, x\rangle; \tag{1}$$

$$NEG := \lambda x.\langle\overline{4}, x\rangle;\ \ AND := \lambda xy.\langle\overline{5}, \langle x, y\rangle\rangle;\ \ ALL := \lambda x.\langle\overline{6}, x\rangle;$$

they represent the basic predicates and logical operations in $\mathcal{L}_{op}$. However, this choice is immaterial for the development of the theory MF^-. This fact can be seen as follows. We choose an expansion $\mathcal{L}_+$ of the basic language $\mathcal{L}$, in which $\dot{ID}$, $\dot{TR}$, $\dot{NAT}$, $\dot{NEG}$, $\dot{AND}$, $\dot{ALL}$ are new *primitive individual constants*. The new constants are supposed to satisfy the set LOG of finitely many axioms, obviously suggested by 7.1.1:

LOG1 $L_1 x = L_2 y \rightarrow L_1 = L_2 \wedge x = y$,

where L_1, $L_2 \in LOG_1 = \{\dot{NAT}, \dot{NEG}, \dot{TR}, \dot{ALL}\}$;

LOG2 $G_1xy = G_2x'y' \rightarrow . \; G_1 = G_2 \wedge x = x' \wedge y = y'$,
where $G_1, G_2 \in LOG_2 = \{I\dot{D}, AN\dot{D}\}$;

LOG3 $\neg \; L_1x = L_2yz$, for $L_1 \in LOG_1$, $L_2 \in LOG_2$;

LOG4 $\neg \; L_1 = L_2$, where L_1, L_2 are distinct elements of $LOG_1 \cup LOG_2$.

Now let MFL^- be the theory which is obtained from MF^-:

1) by replacing the terms ID, NAT, TR, NEG, AND, ALL with $\mathfrak{L}_+$-primitive constants $I\dot{D}$, $NA\dot{T}$, $T\dot{R}$, $NE\dot{G}$, $AN\dot{D}$, $AL\dot{L}$ (respectively) in the truth axioms $T.1$-$T.6$ of theorem 7.10;

2) by adding the axioms LOG.

The reader can check that, if the axioms LOG are assumed, all theoretical developments of part A through part C can be carried out without difficulty in MFL^-.

PROPOSITION (*"Change-of-basis"*)

1.1. *We can define a term $\Phi_+ x$ and a map $A \mapsto (A)_+$ from the set of $\mathfrak{L}$-formulas into the set of $\mathfrak{L}_+$-formulas such that:*

(i) *$(A)_+$ is obtained from A by replacing every subformula of the form Tt with $T(\Phi_+ t)$, and hence $(A)_+ = A$ if A does not contain T;*

(ii) *if $\mathrm{MF}^- \vdash A$, then $\mathrm{MFL}^- \vdash (A)_+$ (A arbitrary).*

1.2. *Conversely, we can define a map $A \mapsto (A)_-$ from the set of $\mathfrak{L}_+$-formulas into the set of $\mathfrak{L}$-formulas such that:*
(i) *if $\mathrm{MFL}^- \vdash A$, then $\mathrm{MF}^- \vdash (A)_-$ (A arbitrary);*
(ii) *$A = (A)_-$ if A is an $\mathfrak{L}$-formula.*

PROOF.
Ad 1.1. By applying the fixed point theorem 2.3 and definition by cases on N, we can find a term Φ_+, which satisfies the following recursive conditions:

$$\Phi_+ x = I\dot{D}((x)_2)_1((x)_2)_2 \text{ if } (x)_1 = \overline{1};$$
$$\Phi_+ x = T\dot{R}(\Phi_+((x)_2)) \text{ if } (x)_1 = \overline{2};$$
$$\Phi_+ x = NA\dot{T}((x)_2) \text{ if } (x)_1 = \overline{3};$$
$$\Phi_+ x = NE\dot{G}(\Phi_+((x)_2)) \text{ if } (x)_1 = \overline{4};$$
$$\Phi_+ x = AN\dot{D}(\Phi_+((x)_2)_1)(\Phi_+((x)_2)_2) \text{ if } (x)_1 = \overline{5};$$
$$\Phi_+ x = AL\dot{L}(\lambda u.\Phi_+(((x)_2)u)) \text{ if } (x)_1 = \overline{6}.$$

Then we define $A \mapsto A_+$ by induction on A:

(i) $(Nt)_+ = Nt$ and $(t = s)_+ = (t = s)$; $(Tt)_+ = T(\Phi_+ t)$;
(ii) $(\neg A)_+ = \neg(A_+)$; $(A \wedge C)_+ = A_+ \wedge C_+$; $(\forall x A)_+ = \forall x(A_+)$.

The claim is checked by a straightforward induction on the length on the derivation of A.

Ad 1.2. $(A)_-$ is obtained from A by replacing the constants $I\dot{D}$, $T\dot{R}$, $NA\dot{T}$, $NE\dot{G}$, $AN\dot{D}$, $AL\dot{L}$ with the corresponding terms ID, NAT, TR, NEG, AND, ALL. □

A second route for ensuring the independence of the truth theory on the choice in (1) is (roughly) suggested by the analogy with recursion theory and acceptable Gödel numberings (see Rogers 1967, Odifreddi 1989). One might introduce a notion of acceptable logical basis and show that the truth theories do not depend on the choice of a particular basis.

PART B

TRUTH AND RECURSION THEORY

"Niemand kann eine unendliche Menge anders beschreiben als durch Angabe von Eigenschaften, welche für die Elemente der Menge charakteristisch sind; niemand eine Zuordnung zwischen unendlich vielen Dingen stiften ohne Angabe eines Gesetzes, d.h. eine Relation, welche die zugeordneten Gegenstände miteinander verknüpft." (H.Weyl 1918).

CHAPTER 3

INDUCTIVE MODELS AND DEFINABILITY THEORY

§12. Inductive models and the induction theorem
§13. The envelope of an inductive model
§14. The uniform ordinal comparison theorem for inductive models
§15. Applications of the uniform ordinal comparison theorem

We know from §7 that the minimal fixed points of a simple monotone operator yield natural set-theoretic models of MF^-, the so-called inductive models. On the other hand, there exists a close relation between inductive models and elementary inductive definability on combinatory algebras: not only inductive definitions are crucial for the semantics of reflective truth, but the language of reflective truth offers a natural framework for an intensional approach to non trivial recursion-theoretic facts.

The two-sided link with generalized recursion theory is first illustrated by the equivalence theorems of this chapter. First of all, the fixed points of elementary inductive definitions in the sense of Moschovakis (1974) are naturally represented by terms of our language, as soon as we work within inductive models (§§12-13). Second and more important, we can prove in §14 a uniform ordinal comparison theorem for inductive models, which readily implies that total predicates on a given model $\mathcal{M}$ of OP^- ($= OP$ without N-induction) coincide with the collection of hyperelementary subsets of $\mathcal{M}$ (§15).

The recursion-theoretic investigations naturally lead to an extension of the basic system MF^- with simple axioms on a suitable approximation operator π; π uniformly splits each non-empty property into a $\subseteq$-chain of subclasses, satisfying certain simple compatibility and well-foundedness conditions. The π-axioms are still conservative, even in presence of a full generalized induction schema, as far as number-theoretic induction holds for total predicates only (see the general conservation theorem of 15.5). We shall see that π-axioms are quite effective for obtaining consequences, that usually require ordinal-theoretic arguments. In a sense, this is not surprising: π-axioms are directly responsible for interpreting fragments of set theory in Ch.V.

§12. Inductive models and the induction theorem

$\mathcal{M}$ is a fixed model of OP^-, whose universe is the set M. As in Ch.II, $\mathcal{L}(\mathcal{M})$ ($\mathcal{L}_{op}(\mathcal{M})$) denotes the language of MF (of OP without the truth predicate T respectively) with distinct individual constants, for each element of M. For the sake of simplicity, we stick to the notations and conventions of §7. If Γ is the monotone operator of §7, $\mathbb{O}(\mathcal{M})$ is the $\subseteq$-least fixed point, which is generated from below by transfinite iteration of Γ:

(i) $\mathbb{O}(\mathcal{M},0) = \emptyset$;
(ii) $\mathbb{O}(\mathcal{M},\alpha+1) = \Gamma(\mathbb{O}(\mathcal{M},\alpha))$;
(iii) $\mathbb{O}(\mathcal{M},\lambda) = \bigcup_{\alpha<\lambda} \mathbb{O}(\mathcal{M},\alpha)$ (λ limit).

Once $\mathcal{M}$ is fixed or clear from the context, we set $\mathbb{O} := \mathbb{O}(\mathcal{M})$ and $\mathbb{O}(\alpha) := \mathbb{O}(\mathcal{M},\alpha)$. If $S \subseteq M$, we identify $\langle\mathcal{M},S\rangle$ with S; thus we simply write $S \models A$ instead of $\langle\mathcal{M},S\rangle \models A$. As usual, if t is a closed term of $\mathcal{L}_{op}(\mathcal{M})$, $\mathcal{M}(t)$ is the unique value of t in $\mathcal{M}$.

12.1. LEMMA. *If A is an arbitrary sentence of $\mathcal{L}(\mathcal{M})$,*

$$\mathbb{O}(\alpha+1) \models TA \textit{ implies } \mathbb{O}(\alpha) \models A, \textit{ for every } \alpha. \quad (*)$$

PROOF. It is enough to check $(*)$, by induction on the logical complexity of A. The case where A is an e-atom is trivial. If $A = \neg Tt$, we have, by assumption and the inversion lemma 7.4 (ii)-(iii), $\mathcal{M}(\neg t) \in \mathbb{O}(\alpha)$, whence $\mathcal{M}(t) \notin \mathbb{O}(\alpha)$ by consistency, i.e. $\mathbb{O}(\alpha) \models \neg Tt$. Let $A = B \wedge C$. Again by assumption and inversion, we have $\mathcal{M}([B])$, $\mathcal{M}([C]) \in \mathbb{O}(\alpha) \subseteq \mathbb{O}(\alpha+1)$, which implies $\mathbb{O}(\alpha+1) \models TB$ and $\mathbb{O}(\alpha+1) \models TC$, whence we conclude by IH $\mathbb{O}(\alpha) \models B \wedge C$. The remaining cases are similar. □

12.2. (i) Below we need to consider the expansion of $\mathcal{L}(\mathcal{M})$ by a new unary predicate symbol; we put $\mathcal{L}(\mathcal{M},P) := \mathcal{L}(\mathcal{M}) \cup \{P\}$ (P unary predicate symbol distinct from T). Then a realization of $\mathcal{L}(\mathcal{M},P)$ over $\mathcal{M}$ is given by any pair $\langle S_1,S_2\rangle$ of subsets $\subseteq M$: it is understood that S_1 is the interpretation of T, while S_2 is the interpretation of P.

(ii) Let $A(x,P)$ be any formula of $\mathcal{L}(\mathcal{M},P)$; if $B(x)$ is a formula of $\mathcal{L}(\mathcal{M})$, $A(x,B)$ is the $\mathcal{L}(\mathcal{M})$-formula obtained by replacing each occurrence of Pt in A with $B[x:=t]$ (we assume that the substitution is legitimate). Then, if $S \subseteq M$, we obviously have (induction on A):

12.2.1. FACT

(i) *If $A(x,P)$ is a formula of $\mathcal{L}(\mathcal{M},P)$ with the free variables shown, then $\langle S,\{a : a \in M \text{ and } S \models B(a)\}\rangle \models A(a,P)$ iff $S \models A(a,B)$.*

(ii) *If $A(x,v)$ is an operator (in the variable v) of $\mathcal{L}(\mathcal{M})$ (see 10.3.1),* *and $A(x,P)$ is the $\mathcal{L}(\mathcal{M}, P)$-formula obtained from $A(x,v)$ by replacing each subformula of the form $t\eta v$ with Pt, then $A(x,P)$ is positive in P and T (i.e. its negation normal form contains an even number of negations in front of atoms of the form Tt, Pt). Conversely every $A(x,P)$ of $\mathcal{L}(\mathcal{M}, P)$, which is positive in T and Q, determines an operator $A(x,v)$ of $\mathcal{L}(\mathcal{M})$, obtained by replacing Pt with $t\eta v$ (v being a fresh variable).*

Clearly if $A(x,v)$ is an operator and $B(x)$ is a formula, it is natural to write $A(x,B)$ also for the result of replacing each occurrence of "$t\eta v$" by $B(t)$.

12.3. DEFINITION

(i) If $B(x)$ is a formula of $\mathcal{L}(\mathcal{M})$ and $A(x,v)$ is an operator of $\mathcal{L}(\mathcal{M})$,

$$Clos_A(B) := \forall x(A(x,B) \to B(x)) \; (= \text{"}B\text{ is }A\text{-closed"});$$

(ii) If $B(x) = x\eta b$, we simply write $Clos_A(b)$ (= "b is A-closed").

12.4. THEOREM (*Generalized induction*). *If $A(x,v)$ is an operator of $\mathcal{L}(\mathcal{M})$ and $I(A)$ is the fixed point term of* 10.1, *then*

$$\mathbb{O} \models Clos_A(B) \text{ implies } \mathbb{O} \models \forall x(x\eta I(A) \to B(x)).$$

PROOF. Assume $\mathbb{O} \models Clos_A(B)$. We verify by transfinite induction on α:

$$\text{for every } a \in \mathcal{M},\ \mathbb{O}(\alpha) \models a\eta I(A) \text{ implies } \mathbb{O} \models B(a). \tag{1}$$

If $\alpha = 0$ or α is a limit, (1) is trivially true. Let (1) hold for α, and assume that $\mathbb{O}(\alpha+1) \models a\eta I(A)$, i.e by ($\beta$)-conversion, $\mathbb{O}(\alpha+1) \models TA(a, I(A))$. But Lemma 12.1 implies $\mathbb{O}(\alpha) \models A(a, I(A))$; if P is a new predicate symbol and we set

$$I(A,\alpha) := \{a : a \in \mathcal{M} \text{ and } \mathbb{O}(\alpha) \models a\eta I(A)\},$$

we obtain by 12.2.1:

$$\langle \mathbb{O}(\alpha), I(A,\alpha)\rangle \models A(a,P).$$

By IH, since A depends positively upon the interpretations of T and P,

$$\langle \mathbb{O},\ \{a :\ \in \mathcal{M} \text{ and } \mathbb{O} \models B(a)\}\rangle \models A(a,P);$$

$$\mathbb{O} \models A(a,B) \text{ (by 12.2.1)};$$

hence, by A-closure of B, $\mathbb{O} \models B(a)$. □

12.4.1. GID (= generalized inductive definition) is the schema:

$$Clos_A(B) \to \forall x(x\eta I(A) \to B(x)),$$

where B is an arbitrary formula of $\mathcal{L}$ and $A(x,v)$ is an operator of $\mathcal{L}$.

Theorem 12.4 and 7.10 imply:

12.5. COROLLARY. MF + GID *is consistent.*

§13. The envelope of an inductive model

We fix a structure $\mathcal{M}$ with universe M, which is a model of OP^-. If $S \subseteq M$, each closed term t of $\mathcal{L}(\mathcal{M})$ naturally defines a subset of M in the structure $\langle \mathcal{M}, S\rangle$, namely the set:

13.1 $$t(S) := \{a \colon a \in M \text{ and } S \models a\eta t\}.$$

If $X \subseteq M$ and $X = t(S)$, for some closed term t of $\mathcal{L}(\mathcal{M})$, we say that X is *representable in* $\langle \mathcal{M}, S\rangle$ (or simply in S), and we put:

13.2. $$ENV(\mathcal{M}, S) := \{X \colon X \subseteq M \text{ and } X \text{ is representable in } \langle \mathcal{M}, S\rangle\}.$$

Trivially $S \in ENV(\mathcal{M}, S)$. We call $ENV(\mathcal{M}, S)$ the *envelope of* $\langle \mathcal{M}, S\rangle$; as usual, once $\mathcal{M}$ is clear, we simply neglect its explicit mention and we speak of the envelope of a given $S \subseteq M$. The envelope of $\mathbb{O}(\mathcal{M})$ has a natural characterization in terms of elementary inductive definability in the sense of Moschovakis (1974). The argument is the recursion-theoretic *pendant* of the generalized induction theorem 12.4. We give the basic definitions, suitably adapted to the present context.

13.3. DEFINITION

(i) If P is a predicate symbol distinct from T and $A(u, P)$ is a formula of $\mathcal{L}_{op}(\mathcal{M}, P) := \mathcal{L}_{op}(\mathcal{M}) \cup \{P\}$, which only contains the free variables shown, $A(u, P)$ is *positive elementary* (in P) iff A belongs to the smallest class of $\mathcal{L}_{op}(\mathcal{M}, P)$-formulas, which includes formulas of the form $t = s$, Nt, $\neg t = s$, $\neg Nt$, Pt, and is closed under $\wedge$, $\vee$ and Qx ($Q = \exists, \forall$).

(ii) If $A(u, P)$ is positive elementary in P, then $A(u, P)$ defines a monotone operator $\Gamma_A : \mathcal{P}(M) \to \mathcal{P}(M)$ ($\mathcal{P}(M) =$ power set of M), namely, if $S \subseteq M$ and we interpret P by S,

$$\Gamma_A(S) := \{a \colon a \in M \text{ and } \langle \mathcal{M},\ S\rangle \models A(a, P)\}.$$

For simplicity, it is convenient to call $A(u, P)$ itself *an elementary positive operator* on $\mathcal{M}$ (thus identifying the formula with the operator it defines).

(iii) We write I_A for the smallest fixed point of Γ_A (in short, the fixed point of the given operator; so I_A satisfies $\Gamma_A(I_A) \subseteq I_A$ and $I_A \subseteq X$, whenever $\Gamma_A(X) \subseteq X$).

(iv) A set $X \subseteq M$ is *inductive on* $\mathcal{M}$ iff there are a positive elementary operator A and an element f of M such that, for every $a \in M$, $a \in X$ iff $(f \cdot_M a) \in I_A$ ($\cdot_M$ is the application operation of $\mathcal{M}$).

(v) $IND(\mathcal{M}) :=$ the collection of sets $\subseteq M$, which are inductive on $\mathcal{M}$.

(vi) $X \subseteq M$ is *coinductive on* $\mathcal{M}$ iff $-X(=$ the complement of X in $M)$ is inductive on $\mathcal{M}$.

(vii) $X \subseteq M$ is *hyperelementary on* $\mathcal{M}$ iff X is both inductive and coinductive on $\mathcal{M}$.

(viii) $HYP(\mathcal{M}) :=$ the collection of subsets $\subseteq M$, hyperelementary on $\mathcal{M}$.

13.4. THEOREM. *Let* $\mathcal{M}$ *be a model of* OP^-. *Then:*

(i) $ENV(\mathbb{O}(\mathcal{M})) = IND(\mathcal{M})$;

(ii) $\mathbb{O}(\mathcal{M}) \in IND(\mathcal{M}) - HYP(\mathcal{M})$.

PROOF. (i) $IND(\mathcal{M}) \subseteq ENV(\mathbb{O}(\mathcal{M}))$. It clearly suffices to check that, if $A(u,P)$ is an arbitrary positive elementary operator on $\mathcal{M}$, then I_A is representable in $\mathbb{O}(\mathcal{M})$. Consider the operator $A(u,v)$ in the language $\mathfrak{L}(\mathcal{M})$, obtained from $A(u,P)$ according to 12.2.1 (ii) (v fresh variable). Now $\mathbb{O}(\mathcal{M})$ is a model of MF^- (by 7.10) and MF^- proves by 10.4 that there is an abstract $I(A) = IxvA(x,v)$ such that:

$$\mathbb{O}(\mathcal{M}) \models \forall x(x\eta I(A) \leftrightarrow A(x, I(A))). \tag{1}$$

Hence, if we let $I(A)(\mathbb{O}(\mathcal{M})) := I$ (recall the notation of 13.1), and we notice that the only T-atoms occurring in $A(x, I(A))$ have the form $t\eta I(A)$ by assumption on A, we get by 12.2.1 (i),

$$\mathbb{O}(\mathcal{M}) \models A(a, I(A)) \text{ iff } \langle \emptyset, I \rangle \models A(a,P), \tag{2}$$

which implies by (1),

$$\Gamma_A(I) = I. \tag{3}$$

By minimality of I_A, $I_A \subseteq I$. Conversely, it is straightforward to check, by induction on α with 12.1-12.2:

$$\mathbb{O}(\alpha) \models a\eta I(A) \text{ implies } a \in I_A \text{ (for } a \in M). \tag{4}$$

Thus $I \subseteq I_A$ and $I_A = I$; hence $I \in ENV(\mathbb{O}(\mathcal{M}))$. In order to complete the proof of the theorem, we show:

$$ENV(\mathbb{O}(\mathcal{M})) \subseteq IND(\mathcal{M}).$$

If $X = t(\mathbb{O}(\mathcal{M}))$, for some closed term t of $\mathfrak{L}(\mathcal{M})$ and $f = \mathcal{M}(t) \in M$, we have, for every $a \in M$:

$$a \in X \text{ iff } \mathbb{O}(\mathcal{M})\models a\eta t \text{ iff } (f \cdot_M a) \in \mathbb{O}(\mathcal{M}).$$

Hence X is inductive on $\mathcal{M}$, since $\mathbb{O}(\mathcal{M})$ is the fixed point of the positive elementary operator $\Gamma(u,P)$ of section 7.3 (v).

(ii) If $M - \mathbb{O}(\mathcal{M}) \in IND(\mathcal{M})$, $\{a \in M: \mathbb{O}(\mathcal{M})\models\neg a\eta a\}=\{a \in M: \mathbb{O}(\mathcal{M})\models a\eta t\}$, for some closed term t, and we can derive a contradiction *à la* Russell. □

13.5. DEFINITION. Let $S \subseteq M$:

(i) $SEC(S) := \{X: X \subseteq \mathcal{M}$ and $X = t(S)$, for some closed term t of $\mathcal{L}(\mathcal{M})$ such that $S\models Cl(t)\}$ (= " the section of S");

(ii) $SEC^+(S) := \{X: X \subseteq M$ and $X, -X \in ENV(S)\}$ (= "the $^+$-section of S").

We immediately obtain, with theorem 13.4:

13.6. LEMMA
If S is consistent and $S \in FIX(\Gamma, \mathcal{M})$ (cf. 7.5), $SEC(S) \subseteq SEC^+(S)$ and $SEC^+(S) \subset ENV(S)$. In particular, $SEC^+(\mathbb{O}(\mathcal{M})) = HYP(\mathcal{M})$.

From the ordinal comparison theorem of the next section, it will follow that $SEC^+(\mathbb{O}(\mathcal{M})) = SEC(\mathbb{O}(\mathcal{M})) = HYP(\mathcal{M})$. We conclude with a proof-theoretic application of 13.4-13.6: as far as we restrict number-theoretic induction to total predicates, i.e. classes, we obtain a theory of abstraction and truth, whose arithmetical content does not exceed that of OP, and hence of Peano arithmetic, *even if we add the generalized induction principle of* §12.

13.7. THEOREM. $\mathrm{MF}_c + \mathrm{GID}$ *is a conservative extension of* OP, *i.e. if A is a formula of $\mathcal{L}_{op}$ and* $\mathrm{MF}_c + \mathrm{GID} \vdash A$, *then* $\mathrm{OP} \vdash A$.
(For the definition of MF_c, see 10.7; as to GID, see 12.4.1).

PROOF. The argument uses well-known results from general model theory:

(i) every structure $\mathcal{M}$ for a given elementary language $\mathcal{L}_{op}(\mathcal{M})$ has an elementary extension, which is recursively saturated (see Chang-Keisler 1990[3]);

(ii) if $\mathcal{M}$ is recursively saturated, $HYP(\mathcal{M}) = DEF(\mathcal{M})$ (here $X \in DEF(\mathcal{M})$ iff $X = \{a: a \in M$ and $\mathcal{M}\models B(a)\}$, for some $B(x)$ of $\mathcal{L}_{op}(\mathcal{M})$ (see Barwise 1975, Barwise-Schlipf 1975).

In order to prove the theorem, we assume that $\mathcal{M}$ is a model of $\mathrm{OP} + A$ (A sentence of $\mathcal{L}_{op}(\mathcal{M})$). By 7.10, 12.4, $\mathbb{O}(\mathcal{M})$ is a model of $\mathrm{MF}^- + \mathrm{GID} + A$, plus N-induction for $\mathcal{L}_{op}(\mathcal{M})$-formulas. Assume:

$$\mathbb{O}(\mathcal{M})\models Cl(t) \wedge \forall x(x\eta t \rightarrow (x+1)\eta t) \wedge \bar{0}\eta t \text{ and } X = t(\mathbb{O}(\mathcal{M}));$$

then, by lemma 13.6, $X \in HYP(\mathcal{M})$. By (i), it is not restrictive to suppose that $\mathcal{M}$ is recursively saturated; so by (ii) $X \in DEF(\mathcal{M})$. By hypothesis on t and N-induction for $\mathcal{L}_{op}(\mathcal{M})$–conditions, we have $\mathbb{O}(\mathcal{M}) \models \forall x(Nx \to x\eta t)$. In conclusion, $\mathbb{O}(\mathcal{M}) \models \mathrm{MF}_c + \mathrm{GID} + A$. □

Similar results hold if we add the enumeration axiom EA of §4 or extensionality for operations Ext_{op}, and we replace class N-induction by the weaker $\exists(+)$-N-induction of 4.13. Let:

$$\mathrm{MF}_1 := \mathrm{MF}^- + \exists(+)\text{-NIND}.$$

Then by 4.15 and the appendix to Ch. I, we immediately have:

13.8. THEOREM. *Let Ax be either* EA *or* Ext_{op}.

(i) *If f is a combinator such that* $\mathrm{MF}_c + \mathrm{GID} + Ax \vdash f: N \to N$, *then f defines a number-theoretic function, which is still provably recursive in Peano arithmetic.*

(ii) *If f is a combinator such that* $\mathrm{MF}_1 + \mathrm{GID} + \mathrm{Ax} \vdash f: N \to N$, *then f defines a primitive recursive function.*

We do not know what happens if both EA and Ext_{op} are present; perhaps, it might be useful to check that $\mathbf{CTM}(\omega)$ (cf. §6) is a model of $\mathrm{OP} + \mathrm{Ext}_{op} + \mathrm{EA}$, provably in (some conservative extension of) $\mathrm{MF}_c + \mathrm{GID}$.

§14. The uniform ordinal comparison theorem for inductive models

We further investigate the recursion-theoretic structure of $\mathbb{O}(\mathcal{M})$ ($\mathcal{M}$ fixed model of OP^-), in order to prove, in the notation of §13:

$$HYP(\mathcal{M}) = SEC^+(\mathbb{O}(\mathcal{M})) = SEC(\mathbb{O}(\mathcal{M})). \qquad (*)$$

The main result of this section, from which (*) easily follows, is the extension to $\mathbb{O}(\mathcal{M})$ of a fundamental theorem, originally stated for Kleene's recursion in higher types by R. Gandy (see Hinman 1978). Let $b \in M$: since $\mathbb{O}(\mathcal{M})$ is inductively generated by the operator Γ of 7.3, we apply backwards to b the reductive clauses which specify Γ, just as in the standard generation of a semantic tableau. Except for the trivial case where no Γ-clause is applicable, the reductive procedure produces a "search tree" $\tau(b)$: $\tau(b)$ starts with a "root" (labeled by) b and $\tau(b)$ is well founded (in the usual sense) iff $b \in \mathbb{O}(\mathcal{M})$, i.e. $\mathbb{O}(\mathcal{M}) \models Tb$. Therefore we can naturally associate to b an ordinal, called $|b|$, whenever $\mathbb{O}(\mathcal{M}) \models Tb$: it is the rank of the "direct" (cut-free) proof of Tb. The remarkable fact is that if $\mathbb{O}(\mathcal{M}) \models Ta$ or $\mathbb{O}(\mathcal{M}) \models Tb$, then the ordering relation between $|a|$ and $|b|$ is representable in $\mathbb{O}(\mathcal{M})$: this is the key property for obtaining (*) and selection, reduction and separation principles for $ENV(\mathbb{O}(\mathcal{M}))$.

14.1. *Preliminary notions and conventions.*

(i) $E(z) := \exists x \exists y((\neg PFOR(x) \wedge z = (\neg x)) \vee z = [Nx] \vee z = [\neg Nx] \vee$
$\vee z = [x = y] \vee z = [\neg(x = y)]);$

$S(z) := \exists x(z = [\neg\neg x] \vee z = [Tx] \vee z = [\neg Tx]);$

$\Sigma(z) := \exists x \exists y(z = \neg(\forall x) \vee z = \neg(x \wedge y));$

$\Pi(z) := \exists x \exists y(z = (\forall x) \vee z = (x \wedge y));$

(for the definition of $PFOR(x) :=$ "x is a p-form", cf. 7.3).

(ii) If $b \in M$, we say that b is in E-form (in S-, Σ-, Π-form) iff $\mathcal{M} \models E(b)$ ($\mathcal{M} \models S(b)$, $\Sigma(b)$, $\Pi(b)$ respectively).

(iii) According to 7.3, if b, $c \in M$, we adopt the abbreviations bc, $\neg b$, $\forall b$, $b \wedge c$, instead of the proper $\mathcal{M}(Ap(bc))$, $\mathcal{M}(NEGb)$, $\mathcal{M}(ALLb)$, $\mathcal{M}(ANDbc)$ (respectively); furthermore, $id(b,c)$, $nat(b)$, $tr(b)$ stand for $\mathcal{M}([b = c])$, $\mathcal{M}([Nb]$, $\mathcal{M}([Tb])$ (in the given order).

(iv) We stipulate that $b \ll c$ holds for b, $c \in M$ iff

either $c = \forall d$ and $b = da$, for some $a \in M$, or
$c = \neg(\forall d)$ and $b = \neg(da)$, for some $a \in M$, or
$c = (e \wedge f)$ and $b = e$ or $b = f$, or
$c = \neg(e \wedge f)$ and $b = \neg e$ or $b = \neg f$, or
$c = tr(d)$ and $b = d$, or
$c = \neg(tr(d))$ and $b = \neg d$, or
$c = \neg\neg d$ and $b = d$.

(v) If $c = \neg\neg d$, $tr(d)$, $\neg tr(d)$, we put

$PRD(c) = d =$ "the unique $b \ll c$".

(vi) If $b \ll c$ holds, we say that b *is a predecessor of* c.

Obviously, there is a formula $B(x,y)$ of $\mathcal{L}_{op}$ which defines $\ll$, uniformly in every model of OP^{-}.

14.2. DEFINITION

(i) Let δ^+ be the successor cardinal of δ (δ being the cardinal of M). If $b \in M$ and $b \notin \mathbb{O}(\mathcal{M})$, we put: $|\, b \,| := \delta^+$.

(ii) Assume $b \in \mathbb{O}(\mathcal{M})$; then we define (by induction on the generation of $\mathbb{O}(\mathcal{M})$):

if b is in E-form, $|\, b \,| = 0$;
if b is in S-form, $|\, b \,| = |\, PRD(b) \,| + 1$;
if b is in Π-form, $|\, b \,| = sup\{\, |\, c \,| + 1\colon c \in M$ and $c \ll b\}$;

if b is in Σ-form, $|b| = inf\{|c|+1\colon c \in M \text{ and } c \ll b\}$.

(iii) $b \leq c := (b \in \mathbb{O}(\mathcal{M}) \text{ and } |b| \leq |c|)$;

$b < c := (b \in \mathbb{O}(\mathcal{M}) \text{ and } |b| < |c|)$.

Obviously, $c < b$ iff $c \in \mathbb{O}(\mathcal{M})$ and not ($b \leq c$). $|b|$ is called *the norm of* b. It is straightforward to verify, by definition of norm and $\mathbb{O}(\mathcal{M})$:

14.2.1. $$|b| < \delta^+ \text{ iff } b \in \mathbb{O}(\mathcal{M}).$$

$\leq$, $<$ are well-founded relations on $\mathcal{M}$ and we have, for every b, $c \in M$:

14.2.2. $$b \leq c \text{ iff } b \in \mathbb{O}(\mathcal{M}) \text{ and (if } c \in \mathbb{O}(\mathcal{M}),\ |b| \leq |c|);$$

$$b < c \text{ iff } b \in \mathbb{O}(\mathcal{M}) \text{ and (if } c \in \mathbb{O}(\mathcal{M}),\ |b| < |c|).$$

14.2.3. LEMMA

(i) *The predicates* E, S, Σ, Π *are pairwise disjoint: if* $W \neq V$ *and* W, $V \in \{E, \Pi, \Sigma, S\}$, *then for every* $b \in M$, $\mathcal{M} \models \neg(W(b) \wedge V(b))$.

(ii) *For every* $b \in M$,

$$\mathcal{M} \models (PFOR(b) \wedge \neg E(b)) \leftrightarrow (S(b) \vee \Pi(b) \vee \Sigma(b)).$$

PROOF: apply 7.1.1 and the definition of the logical combinators before 7.1. □

14.2.4. LEMMA. *Let* b, $c \in M$ *and assume*

$$\mathcal{M} \models E(b) \vee E(c) \vee \neg PFOR(b) \vee \neg PFOR(c).$$

Then:

$$b \leq c \textit{ iff } \mathbb{O}(\mathcal{M}) \models Tb \wedge (E(b) \vee (Fc \wedge E(c)) \vee \neg PFOR(c)); \tag{1}$$

$$c < b \textit{ iff } \mathbb{O}(\mathcal{M}) \models Tc \wedge F(Tb \wedge (E(b) \vee (Fc \wedge E(c)) \vee \neg PFOR(c))). \tag{2}$$

PROOF. The restriction axioms RES, $T.1$ and $T.3$ of 7.10 easily imply, for every $d \in M$:

$$\mathbb{O}(\mathcal{M}) \models E(d) \wedge \neg Td \rightarrow Fd. \tag{$*$}$$

Ad (1). Assume $b \leq c$. Then $\mathbb{O}(\mathcal{M}) \models Tb$ and hence b is in p-form. If b is an E-form or c is not in p-form, then the right hand side of (1) holds trivially. Else, c is in E-form and b is not in E-form and we must have $|b| > 0$, whence $|c| > 0$, which implies $\mathbb{O}(\mathcal{M}) \models \neg Tc$ (by 14.2 (ii)), i.e. $\mathbb{O}(\mathcal{M}) \models Fc$ with ($*$). Assume the right hand side of (1). Then clearly $b \in \mathbb{O}(\mathcal{M})$ and b must be in p-form by RES. If b is in E-form, $|b| = 0$ and trivially $b \leq c$. If

c is in E-form and $\mathbb{O}(\mathcal{M})\models Fc$, or c is not in p-form, then $c \notin \mathbb{O}(\mathcal{M})$ by consistency of $\mathbb{O}(\mathcal{M})$: hence $b \leq c$ (by 14.2.2).
Ad (2): similar argument, using the assumption from right to left. □

14.3. THEOREM (*Uniform Ordinal Comparison*). *There exists an operator* $G(u,v)$ *in the language* $\mathfrak{L}$, *such that if* $\mathcal{M}\models \mathrm{OP}^-$, *then for every* b, $c \in M$:

$$b \leq c \text{ iff } \mathbb{O}(\mathcal{M}) \models \langle b,c\rangle \eta I(G); \qquad (*)$$

$$c < b \text{ iff } c \in \mathbb{O}(\mathcal{M}) \text{ and } \mathbb{O}(\mathcal{M}) \models \langle b,c\rangle \overline{\eta} I(G). \qquad (**)$$

($I(G)$ = *the fixed point predicate of* 10.4)

PROOF. We describe a set of inference rules for deriving statements of the form $b \leq c$. In each rule, the conclusion will depend positively on $b' \leq c'$, where b', c' are immediate predecessors of b, c in the sense of $\ll$. $G(u,x)$ will be easily assembled by formalizing the $\leq$-rules. $(*)$-$(**)$ are verified by transfinite induction.

Initial Rules. These rules handle the cases where b, $c \in M$ and

$$\mathcal{M}\models E(b) \vee E(c) \vee \neg PFOR(b) \vee \neg PFOR(c). \qquad (1)$$

I.1 $$\frac{b \in \mathbb{O}(\mathcal{M})\ ,\ b \text{ is in } E\text{-form}}{b \leq c\ (c \text{ arbitrary})}\ ;$$

I.2. $$\frac{b \in \mathbb{O}(\mathcal{M}),\ (\neg c) \in \mathbb{O}(\mathcal{M}) \text{ and } c \text{ is in } E\text{-form}}{b \leq c}\ ;$$

I.3 $$\frac{b \in \mathbb{O}(\mathcal{M}) \text{ and } c \text{ is not in p-form}}{b \leq c}.$$

By 14.2.4, I.1-I.3 suffice.

Inductive Rules (under the assumption that (1) does not hold). By 14.2.3 (ii) $\mathcal{M}\models S(b) \vee \Sigma(b) \vee \Pi(b)$ and $\mathcal{M}\models S(c) \vee \Sigma(c) \vee \Pi(c)$. Hence, we have to specify nine rules; each rule is labeled by a two-letter word $W{*}V$ where W, $V \in \{\Pi,\ \Sigma,\ S\}$. $W{*}V$ denotes the case where $W(b)$, $V(c)$ hold in $\mathcal{M}$. We adopt the abbreviations:

$$(\exists u \ll b)(\ldots) := \exists u(u \ll b \wedge \ldots) \text{ and } (\forall v \ll c)(\ldots) := \forall v(v \ll c \rightarrow \ldots).$$

$\boxed{S{*}S}$ b, c are in S-form:

$$\frac{PRD(b) \leq PRD(c)}{b \leq c}\ ;$$

$\boxed{S{*}\Pi}$ b is in S-form and c is in Π-form:

$$\frac{(\exists c' \ll c)(PRD(b) \leq c')}{b \leq c};$$

$\boxed{\Pi{*}S}$ b is in Π-form and c is in S-form:

$$\frac{(\forall b' \ll b)(b' \leq PRD(c))}{b \leq c};$$

$\boxed{S{*}\Sigma}$ b is in S-form and c is in Σ-form:

$$\frac{(\forall c' \ll c)(PRD(b) \leq c')}{b \leq c};$$

$\boxed{\Sigma{*}S}$ b is in Σ-form and c is in S-form:

$$\frac{(\exists b' \ll b)(b' \leq PRD(c))}{b \leq c};$$

$\boxed{\Pi{*}\Pi}$ b, c are in Π-form:

$$\frac{(\forall b' \ll b)(\exists c' \ll c)(b' \leq c')}{b \leq c};$$

$\boxed{\Sigma{*}\Sigma}$ b, c are in Σ-form:

$$\frac{(\exists b' \ll b)(\forall c' \ll c)(b' \leq c')}{b \leq c};$$

$\boxed{\Pi{*}\Sigma}$ b is in Π-form and c is in Σ-form:

$$\frac{(\forall b' \ll b)(\forall c' \ll c)(b' \leq c')}{b \leq c};$$

$\boxed{\Sigma{*}\Pi}$ b is in Σ-form, c is in Π-form:

$$\frac{(\exists b' \ll b)(\exists c' \ll c)(b' \leq c')}{b \leq c}.$$

This completes the list of the inductive inferences.

Obviously, to each inference $V{*}W$ (where $V, W \in \{S, \Pi, \Sigma\}$) we can associate a T-positive formula $V{*}W(\langle b, c\rangle, v)$ of $\mathfrak{L}$, operative in v, which states that b is in V-form, c is in W-form and formalizes the general clause embodied in the premise of $V{*}W$, by replacing each condition $b' \leq c'$ with $\langle b', c'\rangle \eta v$. Let $G(x, v)$ be the T-positive formula, operative in v, which describes the initial rules and the $V{*}W$-rules; then we can find a fixed point $I(G)$ (by 10.4) such that, for every b, $c \in M$:

$$\mathbb{O}(\mathcal{M}) \models \langle b, c\rangle \eta I(G) \text{ iff } \mathbb{O}(\mathcal{M}) \models G(\langle b, c\rangle, I(G)); \tag{2.1}$$

$$\mathbb{O}(\mathcal{M})\models\langle b,c\rangle\overline{\eta}\,I(G) \text{ iff } \mathbb{O}(\mathcal{M})\models FG(\langle b,c\rangle,I(G)). \tag{2.2}$$

Then

$$b\leq c \text{ implies } \mathbb{O}(\mathcal{M})\models\langle b,c\rangle\eta I(G); \tag{3.1}$$

$$c<b \text{ implies } \mathbb{O}(\mathcal{M})\models\langle b,c\rangle\overline{\eta}\,I(G). \tag{3.2}$$

Indeed, if b, c satisfy (1), (3.1)-(3.2) simply follow from the lemma 14.2.4, (2.1)-(2.2) and the fact that the conditions $E(u)$, $S(u)$, $\Pi(u)$ and $\Sigma(u)$ are mutually incompatible. Otherwise, we can assume that b, c are not in E-form and we establish (3.1)-(3.2) by transfinite induction on $\alpha=min\{\,|\,b\,|\,,|\,c\,|\,\}$, whenever b is in V-form and c is in W-form (V, $W\in\{S,\Sigma,\Pi\}$). In particular, (3.1)-(3.2) can be reduced to an inductive verification on α that if b is in V-form and c is in W-form, then

$$b\leq c \text{ implies } \mathbb{O}(\mathcal{M})\models V{*}W(\langle b,c\rangle,I(G));$$

$$c<b \text{ implies } \mathbb{O}(\mathcal{M})\models F(V{*}W(\langle b,c\rangle,I(G))).$$

We only discuss two significant cases; the remaining ones are obtained by straightforward arguments.

$\Sigma{*}\Sigma$: b, c are in Σ-form, e.g. $b=\neg(\forall e)$, $c=\neg(\forall d)$. If $b\leq c$ holds, then $b\in\mathbb{O}(\mathcal{M})$ and $inf\{\,|\neg(ey)\,|+1:y\in M\}\leq inf\{\,|\neg(dy)\,|+1:y\in M\}$. Since $\mathbb{O}(\mathcal{M})$ is a model of T-axioms and $\mathbb{O}(\mathcal{M})\models F(\forall e)$, we can also assume that there exists a $y\in M$ such that $\mathbb{O}(\mathcal{M})\models Fey$ and $|\neg(ey)\,|\,\leq\,|\neg da\,|$, for every $a\in M$. By IH,

$$\mathbb{O}(\mathcal{M})\models\exists u(u\ll b\wedge\forall v(v\ll c\rightarrow\langle u,v\rangle\eta I(G))),$$

which implies $\mathbb{O}(\mathcal{M})\models\Sigma{*}\Sigma(\langle b,c\rangle,I(G))$.
Hence by (2.1) and logic, $\mathbb{O}(\mathcal{M})\models\langle b,c\rangle\eta I(G)$. This takes care of (3.1). As to (3.2), if we assume $c<b$, we must verify $\mathbb{O}(\mathcal{M})\models F(G(\langle b,c\rangle,I(G)))$; since $\mathbb{O}(\mathcal{M})\models\Sigma(b)\wedge\Sigma(c)$, it is enough to check:

$$\mathbb{O}(\mathcal{M})\models F(\exists u(u\ll b\wedge\forall v(v\ll c\rightarrow\langle u,v\rangle\eta I(G))), \text{ or equivalently,}$$

$$\mathbb{O}(\mathcal{M})\models\forall u(u\ll b\rightarrow\exists v(v\ll c\wedge\langle u,v\rangle\overline{\eta}\,I(G))). \tag{4}$$

By assumption, for some $x\in M$, $\mathbb{O}(\mathcal{M})\models F(dx)$ and $|\neg(dx)\,|\,<\,|\neg(ey)\,|$ hold for arbitrary $y\in M$, whence by IH

$$\mathbb{O}(\mathcal{M})\models\exists v(v\ll c\wedge\forall u(u\ll b\rightarrow\langle u,v\rangle\overline{\eta}\,I(G))), \tag{5}$$

which implies (4).

$\Pi{*}\Sigma$. We assume b is in Π-form and c in Σ-form.

Ad (3.1). If $b\leq c$, then $b'\leq c'$ holds for every $b'\ll b$, $c'\ll c$, whence by IH $\mathbb{O}(\mathcal{M})\models\forall x\forall y(x\ll b\wedge y\ll c\rightarrow\langle x,y\rangle\eta I(G))$, i.e. $\mathbb{O}(\mathcal{M})\models\Pi{*}\Sigma(\langle b,c\rangle,I(G))$, which yields by (2.1) the required conclusion.

Ad (3.2). By assumption we have, for some $c' \ll c$ and some $b' \ll b$, $c' < b'$. It follows by IH that $\mathbb{O}(\mathcal{M})\models \exists x \exists y(x \ll b \wedge y \ll c \wedge \langle x, y\rangle \overline{\eta} I(G))$, which readily implies the validity of $F(\Pi * \Sigma(\langle b, c\rangle, I(G)))$ in $\mathbb{O}(\mathcal{M})$. By (2.2) we obtain $\mathbb{O}(\mathcal{M})\models \langle b, c\rangle \overline{\eta} I(G)$.

We then proceed by observing that, for every ordinal γ:

$$\mathbb{O}(\mathcal{M},\gamma)\models \langle b,c\rangle \eta I(G) \text{ implies } \mathbb{O}(\mathcal{M})\models Tb. \tag{6}$$

$\mathbb{O}(\mathcal{M},\gamma)$ is the γth-stage of $\mathbb{O}(\mathcal{M})$ (see 7.6 or §12); (6) is checked by transfinite induction on γ. If $\gamma = 0$, the claim holds, as the premise is false, and, if γ is a limit, we use IH. Assume $\mathbb{O}(\mathcal{M}, \beta+1)\models \langle b, c\rangle \eta I(G)$; by lemma 12.1 we have:

$$\mathbb{O}(\mathcal{M},\beta)\models G(\langle b,c\rangle, I(G)). \tag{7}$$

As above, we must distinguish several cases according to the form of b, c. If b (or c) is in E-form, (7) implies $\mathbb{O}(\mathcal{M})\models Tb$ (by (1)). In the remaining cases, IH applies and it always entails a sufficient condition to assert $\mathbb{O}(\mathcal{M})\models Tb$ (e.g. if b is in Π-form, we have from ($\Pi * S$), ($\Pi * \Pi$), ($\Pi * \Sigma$) that every $b' \ll b$ holds in $\mathbb{O}(\mathcal{M})$, which implies the conclusion with T-axioms for $\wedge$ and $\forall$).
Then we prove:

$$\mathbb{O}(\mathcal{M})\models \langle b,c\rangle \eta I(G) \text{ implies } b \leq c; \tag{8.1}$$

$$\mathbb{O}(\mathcal{M})\models \langle b,c\rangle \overline{\eta} I(G) \text{ and } \mathbb{O}(\mathcal{M})\models Tc \text{ imply } c < b. \tag{8.2}$$

Ad (8.1). Assume $\mathbb{O}(\mathcal{M})\models \langle b,c\rangle \eta I(G)$. Were $b \leq c$ false, then by (6) $\mathbb{O}(\mathcal{M})\models Tb$ and $|c| < |b|$, whence $\mathbb{O}(\mathcal{M})\models Tc$, i.e. $c < b$, which implies $\mathbb{O}(\mathcal{M})\models \langle b,c\rangle \overline{\eta} I(G)$ by (3.2), against consistency of $\mathbb{O}(\mathcal{M})$.

Ad(9.2). Assume $b < c$ is false and $\mathbb{O}(\mathcal{M})\models \langle b,c\rangle \overline{\eta} I(G)$, $\mathbb{O}(\mathcal{M})\models Tc$. Then $|b| \leq |c|$ and hence $\mathbb{O}(\mathcal{M})\models Tb$; by definition, $b \leq c$ and by (3.1), $\mathbb{O}(\mathcal{M})\models \langle b,c\rangle \eta I(G)$: contradiction!
(8.1)-(8.2), (3.1)-(3.2) prove the theorem. □

§15. Applications of the uniform ordinal comparison theorem

We conclude the recursion-theoretic analysis of §14 by characterizing the section of an arbitrary inductive model. We then introduce an approximation operation π, which splits each non-empty property in a $\subseteq$-chain of subclasses, and we show that π satisfies six simple conditions, the π-approximation axioms. As we shall see in the next chapter, π-axioms suffice to derive *non-trivial ordinal-free consequences* of the ordinal comparison theorem. However, we easily have that π-axioms are still *conservative* over $\mathrm{MF}_c + \mathrm{GID}$ (hence over OP), by an obvious extension of the model-theoretic argument of §13.

The basic step for proving the full characterization theorem is a separation lemma for disjoint predicates, which is easily implied by 14.3.

15.1. LEMMA (*Separation*). *We can find a closed term* $\lambda x \lambda y.SEP(x,y)$ *such that if* $\mathcal{M} \models \mathrm{OP}^-$, b, $c \in M$ *and* $\mathrm{O}(\mathcal{M}) \models \forall u(\neg u\eta b \vee \neg u\eta c)$, *then*

$$\mathrm{O}(\mathcal{M}) \models \forall x(x\eta b \to x\eta SEP(b,c)) \wedge \forall x(x\eta c \to x\overline{\eta} SEP(b,c)).$$

PROOF. Choose $SEP(b,c) := \{x : \langle bx, cx\rangle \eta I(G)\}$, where $I(G)$ is the predicate, whose existence is ensured by the ordinal comparison theorem. If $d \in M$ and $\mathrm{O}(\mathcal{M}) \models d\eta b$, we have by assumption $\mathrm{O}(\mathcal{M}) \models \neg(d\eta c)$; by definition 14.2, $bd \leq cd$ trivially holds and hence $\mathrm{O}(\mathcal{M}) \models \langle bd, cd\rangle \eta I(G)$ by 14.3 (*), i.e. $\mathrm{O}(\mathcal{M}) \models d\eta SEP(b,c)$. On the other hand, if $\mathrm{O}(\mathcal{M}) \models d\eta c$, then $cd < bd$ holds and hence by (**) of 14.3, $\mathrm{O}(\mathcal{M}) \models d\overline{\eta} SEP(b,c)$. □

15.2. THEOREM. *Let* $\mathcal{M}$ *be an arbitrary model of* OP^-; *then*

$$HYP(\mathcal{M}) = SEC(\mathrm{O}(\mathcal{M})) = SEC^+(\mathrm{O}(\mathcal{M})).$$

PROOF. By lemma 13.6 it suffices to verify $SEC^+(\mathrm{O}(\mathcal{M})) \subseteq SEC(\mathrm{O}(\mathcal{M}))$. Let $X \subseteq M$ and assume that

$$X = \{a : a \in M \text{ and } \mathrm{O}(\mathcal{M}) \models a\eta b\},\ -X = \{a : a \in M \text{ and } \mathrm{O}(\mathcal{M}) \models a\eta c\},$$

for some b, $c \in M$. By 15.1 and consistency, $\mathrm{O}(\mathcal{M}) \models d\eta SEP(b,c)$ iff $\mathrm{O}(\mathcal{M}) \models d\eta b$, for every $d \in M$; but $SEP(b,c)$ is a class in $\mathrm{O}(\mathcal{M})$: $\mathrm{O}(\mathcal{M}) \models \neg d\eta SEP(b,c)$ implies $\mathrm{O}(\mathcal{M}) \models d\eta c$, whence $\mathrm{O}(\mathcal{M}) \models d\overline{\eta} SEP(b,c)$ again by 15.1. □

15.2.1. REMARK. If $\mathcal{M}$ is countable, $SEC(\mathrm{O}(\mathcal{M})) = \Delta^1_1(\mathcal{M})$, the collection of subsets of M, which are definable by formulas $\exists Y A(Y,x)$, $\forall Z B(Z,x)$, where $\exists Y$, $\forall Z$ range over subset of M, and A, B are formulas in the language $\mathcal{L}_{op}$ possibly containing atoms Yt, Zt, Xt. This follows from the generalized Suslin-Kleene theorem of Moschovakis (1974).

15.3. DEFINITION

$$\pi := \lambda y \lambda x.\{u : u = x \vee \langle yu, yx\rangle \eta I(G)\}.$$

Henceforth, it is convenient to adopt the more perspicuous abbreviations:

15.3.1. $\quad xu \leq^G yv := \langle xu, yv\rangle \eta I(G);$

15.3.2. $\quad yv <^G xu := v\eta y \wedge \langle xu, yv\rangle \overline{\eta} I(G).$

The intuition behind π is that, if x is in y, then πyx defines those u which satisfy $yu \leq^G yx$, i.e. which fall under the property y at a level not higher than $|yx|$ (see 14.2); it turns out that πyx is a *class* in $\mathbb{O}(\mathcal{M})$.

15.4. THEOREM (*Approximation*). *Let* $\mathcal{M}$ *be a model of* OP^-. *Then* $\mathbb{O}(\mathcal{M})$ *verifies the following sentences:*

$\pi.1.$ $\forall x \forall y (x \eta \pi y x)$;

$\pi.2.$ $\forall x \forall y (x \eta y \rightarrow Cl(\pi y x) \wedge \pi y x \subseteq y)$;

$\pi.3.$ $\forall x \forall y (\neg x \eta y \rightarrow y \subseteq \pi y x)$;

$\pi.4.$ $\forall y \forall u \forall v (u \eta \pi y v \wedge v \eta y \rightarrow \pi y u \subseteq \pi y v)$;

$\pi.5.$ $\forall y \forall v \forall u ((u \eta y \wedge v \eta y) \rightarrow (\pi y u \subseteq \pi y v \vee \pi y v \subset \pi y u))$;

$\pi.6.$ $\forall y \forall x (x \eta y \rightarrow \exists z (z \eta y \wedge \forall v (v \eta y \rightarrow \pi y z \subseteq \pi y v)))$.

PROOF. $\pi.1$ holds trivially. As to the other sentences, we first observe:

$$\mathbb{O}(\mathcal{M}) \models \forall x \forall y \forall u (x \eta y \rightarrow (u \eta \pi y x \leftrightarrow yu \leq^G yx)). \tag{1}$$

($\Leftarrow$: by abstraction and definition of π. $\Rightarrow$: if $x\eta y$, $\leq^G$ is reflexive, and the conclusion follows by abstraction).

Ad $\pi.2$. Assume $b\eta a$, $d\eta\pi ab$ hold in $\mathbb{O}(\mathcal{M})$, where a, b, c, $d \in M$; for typographical reasons, we keep using the same symbols for elements of M and their names. By (1), $\mathbb{O}(\mathcal{M}) \models ad \leq^G ab$, hence by 14.3 (*), $\mathbb{O}(\mathcal{M}) \models d\eta a$, i.e. $\mathbb{O}(\mathcal{M}) \models \pi ab \subseteq a$. Assume now:

$$\mathbb{O}(\mathcal{M}) \models b\eta a \wedge \neg d \eta \pi ab. \tag{2}$$

By definition of π,

$$\mathbb{O}(\mathcal{M}) \models \neg d = b; \tag{3}$$

$$\mathbb{O}(\mathcal{M}) \models \neg ad \leq^G ab. \tag{4}$$

(4) and 14.3 (*) imply $\mathbb{O}(\mathcal{M}) \models \neg d\eta a$, or $|ab| < |ad|$. In both cases (with (1), 14.3 (**)), we have

$$\mathbb{O}(\mathcal{M}) \models ab <^G ad. \tag{5}$$

Hence from (5) and (3), $\mathbb{O}(\mathcal{M}) \models d \overline{\eta} \pi ab$, which implies that πab is a class in $\mathbb{O}(\mathcal{M})$.

Ad $\pi.3$. Assume

$$\mathbb{O}(\mathcal{M}) \models \neg d\eta a \wedge b\eta a. \tag{6}$$

Then $|ab| < |ad|$ (by 14.2 (i)), i.e. trivially $|ab| \leq |ad|$, which yields, by (6) and 14.3 (**), $\mathbb{O}(\mathcal{M})\models ab \leq^G ad$, whence $\mathbb{O}(\mathcal{M})\models b\eta\pi ad$. Since b is arbitrary, $\mathbb{O}(\mathcal{M})\models a \subseteq \pi ad$.

Ad π.4. Assume

$$\mathbb{O}(\mathcal{M})\models d\eta\pi ab; \tag{7.1}$$

$$\mathbb{O}(\mathcal{M})\models b\eta a; \tag{7.2}$$

$$\mathbb{O}(\mathcal{M})\models c\eta\pi ad. \tag{7.3}$$

By (7.1)-(7.3), (1) and ordinal comparison, we obtain $|ad| \leq |ab|$ and $|ac| \leq |ad|$, whence

$$|ac| \leq |ab|. \tag{8}$$

By (7.2) and 14.2.1, we must have

$$\mathbb{O}(\mathcal{M})\models c\eta a. \tag{9}$$

From 14.3 (*), (9) and (8), $\mathbb{O}(\mathcal{M})\models c\eta\pi ab$, whence $\mathbb{O}(\mathcal{M})\models \pi ad \subseteq \pi ab$.

Ad π.5. Assume

$$\mathbb{O}(\mathcal{M})\models b\eta a \wedge c\eta a; \tag{10.1}$$

$$\mathbb{O}(\mathcal{M})\models d\eta\pi ab \wedge \neg d\eta\pi ac, \text{ for some } d \in M; \tag{10.2}$$

$$\mathbb{O}(\mathcal{M})\models e\eta\pi ac. \tag{10.3}$$

By (10.1) and (10.3), (1) and 14.3 (*),

$$|ae| \leq |ac| \text{ and } \mathbb{O}(\mathcal{M})\models e\eta a. \tag{11}$$

From $d\eta\pi ab$, (1) (10.1) we infer

$$\mathbb{O}(\mathcal{M})\models d\eta a \text{ and } |ad| \leq |ab|. \tag{12}$$

But $\neg d\eta\pi ac$, (1), (12) and (10.1) imply $|ac| < |ad|$, whence

$$|ac| < |ab|. \tag{13}$$

(13) and (11) entail $|ae| \leq |ab|$, i.e. with 14.3 (*) $\mathbb{O}(\mathcal{M})\models e\eta\pi ab$.

Ad π.6. Let $\mathbb{O}(\mathcal{M})\models c\eta a$, for some $c \in M$. Then the set of ordinals

$$\{\,|ad| : d \in M,\ \mathbb{O}(\mathcal{M})\models d\eta a \text{ and } |ad| \leq |ac|\,\}$$

is non-empty, and it possesses a least element δ. Choose any d such that $\delta = |ad|$ and suppose $\mathbb{O}(\mathcal{M})\models b\eta a \wedge e\eta\pi ad$; then by ordinal comparison and the choice of d, $|ae| \leq |ab|$ and $\mathbb{O}(\mathcal{M})\models e\eta a$ (because $\mathbb{O}(\mathcal{M})\models b\eta a$), which yields $\mathbb{O}(\mathcal{M})\models e\eta\pi ab$. Since b, c are arbitrary, it follows that $\mathbb{O}(\mathcal{M})\models \forall v(v\eta a \rightarrow \pi ad \subseteq \pi av)$. □

Let ΠAX be the list π.1-π.6 of the approximation theorem 15.4. Then we immediately have, with the conservation results of 13.7-13.8:

15.5. THEOREM (*General Conservation*)

Let $Ax = \mathrm{EA}$ *or* Ext_{op}.

(i) *If* A *is a formula of* $\mathcal{L}_{op}$ such that $\mathrm{MF}_c{+}\mathrm{GID}{+}\Pi\mathrm{AX}(+\mathrm{Ax}) \vdash A$, *then* A *is already provable in* OP $(+\mathrm{Ax})$.

(ii) *If* f *is a combinator such that* $\mathrm{MF}{+}\mathrm{GID}{+}\Pi\mathrm{AX}(+\mathrm{Ax}) \vdash f: N \rightarrow N$, *then* f *defines a number-theoretic function which is provably recursive in Peano arithmetic* PA.

(iii) *If* $\mathrm{MF}_1 + \mathrm{GID} + \Pi\mathrm{AX}(+\mathrm{Ax}) \vdash f: N \rightarrow N$ *and* f *is a combinator, then* f *defines a primitive recursive function* (MF_1 *being the subsystem of* §13).

15.6. FINAL REMARKS

(i) 15.5(i) can be strengthened to a proof-theoretic equivalence by using a refinement of the techniques of part D (see Cantini 1992).

(ii) Set $\hat{\pi} := \lambda y \lambda x.\{u: \langle yu, yx\rangle \eta I(G)\}$. If $\pi.i$ is any sentence of ΠAX (where $i \in \{1, \ldots, 6\}$), let $\hat{\pi}.i$ be the sentence, which is obtained from $\pi.i$ by replacing everywhere π with $\hat{\pi}$. Then we obtain:

if $\mathcal{M}$ *is any model of* OP^-, $\mathbb{O}(\mathcal{M})$ *verifies* $\hat{\pi}.2$–$\hat{\pi}.6$;

$$\mathbb{O}(\mathcal{M}) \models \forall y \forall u \forall v (v \eta y \wedge u \eta y \wedge \hat{\pi} y u \subseteq \hat{\pi} y v \rightarrow u \eta \hat{\pi} y v). \qquad (\hat{\pi}.7)$$

Clearly $\hat{\pi}.4$ and $\hat{\pi}.7$ imply:

$$\mathbb{O}(\mathcal{M}) \models \forall y \forall u (u \eta y \rightarrow u \eta \hat{\pi} y u);$$

$$\mathbb{O}(\mathcal{M}) \models \forall y \forall u \forall v (v \eta y \wedge u \eta y \rightarrow (\hat{\pi} y u \subseteq \hat{\pi} y v \leftrightarrow u \eta \hat{\pi} y v)). \qquad (*)$$

Using $(*)$, we can show:

$$\mathrm{MF}^- + \Pi\mathrm{AX} \textit{ is interpretable into } \mathrm{MF}^- + \{\hat{\pi}.2\text{–}\hat{\pi}.7\}. \qquad (**)$$

$(**)$ is made precise as follows. First of all, let us consider π and $\hat{\pi}$ as new primitive symbols. Then we define a translation $\hat{}$ from the language of $\mathrm{MF}^- + \Pi\mathrm{AX}$ into the language of $\{\hat{\pi}.2\text{–}\hat{\pi}.7\}$: if A is a formula of $\mathrm{MF}^- + \{\pi.1\text{–}\pi.6\}$, $A\hat{}$ is the formula obtained by replacing each occurrence of π with the term $\pi' := \lambda y \lambda x.\hat{\pi} y x \cup \{x\}$. Then, for each $i \in \{1, \ldots, 6\}$,

$$\mathrm{MF}^- + \{\hat{\pi}.2\text{–}\hat{\pi}.7\} \vdash (\pi.i)\hat{}. \qquad (***)$$

The verification of $(***)$ is trivial for $i = 1, 2, 3, 4$; it makes use of $\hat{\pi}.7$ in the cases of $\pi.5$ and $\pi.6$. A consequence of $(**)$ is that we can work with an operator, which satisfies $\hat{\pi}.7$ instead of $\pi.1$, without weakening the approximation structure.

CHAPTER 4

TYPE-FREE ABSTRACTION WITH APPROXIMATION OPERATOR

§16. Approximating properties by classes
§17. The approximation theorem for extensional operations and the fixed point theorem for monotone operations
§18. Topology displayed: basic definitions
§19. The representation theorem for explicitly CL-continuous operators
Appendix: alternative proofs

In chapter III we semantically justified the introduction of a suitable approximation operator π with a simple axiomatic description, and we proved that the resulting extension of the minimal framework MF_c – to be called PW_c below – is conservative over OP, *even in presence of a generalized induction schema* GID *for fixed points of positive operators* (see 15.5). We now investigate in some detail the extended axiomatic system $PW_c + GID$ and we show that it has natural non trivial consequences.

As we expect from the basic intuition underlying π, the relevant consequences are recursion-theoretic and show that the chosen system yields a kind of *axiomatic intensional theory of inductive sets over a combinatory structure.* In §16 we obtain principles of approximation for properties by classes and we get typical results from definability theory. Incidentally, we obtain a strong form of separation for disjoint properties, which is a basic tool for proving the consistency of an interesting property theory, due to Myhill-Flagg (1987). In §17 we exploit the local structure of properties and we prove *a generalized continuity theorem for extensional operations*, an abstract version of the Myhill-Shepherdson theorem. Of course, "extensional" operations are those operations which preserve the membership relation η, associated to reflective truth. As a consequence (together with generalized induction), we can prove uniform internal version of the fixed point theorem for monotone operators; *it also turns out that extensional, positive, monotone and (generalized) continuous operators all coincide*, provably in $PW_c + GID$.

The axiomatic results can be rephrased in denotational style: in §18 we introduce a natural topology, the class topology on the power set of M (M = support of a given OP-model $\mathcal{M}$), and we show that non-trivial facts

about "constructive objects" of the space can be adequately dealt with in the untyped language of reflective truth. Among others, we characterize the "explicitly" open sets of the space and the explicitly continuous operators of the space into itself (§§18-19); we conclude with an analogue of the Kleene "first recursion theorem" for explicitly continuous operators.
The class topology is an analogue of the positive information topology (in the sense of Scott-Ershov), where the role of "finite" (or compact) elements is played by the hyperelementary subsets of M, i.e. the extensions of total predicates in the inductive models.

§16. Approximating properties by classes

In Ch. III, §15, we defined – within inductive models – an operator π, which uniformly assigns to every property a chain of subclasses. π has its source in this informal idea: in order to verify, by "predicative means", that x witnesses a property b, one employs a portion of b, which is a class and can be determined, uniformly in b and x, by looking at the "search tree", which is naturally associated to (the presentation) of b. Formally, the basic consequence is that we can compare two given elements x, y falling under a property b, according to their "order of generation", and this order is well-founded. A major corollary is that there is an operation ξ, which associates to any given non-empty property b a collection of witnesses of b, which is a class (CL-compactness). Below, we explicitly define the generation order and we collect together a few basic facts, inspired by generalized recursion theory.

16.1. DEFINITION

(a) $\mathrm{PW}_c := \mathrm{MF}_c + \{\pi.1\text{–}\pi.6\}$
(NB: MF_c is based on class N-induction of 10.7).

PW is an acronym for "prewellordering"; $\pi.1$–$\pi.6$ are the sentences of the approximation theorem 15.4, here restated for reader's sake:

$\pi.1.$ $\forall x \forall y (x \eta \pi y x)$;

$\pi.2.$ $\forall x \forall y (x \eta y \to Cl(\pi y x) \wedge \pi y x \subseteq y)$;

$\pi.3.$ $\forall x \forall y (\neg x \eta y \to y \subseteq \pi y x)$;

$\pi.4.$ $\forall y \forall u \forall v (u \eta \pi y v \wedge v \eta y \to \pi y u \subseteq \pi y v)$;

$\pi.5.$ $\forall y \forall v \forall u (u \eta y \wedge v \eta y \to \pi y u \subseteq \pi y v \vee \pi y v \subset \pi y u)$;

$\pi.6.$ $\forall y \forall x (x \eta y \to \exists z (z \eta y \wedge \forall v (v \eta y \to \pi y z \subseteq \pi y v)))$.

As usual, PW^- is the subsystem of PW_c without the number-theoretic induction axiom, while PW_p is $\mathrm{PW}^- + \text{P-NIND}$ (= the property induction axiom of 10.7).

(b) $x \leq_z y := x\eta z \wedge \pi z x \subseteq \pi z y$;

(c) $x <_z y := x\eta z \wedge \pi z x \subset \pi z y$, where $z \subset b := z \subseteq b \wedge \exists x(x\eta b \wedge \neg x\eta z)$.

16.2. LEMMA. *The following formulas are provable in* PW^-:

(i) $T(x \leq_z y) \leftrightarrow x \leq_z y$;
(ii) $\neg\, x <_z x$;
(iii) $x \leq_z y \wedge y \leq_z w \rightarrow x \leq_z w$;
(iv) $x \leq_z y \rightarrow x\eta z$;
(v) $x <_z y \rightarrow x \leq_z y$;
(vi) $x\eta z \wedge \neg y\eta z \rightarrow x <_z y$;
(vii) $x\eta z \wedge y\eta z \rightarrow x \leq_z y \vee y <_z x$;
(viii) $x\eta z \rightarrow x \leq_z x$;
(ix) $x\eta z \vee y\eta z \rightarrow x \leq_z y \vee y <_z x$;
(x) $y\eta z \wedge F(x \leq_z y) \leftrightarrow y <_z x$;
(xi) $x\eta z \vee y\eta z \rightarrow Prop(x \leq_z y)$ (*where* $Prop(z) := Tz \vee Fz$);
(xii) $x\eta z \rightarrow \forall y(y <_z x \rightarrow \neg\, x\eta\pi z y)$;
(xiii) $x\eta z \rightarrow \exists y(y\eta z \wedge \forall v(v\eta z \rightarrow y \leq_z v))$.

PROOF. (i) $\Rightarrow$: trivial by T-logic. $\Leftarrow$: if $x \leq_z y$, then $x\eta z$, whence $Cl(\pi z x)$ (by axiom π.2). It follows that $\pi z x \subseteq \pi z y$ implies $T(\pi z x \subseteq \pi z y)$, i.e. $T(x \leq_z y)$.
(ii)-(v): trivial by definition 16.1.
(vi): from $x\eta z$ and $\neg y\eta z$, we have $z \subseteq \pi z y$ (π.3) and $\pi z x \subseteq z$ (π.2). But $y\eta\pi z y$ (π.1); hence $(\pi z x \subset \pi z y) \wedge x\eta z$.
(vii): it is a restatement of π.5.
(viii): trivial.
(ix): apply (vi)-(v), (vii).
(x) $\Rightarrow$: assume $y\eta z \wedge F(x \leq_z y)$. If $\neg x\eta z$, $y <_z x$ (by (vi)); if $x\eta z$, then $\pi z x \subseteq \pi z y$ or $\pi z y \subset \pi z x$ (π.5). But $x\eta z$ implies $\neg F(x\eta z)$, whence it follows $\exists u(u\eta\pi z x \wedge u\overline{\eta}\pi z y)$; then $\pi z y \subset \pi z x$, i.e. $y <_z x$. $\Leftarrow$: if $y <_z x$ is assumed, $\exists u(u\eta\pi z x \wedge u\overline{\eta}\pi z y)$ and $y\eta z$ hold (π.2); hence $y\eta z$ and $F(x \leq_z y)$.
(xi): by (ix)-(x), (i).
(xii): assume $x\eta z$ and $x\eta\pi z y$, $y <_z x$; then $y\eta z$ and $\pi z y \subset \pi z x$. Hence $\pi z x \subseteq \pi z y$ (π.4): contradiction!
(xiii): it is a reformulation of π.6. □

16.3. THEOREM (CL-*compactness*). *We can find a closed term* ξ *such that* PW^- *proves*:

(i) $\xi z \subseteq z$;

(ii) $\exists x(x\eta z) \rightarrow Cl(\xi z) \wedge \exists u(u\eta\xi z)$;
(iii) $\exists x\exists y(x =_e y \wedge \neg\xi x = \xi y)$ (*so* ξ *is non-extensional*).

PROOF. We define $\xi := \lambda v.\{x : \forall y(x \leq_v y)\}$. Clearly, if $u\eta\xi z$ is assumed, then $u\eta z$ holds (by 16.2 (ii)-(iv)). It is also clear that 16.2 (vi)-(v) entail

$$\forall y(x \leq_z y) \text{ iff } x\eta z \text{ and } \forall y(y\eta z \rightarrow x \leq_z y). \tag{1}$$

We assume that z is non-empty and we verify, for every x,

$$F\forall y(x \leq_z y) \text{ iff } \neg\forall y(x \leq_z y). \tag{2}$$

$\Rightarrow$: trivial by T-logic. $\Leftarrow$: suppose that there is an x such that:

$$\exists y \neg(x \leq_z y); \tag{3.1}$$

$$\neg F\forall y(x \leq_z y). \tag{3.2}$$

Fix an element u of z; then (3.2) and 16.2 (xi) imply $x \leq_z u$, i.e. $x\eta z$. Hence (3.1) and 16.2 (ix)-(x) yield $\exists y F(x \leq_z y)$, whence $F\forall y(x \leq_z y)$: by classical logic we are done. By (2) and 16.2 (i), ξz is a class; moreover, if z is non-empty, ξz is non-empty, by (1) and 16.2 (xiii).
We conclude by verifying (iii): we show that, if $\forall x\forall y(x =_e y \rightarrow \xi x =_e \xi y)$ were assumed, then there would exist a class $b =_e r = \{x : \neg x\eta x\}$, which is impossible by 9.3. Choose

$$W(x) := \{y : y = \overline{0} \vee (y = \overline{1} \wedge x\eta r)\},\ V(x) := \{y : y = \overline{1} \vee (y = \overline{0} \wedge x\eta r)\},$$

and let $b = \{x : \xi W(x) =_e \xi V(x)\}$. Since $W(x), V(x)$ are non-empty, $\xi W(x)$ and $\xi V(x)$ are non-empty classes by (ii) and hence b is a class. Assume $\neg x\eta r$: then $W(x) =_e \{\overline{0}\}$ and $V(x) =_e \{\overline{1}\}$. Since $\xi W(x)$ and $\xi V(x)$ are non-empty and $\xi W(x) \subseteq W(x)$, $\xi V(x) \subseteq V(x)$, then $\neg\xi W(x) =_e \xi V(x)$; hence $b \subseteq r$. Conversely, let $x\eta r$: then $W(x) =_e V(x) =_e \{\overline{0}, \overline{1}\}$. So, if ξ preserves $=_e$, $\xi W(x) =_e \xi V(x)$ i.e. $x\eta b$. Hence $r \subseteq b$: contradiction . □

As a consequence of 16.3 (iii), the existence of an approximation operator π, satisfying $\pi.1$–$\pi.6$, is inconsistent with

$$\mathrm{EXT}(\pi) := \forall y\forall z\forall x(x\eta y \wedge y =_e z \rightarrow \pi yx =_e \pi zx).$$

CL-compactness implies a useful result:

16.3.1. COROLLARY (*CL-Reflection Schema*). *Let* $A(x,y)$ *be an arbitrary formula. Then there is a term* $\lambda u\lambda v.\rho(A,u,v)$ *such that, provably in* PW^-:

$$Cl(a) \wedge \forall x\eta a.\exists y\eta b.TA(x,y) \rightarrow$$
$$\rightarrow \exists c(c = \rho(A,a,b) \wedge Cl(c) \wedge c \subseteq b \wedge \forall x\eta a.\exists y\eta c.A(x,y)).$$

PROOF: choose $H(A,x,b) = \{y : y\eta b \wedge A(x,y)\}$ and define

$$\rho(A,a,b) := \bigcup_{x\eta a} \xi H(A,x,b).$$

Assume $Cl(a)$, $\forall x\eta a.\exists y\eta b.TA(x,y)$. Then we have, by abstraction 9.2:

$$\forall x\eta a\exists y(y\eta H(A,x,b)).$$

By 16.3, since $H(A,x,b)\subseteq b$ and $y\eta H(A,x,b)\to A(x,y)$, we conclude:

$$\forall x\eta a.\,(Cl(\xi H(A,x,b))\wedge \xi H(A,x,b)\subseteq b); \quad (1)$$

$$\forall x\eta a.\,\exists y\eta\xi H(A,x,b).\,A(x,y). \quad (2)$$

If we set $c:=\rho(A,a,b)$, then (1), 9.14.6 and the assumption $Cl(a)$ imply:

$$Cl(c)\wedge c\subseteq b;$$

from (2) and the choice of c, it follows $\forall x\eta a.\exists y\eta c.A(x,y)$. □

We now use 16.3 to give natural uniform counterparts of the classical reduction and separation theorems.

16.4. DEFINITION. $RD(\langle y',z'\rangle,\langle y,z\rangle)$ ($=\langle y',z'\rangle$ *reduces* $\langle y,z\rangle$) is the conjunction of the the following sentences:

(i) $y'\subseteq y\wedge z'\subseteq z$;
(ii) $y\cup z\subseteq y'\cup z'$;
(iii) $\neg\exists x(x\eta y'\wedge x\eta z')$.

16.5. THEOREM (*Reduction*). *We can find closed terms* R_1, R_2 *such that* PW^- *proves:*

$$\forall y\forall z.RD(\langle R_1yz,\ R_2yz\rangle,\langle y,\ z\rangle).$$

PROOF. Define $d(y,z,x):=\{i:(i=\overline{0}\vee i=\overline{1})\wedge\langle x,i\rangle\eta(y\oplus z)\}$ (for $\oplus$, cf. 9.14.2). Then we choose:

$$R_1yz:=\{x:\overline{0}\eta\xi(d(y,z,x))\}$$

$$R_2yz:=\{x:\overline{1}\eta\xi(d(y,z,x))\wedge\neg\overline{0}\eta\xi(d(y,z,x))\}.$$

Conditions (i) and (iii) of 16.4 are immediately checked by 16.3. As to (ii), assume $x\eta(y\cup z)$. Then for some $i\in\{\overline{0},\overline{1}\}$, $i\eta d(y,z,u)$; hence $\xi(d(y,z,x))$ is a non-empty class $\subseteq\{0,1\}$ by 16.3 (ii). If $\overline{0}\eta\xi(d(y,z,x))$, $x\eta R_1yz$ and we are done. Else, $\overline{0}\ \overline{\eta}\ \xi(d(y,z,x))$ and hence $\overline{1}\eta\xi(d(y,z,x))$, whence $x\eta R_2yz$. □

16.6. COROLLARY (*CL-Separation*). *We can find a term* $CSP(y,z)$ *such that:*

$$PW^-\vdash\forall x(x\eta y\vee x\eta z)\to(Cl(CSP(y,z))\wedge$$

$$\wedge\,\forall x(\neg x\eta y\to x\eta CSP(y,z))\wedge CSP(y,z)\subseteq z). \quad (*)$$

PROOF. We reduce the pair $\langle y,z\rangle$: then $\forall x(x\eta R_1yz\vee x\eta R_2yz)$ and R_1yz, R_2yz are disjoint. If $\neg x\eta y$ and $\neg x\eta R_2yz$, we should get $x\eta R_1yz\subseteq y$:

absurd. Therefore we have $\forall x(\neg x\eta y \to x\eta R_2 yz)$ and $R_2 yz \subseteq z$. Put:

$$d(y,z,x) := \{u : \langle x,u\rangle \eta R_1 yz \oplus R_2 yz\};$$

then we get: $\forall x \exists ! u.\langle x,u\rangle \eta d(y,z,x)$.
We choose $CSP(y,z) := \{x : \bar{1}\eta\xi d(y,z,x)\}$. Since $\xi d(y,z,x)$ is a class for every x, $CSP(y,z)$ is a class too, and it satisfies $(*)$. □

An interesting consequence of CL-separation is the "Δ-comprehension" schema, which corresponds to hyperarithmetical comprehension in second-order arithmetic:

16.7. COROLLARY. *Let $A(u,x)$, $B(u,y)$ be formulas elementary in x, y respectively. Let $\Delta(A,B) := \forall u(\exists x(Cl(x) \wedge A(u,x)) \leftrightarrow \forall y(Cl(y) \to B(u,y)))$. Then:*

$$\mathrm{PW}^- \vdash \Delta(A,B) \to \exists z(Cl(z) \wedge \forall u(u\eta z \leftrightarrow \exists x(Cl(x) \wedge A(u,x)))).$$

16.8. DEFINITION. *z weakly separates the pair $\langle x,y\rangle$* of disjoint properties (i.e. $x \cap y =_e \emptyset$), iff $x \subseteq z$ and $y \subseteq -z$.

We now prove in the extended axiomatic framework the separation lemma of 15.1:

16.9. PROPOSITION (PW^-). *We can define a term $SEP(x,y)$ such that, if $x \cap y =_e \emptyset$, then $SEP(x,y)$ weakly separates $\langle x,y\rangle$.*

PROOF. Define

$$SEP(x,y) := \{u : u\eta x \leq u\eta y\}$$

where $u\eta x \leq u\eta y := \langle u,x\rangle \eta E \wedge \pi E\langle u,x\rangle \subseteq \pi E\langle u,y\rangle$ and $E := \{\langle u,y\rangle : u\eta y\}$. If $u\eta x$, then $\neg u\eta y$ by assumption: hence by 16.2 (v), $u\eta x \leq u\eta y$, i.e. $u\eta SEP(x,y)$. If $u\eta y$, then $F(u\eta x \leq u\eta y)$ (by 16.2 (vi), (x)), which implies $u\bar{\eta} SEP(x,y)$. □

16.10. COROLLARY (*in* PW^-, *Effective Inseparability*). *If $\langle x,y\rangle$ is any pair of disjoint properties such that $r := \{u : \neg u\eta u\} \subseteq x$, $-r \subseteq y$, then $\neg SEP(x,y)\eta(x \cup y)$ (where SEP is given by* 16.9).

We conclude the section with a nice application of reduction and π-axioms. Each c trivially defines a pair $\langle d,b\rangle$ of disjoint properties with $d =_e c$ and $b =_e -c$ ($-c := \{x : \neg x\eta c\}$; $=_e$ is η-extensional equality); conversely, we may wonder whether to each pair $\langle d,b\rangle$ of disjoint properties we can associate a property c satisfying $d =_e c$, $b =_e -c$. We call such a c, if it exists, *an exact representation* of the pair $\langle d,b\rangle$. The answer to the problem is positive and shows that the present language is complete with respect to exact representability of pairs of disjoint properties. The result generalizes a well-known theorem for recursively enumerable sets of numbers, due to Putnam, Smullyan, Shepherdson (see Smullyan 1961).

16.11. THEOREM. *We can define a closed term* $\lambda y\lambda z.ER(y,z)$, *such that* PW^- *proves:*

$$y \cap z =_e \emptyset \to (ER(y,z) =_e y \wedge -ER(y,z) =_e z).$$

PROOF. Define:

$$r_1(y,z) := \{\langle v,u\rangle : v\eta y \vee \langle v,u\rangle\bar{\eta}u\};$$

$$r_2(y,z) := \{\langle v,u\rangle : v\eta z \vee \langle v,u\rangle\eta u\}.$$

Below we simply write r_1 for $r_1(y,z)$ and r_2 for $r_2(y,z)$. The definition of R_1, R_2 (see 16.5) and CL-compactness imply:

$$\forall u(u\eta r_1 \wedge \neg u\eta r_2 \to u\eta R_1(r_1,r_2)); \qquad (1)$$

$$\forall u(u\eta r_2 \wedge \neg u\eta r_1 \to u\eta R_2(r_1,r_2)).$$

Since $R_1(r_1,r_2) \cap R_2(r_1,r_2) =_e \emptyset$, we find $c = SEP(R_1(r_1,r_2), R_2(r_1,r_2))$ by 16.9, such that:

$$\forall u((u\eta r_1 \wedge \neg u\eta r_2) \to u\eta c); \qquad (2)$$

$$\forall u((u\eta r_2 \wedge \neg u\eta r_1) \to u\bar{\eta}c).$$

Let $ER(y,z) := \{v : \langle v,c\rangle\eta c\}$; we claim

$$v\eta y \leftrightarrow \langle v,c\rangle\eta c. \qquad (3)$$

(3) $\Rightarrow$: assume $v\eta y$ and $\neg\langle v,c\rangle\eta c$. Since $\neg v\eta z$, we get $\neg\langle v,c\rangle\eta r_2$. But $v\eta y$ implies $\langle v,c\rangle\eta r_1$: hence by (2), $\langle v,c\rangle\eta c$ and the conclusion follows by tertium non datur.
(3) $\Leftarrow$: let $\langle v,c\rangle\eta c$ and $\neg v\eta y$. Then $\langle v,c\rangle\eta r_2$; also, by consistency, $\neg\langle v,c\rangle\bar{\eta}c$, whence $\neg\langle v,c\rangle\eta r_1$. By (2) we have $\langle v,c\rangle\bar{\eta}c$: contradiction.
The verification of $\forall v(v\eta z \leftrightarrow \langle v,c\rangle\bar{\eta}c)$ is similar. □

§17. The approximation theorem for extensional operations and the fixed point theorem for monotone operations

We consider the following *generalized continuity problem*: if we have the value of an operation f on a property a, under which condition can fa be approximated by the values of f on the subclasses of a ?
The answer turns out to be natural and simple: f must be *extensional* in the sense of η and the resulting continuity theorem is the analogue of the Myhill-Shepherdson theorem in a general setting. A remarkable consequence

is that the notions of *extensional operation, monotone operation* and *operation induced by a positive operator* (in the sense of 10.3) essentially coincide (modulo η-extensional equality). In particular *the minimal fixed points of monotone operations are already generated by positive operators.*

We derive the main result within PW^-, as a corollary of a generalized Rice-Shapiro theorem; the neat exposition of Fitting (1981) is our guideline.

17.1. DEFINITION

(i) An operation f is extensional$_2$ iff $x =_e y$ implies $fx =_e fy$, for every x, y.

(ii) z is extensional$_1$ iff $x =_e y$ and $y\eta z$ imply $x\eta z$.

17.2. LEMMA (*Upward closure*). *If z is extensional$_1$, then z is $\subseteq$-upward closed. Formally:*

$$\mathrm{MF}^- \vdash Ext_1(z) \rightarrow \forall x \forall y((x \subseteq y \wedge x\eta z) \rightarrow y\eta z)$$

(*where $Ext_1(z)$ is the obvious formula representing extensionality$_1$*).

PROOF: assume $x \subseteq y$ and $x\eta z$. By the fixed point theorem for properties 10.4, we find a term I depending on x, y, z, such that:

$$\forall u(u\eta I \leftrightarrow (u\eta x \vee (u\eta y \wedge I\eta z))). \qquad (*)$$

If $\neg I\eta z$, then by $(*)$, $I =_e x$, i.e. $I\eta z$. Hence $I\eta z$, which implies $I =_e y$ (because $x \subseteq y$), i.e. by extensionality, $y\eta z$. □

17.2.1. REMARK. Observe that $\mathrm{MF}^- \vdash \forall z \forall x(Ext_1(z) \wedge \neg x\eta z \rightarrow \neg\emptyset\eta z)$. If z is an extensional class, so is the complement $-z$; now either $\emptyset\eta z$ or $\emptyset\eta(-z)$, whence either $z =_e V$ or $z =_e \emptyset$; so 10.15 is a corollary of 17.2.

17.3. LEMMA ("Rice-Shapiro·generalized"). *If z is extensional$_1$ and x is in z, then there is a class u in z such that $u \subseteq x$. Formally:*

$$\mathrm{PW}^- \vdash \forall z \forall x(Ext_1(z) \wedge x\eta z \rightarrow \exists u(Cl(u) \wedge u\eta z \wedge u \subseteq x));$$

(*indeed, u is given uniformly in z, x by a term of the language*).

PROOF. Fix x in z and define $t := x \oplus z$ (or, which is equivalent, $x \oplus \pi zx$). By the fixed point theorem for properties, we can find a term J, which depends on x, z, such that:

$$\forall u(u\eta J \leftrightarrow \exists y(y\eta t \wedge \langle u, \overline{0}\rangle \eta \pi ty \wedge \neg\langle J, 1\rangle \eta \pi ty)). \qquad (1)$$

(From right to left, we apply $\pi.2$ to show that $\neg\langle J, 1\rangle\eta\pi ty$ implies $\langle J, 1\rangle\overline{\eta}\, \pi ty$). But (1), the definition of t and $\pi.2$ again entail

$$J \subseteq x. \qquad (2)$$

Then we obtain

$$J\eta z. \tag{3}$$

Indeed, if $\neg J\eta z$ holds, we have by π.2:

$$\forall u(u\eta J \leftrightarrow \exists y(y\eta t \wedge \langle u,\overline{0}\rangle\eta\pi ty)). \tag{4}$$

Moreover, if $u\eta x$ holds, we have $\langle u,\overline{0}\rangle\eta t$ and $\langle u,\overline{0}\rangle\eta\pi t\langle u,\overline{0}\rangle$ (by π.1), i.e. $\exists y(y\eta t \wedge \langle u,\overline{0}\rangle\eta\pi ty)$, whence by (4), $u\eta J$. By (2), $x =_e J$ is true; since $x\eta z$ and z is extensional$_1$, $J\eta z$.
If $d := \{u : \exists y(y\eta\pi t\langle J,\overline{1}\rangle \wedge \langle u,\overline{0}\rangle\eta\pi ty \wedge \neg\langle J,\overline{1}\rangle\eta\pi ty)\}$, we claim:

$$d \text{ is a class such that } d =_e J. \tag{5}$$

Indeed, as to the classhood of d, let $A(u)$ be the defining condition of d. We show that $\neg TA(u)$ implies $T\neg A(u)$. By T-axioms, this amounts to check that for arbitrary y, $\langle J,\overline{1}\rangle\eta\pi ty$ follows from the assumptions:

$$\neg y\eta\pi t\langle J,\overline{1}\rangle \vee \neg\langle u,\overline{0}\rangle\eta\pi ty \vee \neg\langle J,\overline{1}\rangle\overline{\eta}\,\pi ty; \tag{6.1}$$

$$\neg y\overline{\eta}\,\pi t\langle J,\overline{1}\rangle \wedge \neg\langle u,\overline{0}\rangle\overline{\eta}\,\pi ty. \tag{6.2}$$

Since t is non-empty by (3), $\pi t\langle J,\overline{1}\rangle$ is a subclass of t by π.2. Hence by (6.2) $y\eta t$, which implies that πty is a subclass of t (again by π.2). By the definition of class, (6.1)-(6.2) transform into

$$\neg y\eta\pi t\langle J,\overline{1}\rangle \vee \neg\langle u,\overline{0}\rangle\eta\pi ty \vee \langle J,\overline{1}\rangle\eta\pi ty;$$

$$y\eta\pi t\langle J,\overline{1}\rangle \wedge \langle u,\overline{0}\rangle\eta\pi ty,$$

whence $\langle J,\overline{1}\rangle\eta\pi ty$.
As to the claim $d \subseteq J$, assume $u\eta J$. Then $\langle u,\overline{0}\rangle\eta\pi ty$, for some y in t such that $\langle J,\overline{1}\rangle\overline{\eta}\,\pi ty$ (by (1) and π.2). Since J is in z, $\langle J,\overline{1}\rangle\eta t$ and hence $\pi ty \subseteq \pi t\langle J,\overline{1}\rangle$ (by π.1, π.5); hence y is in $\pi t\langle J,\overline{1}\rangle$, i.e. u is in d.
That $d \subseteq J$ follows from $\pi t\langle J,\overline{1}\rangle \subseteq t$ and $J\eta z$. Extensionality$_1$ of z and (2)-(3)-(5) yield $(d\eta z \wedge Cl(d) \wedge d \subseteq x)$. □

17.4. COROLLARY (PW$^-$)

(i) *Every infinite property contains an infinite class.*
(ii) *CL-compactness again: there is a term $\lambda x.\xi x$ such that*

$$\xi z \subseteq z \wedge (\exists x(x\eta z) \rightarrow Cl(\xi z) \wedge \exists u(u\eta\xi z)).$$

PROOF. (i): let INF $:= \{x : \exists f : N \rightarrowtail x\}$ be the property of being infinite (here $f : N \rightarrowtail x := f : N \rightarrow x$ and f is 1-1). Then INF is extensional$_1$ and we can apply 17.3.

(ii) $c := \{x : \exists u(u\eta x)\}$ is extensional$_1$; hence by 17.3 there is a term $\phi(c,z)$, such that

$$x\eta z \rightarrow Cl(\phi(c,z)) \wedge \exists u(u\eta\phi(c,z)) \wedge \phi(c,z) \subseteq z. \qquad (*)$$

Choose (hint by Minari): $\xi z = \{x : x\eta z \wedge x\eta\phi(c,z)\}$. Then apply (*). □

17.5. THEOREM ("Myhill-Shepherdson"). *Let f be extensional$_2$. Then for every $a, fa =_e \bigcup\{fc : Cl(c) \wedge c \subseteq a\}$. Formally:*

$$\mathrm{PW}^- \vdash \forall f \forall x\ (Ext_2(f) \rightarrow fx =_e \bigcup\{fy : Cl(y) \wedge y \subseteq x\}),$$

(where $Ext_2(f) := \forall x \forall y(x =_e y \rightarrow fx =_e fy)$).

PROOF. Define $c(f,u) := \{y : u\eta fy\}$, where u is in fx. By assumption on f, $c(f,u)$ is extensional$_1$. If $u\eta fy$, for some subclass y of x, then $y\eta c(f,u)$ and by lemma 17.2 $x\eta c(f,u)$, i.e. $u\eta fx$. If $u\eta fx$, then $x\eta c(f,u)$: hence 17.3 yields a subclass y of x such that $u\eta fy$. □

17.5.1. REMARK. We already know that every operator $A(x,a)$ gives rise to an extensional$_2$ operation $f_A a := \{x : A(x,a)\}$ (see 10.3-10.4). By 17.5, the converse also holds: *for every extensional$_2$ operation f, we can find a positive operator $A_f(x,a)$ such that for every a, $fa =_e \{x : A_f(x,a)\}$.*
Of course, we have a uniform choice: $A_f(x,a) = \exists c(Cl(c) \wedge c \subseteq a \wedge x\eta fc)$.
It is easy to check that the two operations $f \mapsto A_f$ and $A \mapsto f_A$ are inverse to each other.

17.6. DEFINITION

(i) $Mon(f) := \forall a \forall b(a \subseteq b \rightarrow fa \subseteq fb)$; if $Mon(f)$ is assumed, we say that f is $\subseteq$-monotone, or, simply, monotone.

(ii) $f\text{-}clos(x) := fx \subseteq x$ (in words: x is f-closed).

If $B(x)$ is an arbitrary formula,

$$f\text{-}clos(B) := \forall w \forall u((u\eta fw \wedge Cl(w) \wedge \forall v(v\eta w \rightarrow B(v))) \rightarrow B(u)).$$

(iii) $I(f) := I(A_f)$; (here $I(A) :=$ fixed point of A, see 10.1).

17.7. FACT

(i) *By* 17.5 *an operation f is extensional$_2$ iff f is monotone.*
(ii) *If f is monotone and $B(u) := u\eta x$ (with u, x distinct variables),* 17.5 *implies*

$$f\text{-}clos(x) \leftrightarrow f\text{-}clos(u\eta x).$$

An important application of the continuity theorem and generalized induction (12.4.1) is a uniform version of the fixed point theorem for monotone operations:

17.8. THEOREM ("Knaster-Tarski"). $PW^- + GID$ *proves:*

$$Mon(f) \rightarrow f(I(f)) \subseteq I(f) \wedge (f\text{-}clos(B) \rightarrow \forall u(u\eta I(f) \rightarrow B(u))).$$

In particular, $PW^- + GID \vdash Mon(f) \rightarrow (f\text{-}clos(x) \rightarrow I(f) \subseteq x)$.

PROOF. If f is monotone, 17.5 and $u\eta f(I(f))$ imply:

$$\exists w(w \subseteq I(f) \wedge Cl(w) \wedge u\eta fw). \tag{1}$$

By choice of $I(f)$, we can conclude:

$$u\eta I(f). \tag{2}$$

If B is f-closed, B is by definition A_f-closed and hence the conclusion follows by GID. The second part is a consequence of 17.7 (ii). □

17.8.1. REMARK. The fixed point operator I, which can be explicitly defined by the term $\lambda f.Y(\lambda v.\{x : \exists c(Cl(c) \wedge c \subseteq v \wedge x\eta fc)\})$, is extensional too, in the sense that it preserves the natural pointwise equality on operations. Set $f \sqsubseteq g := \forall x(fx \subseteq gx)$: then $I(f) \subseteq I(g)$ (apply GID).

§18. Topology displayed: basic definitions

We make explicit the topological flavour of the previous results on extensional operations. Thus we try to build a bridge between the present *intensional approach* and *denotational semantics*, which relies on topologically restricted notions of partial function and functional.

After the definability results of Ch. III (see 13.4), we know that the syntax of $\mathfrak{L}$ is strongly adequate for representing inductive sets. But there is a more recondite adequacy, which is embodied in the results of §17, specifically in the MS-theorem 17.5. We prove that, if the power set of the ground combinatory algebra $\mathcal{M}$ is endowed with a suitably natural topology, then "explicitly continuous" operators, mapping the power set of M into itself, can be adequately mirrored by our logical approach.

We work in a fixed model $\mathcal{M}$ of OP^-, while $\mathbb{O}(\mathcal{M})$ is the least inductive model of $PW^- + GID$. Of course, $M := |\mathcal{M}|$ is the support of $\mathcal{M}$, while $\mathcal{P}(M)$ is the power set of M.

18.1. DEFINITION. If $b \in M$, $E(b) := \{a : \mathbb{O}(\mathcal{M}) \models a\eta b\}$ = "the extension of property b". By 13.4 and 15.2, we have $IND(\mathcal{M}) = \{E(a) : a \in M\}$ and $HYP = \{E(b) : b \text{ is a class in } \mathbb{O}(\mathcal{M})\}$.

All the notions introduced depend on $\mathcal{M}$ (e.g. $E := E_{\mathcal{M}}$, $IND(\mathcal{M})$). Since

$\mathcal{M}$ is fixed, we can simply omit the explicit indication of $\mathcal{M}$ and M. Also, we shall speak of classes instead of M-classes.

CONVENTION: X, Y, Z range over $\mathcal{P}(M)$, P, Q, R range over families of subsets of M. We freely use formulas of the language of PW^- to state facts about $\mathbb{O}(\mathcal{M})$; e.g. " assume $a\eta b$" means " assume $\mathbb{O}(\mathcal{M})\models a\eta b$", etc...

We induce a topology on $\mathcal{P}(M)$ and, by relativization, on IND. For the topological notions, see any good reference (e.g. Kelley 1955).

18.2. DEFINITION. If $\mathbb{O}(\mathcal{M})\models Cl(e)$, $V(e) := \{X \subseteq M : E(e) \subseteq X\}$ is called a *basic open* (set).

18.3. LEMMA

(i) *The family* $\{V(e) : \mathbb{O}(\mathcal{M})\models Cl(e)\}$ *is a basis for a topology* Cl-$\mathcal{T}$, *the class topology over* $\mathcal{M}$.

(ii) $\langle \mathcal{P}(M), Cl\text{-}\mathcal{T}\rangle$ *(and hence the subspace* $\langle IND, Cl\text{-}\mathcal{T}\rangle$*) is a* T_0*-space, which is not* T_1.

(iii) *For any class* e, *if* $V(e)$ *is covered by a family of basic opens, then* $V(e)$ *is already covered by a basic open of the given family itself*:

if $V(e) \subseteq \cup\{V(e_k) : k \in I\}$, *then* $V(e) \subseteq V(e_j)$, *for some* $j \in I$.

(iv) $P \subseteq \mathcal{P}(M))$ *is open in the class topology iff*
1. P *is upward closed (i.e.* $Y \in P$ *and* $Y \subseteq X$ *imply* $X \in P$*)*;
2. *whenever* $X \in P$, *then there exists a class* e *such that* $E(e) \in P$ *and* $E(e) \subseteq X$.

The proof is an easy exercise. Clearly $\langle \mathcal{P}(M), Cl\text{-}\mathcal{T}\rangle$ induces a sort of generalized Scott topology on inductive sets (under the Kreisel-Sacks's analogy "finite" = "to be a class"; see Kreisel-Sacks 1965). It is also useful to regard the subspace IND as an analogue of f-spaces in the sense of Ershov (1977), where the subset of *finite elements* is just the collection $\{E(c)$: c is an M-class$\}$.

We now state a simple proposition relating classhood to a constructive abstract version of the compactness property; not surprisingly, the proof requires CL-compactness (16.3).

18.4. THEOREM (*CL-finiteness*). *Let* $X \in IND$. *Then* $X = E(c)$, *for some* M*-class* c *iff whenever* $X \subseteq E(\cup\{fi : i\eta I\})$, *then also* $X \subseteq E(\cup\{fi : i\eta I_0\})$ *for some* M*-subclass* I_0 *of* I *(*I, f *arbitrary)*.

PROOF. $\Rightarrow$. By assumption $X = E(c)$ and we have $\forall x\eta c.\exists i\eta I.x\eta fi$. Set $A(x,i) := i\eta I \wedge x\eta fi$ and choose $I_0 = \rho(A, c, I)$ by CL-reflection 16.3.1.

Then I_0 is a subclass of I which meets the condition of the lemma.
$\Leftarrow$: assume that $X := E(c)$ satisfies the right side of the lemma; trivially $E(c) = E(\cup\{\{x\}: x\eta c\})$ and hence for $f := \lambda x.\{x\}$ and some M-class b with $E(b) \subseteq E(c)$, $E(c) \subseteq E(\cup\{\{x\}: x\eta b\})$, i.e. $E(c) \subseteq E(b)$, for some M-class b, i.e. $X = E(b)$. □

Let $PS(X) := \{E(a): Cl(a) \text{ and } E(a) \subseteq X\}$. Clearly:

$$PS(E(a)) = \{E(c):\ c\eta P^+(a)\},$$

where $P^+(a) := \{x: Cl(x) \wedge x \subseteq a\}$ (see 9.15). It is immediate to verify:

18.5. LEMMA

(i) *$PS(X)$ is $\subseteq$-directed: if Z, $Y \in PS(X)$, there exists $W \in PS(X)$ such that $Z \subseteq W$ and $Y \subseteq W$.*

(ii) *If $Y \in IND$, $Y = \cup PS(Y)$.*

(iii) *$V(b) \cap V(c) = V(e)$, for some class e (b, c classes).*

(iv) *$IND \cap V(\emptyset) = IND$ ($\emptyset$ = empty property).*

IND can be naturally viewed as a sort of constructive "ideal completion" of HYP (in the sense of lattice theory). In particular, ideals in HYP which are explicitly presented by means of *arbitrary operations*, are already principal ideals in IND. The point is made precise below.

18.6. DEFINITION. If $f: I \to CL$ shortens $\forall x\eta I.Cl(fx)$, we let:

$$IDE(f) := \{c: Cl(c) \wedge \exists I(Cl(I) \wedge f: I \to CL \wedge c \subseteq \cup\{fi:\ i\eta I\})\}.$$

18.6.1. FACT. *The following is provable in* MF$^-$:

$$c\eta IDE(f) \leftrightarrow Cl(c) \wedge \exists I(Cl(I) \wedge f: I \to CL \wedge c \subseteq \cup\{fi:\ i\eta I\}).$$

The term $IDE(f)$ represents an ideal in the lattice with support $HYP = \{E(a): a \text{ is a class in } \mathbb{O}(\mathcal{M})\}$ (and with set inclusion as ordering); the definition is justified by the following:

18.7. PROPOSITION

(i) *$IDE(f)$ is $\subseteq$-downward closed:*

$$y\eta IDE(f) \wedge Cl(x) \wedge x \subseteq y \to x\eta IDE(f);$$

(ii) *$IDE(f)$ is closed under $\cup$ over its subclasses:*

$$Cl(b) \wedge b \subseteq IDE(f) \to (\cup b)\eta IDE(f);$$

(iii) *Constructive completeness: we can find a term L such that*

$$E(IDE(f)) = E(P^+(L(f))).$$

PROOF. (i) is trivial.
(ii) Assume that b is a subclass of $IDE(f)$. Then by definition

$$b \subseteq CL = \{x : Cl(x)\} \text{ and } \forall x\eta b.\exists c\eta CL.A(x,c),$$

where $A(x,c) := f : c \to CL \wedge x \subseteq \cup\{fi : i\eta c\}$.
Since $A(x,c)$ is quasi-elementary in x, c, we have by 9.6 (i):

$$\forall x\eta b.\forall c\eta CL.(A(x,c) \leftrightarrow TA(x,c)).$$

Then one can apply CL-reflection 16.3.1 and there exists a class of classes e such that

$$\forall x\eta b.\exists c\eta e.A(x,c). \tag{1}$$

By 9.14 $d = \cup e$ is a class and $\cup b \subseteq \cup\{fi : i\eta d\}$. Indeed:

$$\begin{aligned} u\eta \cup b &\Rightarrow \exists x\eta b.u\eta x; \\ &\Rightarrow \exists x\exists c(x\eta b \wedge c\eta e \wedge u\eta x \wedge x \subseteq \cup\{fi : i\eta c\}); \qquad \text{(by (1))} \\ &\Rightarrow \exists c\eta e.u\eta \cup\{fi : i\eta c\}; \\ &\Rightarrow \exists c\eta e.\exists i\eta c.u\eta fi; \\ &\Rightarrow \exists i\eta d.u\eta fi. \end{aligned}$$

(iii) Let $D(f) := \cup\{c : Cl(c) \wedge f : c \to CL\}$ and $L(f) := \cup\{fi : i\eta D(f)\}$.
$IDE(f) \subseteq P^+(L(f))$ is immediate by definition of $L(f)$ and $IDE(f)$.
As to $P^+(L(f)) \subseteq IDE(f)$, assume $Cl(d)$ and $d \subseteq \cup\{fi : i\eta D(f)\}$. Then by CL-finiteness $d \subseteq \cup\{fi : i\eta c\}$, for some class $c \subseteq D(f)$. But $f : D(f) \to CL$ and a fortiori $f : c \to CL$; hence $d\eta IDE(f)$. □

18.8. DEFINITION

(i) Put $\mathfrak{I}(b) := \{E(a) : a\eta b\}$. Then $P \subseteq \mathfrak{P}(M)$ is an RS-family iff for some extensional$_1$ b, $P = \mathfrak{I}(b)$; (in short $P \in RS$; RS = Rice-Shapiro).

(ii) Let $\mathfrak{O}(b) := \cup\{V(e) : e \in E(b), e \text{ class}\}$. Then $P \subseteq \mathfrak{P}(M)$ is *explicitly open* in the class topology iff for some $b \in M$, $P = \mathfrak{O}(b)$ (b is called *index* of the open set).

(iii) $D(b) := \{x : \exists c(Cl(c) \wedge c\eta b \wedge c \subseteq x\}$.

(iv) ECL-OPEN $:= \{\mathfrak{O}(b) : b \in M\}$; $RS := \{\mathfrak{I}(b) : b \in M,\ b \text{ extensional}_1\}$.

RS-families are important because they characterize the explicitly open sets of IND.

18.9. LEMMA

(i) $D(b)$ *is extensional*$_1$: $x =_e y \wedge y\eta D(b) \to x\eta D(b)$;
(ii) $\lambda x.D(x)$ *is extensional*$_2$: $x =_e y \to D(x) =_e D(y)$;
(iii) $\mathfrak{I}(D(b))$ *is an* RS-*family*;

(iv) *if b is extensional$_1$, then* $\mathcal{I}(b) = \mathcal{I}(D(b))$;
(v) $\mathcal{O}(b) = \mathcal{O}(D(b))$ *and* $\mathcal{O}(b) \cap IND = \mathcal{I}(D(b))$.

PROOF. (i)-(ii) follow from the fact that the formula defining $D(b)$ is positive in x and b; (iii) is immediate from (i), while (iv) is a restatement of lemmas 17.3 and 17.2. Proposition (v) is a straightforward application of the definition of $\mathcal{O}$ and $D(b)$. □

18.10. DEFINITION

$\sigma: ECL$-OPEN $\to RS$ and $\omega: RS \to ECL$-OPEN are the maps defined by:

$$\sigma(\mathcal{O}(b)) = \mathcal{I}(D(b)) \text{ and } \omega(\mathcal{I}(b)) = \mathcal{O}(b).$$

18.11. THEOREM. *RS-families are exactly the intersections of explicitly open sets with IND. More precisely:*

(i) *if $b \in M$ and b is extensional$_1$, then* $\mathcal{O}(b) \cap IND = \mathcal{I}(b)$;

(ii) *for every $b \in M$,* $\omega(\sigma(\mathcal{O}(b)) = \mathcal{O}(b))$;

(iii) *if $b \in M$ and extensional$_1$, then* $\sigma(\omega(\mathcal{I}(b))) = \mathcal{I}(b)$.

PROOF. (i) is a consequence of 18.9 (v)-(iv).
(ii): by definition of ω, σ and 18.9 (v).
(iii): by 18.9 (iv). □

Since $IND \subseteq \mathcal{I}(V)$ and the universe $V = \{x : x = x\}$ is an extensional$_1$ class, then we have that $IND = \bigcup \{\mathcal{I}(b) : b \text{ is extensional}_1\}$; on the other hand, $\mathcal{I}(b \cap c) \subseteq \mathcal{I}(b) \cap \mathcal{I}(c)$: hence RS is a basis for a topology on IND, we simply label RS-topology. Once we observe that the intersection of any basic open of Cl-$\mathfrak{T}$ with IND is an RS-family and we keep in mind 18.11 (i), we obtain:

18.12. COROLLARY. *The RS-topology coincides with the class topology on IND.*

§19. The representation theorem for explicitly *CL*-continuous operators

As usual, an operator $F: \mathcal{P}(M) \to \mathcal{P}(M)$ is continuous with respect to the class topology Cl-$\mathfrak{T}$ iff the inverse image of a basic open of Cl-$\mathfrak{T}$ is open in Cl-$\mathfrak{T}$.

19.1. DEFINITION

(i) $F: \mathcal{P}(M) \to \mathcal{P}(M)$ is *CL-continuous* iff 1) F is $\subseteq$-monotone; 2) whenever $x \in F(X)$ ($X \subseteq M$), then $x \in F(E(c))$, for some class c with $E(c) \subseteq X$;

(ii) $F : \mathcal{P}(M) \to \mathcal{P}(M)$ is ECL (*explicitly CL-continuous*) iff $F = F_f$ where $F_f(X) := \cup \{E(fc) : Cl(c) \wedge E(c) \subseteq X\}$; f is called an *index* for F.

NOTATION. $ECL :=$ space of explicitly continuous operators (from $\mathcal{P}(M)$ to $\mathcal{P}(M)$).

ECL-continuous operators are CL-continuous; as soon as we consider the subspace IND with ECL-operators restricted to IND, CL-continuity and continuity in the class topology yield the same notion.

19.2. LEMMA

(i) *Assume that $F : \mathcal{P}(M) \to \mathcal{P}(M)$ is continuous with respect to the class topology . Then F is CL-continuous.*

(ii) *If F is ECL-continuous, then F is CL-continuous and the restriction of F to IND is continuous in the class topology relativized to IND.*

PROOF. (i): straightforward with 18.3 (iv).
(ii) Let $F = F_f$:

$a \in F(X) \Rightarrow$ for some class c with $E(c) \subseteq X$, $a \in E(fc)$,

$\Rightarrow$ there are classes c, d with $E(d) \subseteq E(c) \subseteq X$, $a \in E(fd)$,

$\Rightarrow$ for some class c with $E(c) \subseteq X$, $a \in F(E(c))$.

Monotonicity is also immediate by definition. As to the second claim, if e is a class, f is a index for F and F is restricted to IND,

$$F^{-1}(V(e) \cap IND) = \{E(a) : \exists b(E(e) \subseteq E(b) \wedge F(E(a)) = E(b)\};$$
$$= \{E(a) : e \subseteq \cup \{fc : Cl(c) \wedge c \subseteq a\}\} \text{ (by } ECL\text{-continuity)};$$
$$= \mathcal{S}(J(e)),$$

where $J(e) := \{x : e \subseteq \cup \{fc : Cl(c) \wedge c \subseteq x\}\}$. But $J(e)$ is extensional$_1$ and hence $F^{-1}(V(e) \cap IND)$ is an RS-family, i.e. open in the relativized class topology by 18.11. □

We now see that the restrictions of ECL-operators to IND have the expected representation, as effective counterparts of RS-continuous operators on IND; we write $F \lceil P$ for the function F, restricted to the domain P.

19.3. DEFINITION. An operator $F : IND \to IND$ is *effective* (in short $F \in EFF$) iff $F(E(a)) = E(fa)$, for some extensional$_2$ f and every $a \in M$. F is "generated by f" and simply denoted by E_f; thus $E_f(E(a)) = E(fa)$.

EFF will denote the the space of effective operators.

19.4. LEMMA. *Assume $F \in EFF$. Then F is RS-continuous and hence CL-continuous in the class topology relativized to IND.*

PROOF. Observe that if $F = E_f$, f is extensional$_2$ and b is extensional$_1$, $F^{-1}(\{E(a): a \in E(b)\}) = \{E(c): (fc)\eta b\}$ and $\{c: (fc)\eta b\}$ is extensional$_1$; then apply 18.12. □

19.5. DEFINITION. $f^*a := \cup\{fc : c \subseteq a \wedge Cl(c)\}$.

19.6. LEMMA

(i) *f^* is extensional$_2$; hence $E_{f^*} \in EFF$.*

(ii) *If f is extensional$_2$, $f^*a =_e fa$; hence $f^*a =_e (f^*)^*a$ (a arbitrary).*

(iii) $F_f = F_{f^*}$.

PROOF. (i) f^* is extensional$_2$ because f^*a depends positively on a.
(ii) $fa \subseteq f^*a$ is simply the main theorem 17.5, while $f^*a \subseteq fa$ follows by $\subseteq$-monotonicity of extensional$_2$ operations (17.7).
(iii) : by definition of F_f and f^*. □

We now see that the restrictions to IND of explicitly CL-continuous operators are effective operators; viceversa, each effective operator has its source in a uniquely determined element of ECL.

19.7. DEFINITION

(i) If $E_f \in EFF$, let $\mathcal{C}(E_f) := F_f$.

(ii) If $F_f \in ECL$, let $\mathcal{E}(F_f) := E_{f^*}$.

19.8. THEOREM

(i) *Let $F_f \in ECL$. Then $\mathcal{E}(F_f) \in EFF$ and*

$$\mathcal{E}(F_f) = F_f\lceil IND = F_{f^*}\lceil IND.$$

(ii) *Let f be extensional$_2$. Then*

$$\mathcal{C}(E_f) \in ECL \text{ and } \mathcal{C}(E_f)\lceil IND = E_f.$$

(iii) *Moreover:*

$$\text{if } f \text{ is extensional}_2,\ \mathcal{E}(\mathcal{C}(E_f)) = E_f; \tag{$*$}$$

$$\text{if } F_f \in ECL,\ \mathcal{C}(\mathcal{E}(F_f)) = F_f. \tag{$**$}$$

PROOF. (i) $\mathcal{E}(F_f) \in EFF$ by 19.6 (i). Moreover, $F_f(E(a)) = E(f^*a)$ is immediate by definition of f^* and explicit CL-continuity with index f. On the other hand, $F_{f^*}(E(a)) = E(f^{**}a) = E(f^*a)$ by 19.6 (ii).

(ii) Let $G := \mathfrak{C}(E_f)$. Then we have $G \in ECL$ by lemma 19.2 (ii). $G(E(a)) = E(fa)$ is a restatement of theorem 17.5.

(iii) Let $F := E_f \in EFF$: $\mathfrak{E}(\mathfrak{C}(F)) = E_{f^*}$ as $\mathfrak{C}(F)$ has index f; but $E_{f^*} = E_f$ by 19.6 (ii). Let $F_f \in ECL$: then $\mathfrak{C}(\mathfrak{E}(F_f)) = F_{f^*} = F_f$ by 19.6 (iii). □

If F is a monotone operator from $\mathcal{P}(M)$ to $\mathcal{P}(M)$, let $\mathfrak{I}(F)$ denote the (set-theoretic) least fixed point of F. Then we obtain the fundamental:

19.9. THEOREM (*"First Recursion Theorem"*). *If $F_f \in ECL$, then*

$$\mathfrak{I}(F_f) = E(I(f)).$$

PROOF. Let F_f be ECL: then

$a \in F_f(E(I(f))) \Rightarrow \mathbb{O}(\mathcal{M}) \models A_f(a, I(f))$ (CL-continuity);

$\Rightarrow \mathbb{O}(\mathcal{M}) \models a\eta I(f)$, i.e. $a \in E(I(f))$ (by fixed point 10.1);

hence $E(I(f))$ is F_f-closed and $\mathfrak{I}(F_f) \subseteq E(I(f))$. On the other hand,

$$\{a \in M : \langle \mathbb{O}(\mathcal{M}), \mathfrak{I}(F_f)\rangle \models A_f(a, X)\} \subseteq \mathfrak{I}(F_f).$$

But $E(I(f))$ is the $\subseteq$-least A_f-closed subset of M by the induction theorem of §12: hence $E(I(f)) \subseteq \mathfrak{I}(F_f)$. □

Theorem 19.9 is *a completeness result*: as far as we consider continuous operators, which are explicitly given via internal maps of the underlying combinatory algebra, the restriction to the subspace IND is unessential and we are still able to capture *all the explicitly "recursive" objects we have in the full space.* Theorems 18.11 and 19.9 establish the adequacy of a syntax – the language of operations and reflective truth – to the "semantics" of ECL-opens and ECL-operators. This should add some evidence in favour of the approximation axioms as reasonable choices.

19.10. REMARK

(i) Define:

$$FUN^*(c)(X) := \{a \in M : \text{for some class } d, \text{ with } E(d) \subseteq X,\ \langle d, a\rangle \in E(c)\}.$$

It is easy to see that $\{FUN^*(c) : c \in M\} = \{F_f : f \in M\}$; so one has an alternative definition for explicitly CL-continuous operators. Also, if $F, G : \mathcal{P}(M) \to \mathcal{P}(M)$, $F \subseteq G := F(X) \subseteq G(X)$ for every $X \subseteq M$.
Finally, we define:

$$V^1(c) := \{F : (F : \mathcal{P}(M) \to \mathcal{P}(M)) \text{ and } FUN^*(c) \subseteq F\};$$

then $CLF\text{-}\mathcal{K} := \{V^1(c) : c \text{ class}\}$ is a basis for a topology on $\{F : F$ operator from $\mathcal{P}(M)$ to $\mathcal{P}(M)\}$. One checks that ECL and EFF are homeomorphic as function subspaces via $\mathfrak{E}$ and $\mathfrak{C}$.

(ii) Define:

$$\text{Graph}^+(f) := \{\langle c, x\rangle : Cl(c) \wedge x\eta fc\};$$

$$\text{Fun}^+(a) := \lambda u.\{x : \exists y(y \subseteq u \wedge Cl(y) \wedge \langle y, x\rangle \eta a)\};$$

$$a{*}b := (\text{Fun}^+(a))b.$$

Then:

if $f \subseteq g$, then $\text{Graph}^+(f) \subseteq \text{Graph}^+(g)$;

if f is extensional$_2$, then for every a, $fa =_e (\text{Graph}^+(f)){*}a$.

IND becomes a λ-model if we interpret application via $*$ and λ-abstraction via Graph$^+$.

(iii) By 17.8.1, the internal fixed point operation I preserves the natural pointwise equality on operation. This suggests the definition of a hierarchy of extensional operations, in analogy with the classical hierarchy of hereditarily effective operations HEO (see Troelstra 1973). One may wonder whether this hierarchy mirrors the corresponding natural hierarchy based on class topology at the ground type $\mathcal{P}(M) = \{X : X \subseteq M\}$ and extending upwards via pointwise convergence.

Perhaps, it might be useful to generalize to the present theory Ershov's notion of *numeration*. Ershov's basic idea is to consider *enumerated sets*, i.e. pairs (X, ν), where X is any set and ν is a map from natural numbers onto X. Enumerated sets with an appropriate notion of (computable) morphism give rise to a category, which forms a suitable environment for developing a theory of partial computable functionals in higher types.

In this respect, the reader, who seeks for possible generalizations, could profitably read the interesting paper of Ershov (1985): there, the author constructs a theory of Σ-predicates of higher types over any admissible set $\mathbb{A}$. In Ershov's approach, *sets* in $\mathbb{A}$ play the role of *classes* and one can introduce a topology on Σ-definable families of Σ-sets of $\mathbb{A}$, which appears an analogue of the class topology. The topological structure is then applied, in order to define morphisms between enumerated sets (in the sense of $\mathbb{A}$), and to induce an appropriate notion of approximation.

Appendix: alternative proofs

The proof of the generalized Rice-Shapiro, given in §17, depends upon axioms $\pi.1$, $\pi.2$ and $\pi.5$. However, during the final proof reading of the book, P. Minari found a slick variant $\mathbb{J}$ of the term J of 17.3. The resulting argument relies on $\pi.2$ and $\pi.3$ and $\mathbb{J}$ satisfies an additional condition.

Alternative proof of 17.3. By the fixed point theorem for properties, we can find $\mathbb{J}$ such that, if we set $J := \mathbb{J}zx$ and $E = \{\langle x,y\rangle : x\eta y\}$, then

$$\forall u(u\eta J \leftrightarrow \langle u,x\rangle \eta \pi E\langle J,z\rangle). \tag{1}$$

Assume $Ext_1(z)$ and $x\eta z$. If $\neg J\eta z$, by $\pi.3$, $E \subseteq \pi E\langle J,z\rangle$; hence, if $u\eta x$, $\langle u,x\rangle \eta E \subseteq \pi E\langle J,z\rangle$. So $x \subseteq J$, which implies $J\eta z$ by 17.2. By classical logic, we conclude:

$$J\eta z. \tag{2}$$

By (2) and $\pi.2$, $\pi E\langle J,z\rangle$ is a class: hence

$$Cl(J). \tag{3}$$

Again (2) and $\pi.2$ imply $\pi E\langle J,z\rangle \subseteq E$. If $u\eta J$, then $\langle u,x\rangle \eta\pi E\langle J,z\rangle)$ and hence $u\eta x$; hence

$$J \subseteq x \tag{4}$$

(2)-(4) verify $\forall z\forall x(Ext_1(z) \wedge x\eta z \rightarrow (Cl(\mathbb{J}zx) \wedge (\mathbb{J}zx)\eta z \wedge \mathbb{J}zx \subseteq x)$.

We can also prove, recalling that that $J := \mathbb{J}zx$:

$$Ext_1(z) \wedge J\eta z \rightarrow x\eta z.$$

Indeed, assume $Ext_1(z)$, $\neg x\eta z$ and $J\eta z$. Then by (4) $J \subseteq x$, whence $x\eta z$ by extensionality$_1$ of z: contradiction. □

It is also clear that CL-compactness becomes provable in $\mathrm{MF}^- + \{\pi.2, \pi.3\}$. The separation property 16.9 can be proved on the ground of $\pi.2$ and $\pi.3$.

PROPOSITION ($\mathrm{MF}^- + \{\pi.2, \pi.3\}$). *There is a term $SEP(x,y)$ such that, if $x \cap y =_e \emptyset$, then $SEP(x,y)$ weakly separates $\langle x,y\rangle$.*

PROOF. Choose $SEP(x,y) := \{u : \langle u,x\rangle \eta \pi E\langle u,y\rangle\}$. □

Since the developments of the chapters to come, only depend upon class compactness and exact representation 16.11 (which only requires 16.9 and CL-compactness), we might replace PW^- by $\mathrm{MF}^- + \{\pi.2, \pi.3\}$.
However, the proofs of §§16-17 illustrate a different way of handling stage arguments, that might perhaps be useful elsewhere. Moreover, the definition

of ξ, given in 16.3, is easier to understand, because it reflects a natural set-theoretic interpretation of ξa, as the collection containing the elements of a with minimal stage.

The previous remarks naturally raise the following problems:

1. find models of $\pi.2$ and $\pi.3$, which are not inductively generated;
2. prove $\pi.2$ and $\pi.3$ from some of their notable consequences (say class compactness) and MF^--axioms;
3. prove that the system of the π-axioms is independent.
4. find significant consequences of $\pi.4$, $\pi.5$ (if the previous problem has positive solution), which cannot be obtained by $\pi.2$, $\pi.3$.

CHAPTER 5

TYPE-FREE ABSTRACTION, CHOICE AND SETS

§20. Choice principles and the distinction between operations and functions
§21. Admissible hulls: elementary facts
§22. A model of admissible set theory
§23. The boundedness theorem

This part concludes the axiomatic investigation of the recursion-theoretic properties of inductive models for reflective truth. We are interested in relating the present framework, extended with approximation axioms and generalized induction, to classical systems, namely theories of sets. Thus the main bulk of this chapter is devoted to the *construction of a model for admissible set theory* within $PW_c + GID$.

In §20 we investigate choice principles, under the assumption of the enumeration axiom 4.14, and we observe that they have a problematic role in the present theory. Inconsistencies underline the difference between the intensional notion of operation and the set-theoretic one.

By the results of §§21-22, an *iterative notion of set* can be adequately defined within the theory, extended with approximation axioms and the generalized induction principle GID. Indeed, we uniformly associate to each property U a structure $AD(U)$, *the admissible hull of* U, which is a model of admissible set theory with urelements in U.

In the final section §23 we show that two natural ways of characterizing the minimal fixed point of a monotone operator are extensionally equivalent ("boundedness" theorem), and it is quite surprising to see that the equivalence result holds in a proof theoretically weak system. The theorem exploits the generalized continuity property of 17.5 and the uniform Knaster-Tarski theorem from the previous chapter, together with the theory of ordinals available from inner set-theoretic models.

The boundedness theorem should add some evidence in favour of the present axiomatic choice: the extended theory is *natural,* because there is a certain harmony between set theory and recursion theory, predicative and impredicative definitions.

§20. Choice principles and the distinction between operations and functions

In this section we investigate the problem of consistently refining the approximation operator π of §16 to a choice operator, in presence of suitable well-orderings of the universe, in particular when the enumeration axiom of §4 is assumed. This problem leads to consider how far the operational axioms can proceed in witnessing "implicitly defined" functions. We shall see that it is essential to distinguish between *operations* and *functions-as-graphs*, and to take into account *the extensional character of the operations involved.*

If the approximation operation π is applied to properties of natural numbers, π can be obviously strengthened to a selection operation and the same happens if the enumeration axiom EA (cf. 4.14 and following) is assumed. We recall that EA is the sentence $\exists f \forall x \exists y(Ny \wedge fy = x)$ and that EA is consistent with PW_c (consider the inductive models $\mathbb{O}(\mathbf{CTM})$ or $\mathbb{O}(RE)$, where **CTM** and RE are introduced in Ch. I).

20.1. DEFINITION

(i) Let w encode a binary relation, i.e. let w be a property of ordered pairs. We keep using the infix notation $x \prec_w y$ in place of $\langle x,y\rangle \eta w$. $Field(\prec_w)$ is the term $\{x : \exists z(x \prec_w z \vee z \prec_w x)\}$ representing the field of $\prec_w$, while the x-segment of $\prec_w$ determined by x is defined by the term $\prec_w \lceil x := \{u : u \prec_w x\}$.

(ii) $$LO(\prec_w) := \forall x \forall y \forall z (\neg(x \prec_w x) \wedge \wedge (x \prec_w y \wedge y \prec_w z \rightarrow x \prec_w z) \wedge Conn(\prec_w)),$$

where $$Conn(\prec_w) := \forall x \forall y (x \eta Field(\prec_w) \wedge \wedge y \eta Field(\prec_w) \rightarrow (x \prec_w y \vee x = y \vee y \prec_w x)).$$

$LO(\prec_w)$ states that $\prec_w$ is a linear ordering.

(iii) $Progr(b, \prec_w) := \forall x (x \eta Field(\prec_w) \wedge \forall y (y \prec_w x \rightarrow y \eta b) \rightarrow x \eta b)$.

$Progr(b, \prec_w)$ is to be read "b is progressive" (relative to $\prec_w$).

We also define:

$$TI(\prec_w, b) := Progr(b, \prec_w) \rightarrow Field(\prec_w) \subseteq b.$$

(iv) A linear ordering $\prec_w$ is called a *pseudo-well-ordering* – in symbols $PWO(\prec_w)$, and, in short, $\prec_w$ is a *pwo* – iff $\forall b(Cl(b) \rightarrow TI(\prec_w, b))$.

(v) A *pwo* $\prec_w$ is *acceptable* iff $\prec_w$ is a class.

(vi) In PW_c the standard ordering $<$ of $\mathbb{N}$ is a pseudo-well-ordering; if we assume the axiom EA and f is any surjective operation from $\mathbb{N}$ onto

$V = \{x : x = x\}$ given by EA, we can define $x <_f y := |x|_f < |y|_f$, where $|x|_f =$ the least k in $\mathbb{N}$ such that $fk = x$.

20.1.1. CONVENTION. It is understood that all the constructions below, which rely upon EA, will be uniform in some fixed witness E for EA, such that

$$\forall x \exists y (Ny \wedge Ey = x).$$

For simplicity, we leave dependence on E, usually implicit.

Of course, if E satisfies the enumeration axiom, it is easy to get:

20.1.2. FACT. $<_E$ is an acceptable *pwo*, provably in $\mathrm{PW}_c + \mathrm{EA}$.

On the other side, we can verify:

20.1.3. FACT (*provable in* PW^-). *$\prec_w$ is an acceptable pwo iff $\prec_w$ is a linear ordering such that*

(i) *$\{\langle x, y\rangle : x \prec_w y\}$ is a class;*

(ii) *every non-empty class $\subseteq Field(\prec_w)$ has a $\prec_w$-least element.*

For the proof of 20.1.3, one applies closure of CL under complement, the assumption that $\prec_w$ is a class and elementary comprehension from 9.7.

20.2. LEMMA (*Selection, provable in* PW^-).
We can define a term $\delta(a, \prec_w)$ such that, if $\prec_w$ is an acceptable pwo and $a \subseteq Field(\prec_w)$, then $\delta(a, \prec_w) \subseteq a$ and, if a is non-empty, $\delta(\prec_w, a)$ is a class, which contains exactly one element of a (i.e. $x\eta\delta(\prec_w, a)$, $y\eta\delta(\prec_w, a)$ imply $x = y$).

PROOF. Set

$$\delta(\prec_w, a) := \{x : x\eta\xi a \wedge \forall y (y \prec_w x \rightarrow \neg y\eta\xi a)\};$$

CL-compactness (16.3) and elementary comprehension 9.7 complete the proof. □

If we apply the selection lemma, binary relations can be uniformized via functions-as-graphs, provably in $\mathrm{PW}_c + \mathrm{EA}$.

20.3. DEFINITION. Let:

$$Funrel(f) := \ Cl(f) \wedge \forall u (u\eta f \rightarrow \exists x \exists y (u = \langle x, y\rangle)) \wedge$$
$$\wedge \forall x \forall y \forall z (\langle x, y\rangle \eta f \wedge \langle x, z\rangle \eta f \rightarrow y = z);$$

Funrel(f) is read " f is (or encodes) a functional binary relation ".

20.4. THEOREM. *We can find a term $Sel(r, b) := Sel(r, b, <_E)$ (cf. 20.1.1 for notations) such that, provably in* $\mathrm{PW}_c + \mathrm{EA}$, *if b is a class and r is a* *binary relation defined on b, then Sel(r,b) is a functional relation which uniformizes r on b; formally*

$$Cl(b) \wedge \forall x(x\eta b \rightarrow \exists y(\langle x,y\rangle \eta r)) \rightarrow$$
$$\rightarrow Sel(r,b) \subseteq r \wedge \forall u\eta b.\exists! y(\langle u,y\rangle \eta Sel(r,b)).$$

PROOF. Define $U(r,x) := \{y : \langle x,y\rangle \eta r\}$ and choose

$$Sel(r,b) := \{u : \exists x \exists y(x\eta b \wedge y\eta\delta(U(r,x), <_E) \wedge u = \langle x,y\rangle)\}. \qquad (*)$$

In $(*)$ $<_E$ is the pseudo-well-ordering which exists by EA, while δ is the operator of lemma 20.2. □

It is natural to ask whether the theorem 20.4 can be strengthened. If $\forall x\eta b.\exists y.A(x,y)$ is assumed, is there *a choice operation* g such that $A(x,gx)$ for every x in b ? The answer is easily seen to be negative in general.

20.5. THEOREM. (i) *The choice schema* $AC_V(op)$ *over the universe* V

$$\forall x \exists y A(x,y) \rightarrow \exists f \forall x A(x,fx) \ (A \text{ in the language of OP})$$

is inconsistent with OP^- (= OP *without N-induction).*

(ii) *The choice axiom for classes* Cl-AC(op)

$$Cl(a) \wedge Cl(r) \wedge \forall x\eta a.\, \exists y\, \langle x,y\rangle \eta r \rightarrow \exists g \forall x\eta a.(\langle x,gx\rangle \eta r)$$

is inconsistent with any theory, which includes elementary comprehension EC and OP^-.

PROOF. (i) By intuitionistic logic plus $(x = y \vee \neg x = y)$, we can prove in OP^-:

$$\forall x \exists! y((y = \bar{1} \wedge x = \bar{0}) \vee (y = \bar{0} \wedge \neg x = \bar{0})). \qquad (1)$$

Were a choice operation f available in OP^-, we should have:

$$\forall x((fx = \bar{1} \wedge x = \bar{0}) \vee (fx = \bar{1} \wedge \neg\, x = \bar{0})). \qquad (2)$$

But (2) yields a "global test for zero" and it leads to an inconsistency via paradoxical combinator (see 3.9).
(ii): apply (i) and elementary comprehension 9.7. □

20.5.1. REMARK. The argument of (i) essentially depends on classical logic (i.e. the decidability of $=$) and on the use of type-free operations; in fact, Barendregt (1973) shows that $AC_V(op)$ is consistent with combinatory logic based on intuitionistic logic. Part (ii) also holds without classical logic: the trick can be found in Beeson (1985), where a survey of consistency results in constructive type-free systems is given (see also Troelstra-van Dalen 1989).

The inconsistency of theorem 20.5 has, however, a conceptual significance: it underlines the limits of the interplay between type-free operations and type-free predicate abstraction. Nevertheless, we can produce choice principles, consistent with the present framework, as soon as we consider *extensional operations*.

First of all, recall that m, n, k range over number-theoretic variables (hence $\forall nA$ and $\exists nA$ stand for $\forall x(Nx \rightarrow A)$, $\exists x(Nx \wedge A)$).

20.6. DEFINITION

(i) $Ext_2(f) := \forall x \forall y(x =_e y \rightarrow fx =_e fy)$;

$$2\text{-}Ext(f) := \forall x \forall y \forall z \forall w(x =_e z \wedge y =_e w \rightarrow fxy =_e fzw).$$

(Here $x =_e y := \forall u(u\eta x \leftrightarrow u\eta y)$); $Ext_2(f) := f$ is extensional$_2$ in the sense of §17); $2\text{-}Ext(f) := f$ is 2-extensional or extensional as a function of two variables).

(ii) Ext-AC is the axiom :

$$\forall b \exists h \forall f(Ext_2(f) \wedge \forall x\eta b.\exists y(Cl(y) \wedge x\eta fy) \rightarrow \forall x\eta b(Cl(hfx) \wedge x\eta f(hfx))).$$

(iii) Ext-DC is the axiom:

$$\exists h \forall f(2\text{-}Ext(f) \wedge \forall n \forall x(Cl(x) \rightarrow \exists y(Cl(y) \wedge n\eta fxy)) \rightarrow$$
$$\rightarrow \forall x(Cl(x) \rightarrow (hfx\bar{0} = x \wedge \forall n(Cl(hfxn) \wedge n\eta f(hfxn)(hfx(n+1)))))).$$

20.7. THEOREM ("*Extensional choice*"). $PW_c + EA \vdash Ext$-AC.

PROOF. Assume that f is extensional$_2$ and let $\forall x\eta b.\exists y(Cl(y) \wedge x\eta fy)$. Put

$$t(f,x) := \{y : Cl(y) \wedge x\eta fy\}.$$

Then, by assumption, we have $\forall x\eta b.\exists y(y\eta t(f,x))$; thus, if we choose $r(f,x) := \delta(t(f,x), <_E)$, where $<_E$ is the *pwo* induced by EA on the universe, lemma 20.2 implies in $PW_c + EA$:

$$r(f,x) \subseteq t(f,x) \wedge \forall x\eta b.(Cl(r(f,x)) \wedge \exists! y(y\eta r(f,x))). \quad (1)$$

Hence $r(f,x)$ is a non-empty class of classes, for every x in b, and if we choose $hfx = \bigcup r(f,x)$ we obtain $\forall x\eta b.Cl(hfx)$ by 9.14. Observe now that, for x in b:

$$y\eta r(f,x) \rightarrow y =_e hfx \quad (2)$$

(apply the fact that $r(f,x)$ has a unique element). If $x\eta b$, (1)-(2) imply $x\eta fy$ for some $y =_e hfx$, hence $Ext_2(f)$ yields $x\eta f(hfx)$. □

20.8. THEOREM("*Extensional dependent choices*"). $PW_p + EA \vdash Ext$-DC.

PROOF. Assume the antecedent of Ext-DC, and let f be 2-extensional.

Then, if $R(n,x,f) := \{y\colon Cl(y) \wedge n\eta(fxy)\}$:

$$\forall n \forall x(Cl(x) \to \exists y(y\eta R(n,x,f))). \tag{1}$$

Then we can find $S(n,x,f) := \delta(R(n,x,f), <_E)$, such that, provably in $\mathrm{PW}_c + \mathrm{EA}$:

$$\forall n \forall x(Cl(x) \to (S(n,x,f) \subseteq R(n,x,f) \wedge$$
$$\wedge\, Cl(S(n,x,f)) \wedge \exists ! y(Cl(y) \wedge y\eta S(n,x,f)))). \tag{2}$$

Then we can find, by primitive recursion on N (cf. 3.2) with parameters x, f, an operation h such that:

$$hfx\bar{0} = x \text{ and } hfx(m+1) = \cup S(m, hfxm, f) \tag{3}$$

($\cup$ = generalized union of 9.14). By N-induction for properties, it easily follows with (2), if x is a class:

$$\forall m(Cl(hfxm) \wedge S(m, hfxm, f) \subseteq CL); \tag{4}$$

(4) and the uniqueness requirement of (2) also yield:

$$\forall y(y\eta S(m, hfxm, f) \to y =_e hfx(m+1)). \tag{5}$$

Hence by property N-induction, definition of h, (5) and 2-extensionality, we can conclude:

$$hfx\bar{0} =_e x \wedge \forall n(n\eta f(hfxn)(hfx(n+1))). \ \square \tag{6}$$

The uniform choice axioms *Ext*-AC and *Ext*-DC entail two schemata for elementary conditions, which are extensional in the relevant parameters. These schemata are significant for interpreting fragments of second-order arithmetic (see Ch. VIII).

20.9. DEFINITION

(a) A formula A of $\mathfrak{L}$ is *elementary extensional* in $x_1, \ldots, x_n$ iff A belongs to the least class of formulas inductively generated by means of $\wedge$, $\neg$, $\forall y$ (y distinct from $x_1, \ldots, x_n$) from atoms of the form $t = s$, Nt, $t\eta x_i$, provided $x_1, \ldots, x_n$ do not occur in t, s (compare with 9.5).

(b) Let $z_x := \{u\colon \langle x, u\rangle \eta z\}$. EAC (= the elementary choice schema):

$$Cl(b) \wedge \forall x\eta b.\ \exists y(Cl(y) \wedge A(x,y)) \to \exists z(Cl(z) \wedge \forall x\eta b. A(x, z_x)),$$

for $A(x,y)$ elementary extensional in y;

(c) EDC(= the elementary dependent choice schema)):

$$\forall n \forall x(Cl(x) \to \exists y(Cl(y) \wedge A(n,x,y))) \to$$
$$\to \forall x(Cl(x) \to \exists z(Cl(z) \wedge z_0 =_e x \wedge \forall n A(n, z_n, z_{n+1}))),$$

for every $A(u,x,y)$ elementary extensional in x,y.

20.9.1. FACT. If $A(u,x_1,\ldots,x_n)$ is elementary extensional in $x_1,\ldots,x_n$ and $=_e$ is extensional equality with respect to η (see 9.11), we can prove in pure logic, for $1 \leq i \leq n$:

$$A(u,x_1,\ldots,x_n) \wedge x_i =_e y_i \to A(u,x_1,\ldots,x_{i-1},y_i,x_{i+1},\ldots,x_n).$$

20.10. COROLLARY

(i) $\mathrm{PW}_c + \mathrm{EA} \vdash \mathrm{EAC}$;
(ii) $\mathrm{PW}_p + \mathrm{EA} \vdash \mathrm{EDC}$.

PROOF. (i) Assume $\forall x\eta b.\exists y(Cl(y) \wedge A(x,y))$, where $A(x,y)$ is elementary extensional in y; then $fy := \{x: A(x,y)\}$ is extensional and by *Ext*-AC there is an operation h such that $Cl(hx)$, whenever $x\eta b$, and such that $A(x,hx)$. Choose $z = \Sigma(b,h)$: then z is a class by join principle 9.9 and $\forall x A(x,z_x)$ (use extensionality of A in y).
(ii): similar argument (apply *Ext*-DC). □

20.11. FINAL REMARK. It may be of interest to restrict the investigation to the special class $b = \mathsf{N}$. Then we can apply 5.13 and we can show that MF_c is consistent with the full N-choice schema AC_{N}:

$$\forall x(x\eta\mathsf{N} \to \exists y A(x,y)) \to \exists f \forall x(x\eta\mathsf{N} \to A(x,fx)) \ (A \text{ arbitrary}).$$

The reader can readily verify in $\mathrm{MF}_c + \mathrm{AC}_{\mathsf{N}}$:

20.11.1. *Let* $Pd(\mathsf{N}) := \{\{x: gx = \overline{0}\} : g\eta 2^{\mathsf{N}}\}$. *Then for every* A, *there exists an element* b *of* $Pd(\mathsf{N})$ *such that* $\forall x(x\eta\mathsf{N} \to (x\eta b \leftrightarrow A(x)))$.

20.11.2. *For every* $a \subseteq \mathsf{N}$ *there is a class* b *in* $Pd(\mathsf{N})$ *such that* $a =_e b$.

§21. Admissible hulls : elementary facts

In the previous sections, we pointed out the fundamental aspects of a theory of abstraction in a world endowed with an approximation structure. We now apply the strengthened machinery to the reconstruction of standard concepts, namely sets.

Henceforth, the general frame theory is $\mathrm{PW}_c + \mathrm{GID}$, which is conservative over Peano arithmetic. Our basic aim is to define models of admissible set theory with urelements (or atoms). Indeed, we shall introduce an extensional operation AD such that $AD(U)$ is a canonical admissible structure over U; $AD(U)$ is called the *admissible hull* of U. If U is a class (in our sense), U will correspond to a set in $AD(U)$. If $U = \mathsf{N}$, we have the counterpart of the "next admissible structure" above natural numbers in the sense of Barwise (1975) (see also Barwise, Gandy, Moschovakis 1971, Moschovakis 1974).

The idea is that atoms are just pairs $\langle x,\overline{0}\rangle$ with x in U, while U-sets have the form $\langle x,\overline{1}\rangle$ where x is a class $\subseteq AD(U)$. Thus $AD(U)$ can be regarded as the solution of natural inductive conditions. As we shall see, *the construction is monotone in* U, *and* AD is actually *functorial*: every injective operation f from U into W can be canonically extended to an embedding $AD(f)$ of $AD(U)$ into $AD(W)$.

21.1. DEFINITION

(i) Let $PAR(a) := (a = \langle (a)_1, (a)_2\rangle)$: we define

$$A_U(x,v) := PAR(x) \wedge [((x)_2 = \overline{0} \ \wedge \ (x)_1 \eta U) \vee ((x)_2 = \overline{1} \wedge \wedge Cl((x)_1) \wedge \forall u(u\overline{\eta}(x)_1 \vee u\eta v))].$$

(ii) Clearly, $A_U(x,v)$ is an operator in v and we can choose, by 10.4:

$AD(U) := Ixv.A_U(x,v)$ (in short $I(A_U)$;

$U\text{-}AT := \{x : PAR(x) \wedge (x)_1 \eta U \wedge (x)_2 = \overline{0}\}$;

$U\text{-}SET := \{x : x\eta AD(U) \wedge (x)_2 = \overline{1}\}$.

If x is in U-AT (U-SET, $AD(U)$), we say that x is an U-atom (U-set, U-object).

21.2. LEMMA (MF^-)

(i) $Cl(U) \rightarrow Cl(U\text{-}AT) \wedge \langle U\text{-}AT, \overline{1}\rangle \eta AD(U)$.

(ii) $\forall x(x\eta AD(U) \leftrightarrow PAR(x) \wedge (((x)_1 \eta U \wedge (x)_2 = \overline{0}) \vee (Cl((x)_1) \wedge \wedge \forall y\eta(x)_1 .\ y\eta AD(U) \wedge (x)_2 = \overline{1})))$.

(iii) $\forall x(x\eta AD(U) \leftrightarrow (x\eta U\text{-}AT \vee x\eta U\text{-}SET))$;

(iv) $\neg(x\eta U\text{-}AT \wedge x\eta U\text{-}SET)$.

(i)-(iv) are straightforward by choice of $AD(U)$ and 10.1.

We now proceed by inductively defining an equality relation $\equiv_U$ on $AD(U)$; $\equiv_U$ is extensional on sets, together with its dual "internal" version $\not\equiv_U$. We put:

$$E^1(u,v,z) := \forall x(x\overline{\eta}(u)_1 \vee \exists y\eta(v)_1 . \langle x,y\rangle \eta z);$$

$$E(u,v,z) := E^1(u,v,z) \wedge E^1(v,u,z).$$

Clearly $E(u,v,z)$ determines an operator in z; if $(u)_1$, $(v)_1$ are classes, we can prove by T-logic:

21.3. $E(u,v,z) \leftrightarrow (\forall x\eta(u)_1.\exists y\eta(v)_1.\langle x,y\rangle\eta z) \wedge (\forall x\eta(v)_1.\exists y\eta(u)_1.\langle x,y\rangle\eta z)$.

Let us define the following formula, which uniformly depends on U and is an operator in z:

$$B_U(u,z) := PAR(u) \wedge (((u)_1\eta U\text{-}SET \wedge (u)_2\eta U\text{-}SET \wedge E((u)_1,(u)_2,z)) \vee$$
$$\vee ((u)_1\eta U\text{-}AT \wedge (u)_2\eta U\text{-}AT \wedge (u)_1 = (u)_2)).$$

Then we find a fixed point $I(B_U) := Iuz.B_U(u,z)$ (by 10.1).

21.4. DEFINITION

(i) $u \equiv_U v := \langle u,v\rangle\eta I(B_U)$; $E_U(u,v) := E(u,v,I(B_U))$;

(ii) $u \not\equiv_U v := u\eta AD(U) \wedge v\eta AD(U) \wedge F(u \equiv_U v)$;

(iii) $DE_U(u,v) := \exists x\eta(u)_1.\forall y(y\overline{\eta}(v)_1 \vee x \not\equiv_U y) \vee$
$\vee \exists x\eta(v)_1.\forall y(y\overline{\eta}(u)_1 \vee x \not\equiv_U y)$.

21.5. LEMMA (MF^-)

(i) $\neg(u \equiv_U v \wedge u \not\equiv_U v)$;

(ii) $u \equiv_U v \leftrightarrow (u\eta U\text{-}SET \wedge v\eta\, U\text{-}SET \wedge E_U(u,v)) \vee$
$\vee (u\eta U\text{-}AT \wedge v\eta U\text{-}AT \wedge u = v)$.

(iii) $u \not\equiv_U v \leftrightarrow (u\eta AD(U) \wedge v\eta AD(U) \wedge (u)_2 \neq (v)_2) \vee$
$\vee (u\eta U\text{-}AT \wedge v\eta U\text{-}AT \wedge u \neq v) \vee$
$\vee (u\eta U\text{-}SET \wedge v\eta U\text{-}SET \wedge DE_U(u,v))$.

PROOF. (i): by T-consistency.
(ii): apply 10.1, 21.3 and the fact that, if u is an $U\text{-}SET$, $(u)_1$ is a class.
(iii) $\Rightarrow$: By T-logic, $F(u \equiv_U v)$ requires distinction of some cases. We only consider the case where u and v are U-sets such that $F(E_U(u,v))$ holds. To be definite, we also suppose $\forall y(y\overline{\eta}(v)_1 \vee F(x \equiv_U y))$, for some x in $(u)_1$. By hypothesis on u and v with 21.2 (iii), $(v)_1$ is a class and x is in $AD(U)$. Hence if $\neg y\overline{\eta}(v)_1$ holds, we also have $y\eta AD(U)$ and $F(x \equiv_U y)$, which imply $x \not\equiv_U y$, i.e. $DE_U(u,v)$.
$\Leftarrow$: Assume that u, v are U-sets and $DE_U(u,v)$. Then a fortiori $F(u\eta AD(U) \wedge (u)_2 = \overline{0})$, i.e. $F(u\eta U\text{-}AT)$, which entails:

$$F(u\eta U\text{-}AT \wedge v\eta U\text{-}AT \wedge u = v). \qquad (1)$$

But $DE_U(u,v)$ yields $FE_U(u,v)$, i.e.

$$F(u\eta U\text{-}SET \wedge v\eta U\text{-}SET \wedge E_U(u,v)). \tag{2}$$

(1)-(2) imply $u \not\equiv_U v$. The other cases are left to the reader. □

21.6. PROPOSITION ($\mathrm{MF}^- + \mathrm{GID}$)

(i) $a\eta AD(U) \rightarrow a \equiv_U a$;

(ii) $\neg a \not\equiv_U a$;

(iii) $(a \equiv_U b \rightarrow b \equiv_U a) \wedge (a \not\equiv_U b \rightarrow b \not\equiv_U a)$;

(iv) $a\eta AD(U) \wedge b\eta AD(U) \rightarrow a \equiv_U b \vee a \not\equiv_U b$;

(v) $a \equiv_U b \wedge b \equiv_U c \rightarrow a \equiv_U c$;

(vi) $a \equiv_U b \wedge b \not\equiv_U c \rightarrow a \not\equiv_U c$.

PROOF. (i): apply the generalized induction schema for $AD(U)$ to the condition $x \equiv_U x$ and 21.2.
(ii): $a \not\equiv_U a$ implies $a\eta AD(U)$, whence $a \equiv_U a$ by (i), which contradicts 21.5 (i).
(iii): trivial by inspection of 21.5 (iii).
(iv): we apply GID on $AD(U)$ to the formula

$$C(a) := \forall b(b\eta AD(U) \rightarrow a \equiv_U b \vee a \not\equiv_U b).$$

Thus we assume $A_U(a, C)$, $b\eta AD(U)$ and we verify $a \equiv_U b$ or $a \not\equiv_U b$. We repeatedly use 21.5 (ii)-(iii). If a is an U-atom and b is an U-atom, we have $a \equiv_U b$ or $a \not\equiv_U b$, according to $a = b$ or $a \neq b$; if a is an U-atom and b is an U-set (or symmetrically), we get $a \not\equiv_U b$. If a, b are U-sets, then $(a)_1$, $(b)_1$ are classes and the following conditions hold:

$$\forall x\eta(a)_1.\forall v(v\eta AD(U) \rightarrow x \equiv_U v \vee x \not\equiv_U v); \tag{1}$$

$$\forall y\eta(b)_1.y\eta AD(U). \tag{2}$$

If $\neg a \equiv_U b$ holds, we have by 21.5 (ii) $\neg E_U(a,b)$; for instance, suppose that there is $x\eta(a)_1$ such that, for every y, either $\neg y\eta(b)_1$ or $\neg x \equiv_U y$. In the first case we obtain $y\overline{\eta}(b)_1$ (classhood of $(b)_1$); if y is in $(b)_1$, $\neg x \equiv_U y$ holds, whence $x \not\equiv_U y$ by (1)-(2). As a consequence, $DE_U(a,b)$ holds and finally $a \not\equiv_U b$.
(v)-(vi): argue by GID, 21.5, (iii)-(iv) above. □

21.7. DEFINITION (Extensional membership and its dual relation).

$$x \in_U y := \exists v(v \equiv_U x \wedge v\eta(y)_1 \wedge (y)_2 = \overline{1});$$

$$x \mathrel{\overline{\in}}_U y := x\eta AD(U) \wedge \forall v(v \not\equiv_U x \vee v\overline{\eta}(y)_1 \vee (y)_2 \neq \overline{1}).$$

If we systematically apply 21.6 (e.g. you need 21.6 (iv) to check (v) below), we have that $\equiv_U$ is a congruence with respect to $\in_U, \overline{\in}_U$ and that $\lambda xy.[x \in_U y]$ is a propositional function, if it is restricted to elements of $AD(U)$:

21.8. LEMMA ($\mathrm{MF}^- + \mathrm{GID}$).

(i) $\neg(x \in_U y \wedge x \overline{\in}_U y)$;

(ii) $x\eta(y)_1 \wedge (y)_2 = \bar{1} \wedge (x\eta AD(U) \vee y\eta AD(U)) \to x \in_U y$;

(iii) $x \in_U y \wedge y \equiv_U z \to x \in_U z$; $\quad x \in_U y \wedge x \equiv_U z \to z \in_U y$;

(iv) $x \overline{\in}_U y \wedge y \equiv_U z \to x \overline{\in}_U z$; $\quad x \overline{\in}_U y \wedge x \equiv_U z \to z \overline{\in}_U y$;

(v) $x\eta AD(U) \wedge y\eta AD(U) \to x \in_U y \vee x \overline{\in}_U y$;

(vi) $y\eta AD(U) \wedge a \in_U y \to y\eta U\text{-}SET$.

As one might expect, $\equiv_U$ and $\in_U$-extensional equality coincide on U-sets: if we define $x \subseteq_U y := \forall w \in_U x.\ w \in_U y$, we can prove:

21.9. LEMMA (*Extensionality*)

$$\mathrm{MF}^- + \mathrm{GID} \vdash (x\eta U\text{-}SET \wedge y\eta U\text{-}SET \wedge x \subseteq_U y \wedge y \subseteq_U x) \to x \equiv_U y.$$

PROOF: let w in $(x)_1$; then w is in $AD(U)$ and hence $w \equiv_U w$ (21.6 (i)), which implies $w \in_U x$. By assumption $w \in_U y$, whence $w\eta(y)_1$; with similar arguments, we obtain $\forall u\eta(y)_1.\exists v\eta(x)_1.u \equiv_U v$ and hence $E_U(x,y)$. Then $x \equiv_U y$ follows from 21.5 (ii). □

21.10. PROPOSITION ($\mathrm{PW}^- + \mathrm{GID}$)

If $\mathfrak{I}(U) := AD(U), \equiv_U, \not\equiv_U, \in_U, \overline{\in}_U$,

$$U \subseteq W \to \mathfrak{I}(U) \subseteq \mathfrak{I}(W). \qquad (*)$$

In particular:

$$\forall x(x\eta AD(U) \to \exists c(Cl(c) \wedge c \subseteq U \wedge x\eta AD(c)). \qquad (**)$$

PROOF. (**) is an application of the approximation theorem 17.5 and (*); (*) is verified by GID on the definition of $AD(U)$ (also use 21.2 (ii) and 21.5 (ii)-(iii)). □

According to 21.10, it is not restrictive to consider only *classes of urelements*; if U is not a class, no new set is generated in $AD(U)$, unless it is not already in some $AD(W)$, where W is a class $\subseteq U$. More generally, in order to check $x \in_U y$ (where also y is in $AD(U)$) or $x \equiv_U y$, we only need a collection c of U-atoms, which contains x, y and is a class.

21.11. DEFINITION

(i) $Inj(f,U,W) :=$ f is an injective operation from U into W $:=$

$$= f : U \to W \wedge \forall x \forall y (x \eta U \wedge y \eta U \wedge fx = fy \to x = y).$$

(ii) $Embed(f, AD(U), AD(W)) :=$ f embeds $AD(U)$ into $AD(W)$ $:=$

$$= \forall x \forall y ((x \equiv_U y \to fx \equiv_W fy) \wedge (x \not\equiv_U y \to fx \not\equiv_W fy)) \wedge$$

$$\wedge \forall x \forall y (y \eta AD(U) \to (x \in_U y \to fx \in_W fy) \wedge (x \bar{\in}_U y \to fx \bar{\in}_W fy)) \wedge$$

$$\wedge \forall x ((x \eta U\text{-}SET \to (fx) \eta W\text{-}SET) \wedge (x \eta U\text{-}AT \to (fx) \eta W\text{-}AT)).$$

(iii) $in(x) := \langle x, \overline{0} \rangle$.

Clearly $\lambda x.in(x)$ is an injective operation from U into $AD(U)$.

21.11.1. REMARK. If f is an embedding of $AD(U)$ into $AD(W)$, we can reverse the arrows in 21.11 (ii), as a consequence of 21.8 (v), 21.6 (iv), 21.2 (iii); for instance, we have:

$$fx \equiv_W fy \to x \equiv_U y \text{ , whenever } x, y \text{ are in } AD(U).$$

Now every injective map f of U into W can be canonically extended to an embedding $AD(f)$ between the corresponding admissible hulls; $AD(f)$ is completely determined by f and its defining recursive conditions. Indeed, let

$$Recur(h, U, W) := \forall a (a \eta U\text{-}SET \to ha \equiv_W \langle \{hy : y \eta (a)_1\}, \overline{1} \rangle);$$

$f \circ g$ stands for the composition $\lambda x (f(gx))$ of f and g; id_U is the restriction of the identity map to U. Then we obtain:

21.12. THEOREM ($PW^- + GID$). *We can define a closed term $\lambda f.AD(f)$ such that:*

(i) $Inj(f,U,W) \to Embed(AD(f), AD(U), AD(W)) \wedge$

$$\wedge \forall x (x \eta U \to AD(f)(in(x)) = in(fx));$$

(ii) $Inj(f,U,W) \wedge Recur(h,U,W) \wedge \forall a (a \eta U \to h(in(a)) \equiv_W in(fa)). \to$

$$\to \forall x (x \eta AD(U) \to ha \equiv_W AD(f)(a)).$$

(iii) *AD is functorial, i.e. AD preserves identity maps and composition:*

$$\forall a (a \eta AD(U) \to AD(id_U)(a) \equiv_U id_{AD(U)}(a));$$

$$Inj(g,U,W) \wedge Inj(f,W,Z) \to Embed(AD(f \circ g), AD(U), AD(Z)) \wedge$$

$$\wedge \forall a (a \in AD(U) \to AD(f \circ g)(a) \equiv_Z (AD(f) \circ AD(g))(a)).$$

PROOF. (i) By the fixed point theorem 2.3 and definition by cases on $\mathbb{N}$, we can find an operation AD such that

$$\text{if } a \in U\text{-}AT,\ (ADf)a = \langle f(a)_1, \overline{0}\rangle; \tag{1.1}$$

$$\text{if } a \in U\text{-}SET,\ (ADf)a = \langle \{(ADf)y : y\eta(a)_1\}, \overline{1}\rangle. \tag{1.2}$$

Let $f: U \to W$ be injective; we have to check that $AD(f) := \lambda x.(ADf)x$ is an embedding of $AD(U)$ into $AD(W)$. First of all, GID with 21.2 (ii) and (1.1)-(1.2) immediately yields that U-atoms (U-sets) are sent by $AD(f)$ into W-atoms (W-sets). $a \not\equiv_U b$ implies $AD(f)a \not\equiv_W AD(f)b$ (a, $b \in AD(U)$). In fact, let

$$C(a) := \forall b(b\eta AD(U) \wedge (((a)_2 \neq (b)_2) \vee ((a)_2 = (b)_2 \wedge a \neq b) \vee$$
$$\vee ((a)_2 = (b)_2 = \overline{1} \wedge DE_U(a,b))) \to AD(f)(a) \not\equiv_W AD(f)b).$$

By GID, it is enough to prove that $C(a)$ follows from the assumption of $A_U(a, C)$ (A_U being the operator defining $AD(U)$).

If b is in $AD(U)$ and $(a)_2 \neq (b)_2$, $AD(f)(a) \not\equiv_U AD(f)(b)$ holds since $AD(f)$ preserves "category" by (1.1)-(1.2). If $(a)_2 = (b)_2 = \overline{0}$ and $a \neq b$, then $(a)_1 \neq (b)_1$ and hence $f(a)_1 \neq f(b)_1$ by injectivity, which implies by definition $AD(f)(a) \neq AD(f)(b)$. Since $AD(f)(a)$ and $AD(f)(b)$ are W-atoms, we can conclude $AD(f)(a) \not\equiv_W AD(f)(b)$.

If a and b are U-sets and $DE_U(a,b)$ holds, let us suppose (for instance) that for some v in $(b)_1$, either $u\overline{\eta}(a)_1$ or $v \not\equiv_U u$ (u arbitrary).
Since $(a)_1$ is a class, $C(u)$ and $v \not\equiv_U u$, for every u in $(a)_1$, which implies $AD(f)(u) \not\equiv_W AD(f)(v)$; hence, if we choose $z = AD(f)(v)$, we conclude $\forall y(y\overline{\eta} AD(f)(a) \vee z \not\equiv_W y)$, i.e. $AD(f)(a) \not\equiv_W AD(f)(b)$. The remaining conditions are left as exercise.

(ii)-(iii): straightforward by GID and the basic properties of $\equiv_U$ (see 21.5-21.6). □

§22. A model of admissible set theory

We show – within PW_c + GID – that $AD(V)$, where V is the universal class, is a model of *admissible set theory above the ground combinatory structure*. An admissible set above a given model $\mathcal{M}$ of OP is basically a two-sorted structure with M, the universe of $\mathcal{M}$, as *domain of urelements*, while the collections of sets always includes M itself and is closed under pairing, union, bounded separation and bounded collection (see below for details). Sets are well-founded, in the sense that they satisfy forms of $\in$-induction and number-theoretic induction.

In order to axiomatize admissible structures of this sort, it is convenient to consider a purely relational variant $\mathcal{L}_r$ of $\mathcal{L}_{op}$, the language of OP without T-predicate (see §1), which coincides with $\mathcal{L}_{op}$, except for replacing the application symbol Ap with a new 3-ary relation symbol App. Thus the terms of $\mathcal{L}_r$ are simply individual variables and individual constants of $\mathcal{L}_{op}$. In addition to Nt, $t = s$, $\mathcal{L}_r$ has the new atom $App(t, s, r)$, with the same meaning as the former $ts = r$. We then extend $\mathcal{L}_r$ to a new language $\mathcal{L}_s$ with a binary relation symbol for set-theoretic membership, to be denoted with $\in$, and two unary predicate symbols Set, Ur, which classify sets and urelements (respectively). Of course, $\mathcal{L}_s$ has atoms of the form $t \in s$, $Set(t)$, $Ur(t)$, besides those of $\mathcal{L}_r$; $\mathcal{L}_s$-terms coincide with $\mathcal{L}_r$-terms and formulas are inductively generated from atoms by means of classical connectives and quantifiers. We systematically adopt the abbreviations:

$$\forall x \in t.\, A := \forall x(x \in t \rightarrow A); \quad \exists x \in t.A := \exists x(x \in t \wedge A).$$

The basic equality $=$ is well-defined on urelements, while extensional equality is adopted for sets; therefore we can introduce a *general equality relation*:

22.1 $$x \equiv y := (Ur(x) \wedge Ur(y) \wedge x = y) \vee$$
$$\vee (Set(x) \wedge Set(y) \wedge \forall u \in x\,.\, u \in y \wedge \forall u \in y.\, u \in x).$$

For convenience, we keep using $\mathcal{L}_{op}$-terms (in short *applicative terms*), as metamathematical abbreviations. Applicative terms (denoted by t, s, r) can be explained away according to the following contextual definitions:

22.2. $$(xy = z) := App(x, y, z);$$
$$(x = ts) := \exists u \exists v(u = t \wedge v = s \wedge x = uv);$$
$$(t = s) := \exists u(u = t \wedge u = s); \quad A(t) := \exists u(u = t \wedge A(u)).$$

22.3

(a) The collection of bounded formulas (of $\mathcal{L}_s$) is the smallest collection $\mathcal{K}$ containing $\mathcal{L}_s$-atoms and closed under the clauses:

(i) if A is in $\mathcal{K}$, so is $\neg A$;
(ii) if A, B are in $\mathcal{K}$, so is $A \wedge B$;
(iii) if A is in $\mathcal{K}$, then so are $\forall x \in t.\, A$ and $\exists x \in t.\, A$ (t arbitrary term of $\mathcal{L}_s$).

(b) If $B(x)$ is a formula with only the free variable shown and A is an arbitrary formula (B, A in $\mathcal{L}_s$), the *relativization* A^B of A to B is obtained by replacing each unbounded quantifier Qx occurring in A with Qx^B, where

$$\exists x^B C := \exists x(B(x) \wedge C),\ \forall x^B C := \forall x(B(x) \rightarrow C).$$

22.4. *The theory* KPU(op) (*Kripke-Platek set theory above a model of* OP)

It consists of elementary classical logic and the universal closures of the following:

(General equality)

$$a \equiv b \rightarrow A[x := a] \rightarrow A[x := b] \ (A\ \mathcal{L}_s\text{-atom});$$

(Ontological axioms)

$$Set(x) \leftrightarrow \neg Ur(x);$$

$$(Nx \rightarrow Ur(x)) \wedge (x = y \rightarrow Ur(x) \wedge Ur(y)) \wedge (x \in y \rightarrow Set(y));$$

$$App(x, y, z) \rightarrow Ur(x) \wedge Ur(y) \wedge Ur(z);$$

$Ur(c)$, for each individual constant c of $\mathcal{L}_{op}$;

$$\forall x \forall y (Ur(x) \wedge Ur(y) \rightarrow \exists z (Ur(z) \wedge App(x, y, z)));$$

$$\forall x \forall y \forall z \forall w (App(x, y, z) \wedge App(x, y, w) \rightarrow z = w).$$

(Combinatory axioms)

A^{Ur}, for each axiom A of OP^-;

($\mathrm{OP}^- = \mathrm{OP}$ without N-induction, see §1; observe that, by the convention on applicative terms, each axiom of OP can be regarded as a formula of $\mathcal{L}_r$).

(V is a set) $\exists a (Set(a) \wedge \forall x (x \in a \leftrightarrow Ur(x)));$

(Pair) $\forall x \forall y \exists a (Set(a) \wedge x \in a \wedge y \in a);$

(Union) $\forall a (Set(a) \rightarrow \exists b (Set(b) \wedge \forall x \in a. \forall y \in x. y \in b));$

(Bounded separation)

$$\forall a (Set(a) \rightarrow \exists b (Set(b) \wedge \forall x (x \in b \leftrightarrow x \in a \wedge A(x)))),$$
for all bounded A, in which b is not free;

(Bounded collection)

$$Set(a) \wedge \forall x \in a. \exists y A(x, y) \rightarrow \exists b (Set(b) \wedge \forall x \in a. \exists y \in b. A(x, y)),$$
for all bounded A, in which b is not free;

($\in$-transfinite induction)

$$\forall x (\forall y \in x. A(y) \rightarrow A(x)) \rightarrow \forall x A(x) \ (A \text{ arbitrary});$$

(Bounded number-theoretic induction)

$$\forall x (N(x) \wedge A(x) \rightarrow A(x+1)) \wedge A(\overline{0}) \rightarrow \forall x (Nx \rightarrow A(x)),$$
for all bounded A.

22.5. *Translation of $\mathcal{L}_s$ into $\mathcal{L}$.* We inductively define a map ${}^\tau : \mathcal{L}_s \to \mathcal{L}$:

(i) if c is an individual constant of $\mathcal{L}_s$ (hence of $\mathcal{L}_{op}$), $(c)^\tau = \langle c, \overline{0} \rangle$;

(ii) the map ${}^\tau$ is the identity map on variables;

(iii) $(t = s)^\tau = t^\tau \eta V\text{-}AT \wedge s^\tau \eta V\text{-}AT \wedge \ (t^\tau)_1 = (s^\tau)_1$;

(iv) $(Ur(t))^\tau = t^\tau \eta V\text{-}AT$;

(v) $(Nt)^\tau = N(t^\tau)_1 \wedge t^\tau \eta V\text{-}AT$;

(vi) $(Set(s))^\tau = s^\tau \eta V\text{-}SET$ (see 21.1);

(vii) $(t \in s)^\tau = t^\tau \in_V s^\tau$;

(viii) $(App(t, r, s)^\tau = t^\tau \eta V\text{-}AT \wedge s^\tau \eta V\text{-}AT \wedge$
$\wedge\, r^\tau \eta V\text{-}AT \wedge (t^\tau)_1 (r^\tau)_1 = (s^\tau)_1$;

(ix) $(\neg A)^\tau = \neg (A)^\tau$; $\quad (A \wedge B)^\tau = (A)^\tau \wedge (B)^\tau$;

(x) $(\forall x\, A)^\tau = \forall x (x \eta AD(V) \to A^\tau)$; $(\exists x A)^\tau = \exists x (x \eta AD(V) \wedge A^\tau)$.

In 22.5, $AD(V)$, $V\text{-}AT$, $V\text{-}SET$, $\in_V$ are the notions defined in the previous section.

We can now state the main interpretation result:

22.6. THEOREM. *If* KPU(op) $\vdash A(x_1, \ldots, x_n)$ *and the free distinct variables of A occur in the list $x_1, \ldots, x_n$, then*

$$\mathrm{PW}_c + \mathrm{GID} \vdash x_1 \eta AD(V) \wedge \ldots \wedge x_n \eta AD(V) \to A^\tau(x_1, \ldots, x_n).$$

In words, $AD(V)$ is a model of KPU(op), *provably in* $\mathrm{PW}_c + \mathrm{GID}$.

The proof of 22.6 requires two preliminary lemmata.

22.6.1. LEMMA (*Soundness of the equality axioms of* KPU(op))

If A is a formula of KPU(op), $\mathrm{MF}^- + \mathrm{GID}$ *proves:*

$$x \eta AD(V) \wedge y \eta AD(V) \wedge (x \equiv y)^\tau \to (A^\tau[u := x] \leftrightarrow A^\tau[u := y])$$
(x, y free for u in A).

PROOF: by induction on A. We first assume that A is an atom of $\mathcal{L}_s$.
Let $A[u := x]^\tau = (x \in_V a)$ and $(x \equiv y)^\tau$.
Case 1: x, y are V-atoms and $(x)_1 = (y)_1$, i.e. $x = y$. Then by 21.5 (ii) $x \equiv_V y$, whence $A[u := y]^\tau = (y \in_V a)$ by 21.8 (iii).
Case 2: x, y are V-sets and $x \subseteq_V y$, $y \subseteq_V x$ (where in general $x \subseteq_V y :=$ $\forall u \in_V x.\, u \in_V y$). By extensionality 21.9, $x \equiv_V y$, whence again by 21.8 (iii) we get $x \in_V a$. The argument for $A = (a \in_V x)$ is similar.

Let $A[u := x]^\tau = ((x)_1(u)_1 = (v)_1)$ and let x, u, v be V-atoms. Then the assumption $(x \equiv y)^\tau$ implies $(x)_2 = (y)_2$ and $(x)_1 = (y)_1$, whence y is a V-atom and by identity logic $((y)_1(u)_1 = (v)_1)$, i.e. $A[u := y]^\tau$.
The remaining atomic cases are left as exercise; the induction step follows without difficulty with induction hypothesis. □

If A is a bounded formula of $\mathcal{L}_s$, let A^τ_+ be the $\mathcal{L}$-formula which results from A^τ by replacing each bounded quantifier

$$\forall x \in_V a \;\text{ with } \forall x(x\eta(a)_1 \wedge (a)_2 = \bar{1} \rightarrow \ldots)$$

$$\{\text{respectively } \exists x \in_V a \;\text{ with } \exists x(x\eta(a)_1 \wedge (a)_2 = \bar{1} \wedge \ldots)\}.$$

Then we obtain:

22.6.2. LEMMA. *For every bounded $\mathcal{L}_s$-formula A,* $\mathrm{MF}^- + \mathrm{GID}$ *proves:*

(i) $x_1\eta AD(V) \wedge \ldots \wedge x_n\eta AD(V) \rightarrow TA^\tau_+(x_1, \ldots, x_n) \vee FA^\tau_+(x_1, \ldots, x_n)$;

(ii) $x_1\eta AD(V) \wedge \ldots \wedge x_n\eta AD(V) \rightarrow A^\tau_+(x_1, \ldots, x_n) \leftrightarrow A^\tau(x_1, \ldots, x_n)$.

PROOF

(i): by induction on the definition of bounded formula of $\mathcal{L}_s$. If A is an atom, $A^\tau = A^\tau_+$; we must check that each A^τ is always well-defined on elements of $AD(V)$. We only consider two cases.
If A has the form $App(x, y, z)$, assume that x, y and z are in $AD(V)$ and $\neg T(App(x, y, z)^\tau)$. Then, by T-logic and definition of the $^\tau$-translation, we must have:

$$\neg x\eta V\text{-}AT \vee \neg y\eta V\text{-}AT \vee \neg z\eta V\text{-}AT \vee \neg(x)_1(y)_1 = (z)_1,$$

which implies, again by T-logic,

$$F(x\eta AD(V) \wedge y\eta AD(V) \wedge z\eta AD(V) \wedge (x)_1(y)_1 = (z)_1) := F(App(x, y, z)^\tau.$$

If $A = (x \in y)$, x, y are in $AD(V)$ and $\neg T(x \in_V y)$, we have, for arbitrary v:

$$\neg v \equiv_V x \vee \neg v\eta(y)_1 \vee \neg(y)_1 = \bar{1}.$$

If $\neg(y)_2 = \bar{1}$, also $F((y)_2 = \bar{1})$ and hence $F(x \in_V y)$ by T-logic. Therefore we can assume that y is a V-set and hence that $(y)_1$ is a class. As a consequence, $\neg v\eta(y)_1$ implies $F(x \in_V y)$. Otherwise, if $v\eta(y)_1$ and $(y)_2 = \bar{1}$, v is in $AD(V)$; hence $\neg v \equiv_V x$ and $v\eta AD(V)$ imply by 21.6 (iv) $F(u \equiv_V x)$, i.e. $F(x \in_V y)$.
Let A be of the form $\forall x \in y.\, B(x, a)$ and assume that y and a are in $AD(V)$, and

$$\neg TA^\tau_+(y, a) = \neg T(\forall x(x\eta(y)_1 \wedge (y)_2 = \bar{1} \rightarrow B^\tau_+(x, a))). \qquad (*)$$

By $(*)$, T-logic and the fact that $(y)_1$ is a class, there is an x such that $(y)_2 = \bar{1}$, $\neg x\bar{\eta}(y)_1$ and $\neg TB^{\tau}_{+}(x,a)$. Then $x\eta(y)_1$ and hence, as y is a V-set, also $x\eta AD(V)$. Since $a\eta AD(V)$, we have by IH $FB^{\tau}_{+}(x,a)$, and by T-logic we can infer $FA^{\tau}_{+}(y,a)$. If A is a conjunction or a negation, the conclusion is an immediate consequence of the induction hypotheses and T-logic.

(ii) is trivial in the atomic case. Again, let A be of the form $\forall x \in y.\ B(x)$. First assume that $(\forall x \in y. B(x))^{\tau} = \forall x \in_V y. B^{\tau}(x)$ where y is in $AD(V)$ and let $x\eta(y)_1$, $(y)_2 = \bar{1}$. Since x is in $AD(V)$, $x \equiv_V x$ holds (21.6 (i)) and hence $x \in_V y$ (definition 21.7). By assumption, we have $B^{\tau}(x)$. Conversely, assume $(\forall x \in y.\ B(x))^{\tau}_{+}$ and let v such that $v \equiv_V x$, $v\eta(y)_1$ and $(y)_2 = \bar{1}$; then we get $B^{\tau}_{+}(v)$ and by IH also $B^{\tau}(v)$, whence $B^{\tau}(x)$ by the equality lemma 22.6.1. □

PROOF of Theorem 22.6. The verification of the ontological and combinatory axioms is straightforward by definition of the $^{\tau}$-translation and by the basic properties of $AD(V)$, $V\text{-}SET$ and $V\text{-}AT$ (see 21.2 (iii)). The equality axiom is already justified by 22.6.1.

Pairing axiom. Let x, y in $AD(V)$ and choose $b = \langle\{u : u = x \vee u = y\}, \bar{1}\rangle$. Clearly $(b)_1$ is a class of elements of $AD(V)$ and hence b is in $AD(V)$ (21.2 (ii)). On the other hand, x, y are in $(b)_1$ and $x \equiv_V x$, $y \equiv_V y$, whence $x \in_V b$ and $y \in_V b$.

Union axiom. If a is a V-set, choose $b = \langle\{u : \exists y\eta(a)_1. x\eta(y)_1 \wedge (y)_2 = \bar{1}\}, \bar{1}\rangle$. Clearly b is a V-set; if $x \in_V y$ and $y \in_V a$, it is immediate to see that there is a $v \equiv_V x$ such that $v\eta(b)_1$ (apply the definition of $\in_V$, 21.8 (iii)).

Bounded separation. If c is a V-set and $A(x)$ is a bounded formula of $\mathfrak{L}_s$, we choose $b = \langle\{x : x\eta(c)_1 \wedge A^{\tau}_{+}(x)\}, \bar{1}\rangle$.
$(b)_1$ is a class by lemma 22.6.2 (i) and it only contains elements of $AD(V)$: hence b is a V-set. If x is in $AD(V)$, $x \in_V b \leftrightarrow x \in_V c \wedge A^{\tau}(x)$ holds by 22.6.2 (ii) and 22.6.1.

Bounded collection. Let A be bounded and assume

$$\forall x \in_V a. \exists y(y\eta AD(V) \wedge A^{\tau}(x, y)), \text{ where } a\eta AD(V). \tag{1}$$

We can also suppose that a is a V-set; hence

$$\forall x\eta\,(a)_1. \exists y(y\eta AD(V) \wedge A^{\tau}(x, y));$$

by 22.6.2 (i)-(ii),

$$\forall x\eta\,(a)_1.\ \exists y(y\eta AD(V) \wedge TA^{\tau}_{+}(x,y)). \tag{2}$$

As a consequence of CL-reflection 16.3.1, there exists a class $b \subseteq AD(V)$, such that, again by 22.6,

$$\forall x \eta (a)_1 . \exists y \eta b. A^\tau(x,y).$$

Then, by construction, $c = \langle b, \overline{1} \rangle$ is a V-set such that:

$$\forall x \in_V a. \exists y \in_V c. A^\tau(x,y).$$

Bounded number-theoretic induction. Let A be a bounded condition of $\mathcal{L}_s$, such that $(A(\overline{0}) \wedge \forall x(Nx \wedge A(x) \to A(x{+}1)))^\tau$: then by definition of $^\tau$-translation with $V\text{-}AT \subseteq AD(V)$, we obtain

$$A^\tau(\langle \overline{0}, \overline{0} \rangle) \wedge \forall x (x \eta V\text{-}AT \wedge N(x)_1 \wedge A^\tau(\langle x, \overline{0} \rangle) \to A^\tau(\langle x{+}1, \overline{0} \rangle)) \qquad (3)$$

Consider $c = \langle \{x : x \eta V\text{-}AT \wedge N(x)_1 \wedge A^\tau_+(x)\}, \overline{1} \rangle$; then $(c)_1$ is a class by 22.6.2 (i) and every element of $(c)_1$ is a V-atom. Hence we can apply number-theoretic induction in MF_c to the class $\{x : \langle x, \overline{0} \rangle \eta (c)_1\}$ and we easily obtain, with (3) and 22.6.2 (ii),

$$\forall x (Nx \to \langle x, \overline{0} \rangle \eta (c)_1),$$

which immediately implies $(\forall x(Nx \to A(x))^\tau$.

Existence of V as a set of urelements: $(\exists b(Set(b) \wedge \forall x(x \in b \leftrightarrow Ur(x))))^\tau$ holds by 21.2 (i) (choose $b = \langle V\text{-}AT, \overline{1} \rangle$).

$\in$-Transfinite induction schema. Assume $(\forall x(\forall y \in x. B(y) \to B(x)))^\tau$, where B is arbitrary; then we have:

$$\text{for every } x \text{ in } AD(V),\ \forall y (y \eta AD(V) \to (y \in_V x \to B^\tau(y)) \to B^\tau(x)). \qquad (4)$$

We apply GID to $AD(V)$; if we put $C(x) := x \eta AD(V) \to B^\tau(x)$, it is enough to verify that $A_V(x, C)$ implies $B^\tau(x)$ under the assumption $x \eta AD(V)$ (here A_V is the operator of 20.1). If $(x)_1 = \overline{0}$, x is a V-atom and hence x is in $AD(V)$; by definition 21.7, we also have $\neg y \in_V x$ and (4) trivially yields $B^\tau(x)$ and $C(x)$. If $(x)_2 = \overline{1}$, $(x)_1$ is a class and $\forall y(y \eta (x)_1 \to C(y))$ holds, we also have $\forall y \in_V x. B^\tau(y)$ (apply 21.6 (i), 21.7, 22.6.1); hence by (4) $B^\tau(x)$ holds. □

The previous theorem easily implies a kind of conservation result of KPU(op) over OP. To this aim, we identify in the set theoretic language of KPU(op) the formulas, which correspond to usual $\mathcal{L}_{op}$-formulas.

22.7. DEFINITION. $\mathrm{FOR}_{op}(\mathcal{L}_s)$ is the smallest collection of $\mathcal{L}_s$-formulas, which is generated from atoms of the form $Ur(t)$, Nt, $App(t,r,s)$, $t = s$ by application of $\neg$, $\wedge$ and quantifiers restricted to urelements $\forall x(Ur(x) \to \ldots)$.

22.7.1. FACT. If $A \in \mathrm{FOR}_{op}(\mathcal{L}_s)$, A^τ is provably equivalent (say, in MF^-) to a formula $A^{\tau_0} \in \mathcal{L}_{op}$.

PROOF: by definition 22.5 , observing that $V\text{-}AT$ is definable is $\mathcal{L}_{op}$. □

Hence, if we apply 22.6 and the general conservation theorem 15.5, we obtain:

22.8. COROLLARY. *If* $A \in FOR_{op}(\mathcal{L}_s)$ *and* KPU(op) $\vdash A$, *then* OP $\vdash A^{\tau_0}$.

22.9. REMARK

(a) 22.8 can be refined and one can show that KPU(op) and OP have the same proof theoretic strength. As to related results, Jäger (1984) shows that the theory KPU^r of admissible sets above natural numbers, with number-theoretic induction and $\in$-transfinite induction restricted to sets, is conservative over Peano arithmetic.

(b) The AD-construction can be used to calibrate the strength of several intermediate options, insofar as we strengthen N-induction and we restrict GID. For instance, if we replace number-theoretic induction for classes with the corresponding axiom for properties, $AD(V)$ validates the existence of an infinite ordinal and a schema of N-dependent choices for Σ_1-formulas (Σ_1-formulas have the form $\exists x A$, where A is bounded). If we only accept the schema GID for conditions of the form $x\eta a$, $AD(V)$ verifies $\in$-induction only for Σ_1-conditions (for more information, see Jäger 1980, 1982, 1986; Simpson 1982; Cantini 1982, 1983, 1985; Rathjen 1992). If we extend OP with the enumeration axiom, $AD(V)$ verifies a choice schema for relations defined by bounded conditions.

§23. The boundedness theorem

In classical mathematics, there are two equivalent ways of handling inductively defined sets. According to the Frege-Dedekind impredicative style, if we are given a monotone operator $F: \mathcal{P}(U) \rightarrow \mathcal{P}(U)$ (where $\mathcal{P}(U) =$ power set of a given non-empty set U), $I(F)$, the set inductively generated by F, is identified with $\cap\{X \subseteq U : F(X) \subseteq X\}$. On the other hand, $I(F)$ can be obtained from below by iterating F on ordinal numbers, i.e. $I(F) = \cup\{I(F,\alpha) : \alpha \in ON\}$, where $I(F,\alpha) = F(\bigcup_{\beta < \alpha} I(F,\beta))$.

This situation has a satisfactory counterpart in $\mathrm{PW}^- + \mathrm{GID}$, as a byproduct of the continuity theorem, the generalized induction principle GID and the fact that a non-trivial theory of ordinals is available in the system, by the main result of the preceding section.

First of all, we single out the collection ON of ordinals within the model $AD(\emptyset)$ of *pure* admissible set theory (see 21.5-22.6). Without urelements, a simpler definition of the notion of set suffices (see 10.2):

23.1. LEMMA. *We can find a term* IS *such that* $\mathrm{MF}^- + \mathrm{GID}$ *proves:*

(i) $$\forall x(x\eta IS \leftrightarrow Cl(x) \wedge \forall y(y\eta x \rightarrow y\eta IS));$$

(ii) *if* $B(x)$ *is an arbitrary formula,*

$$\forall x(Cl(x) \wedge \forall y(y\eta x \rightarrow B(y)) \rightarrow B(x)). \rightarrow \forall x(x\eta IS \rightarrow B(x)).$$

Now let KP^- be the set theory, which includes extensionality, pairing, union, full $\in$-induction schema, Δ_0-separation and Δ_0-collection schemes (in the pure set-theoretic language with only membership as primitive predicate). By a straightforward adaptation of 21.4-21.7 and 22.6, we can find a term $\equiv$, which inductively defines extensional equality on IS: $\equiv$ is the least relation such that, provably in $\mathrm{PW}^- + \mathrm{GID}$:

23.2. $$x \equiv y \leftrightarrow (x\eta IS \wedge y\eta IS \wedge \forall u\eta x. \exists v\eta y.\, u \equiv v \wedge \forall u\eta y. \exists v\eta x.\, u \equiv v).$$

If $x \in y := \exists u(u\eta y \wedge u \equiv x)$, we can rephrase 22.6 as follows:

23.3. PROPOSITION. *The structure* $\langle IS, \in, \equiv \rangle$ *is a model of* KP^-, *provably in* $\mathrm{PW}^- + \mathrm{GID}$.

Now let:

$$Trans(a) := \forall u \in a. \forall v \in u.\, v \in a;$$

$$Ord(x) := x\eta IS \wedge \forall u\eta x. \forall v\eta u.\, v \in x \wedge \forall u\eta x. \forall v\eta u. \forall w\eta v.\, w \in u;$$

$$ON := \{x : Ord(x)\}.$$

23.4. LEMMA. $\mathrm{PW}^- + \mathrm{GID}$ *proves:*

(i) $x\eta ON \leftrightarrow (x\eta IS \wedge Trans(x) \wedge \forall y \in x. Trans(y))$;

(ii) $x\eta ON \wedge y \equiv x \rightarrow y\eta ON$; $\quad x\eta ON \wedge y \in x \rightarrow y\eta ON$;

(iii) $TI(B, ON)$, *for arbitrary* B, *where* $TI(B, ON) :=$
$:= \forall x(x\eta ON \wedge \forall y \in x. B(y) \rightarrow B(x)) \rightarrow \forall x(x\eta ON \rightarrow B(x))$;

(iv) $x\eta ON \wedge y\eta ON \rightarrow (x \equiv x) \wedge (x \equiv y \vee x \in y \vee y \in x)$;

(v) *if* $\cup a := \{x : \exists y\eta a.\, x\eta y\}$, *then:*

$Cl(a) \wedge \forall x(x\eta a \rightarrow x\eta ON) \rightarrow$

$\rightarrow (\cup a)\eta ON \wedge \forall x(x \in a \rightarrow x \in (\cup a) \vee x \equiv \cup a)$;

(vi) *if* $a{+}1 := a \cup \{a\}$,

$a\eta ON \rightarrow (a{+}1)\eta ON \wedge \forall x(x\eta ON \wedge a \in x \rightarrow x \equiv (a{+}1) \vee (a{+}1) \in x)$.

PROOF. (i): $Ord(x) \to TOrd(x)$, because the negative occurrences of η in $Ord(x)$ are applied to classes (being x in IS). By definition of $\in$, $\equiv$ and 21.8, the right member of (i) is equivalent to $Ord(x)$.
(ii) See 21.8.
(iii) Assume:

$$\forall x(x\eta ON \wedge \forall y \in x.B(y) \to B(x)). \tag{1}$$

It is enough to verify

$$\forall x(x\eta IS \to C(x)), \tag{2}$$

where $C(x) := x\eta ON \to \forall u(u \equiv x \to B(u))$. Then, using $ON \subseteq IS$ and 21.6 (i), (2) implies the required conclusion $\forall x(x\eta ON \to B(x))$. In order to check (2), we apply 23.1. So, we assume:

$$Cl(x),\ \forall y\eta x. C(y),\ x\eta ON \text{ and } x \equiv u, \tag{3}$$

and we prove $B(u)$. Since $u\eta ON$ by (3) and 23.4 (ii), and (1) implies $\forall y \in u. B(y) \to B(u)$, it is sufficient to prove $\forall y \in u. B(y)$, or, which is equivalent, since $x \equiv u$, $\forall y \in x. B(y)$. Let $y \in x$, i.e. $y' \equiv y$, for some $y'\eta x$. By (3), it follows $C(y')$, i.e.

$$y'\eta ON \to \forall u(u \equiv y' \to B(u)). \tag{4}$$

Now, as $x\eta ON$, then $y\eta ON$ and hence $y'\eta ON$ by 23.4 (ii). Thus by (4) we obtain $\forall u(u \equiv y' \to B(u))$; if we choose $u = y$, since $y \equiv y'$, we can conclude $B(y)$, and we are done.
(iv)-(vi): exercise (one essentially applies 23.3). □

23.5. CONVENTION. We temporarily adopt small Greek letters as syntactic variables for ordinals; we also define:

$$\alpha < \beta := \alpha \in \beta; \quad \alpha \leq \beta := \alpha \equiv \beta \vee \alpha < \beta;$$

$$\forall \alpha A(\alpha) := \forall x(x\eta ON \to A(x)); \quad \exists \alpha A(\alpha) := \exists x(x\eta ON \wedge A(x));$$

$$\forall \alpha < \beta. A := \forall \alpha(\alpha < \beta \to A); \quad \exists \alpha < \beta. A := \exists \alpha(\alpha < \beta \wedge A).$$

23.6. LEMMA. *We can find an operation Rec such that, provably in* $\mathrm{PW}^- + \mathrm{GID}$:

(i) *if* $(Recf)(<\alpha) := \{x : \exists \beta < \alpha. x\eta(Recf)\beta\}$,

$(Recf)\alpha = f((Recf)(<\alpha))$;

(ii) $Mon(f) \to (\alpha \leq \beta \to (Recf)\alpha \subseteq (Recf)\beta)$;

(*where* $Mon(f) := f$ *is monotone*);

(iii) $Mon(f) \to . \forall \alpha(H\alpha = f\{x : \exists \beta < \alpha. x\eta H\beta\}) \to \forall \alpha(H\alpha =_e (Recf)\alpha)$.

PROOF. (i): choose $REC := \lambda f.FP(\lambda h\lambda y.f\{x : \exists z \in y.x\eta(hz)\})$, by the fixed point theorem for operations.
(ii) If $\alpha \leq \beta$, $(Recf)(<\alpha) \subseteq (Recf)(<\beta)$ by 23.4 (i) and $<$-transitivity; hence $Mon(f)$ and (i) imply $(Recf)\alpha \subseteq (Recf)\beta$.
(iii): immediate by transfinite induction on α. □

23.6.1. REMARK. By (ii) above, $Recf$ is *invariant under* $\equiv$: if $\alpha \equiv \beta$, then $(Recf)\alpha =_e (Recf)\beta$.

23.7. DEFINITION. $RC(f) := \{x : \exists\alpha.x\eta(Recf)\alpha\}$.

23.8. THEOREM (*Boundedness + Covering*). *We can define an operation* $\lambda c\lambda f.Bound(c,f)$ *such that* PW$^-$ + GID *proves:*

$$Cl(c) \wedge c \subseteq RC(f) \rightarrow Bound(c,f)\eta ON \wedge c \subseteq (Recf)(Bound(c,f)). \qquad (*)$$

PROOF. Let c be a class which satisfies the antecedent of $(*)$. By CL-reflection 16.3, there is a term $S(f,c)$ representing a class $\subseteq ON$ such that:

$$\forall x\eta c.\exists\alpha\eta S(f,c).x\eta(Recf)\alpha.$$

Hence $\beta(c,f) := \cup S(f,c)$ is an ordinal by 23.4 (v) and

$$\forall x\eta c.\exists\alpha \leq \beta(c,f).x\eta(Recf)\alpha.$$

Thus we choose $Bound(c,f) := \beta(c,f) + 1 (=$ ordinal successor$)$. □

23.9. COROLLARY (*Closure*). PW$^-$ + GID *proves*:

$$Mon(f) \rightarrow f(RC(f)) \subseteq RC(f).$$

PROOF. If $x\eta f(RC(f))$, there is a class $c \subseteq RC(f)$ with $x\eta fc$ by the generalized Myhill-Shepherdson theorem 17.5; hence $c \subseteq (Recf)\alpha$, for some α, and trivially $c \subseteq (Recf)(<\alpha{+}1)$; by f-monotonicity and 23.6 (i):

$$x\eta fc \subseteq (Recf)(\alpha{+}1) \subseteq RC(f). \ \square$$

23.10. COROLLARY. PW$^-$ + GID $\vdash Mon(f) \rightarrow I(f) =_e RC(f)$.

PROOF: $I(f) \subseteq RC(f)$ is a consequence of 23.9 and GID.
As to $RC(f) \subseteq I(f)$, let $C(\alpha) := (Recf)\alpha \subseteq I(f)$. If we assume by IH $\forall\beta < \alpha.C(\beta)$, then f-closure of $I(f)$, f-monotonicity, and 23.6 (i) imply $C(\alpha)$, whence by 23.4 (iii):

$$\forall\alpha((Recf)\alpha \subseteq I(f)). \qquad \square$$

PART C

SELECTED TOPICS

"Nur wenn man nicht auf den Nutzen nach aussen sieht, sondern in der Mathematik selbst auf das Verhältnis der unbenutzten Teile, bemerkt man das andere und eigentliche Gesicht dieser Wissenschaft. Es ist nicht zweckbedacht, sondern unökonomisch und leidenschaftlich. [...]
Die Mathematik ist Tapferkeitsluxus der reinen Ratio, einer der wenigen, die es heute gibt." (R. Musil 1913)

CHAPTER 6

LEVELS OF IMPLICATION AND INTENSIONAL LOGICAL EQUIVALENCE

§24. Myhill's levels of implication
§25. Formal deducibility based on levels of implication and its proof-theoretic strength
§26. Introducing an intensional equivalence relation
§27. The infinitary reduction relation $\Rightarrow$
§28. The Church-Rosser theorem for $\Rightarrow$
§29. A model of type-free logic based on intensional equivalence

We know from part B that the theory $PW_c + GID$ of reflective truth, extended by generalized induction and approximation principles, has several interesting features from the recursion-theoretic point of view; as to the relation with *classical* notions, we saw how to construct an interpretation of a predicative set theory. We now pursue the *opposite* aim: we relate $PW_c + GID + EA$ (= the enumeration axiom) to *non-classical* systems which allow type-free abstraction. The results we obtain suggest that *the present framework can profitably be adopted as a metatheory* for a variety of formal systems, whose proof-theoretic strength does not go beyond that of Peano arithmetic. We consider two systems stemming from intuitions, which are far from each other and from the theory of reflective truth introduced in part A.

The first proposal – due to Myhill (1972, 1984) – originally moves from a reappraisal of Curry's paradox and explicitly proposes a weakening of the implication introduction rule. According to Myhill's approach, such a rule should be regarded as an "ideal limit" and should be replaced by an infinite sequence of distinct approximations. Carrying out the project in details yields a sort of *ramified logic,* with respect to *the level of implication* assumed. The system turns out to be consistent by essential use of the dual representability theorem of §16 (this is the main idea of the Myhill-Flagg consistency proof).

The second system gives credit to an idea of Behmann (Behmann 1931), according to which we must ascribe the responsibility for the Russell paradox to *the very definition mechanism.* From this perspective, Feferman proposed in the seventies a logic based on a concept of definitional equivalence, which should play for the theory of properties a role, similar to that of the conversion relation in combinatory logic. Feferman and Aczel (1980) offered a consistency proof of the related logic of abstraction by means of a non-trivial adaptation of the Church-Rosser theorem. Rather surprisingly, we show that *a model of an even stronger theory exists just within* $PW_c + GID$, *plus the enumeration axiom*: the proof essentially hinges upon the boundedness theorem of §23 and the strong selection theorem of §20.

§24. Myhill's levels of implication

We are going to prove that the minimal frame system MF_c, expanded with approximation axioms, yields a natural environment for interpreting a theory of properties, that was proposed by Myhill (1972, 1984) and later fully developed in Myhill-Flagg (1987). Myhill's system stems from a criticism of Fitch's "Extended Basic Logic" (Fitch 1948).

As we know from the general introduction, Fitch's ideas are reflected in our frame theory MF_c; thus Myhill's objections to Fitch are easily rephrased as pointing out a crucial weakness of the truth predicate T. In essence, the standard hypothetical reasoning does not work in its full generality with T; the inference from $A \to TB$ to $T(A \to B)$, which corresponds to the deduction theorem or to the usual implication introduction rule, is sound only if we possess *the additional information that* A *is a proposition in the sense of* T (recall 8.6 (iii)). As a consequence, we cannot assert the internal truth of statements of the form $\forall x(Cl(x) \to TA)$, since CL is provably not a class by 9.3. In particular, if we identify Dedekind reals with suitable *classes* of rationals, quantification over the reals $\mathbb{R}$ will not be preserved in general by T. In sum, one would like to have at hand some kind of *internal implication.*

Such a difficulty led Myhill to explore the introduction of a chain of implications ${}_1\supset$, ${}_2\supset, \ldots$, which *reflect various degrees of deducibility in the context of type-free systems à la Fitch.* According to Myhill's proposal, $A\ {}_1\supset B$ is true iff B is deducible from A without using any implication whatsoever; $A\ {}_2\supset B$ is true iff B is deducible from A, using (at most) the inference rules for ${}_1\supset$, and so on.

Flagg and Myhill (1987) propose a system Σ_∞ of illative combinatory logic, which extends (a classical version of) Aczel's theory of Frege' s structures by means of ${}_1\supset, {}_2\supset, {}_3\supset$, etc. The main tool for establishing the consistency of Σ_∞ is given by the theorem 16.11 about the exact representability of pairs of disjoint properties; so we are naturally driven to investigate the relation of Σ_∞ with our theories. Indeed, we adopt

$PW_c + GID$, as a metatheory for developing the semantics of Σ_∞. In the following, we introduce a generalized version of Myhill's hierarchy of implications and we show that the basic construction takes place in $PW_c + GID$; this additional information will be applied in the next section to the characterization of proof-theoretic strength of the corresponding formal systems.

CONVENTION: below we adopt a, b, c as syntactical variables besides the usual x, y, z.

First of all, recall that we have: 1) a closed λ-term of $\mathcal{L}_{op}$, for each predicate symbol and logical operator of $\mathcal{L}$; 2) a canonical term $[A]$ with the same free variables of A, for each $\mathcal{L}$-formula A. $[A]$ expresses the function defined by A itself. For convenience, we agree to identify A and $[A]$, when no ambiguity can arise; thus we still write Nt, $t = s$, $\forall xA$, etc. instead of the corresponding $[-]$-terms; we keep using the standard logical symbols $\neg$, $\wedge$, $\forall$, etc., for the (terms defining in $\mathcal{L}_{op}$ the) corresponding operations. Below, we consider properties $R \subseteq V^2$, which represent abstract infinitary deducibility relations; hence, we shall use the more suggestive notation

$$\Gamma \vdash^R a,$$

instead of the proper $\langle \Gamma, a\rangle \eta R$. We let Γ, Δ range over arbitrary properties, but the informal idea is that Γ, Δ are collections of formula-objects, i.e. denotations of terms of the form $[A]$, where A is an arbitrary formula. $\cup$, $\{\ldots\}$ denote the corresponding operations on properties (see §9). If $\Gamma = \emptyset$, we simply write "$\vdash^R a$" instead of "$\Gamma \vdash^R a$"; if $\Gamma = \{a\}$, "$a \vdash^R b$" is a shortening for "$\{a\} \vdash^R b$". We use the abbreviations:

$$T\Gamma := \forall b(b\eta\Gamma \to Tb) \text{ and } F\Gamma := \exists c(c\eta\Gamma \wedge Fc);$$
$$\Gamma \vdash^R \Delta := \forall a(a\eta\Delta \to \Gamma \vdash^R a).$$

24.1. DEFINITION

(i) $Adr(R) :=$ R is an *abstract deducibility relation* iff R satisfies the following conditions:

Soundness: $\forall\Gamma\forall a(\Gamma \vdash^R a \to (T\Gamma \to Ta))$;

T-completeness: $\forall\Gamma\forall a(F\Gamma \vee Ta \to \Gamma \vdash^R a)$;

Closure under generalized cut:

$$\forall\Gamma\forall\Delta\forall a((Cl(\Delta) \wedge \Gamma \vdash^R \Delta \wedge \Gamma \cup \Delta \vdash^R a) \to \Gamma \vdash^R a).$$

Monotonicity:

$$\forall\Gamma\forall\Delta\forall a((\Gamma \vdash^R a \wedge \Gamma \subseteq \Delta) \to \Delta \vdash^R a).$$

(ii) *R-implication*: if ER is the operation of exact representation, given by theorem 16.11,

$$_R\supset \; := ER(R, \overline{Ded}),$$

where $\overline{Ded} := \{\langle\{a\}, b\rangle : Ta \wedge Fb\}$. Henceforth, we let

$$a\ _R\supset b := (ER(R, \overline{Ded}))\langle\{a\}, b\rangle.$$

(iii) $$Ded_0 := \{\langle\Gamma, a\rangle : F\Gamma \vee Ta\}.$$

Observe that Ded_0 is the $\subseteq$-least abstract deducibility relation.

24.2. LEMMA (*Existence of R-implication*). PW$^-$ *proves:*

$$Adr(R) \rightarrow \forall a \forall b((T(a\ _R\supset b) \leftrightarrow a \vdash^R b) \wedge (F(a\ _R\supset b) \leftrightarrow Ta \wedge Fb)).$$

PROOF: by soundness of R and consistency, $\overline{Ded}$ and R are disjoint and hence 16.11 applies. □

24.3. LEMMA. PW$^-$ *proves that every abstract deducibility relation is CL-compact:*

$$Adr(R) \rightarrow \forall\Gamma\forall\Delta\forall a(\Gamma \vdash^R a \rightarrow \exists\Delta(\Delta \subseteq \Gamma \wedge Cl(\Delta) \wedge \Delta \vdash^R a)).$$

PROOF: $Prem(a, \vdash^R) := \{\Gamma : \Gamma \vdash^R a\}$ is extensional by monotonicity; hence the result follows from the Rice-Shapiro theorem 17.3. □

Every abstract deducibility relation $\vdash^R$ can be extended to a larger deducibility relation $\vdash^R_*$ which is closed under logical inferences and natural rules for $_R\supset$.

24.4. *The basic deducibility relation* $\vdash^R_*$. $\vdash^R_*$ is the least binary relation, which is closed under the rules below.

Initial Rules

$$Hyp: \quad \frac{a\eta\Gamma}{\Gamma \vdash^R_* a}; \qquad Lift(R): \quad \frac{\Gamma \vdash^R a}{\Gamma \vdash^R_* a}.$$

Induction Rule CL-INDR for total predicates

$$\frac{\Gamma \vdash^R_* b\bar{0}; \quad \Gamma \vdash^R_* \forall x(bx \vee \neg bx); \quad \Gamma \vdash^R_* Nc; \quad \text{for every } n,\ \Gamma \cup \{bn\} \vdash^R_* b(n+1)}{\Gamma \vdash^R_* bc}$$

Logical Rules (*we recall that the logical primitives are* $\neg$, $\wedge$, $\forall$)

$$[\mathrm{I}\wedge]:\ \frac{\Gamma\vdash^R_* b_1 \quad \Gamma\vdash^R_* b_2}{\Gamma\vdash^R_* b_1\wedge b_2};\qquad [\mathrm{I}\neg\wedge i]:\ \frac{\Gamma\vdash^R_* \neg b_i\ \ (i=1,2)}{\Gamma\vdash^R_* \neg(b_1\wedge b_2)};$$

$$[\mathrm{E}\wedge i]:\ \frac{\Gamma\vdash^R_* b_1\wedge b_2}{\Gamma\vdash^R_* b_i\ (i=1,2)};$$

$$[\mathrm{E}\neg\wedge]:\ \frac{\Gamma\vdash^R_* \neg(b_1\wedge b_2)\quad \Gamma\cup\{\neg b_1\}\vdash^R_* c\quad \Gamma\cup\{\neg b_2\}\vdash^R_* c}{\Gamma\vdash^R_* c};$$

$$[\mathrm{I}\neg\neg]:\ \frac{\Gamma\vdash^R_* b}{\Gamma\vdash^R_* \neg\neg b};\qquad [\mathrm{E}\neg\neg]:\ \frac{\Gamma\vdash^R_* \neg\neg b}{\Gamma\vdash^R_* b};$$

$$[\mathrm{I}\forall]:\ \frac{\text{for every } c,\ \Gamma\vdash^R_* bc}{\Gamma\vdash^R_* \forall b};\qquad [\mathrm{E}\forall]:\ \frac{\Gamma\vdash^R_* \forall b}{\Gamma\vdash^R_* bc,\ \text{for arbitrary } c};$$

$$[\mathrm{I}\neg\forall]:\ \frac{\Gamma\vdash^R_* \neg\, bc,\ \text{for some } c}{\Gamma\vdash^R_* \neg\forall b};$$

$$[\mathrm{E}\neg\forall]:\ \frac{\Gamma\vdash^R_* \neg\forall b,\ \ \Gamma\cup\{\neg bc\}\vdash^R_* d,\ \text{for arbitrary } c}{\Gamma\vdash^R_* d};$$

$$[\bot]:\ \frac{\Gamma\vdash^R_* b\quad \Gamma\vdash^R_* \neg b}{\Gamma\vdash^R_* c};\qquad [\forall\vee]:\ \frac{\Gamma\vdash^R_* (bc)\vee d,\ \text{for every } c}{\Gamma\vdash^R_* (\forall b)\vee d};$$

$$[\mathrm{I}_R\supset]:\ \frac{\Gamma\cup\{a\}\vdash^R b}{\Gamma\vdash^R_* a\ {}_R\!\supset b};\qquad [\mathrm{E}_R\supset]:\ \frac{\Gamma\vdash^R_* a\quad \Gamma\vdash^R_* a\ {}_R\!\supset b}{\Gamma\vdash^R_* b};$$

$$[\mathrm{I}\neg_R\supset]:\ \frac{\Gamma\vdash^R_* a\qquad \Gamma\vdash^R_* \neg b}{\Gamma\vdash^R_* \neg(a\ {}_R\!\supset b)};$$

$$[\mathrm{E}\neg_R\supset]:\ \frac{\Gamma\vdash^R_* \neg(a\ {}_R\!\supset b)}{\Gamma\vdash^R_* a};\qquad \frac{\Gamma\vdash^R_* \neg(a\ {}_R\!\supset b)}{\Gamma\vdash^R_* \neg b};$$

$$\text{Cut:}\ \frac{\Gamma\vdash^R_* \Delta\quad Cl(\Delta)\quad \Gamma\cup\Delta\vdash^R_* b}{\Gamma\vdash^R_* b}.$$

Equality rules

$$\text{Eq.1:}\ \frac{\Gamma\vdash^R_* a\quad \Gamma\vdash^R_* a=b}{\Gamma\vdash^R_* b};\qquad \text{Eq.2:}\ \frac{\Gamma\vdash^R_* a=b\quad \Gamma\vdash^R_* a'=b'}{\Gamma\vdash^R_* aa'=bb'};$$

$$\text{Eq.3: } \frac{\Gamma \vdash^R_* a = b}{\Gamma \vdash^R_* b = a}; \quad \text{Eq.4: } \frac{\Gamma \vdash^R_* a = b \quad \Gamma \vdash^R_* b = c}{\Gamma \vdash^R_* a = c}.$$

24.4.1. REMARK

(i) $\vdash^R_*$ differs from Myhill's original notion in the following points: 1) we do allow possibly infinite collections of premises (Γ is a finite set in Myhill-Flagg 1987); 2) our cut rule is infinitary; 3) the N-induction rule is restricted to total predicates.

(ii) If $Adr(R)$ holds, the relation $\vdash^R_*$ is closed under the rule below (apply T-completeness of R and $Lift(R)$):

$$True: \ \frac{Ta \text{ (or } F\Gamma)}{\Gamma \vdash^R_* a}.$$

By inspection of the inductive definition 24.4, we can obviously find a T-positive formula $A(R, x, v)$, operative in v (in the sense of 10.3), such that its fixed point $Ixv.A(R, x, v) := I(A_R)$ (cf. 10.4) is closed under the clauses above, provably in MF^-; hence, if we let $\vdash^R_* := I(A_R)$, we have:

24.4.2. LEMMA

(i) $\mathrm{MF}^- \vdash \forall\Gamma\forall b(A(R, \langle\Gamma, b\rangle, \vdash^R_*) \leftrightarrow \Gamma \vdash^R_* b)$;

(ii) $\mathrm{MF}^- + \mathrm{GID} \vdash \forall x(A(R, x, B) \rightarrow B(\Gamma, b)) \rightarrow$

$\rightarrow \forall\Gamma\forall b(\Gamma \vdash^R_* b \rightarrow B(\Gamma, b))$ (B arbitrary).

24.5. THEOREM. (i) $\mathrm{PW}_c + \mathrm{GID} \vdash Adr(\vdash^R) \rightarrow Adr(\vdash^R_*)$.

(ii) *Conservation:*

$\mathrm{PW}_c + \mathrm{GID} \vdash Adr(\vdash^R) \rightarrow \forall a(Ta \leftrightarrow (\vdash^R_* a) \leftrightarrow (\vdash^R a))$.

(iii) *Consistency:* $\mathrm{PW}_c + \mathrm{GID} \vdash Adr(\vdash^R) \rightarrow \neg\exists b(\vdash^R_* b \wedge \vdash^R_* \neg b)$.

PROOF. (i) Monotonicity of $\vdash^R_*$ follows from the corresponding property of $\vdash^R$ by generalized induction (24.4.2 (ii)). As to T-completeness, it is an immediate consequence of the rule *True* of 24.4.1 (ii); closure under cut is immediate by definition of $\vdash^R_*$ and 24.4.2 (i).
It remains to check soundness (again by generalized induction on $\vdash^R_*$). The N-induction axiom for classes (see 10.7) and the T-axioms of §7 take care of $INDR$ and the logical inferences, while $Lift(R)$ preserves soundness, as $Adr(R)$ is assumed.

Let us consider the rules for R-implication.

$[\mathrm{I}_R \supset]$: assume $\Gamma \cup \{b\} \vdash^R c$ and $T\Gamma$. By 24.3, $\Delta \cup \{b\} \vdash^R c$, for some class $\Delta \subseteq \Gamma$; but $T\Gamma$ implies $T\Delta$, whence by *True* and monotonicity, $\{b\} \vdash^R \Delta$. Cut implies $b \vdash^R c$ and hence $T(b\ {}_R\!\supset c)$ (by 24.2).

$[\mathrm{I}\neg_R \supset]$: assume $\Gamma \vdash^R_* a$ and $\Gamma \vdash^R_* \neg b$; then $T\Gamma \rightarrow Ta \wedge T\neg b$ (by IH). If we assume $T\Gamma$, 24.2 implies $T(\neg(a\ {}_R\!\supset b))$.

$[\mathrm{E}_R \supset]$: assume $\Gamma \vdash^R_* a$, $\Gamma \vdash^R_* a\ {}_R\!\supset b$ and $T\Gamma$; by IH, we obtain Ta, $T(a\ {}_R\!\supset b)$, i.e. by 24.2 $a \vdash^R b$, which implies by $\vdash^R$-soundness $Ta \rightarrow Tb$, whence Tb.

$[\mathrm{E}\neg_R \supset]$-rules are also straightforward by IH and 24.2.

(ii) follows from (i) and application of *Lift* and *True*; (iii) is a consequence of T-consistency and (ii). □

The main consequence of 24.5 is that if we have a deducibility relation $\vdash^R$ and hence a corresponding R-implication ${}_R\!\supset$, we can conservatively extend $\vdash^R$ with the rules for ${}_R\!\supset$; but the enlarged deducibility relation induces a new implication and this suggests a natural iteration.

24.6. PROPOSITION. *If A is a formula of $\mathcal{L}_{op}$ and* $\mathrm{OP} \vdash A$, *then:*

$$\mathrm{MF}_c + \mathrm{GID} \vdash Adr(R) \rightarrow (\vdash^R_* A).$$

The proof requires a straightforward metamathematical induction on the derivation of A in OP; the point is that, if A is an axiom of OP, then $\mathrm{MF}_c \vdash TA$ and we can apply the rule *True* of 24.4.1.

24.7. DEFINITION. We introduce a deducibility relation $\vdash^n$ by induction on $n \in \omega$.

(i) *Deducibility of level* 0: $\vdash^0$ is the $\subseteq$-least relation, which is closed under *True* and the rules of 24.4, except $Lift(R)$ and the ${}_R\!\supset$-inferences.

(ii) $\vdash^{n+1} := IxvA(\vdash^n, x, v)$, where $A(\vdash^n, x, v)$ results from the formula $A(R, x, v)$ of 24.4.2 by replacing everywhere R with $\vdash^n$.

(iii) $\qquad {}_n\!\supset\ := {}_R\!\supset$, where $R := \vdash^n$.

Obviously, if $R := \vdash^n$, $\vdash^{n+1} := \vdash^R_*$: hence theorem 24.5 (i) ensures $Adr(\vdash^{n+1})$ under the assumption that $Adr(\vdash^n)$. Note also that $\vdash^0$ can formally be defined by the term $Ixv.A^-(Ded_0, x, v)$, where $A^-(Ded_0, x, v)$ is obtained from $A(R, x, v)$ by replacing R with Ded_0 (see 24.1 (iii)) and by simply omitting the clauses for R-implication. Since $A^-(Ded_0, x, v)$ is T-positive and operative in v, we can rely upon the proof of 24.5 (the cases of R-implication being omitted) and $\vdash^0$ is an abstract deducibility relation.

Hence, if we put $A_0(x,v) := A^-(Ded_0, x, v)$ and $A_{n+1}(x,v) := A(\vdash^n, x, v)$, we can verify by metamathematical induction:

24.8. THEOREM. *For each* $n \in \omega$,

(i) $\mathrm{PW}_c + \mathrm{GID} \vdash Adr\ (\vdash^n)$;

(ii) $\mathrm{PW}_c + \mathrm{GID} \vdash \forall a(Ta \leftrightarrow (\vdash^n a))$;

(iii) $\mathrm{PW}_c + \mathrm{GID} \vdash \neg\exists a(\vdash^n a \wedge \vdash^n (\neg a))$;

(iv) $\mathrm{PW}_c + \mathrm{GID} \vdash \forall\Gamma\forall b(A_n(\langle\Gamma, b\rangle, \vdash^n) \leftrightarrow \Gamma \vdash^n b)$;

(v) *if B is arbitrary,*

$$\mathrm{PW}_c + \mathrm{GID} \vdash \forall\Gamma\forall b(A_n(\langle\Gamma, b\rangle, B) \rightarrow B(\Gamma, b)) \rightarrow \forall\Gamma\forall b(\Gamma \vdash^n b \rightarrow B(\Gamma, b)).$$

24.8.1. REMARK. The internal version of 24.8, i.e. $\forall x(Nx \rightarrow Adr(\vdash^x))$ can be formally verified in PW + GID, where PW contains the full schema of number-theoretic induction.

§25. Formal deducibility based on levels of implication and its proof-theoretic strength

We describe a formal sequent calculus Σ_∞, which formalizes the deducibility relations of §24, and we show that the elementary judgments of Σ_∞ coincide with the provable sentences of OP, the basic theory of operations of Ch.I. First of all, Σ_∞ is formalized in the term fragment of the language $\mathcal{L}_{op}$ for OP, expanded by a formal deducibility symbol $\vdash^n$, for each $n \in \omega$. Besides the usual closed terms representing $\neg$, $\forall$, $\wedge$, $=$, N, we fix a closed term ${}_n\supset$, for each natural number $n \in \omega$. For convenience, we identify ${}_n\supset$ with the term introduced in 24.7 (iii) (NB: the term belongs to the language $\mathcal{L}_{op}$).
The calculus Σ_∞ is a natural deduction system, whose judgements have the form $\Gamma \vdash^n t$, where Γ is a finite (possibly empty) sequence of $\mathcal{L}_{op}$-terms and t is a term of $\mathcal{L}_{op}$. The inference rules of Σ_∞ are finite formal counterparts of the clauses defining $\vdash^R_*$.

25.1. *Inductive definition of the* Σ_∞*-derivability predicate*

$\Gamma \vdash^n t$ is Σ_∞-derivable (in short $\Sigma_\infty \vdash (\Gamma \vdash^n t)$) iff $\Gamma \vdash^n t$ belongs to the $\subseteq$-least collection of judgments which contains the axioms (or initial rules) to be given below, and is closed under the following inferences:

(i) the propositional rules [I $\wedge$], [I$\neg\wedge$], [E $\wedge$ i] ($i = 1, 2$), [E$\neg\wedge$], [I$\neg\neg$], [E$\neg\neg$], [$\perp$];

(ii) the quantifier rules [I$\forall$], [E$\forall$], [I$\neg\forall$], [E$\neg\forall$];

(iii) $[Cut]$ and the structural inferences $[K]$, $[W]$, $[C]$;
(iv) the rule $CL\text{-}INDR$ of N-induction for classes;
(v) the rules $[Lift]$, $[Red]$;
(vi) the implication rules $[\mathrm{I}_n \supset]$, $[\mathrm{E}_n \supset]$, $[\mathrm{I}\neg_n \supset]$, $[\mathrm{E}\neg_n \supset]$;
(vii) the equality rules EQ.1-EQ.4.

$CL\text{-}INDR$, the equality rules and the propositional rules (for $\wedge$, $\perp$, $\neg$) are obtained from the corresponding inferences of 24.4, by replacing everywhere $\vdash^R_*$ with $\vdash^n$ and by reinterpreting Γ as ranging over finite sequences of $\mathcal{L}_{op}$-terms; the remaining inferences are listed below.

Initial Rules

Hyp/TND: $t \vdash^0 t$; $\vdash^0 A \vee \neg A$, if $A = (t = s)$, Nt;

$(K\text{-}S)$: $\vdash^0 Kts = t$; $\vdash^0 Stsr = tr(sr)$; $\vdash^0 \neg(K = S)$;

(P_i): $\vdash^0 (\langle t_1, t_2 \rangle)_i = t_i$ (for $i = 1, 2$);

$(D.1)$: $Nt,\ Ns,\ t = s \vdash^0 Dtsrq = r$;

$(D.2)$: $Nt,\ Ns,\ \neg t = s \vdash^0 Dtsrq = q$;

$(N.1)$: $Nt \vdash^0 \neg(t{+}1) = \bar{0}$;

$(N.2)$: $\vdash^0 N\bar{0}$; $Nt \vdash^0 N(t{+}1)$;

$(N.3)$: $Nt \vdash^0 PRED(t{+}1) = t$;

(Id): $\vdash^0 t = t$.

Quantifier rules

$[\mathrm{I}\forall]$: $\dfrac{\Gamma \vdash^n sx}{\Gamma \vdash^n \forall s}$; $[\mathrm{E}\forall]$: $\dfrac{\Gamma \vdash^n \forall s}{\Gamma \vdash^n st}$;

(proviso for $[\mathrm{I}\forall]$: x not free in Γ, s).

$[\mathrm{I}\neg\forall]$: $\dfrac{\Gamma \vdash^n \neg\, st}{\Gamma \vdash^n \neg\forall s}$; $[\mathrm{E}\neg\forall]$: $\dfrac{\Gamma \vdash^n \neg\forall s \quad \Gamma, \neg sx \vdash^n r}{\Gamma \vdash^n r}$;

(proviso for $[\mathrm{E}\neg\forall]$: x not free in Γ, s, r).

$[\forall \vee]$: $\dfrac{\Gamma \vdash^n \forall(\lambda x.\, sx \vee t)}{\Gamma \vdash^n (\forall s) \vee t}$;

(proviso for $[\forall \vee]$: x not free in t).

Cut and structural rules

$[Cut]$: $\dfrac{\Gamma \vdash^n t \quad \Gamma, t \vdash^n s}{\Gamma \vdash^n s}$; $[K]$: $\dfrac{\Gamma \vdash^n t}{\Gamma, s \vdash^n t}$;

[W]: $\dfrac{\Gamma,\ t,\ t \vdash^{n} s}{\Gamma,\ t \vdash^{n} s}$; [$C$]: $\dfrac{\Gamma,\ t,\ s,\ \Delta \vdash^{n} r}{\Gamma,\ s,\ t,\ \Delta \vdash^{n} r}$.

Rules for implication of level n

[$\mathrm{I}_{n+1} \supset$]: $\dfrac{\Gamma,\ t \vdash^{n} s}{\Gamma \vdash^{n+1} t\ {}_{n}\!\supset s}$; [$\mathrm{I}\neg_{n} \supset$]: $\dfrac{\Gamma \vdash^{n+1} t \quad \Gamma \vdash^{n+1} \neg s}{\Gamma \vdash^{n+1} \neg(t\ {}_{n}\!\supset s)}$;

[$\mathrm{E}_{n} \supset$]: $\dfrac{\Gamma \vdash^{n+1} t \quad \Gamma \vdash^{n+1} t\ {}_{n}\!\supset s}{\Gamma \vdash^{n+1} s}$;

[$\mathrm{E}\neg_{n} \supset$]: $\dfrac{\Gamma \vdash^{n+1} \neg(t\ {}_{n}\!\supset s)}{\Gamma \vdash^{n+1} t}$; $\dfrac{\Gamma \vdash^{n+1} \neg(t\ {}_{n}\!\supset s)}{\Gamma \vdash^{n+1} \neg s}$.

Level Rules

[*Red*]: $\dfrac{\vdash^{n+1} t}{\vdash^{n} t}$; [*Lift*]: $\dfrac{\Gamma \vdash^{n} t}{\Gamma \vdash^{n+1} t}$.

Notice that [*Red*] states a reducibility rule for *unconditional statements*; [*Red*] is not present in Myhill's calculi, but we include it here, since it is sound under the interpretation of the previous section. Clearly, we can interpret $\Gamma \vdash^{n} t$, where Γ is the finite sequence $t_1, \ldots, t_i$, as the formula of $\mathcal{L}$ stating that the abstract deducibility relation $\vdash^{n}$ of 24.8 holds of the pair $\langle \Gamma, t \rangle$, (where Γ is now $\{x : x = t_1 \vee \ldots \vee x = t_i\}$). Under the given interpretation, we show:

25.2. PROPOSITION (*Soundness of* Σ_∞). *For each* $n \in \omega$,

$$\Sigma_\infty \vdash (\Gamma \vdash^{n} t) \textit{ implies } \mathrm{PW}_c + \mathrm{GID} \vdash (\Gamma \vdash^{n} t).$$

PROOF: we argue by induction on the definition of Σ_∞-derivability. We systematically apply the closure properties of the corresponding relations $\vdash^{n}$ of 24.7.

$n = 0$. Initial rules: we apply OP-axioms, the rules *True*, [E$\neg\wedge$], [$\perp$] and $\vdash^{0}$-monotonicity (hence 24.8 (i)). In particular, *TND* follows by *True* and axiom *T*.1 of MF_c. The logical inferences and *CL-INDR* are disposed of by means of the corresponding closure conditions for $\vdash^{0}$, which hold by 24.8 (iv). The structural rules are sound because $\vdash^{n}$ is extensional in Γ; the cut rule of Σ_∞ is a special case of the general cut rule of 24.4.

$n = m+1$. As to [*Red*], assume

$$\Sigma_\infty \vdash (\vdash^{m+1} t) \text{ and } \mathrm{PW}_c + \mathrm{GID} \vdash (\vdash^{m+1} t). \qquad (*)$$

The second part of (∗) implies Tt, provably in $\mathrm{PW}_c + \mathrm{GID}$, by theorem

24.8 (ii). Hence by *True*, $\mathrm{PW}_c + \mathrm{GID} \vdash (\vdash^m t)$, which is exactly the interpretation of the conclusion of [*Red*]. As to the soundness of the ${}_{m+1}\supset$-inferences, it is again granted by 24.5 and 24.8. □

25.3. LEMMA. *If* $\mathrm{OP} \vdash A$, *then* $\Sigma_\infty \vdash (\vdash^0 A)$.

PROOF: by straightforward induction on the length of the derivation of A; it is essential to observe that $\vdash^0 A \vee \neg A$, if A is in $\mathcal{L}_{op}$ (see *TND*). □

If Γ is a finite sequence $B_1, \ldots, B_i$, let $T\Gamma := B_1 \wedge \ldots \wedge B_i$ (if Γ is empty, we can choose $T\Gamma := (\overline{0} = \overline{0})$). Then we conclude by 25.2, 24.8 (i) and the general conservation theorem 15.5:

25.4. COROLLARY

(i) *If* $\Sigma_\infty \vdash (\Gamma \vdash^n A)$, *for some* $n \in \omega$, *then* $\mathrm{PW}_c + \mathrm{GID} \vdash T\Gamma \to TA$.

(ii) *In the same hypothesis of* (i), *if* $\Gamma \cup \{A\}$ *contains only terms of the form* $[B]$, *where* B *is a formula of* $\mathcal{L}_{op}$, *we have:*

$$\mathrm{OP} \vdash \wedge\Gamma \to A.$$

It follows that, if Σ_∞ proves that b is a class, b is a class provably in $\mathrm{PW}_c + \mathrm{GID}$; also, 25.4-25.3 entail:

25.5. COROLLARY. *If A is a formula of* $\mathcal{L}_{op}$,

$$\mathrm{OP} \vdash A \text{ iff } \Sigma_\infty \vdash (\vdash^n A).$$

By the conservation requirement of the previous section (see 24.8 (ii)) and by 25.4, we cannot expect new internal truths out of $\vdash^n$; however, we may expect that the new implications yield useful approximations of classical set-theoretic objects. For instance, we already know that there is no way to distinguish extensionally $[CL \to CL]$ from $[V \to CL]$ (since $\mathrm{MF}^- \vdash \neg FCl(x)$; see 9.3). However, we can define, if $n > 0$,

25.6 $$[a \to b]_n := \{f : \forall x(x \eta a \,{}_n\!\supset (fx)\eta b)\};$$

$$\mathrm{Pow}_n(a) := \{x : \forall y(y \eta x \,{}_n\!\supset y \eta a)\}.$$

Then, even if a is not a class, we do have, provably in $\mathrm{PW}_c + \mathrm{GID}$, by 24.2:

25.6.1. $$\forall f(f\eta[a \to b]_n \leftrightarrow \forall x(x\eta a \,{}_n\!\supset (fx)\eta b));$$

$$\forall b(b\eta \mathrm{Pow}_n(a) \leftrightarrow \forall x(x \eta b \,{}_n\!\supset x\eta a)).$$

It is also easy to see that the function space and the power set operator of 25.6 are monotone in n:

25.6.2. if $n < k$, $\mathrm{Pow}_n(a) \subseteq \mathrm{Pow}_k(a)$ and $[a \rightarrow b]_n \subseteq [a \rightarrow b]_k$.

The previous notions turn out to be interesting, insofar as one can profitably work with the restricted logic of $\vdash^n$ and hence in Σ_n; but it is not clear whether these implications can really have significant applications. However, Flagg and Myhill (1987) show that analysis based on the hierarchy of implications can partially avoid some pathologies, which are typical of recursive analysis.

§26. Introducing an intensional equivalence relation

According to Behmann (1931), (1937), (1959), the Russell paradox concerns *definitions*, not *assertions*: the contradiction shows that we can define notions (hence introduce symbols) by means of the abstraction operator $\{x : A\}$, which *cannot be eliminated by replacing the definiendum with the definiens.* The abstraction principle must be regarded as a conversion or definition schema, which explains the meaning of $t \in \{x : A\}$ as $A[x := t]$; hence the relation of $t \in \{x : A\}$ with $A[x := t]$ is a sort of *definitional equivalence* and not *standard logical equivalence.* If we adopt this view, the Russell argument simply teaches us that we can define a term $R := \{x : \neg x \in x\}$, such that $R \in R$ and $\neg R \in R$ become equivalent in consequence of the given definitions or stipulations; but this fact does not logically entail $R \in R \leftrightarrow (\neg R \in R)$. The Russell paradox only follows, if we know that $R \in R$ represents a genuine proposition with a definite truth value. This situation is familiar from combinatory logic, where we have a great freedom in defining operations, but we must investigate separately their specific properties (e.g. convergence on a given domain, normal form, etc.).

Following this line of thought, Aczel-Feferman (1980) and Feferman (1984) proposed *an extension of classical logic by means of an intensional equivalence operator* $\equiv$, where $A \equiv B$ is to be read as "*A is equivalent to B in consequence of given basic definitions*". In the extended framework, the naive abstraction principle can be consistently rephrased in the sense of Behmann:

AF $\quad \forall u(u \in \{x : A\} \equiv A[x := u])$ (A arbitrary, u free for x in A).

It turns out that the consistency of AF is by no means obvious and is obtained in Aczel-Feferman(1980), using an extension of the Church-Rosser theorem for infinitary calculi. At a first sight, it is not clear how to relate the Aczel-Feferman logic AF with the present set-up.

In the following, we present a suitable generalization BL_c(= Behmann's logic with class induction) of the AF-logic; we then show in the remaining sections that BL_c has a model within PW_c + GID + EA, EA being the enumeration axiom of Ch. I.

The crucial step requires a careful extension to BL_c of the Aczel-Feferman construction and of its underlying reduction relation. As we shall see, an essential role in our proof is played by the fundamental boundedness theorem of §23 and the selection properties of §20. In contrast to Aczel-Feferman's proposal, we formalize Behmann's logic in pure classical predicate calculus: *definitional equivalence is treated as a binary predicate and not as a logical connective.*

We now proceed to a description of the system BL_c, which justifies the principle AF; for convenience, the system is formalized in the extension $\mathcal{L}_B$ of $\mathcal{L}_{op}$ (= the language of OP without T), which contains the new binary predicates $\equiv$, $\in$. It is also convenient to include in $\mathcal{L}_B$:

(i) distinct primitive constants $NAT\hat{}$, $ID\hat{}$, $NEG\hat{}$, $AND\hat{}$, $ALL\hat{}$, corresponding to the $\mathcal{L}_{op}$-terms of 7.1;

(ii) three new constants $EQ\hat{}$, $\mathbb{1}\hat{}$, $\mathbb{O}\hat{}$ representing the predicate $\equiv$ and the boolean values "true" and "false" (respectively).

$\mathcal{L}_B$-terms are inductively generated by application from variables, the constants of $\mathcal{L}_{op}$ and the new constants; $\mathcal{L}_B$-atoms also include expressions of the form $t \in s$, $t \equiv s$ (t, s terms), besides $\mathcal{L}_{op}$-atoms. $\mathcal{L}_B$-formulas are inductively generated using $\forall$, $\neg$, $\wedge$. The [–]-operation of 7.1 which embeds formulas into terms, can obviously be extended to arbitrary formulas of $\mathcal{L}_B$ by stipulating:

26.1 $[Nt] = NAT\hat{}t$; $[t = s] = ID\hat{}ts$; $[t \equiv s] = (EQ\hat{})ts$;

$[t \in s] = st$; $[\neg A] = NEG\hat{}[A]$; $[A \wedge B] = AND\hat{}[A][B]$;

$[\forall x A] = ALL\hat{}(\lambda x.[A])$.

Thus it makes sense to adopt the standard notation $\{x: A\}$, instead of $\lambda x.[A]$, for arbitrary A of $\mathcal{L}_B$. We tend to omit [–], once the given context is clear and no ambiguity arises. In particular, we write $(x \equiv y) \equiv z$, $(x \equiv y) \equiv (u \equiv v)$ for $(EQ\hat{})xy \equiv z$ and $(EQ\hat{})xy \equiv (EQ\hat{})uv$ (respectively); we read $a \equiv \mathbb{O}\hat{}$ ($a \equiv \mathbb{1}\hat{}$) as "$a$ is false (true) according to the basic definitions"; we also let $Da := (a \equiv \mathbb{O}\hat{} \vee a \equiv \mathbb{1}\hat{}) :=$ "a has a definite truth value", $Cl(a) := \forall x D(ax) :=$ "a is a class". We let $A \equiv B$ stand for the proper $[A] \equiv [B]$.

26.2. BL_c contains the system OP^- of §2 (together with classical logic plus equality formalized in $\mathcal{L}_B$), number-theoretic induction for classes, axioms stating that the new constants $EQ\hat{}$, $NAT\hat{}$, $NEG\hat{}$, $ALL\hat{}$, $ID\hat{}$, $AND\hat{}$, $\mathbb{O}\hat{}$, $\mathbb{1}\hat{}$ have the obvious independence properties, and a list of proper axioms on definitional equivalence. The axioms are listed below in three groups.

1. *Independence axioms:*

Cost1 $L_1x = L_2y \rightarrow L_1 = L_2 \wedge x = y$,
where $L_1, L_2 \in SYMB_1 := \{NAT\hat{}, NEG\hat{}, ALL\hat{}\}$;

Cost2 $G_1xy = G_2x'y' \rightarrow G_1 = G_2 \wedge x = x' \wedge y = y'$,
where $G_1, G_2 \in SYMB_2 := \{ID\hat{}, AND\hat{}, EQ\hat{}\}$;

Cost3 $\neg L_1x = L_2yz$, for $L_1 \in SYMB_1$, $L_2 \in SYMB_2$;

Cost4 $\neg L_1 = L_2$, where $L_1, L_2 \in SYMB_1 \cup SYMB_2 \cup \{\mathbb{O}\hat{}, \mathbb{I}\hat{}\}$
and $L_1 \neq L_2$.

2. *Equivalence axioms* ($\equiv$ is an equivalence relation, which preserves itself and the logical operators):

E.1 $(a \equiv a) \wedge (a \equiv b \rightarrow b \equiv a) \wedge (a \equiv b \wedge b \equiv c \rightarrow a \equiv c)$;

E.2.1 $a \equiv a' \wedge b \equiv b' \rightarrow ((a \wedge b) \equiv (a' \wedge b') \wedge (a \equiv b) \equiv (a' \equiv b'))$;

E.2.2 $(\forall x(ax \equiv a'x) \rightarrow (\forall a) \equiv (\forall a)) \wedge (a \equiv a' \rightarrow (\neg a) \equiv (\neg a'))$.

3. *Axioms for the internal logic of* $\equiv$:

E.3.1 $(A \equiv \mathbb{I}\hat{}) \leftrightarrow A$, if $A = Nx$, $(x = y)$, $(x \equiv y)$, $x \in y$;

E.3.2 $(A \equiv \mathbb{O}\hat{}) \leftrightarrow (\neg A)$, if $A = Nx$, $(x = y)$;

E.4 $(\neg a) \equiv \mathbb{I}\hat{} \leftrightarrow a \equiv \mathbb{O}\hat{}$; $(\neg a) \equiv \mathbb{O}\hat{} \leftrightarrow a \equiv \mathbb{I}\hat{}$;

E.5 $(a \wedge b) \equiv \mathbb{I}\hat{} \leftrightarrow (a \equiv \mathbb{I}\hat{} \wedge a \equiv \mathbb{I}\hat{})$; $(a \wedge b) \equiv \mathbb{O}\hat{} \leftrightarrow (a \equiv \mathbb{O}\hat{} \vee b \equiv \mathbb{O}\hat{})$;

E.6 $(\forall a) \equiv \mathbb{I}\hat{} \leftrightarrow \forall x(ax \equiv \mathbb{I}\hat{})$; $(\forall a) \equiv \mathbb{O}\hat{} \leftrightarrow \exists x(ax \equiv \mathbb{O}\hat{})$;

E.7 $(a \equiv b) \equiv \mathbb{O}\hat{} \leftrightarrow (Da \wedge Db \wedge \neg(a \equiv b))$.

Since $\{x : A\}$ is $\lambda x.[A]$ and $[a \in b]$ shortens (ba), it makes sense to state:

26.3. LEMMA
(i) $\mathrm{BL}_c \vdash \forall a(a \in \{x : A\} \equiv A[x := a])$ (*A arbitrary*);

(ii) $\mathrm{BL}_c \vdash (A \equiv \mathbb{I}\hat{}) \rightarrow A. \wedge .(A \equiv \mathbb{O}\hat{}) \rightarrow \neg A$ (*A arbitrary*).

PROOF. (i) AF is a trivial consequence of β-conversion, $\equiv$-reflexivity and $[A[x := t]] = [A][x := t]$, while (ii) follows by induction on A. □

It is also clear that BL_c can be regarded as an extension of MF_c, once we consider MF_c formalized with primitive constants (see Ch. II, appendix III); if we define $Ta := (a \equiv \mathbb{I}\hat{})$ and $TR := \lambda x.[x \equiv \mathbb{I}\hat{}]$, the language $\mathcal{L}$ of MF_c becomes a sublanguage of $\mathcal{L}_B$ and we have:

26.4.PROPOSITION. *For every formula A, if* $\mathrm{MF}_c \vdash A$, $\mathrm{BL}_c \vdash A$.

PROOF. It suffices to check (the translation of the) T-axioms. Axioms $T.1$, $T.2.1$, $T.3$, $T.4$, $T.5$ immediately follow from E.3-E.6; E.3 also implies $\neg(\mathbb{I}\check{} \equiv \mathbb{O}\check{})$, which yields the consistency axiom $T.6$ (via symmetry and transitivity of $\equiv$). As to $T.2.2$, observe:

$$(\neg(a \equiv \mathbb{I}\check{})) \equiv \mathbb{I}\check{} \leftrightarrow (a \equiv \mathbb{I}\check{}) \equiv \mathbb{O}\check{} \leftrightarrow (a \equiv \mathbb{O}\check{}) \leftrightarrow (\neg a) \equiv \mathbb{I}\check{}$$

(apply E.4, E.7 and the equivalence

$$(Da \wedge Db \wedge \neg(a \equiv b)) \leftrightarrow ((a \equiv \mathbb{O}\check{} \wedge b \equiv \mathbb{I}\check{}) \vee (a \equiv \mathbb{I}\check{} \wedge b \equiv \mathbb{O}\check{})). \qquad \square$$

Conversely, BL_c has a model in $\mathrm{PW}_c + \mathrm{GID} + \mathrm{EA}$. In order to prove this theorem, we have to give a detailed analysis of a generalized reduction relation.

§27. The infinitary reduction relation $\Rightarrow$

The naive idea is to define a suitable reduction relation $a \Rightarrow b$ and to interpret $a \equiv b$ as "$a \Rightarrow c$ and $b \Rightarrow c$, for some c". The basic reduction clauses embody natural truth and congruence requirements and $\Rightarrow$ turns out to be confluent (it has the Church-Rosser property; see Ch. I).

Technically, the problem is that if we stick to the obvious reduction clauses for $\forall$, we do get into troubles with the confluence proof, unless we generalize the language to an infinitary propositional language. This move causes additional difficulties which are typical of the Church-Rosser theorem for systems with infinitary terms (due to Girard and Maass). A solution can be found following the layered reduction technique of Aczel-Feferman(1980), which makes essential use of the explicit ordinal analysis of inductive definitions. In what follows, we show that $\mathrm{PW}_c + \mathrm{GID} + \mathrm{EA}$ is powerful enough to grant the existence of a model for a generalization of BL_c.

First of all, besides the terms ID, NAT, AND, NEG, ALL of 7.1, we define four combinators for interpreting the constants $EQ\check{}$, $\mathbb{O}\check{}$, $\mathbb{I}\check{}$ of the previous section and a sort of uniform infinitary conjunction $\sqcap$:

27.1. $\quad \mathbb{I} := \langle \bar{0}, \bar{1} \rangle; \quad \mathbb{O} := \langle \bar{0}, \bar{0} \rangle;$

$$DE := \lambda x \lambda y. \langle \overline{11}, \langle x, y \rangle \rangle; \quad \sqcap := \lambda x. \langle \overline{13}, x \rangle.$$

For convenience, in §§27-28 we keep using $t \equiv s$ for the term $DEts$ (DE is reminiscent of "*definitional equivalent*"); remind that $\neg t := (NEG)t$, $[t = s] := (IDt)s$, $[Nt] := (NAT)s$ (cf. §7).
By choice, DE and $\sqcap$ satisfy the obvious independence axioms (cf. 26.2) provably in OP^-.

27.2. DEFINITION. $Funcl(f) := Cl(f) \wedge Fun(f)$, where $Fun(f)$ is

$$f \subseteq V^2 \wedge \forall x \forall y \forall z(\langle x, y\rangle \eta f \wedge \langle x, z\rangle \eta f \to y = z) \wedge \forall x \exists y(\langle x, y\rangle \eta f).$$

($f \subseteq V^2$ states that f is a property of ordered pairs; see Ch. II, Appendix II, lemma 1). If $Funcl(f)$ is assumed, we say that f is a *function-class*. To simplify notations below, when $Funcl(f)$ is assumed, we tend to write $A(f(u))$ instead of the proper $\exists y(\langle u, y\rangle \eta f \wedge A(y))$.

27.3. DEFINITION. $PTER(x)$ ($= x$ is a pseudo-term) is the formula

$$(x = \mathbb{O} \vee x = \mathbb{I}) \vee \exists y \exists z(x = [Ny] \vee x = [y = z]) \vee$$
$$\vee \exists y \exists z(x = (y \equiv z)) \vee \exists y(x = (\neg y)) \vee \exists z(x = \sqcap z).$$

$PTER(x)$ says that x has the appropriate form, to be processed by means of the reduction relation below (recall the definition of pseudo-form in §7).

27.4. *(Informal) Definition of the reduction relation* $\Rightarrow$

$\Rightarrow$ is the $\subseteq$-least reflexive, transitive relation, which is closed under the following rules:

AT1 : $a \Rightarrow \mathbb{I}$ ($\mathbb{O}$), provided $a := [Nt]$, $[t = s]$ and Nt, $t = s$ are true (false);

NEG.1: from $a \Rightarrow b$ infer $\neg a \Rightarrow \neg b$;

NEG.2: from $a \Rightarrow \mathbb{O}$ ($\mathbb{I}$) infer $\neg a \Rightarrow \mathbb{I}$ ($\mathbb{O}$);

CONJ.1: from $Funcl(f)$, $Funcl(g)$ and $\forall x(f(x) \Rightarrow g(x))$ infer $\sqcap f \Rightarrow \sqcap g$;

CONJ.2: from $Funcl(f)$ and $\forall x(f(x) \Rightarrow \mathbb{I})$ infer $\sqcap f \Rightarrow \mathbb{I}$;

CONJ.3: from $Funcl(f)$ and $\exists x(f(x) \Rightarrow \mathbb{O})$ infer $\sqcap f \Rightarrow \mathbb{O}$;

DE.1: from $a_1 \Rightarrow b_1$ and $a_2 \Rightarrow b_2$ infer $(a_1 \equiv a_2) \Rightarrow (b_1 \equiv b_2)$;

DE.2: from $a \Rightarrow c$ and $b \Rightarrow c$, for some c, infer $(a \equiv b) \Rightarrow \mathbb{I}$;

DE.3: from $a \Rightarrow \mathbb{O}$ ($\mathbb{I}$) and $b \Rightarrow \mathbb{I}$ ($\mathbb{O}$) infer $(a \equiv b) \Rightarrow \mathbb{O}$.

27.5. REMARK. CONJ.3 makes clear that our interpretation of infinitary conjunction $\sqcap$ is *non-strict*, contrary to what happens in the original proposal of Aczel-Feferman (1980). We also stress that $\Rightarrow$ *generates an equivalence relation* $\equiv$, *defined on arbitrary objects, and not a new connective on formulas.*

Before we proceed to a stage-by-stage definition of $\Rightarrow$, we introduce the notion of *reflexive transitive closure $RT(r)$ of a given binary relation r.*

27.6. LEMMA. *We can find a term RT of $\mathfrak{L}$, such that, provably in* $\mathrm{MF}^- + \mathrm{GID}$, *if $r \subseteq V^2$, then $RT(r) \subseteq V^2$; moreover:*

(i) $\forall x \forall y(\langle x,y\rangle \eta RT(r) \leftrightarrow$

$\leftrightarrow ((x=y) \vee (\langle x,y\rangle \eta r) \vee \exists z(\langle x,z\rangle \eta RT(r) \wedge \langle z,y\rangle \eta RT(r))))$.

(ii) *The principle of RT-induction: for each $B(x,y)$,*

$[\forall x B(x,x) \wedge \forall x \forall y(\langle x,y\rangle \eta r \to B(x,y)) \wedge$

$\wedge \forall x \forall y \forall z(B(x,y) \wedge B(y,z) \to B(x,z))] \to \forall x \forall y(\langle x,y\rangle \eta RT(r) \to B(x,y))$.

(iii) $(r \subseteq V^2 \to r \subseteq RT(r)) \wedge (r \subseteq r' \to RT(r) \subseteq RT(r'))$.

PROOF. The formula

$$A_r(u,a) := \exists x \exists y(u = \langle x,y\rangle \wedge (x = y \vee \langle x,y\rangle \eta r \vee \exists z(\langle x,z\rangle \eta a \wedge \langle z,y\rangle \eta a)))$$

is T-positive and operative in a. Hence we can choose $RT(r) := I(A_r)$ by 10.4, and $I(A_r)$ is the $\subseteq$-least fixed point of $A_r(u,a)$ by GID. Clearly, if r is a binary relation, so is $RT(r)$.
(iii) is an immediate application of (ii) and (i). □

As we can guess from Ch. I (see 4.6), RT preserves confluence, provably in $\mathrm{MF}^- + \mathrm{GID}$.

27.7. DEFINITION. $CR(r) :=$ "r is Church Rosser" stands for:

$$r \subseteq V^2 \wedge \forall x \forall y \forall z(\langle x,y\rangle \eta r \wedge \langle x,z\rangle \eta r \to \exists w(\langle y,w\rangle \eta r \wedge \langle z,w\rangle \eta r)).$$

27.8. LEMMA. $\mathrm{MF}^- + \mathrm{GID} \vdash CR(r) \to CR(RT(r))$.

PROOF. We apply 27.6. Let:

$$A(a,b) := \forall b(\langle a,b'\rangle \eta RT(r) \to \exists d(\langle b,d\rangle \eta RT(r) \wedge \langle b',d\rangle \eta RT(r))).$$

Then we easily obtain $A(a,a)$ and $A(a,c) \wedge A(c,b) \to A(a,b)$ by closure of $RT(r)$ (see 27.6 (i)). $CR(r)$ and RT-induction imply:

$$\forall a \forall b'(\langle a,b'\rangle \eta RT(r) \to \forall b(\langle a,b\rangle \eta r \to \exists d(\langle b',d\rangle \eta r \wedge \langle b,d\rangle \eta RT(r)))),$$

which yields $\forall a \forall b(\langle a,b\rangle \eta r \to A(a,b))$. A final application of RT-induction yields $CR(RT(r))$. □

We now show that $\Rightarrow$-reduction is resolvable into a non decreasing sequence of approximations $(\Rightarrow_0) \subseteq (\Rightarrow_1) \subseteq \ldots (\Rightarrow_\alpha) \subseteq \ldots$(for α in ON), which converges to $\Rightarrow$ and is suitable for the proof-theoretic verification of $CR(\Rightarrow)$. The basic tool is ordinal analysis of inductive notions.

First of all, we introduce an operator H of 1-*step reduction*:

27.9. DEFINITION. (i) $H(r) := \{\langle x,y\rangle : H(\langle x,y\rangle, r)\}$,
where $H(u,r) := \exists x \exists y(u = \langle x,y\rangle \wedge (\langle x,y\rangle \eta r \vee H_0(x,y,r)))$ and $H_0(x,y,r)$ is the straightforward formalization of the nine inductive clauses defining $\Rightarrow$ in 27.4, except that reflexivity and transitivity are omitted.

(ii) We then define the operator LR of "layered reduction":

$$LR := \lambda r\, H(RT(r)).$$

We won't bother the reader with the explicit formula for $H_0(x,y,r)$; for instance, a typical clause corresponding to CONJ.1 has the form:

$$\exists f \exists g(Funcl(f) \wedge Funcl(g) \wedge x = \sqcap f \wedge y = \sqcap g \wedge \forall x(\langle f(x), g(x)\rangle \eta r)).$$

Of course, it is essential to observe that $H(u,r)$ is an operator in the sense of 10.3. Then we have:

27.10. LEMMA. $\mathrm{MF}^- + \mathrm{GID}$ *proves* :

(i) $\forall r \forall u(u \eta H(r) \leftrightarrow \exists x \exists y(u = \langle x,y\rangle \wedge (\langle x,y\rangle \eta r \vee H_0(x,y,r))))$;

(ii) $(a \subseteq V^2 \rightarrow a \subseteq H(a)) \wedge (a \subseteq b \rightarrow H(a) \subseteq H(b))$;

(iii) $(a \subseteq V^2 \rightarrow a \subseteq LR(a)) \wedge (a \subseteq b \rightarrow LR(a) \subseteq LR(b))$.

Note that (i) essentially requires the hypothesis that f, g are classes, whenever $x = \sqcap f$, $y = \sqcap g$; (iii) follows with (ii) and 27.6 (iii).
Since LR is monotone, we can apply the bottom-up characterization of inductively defined predicates of § 23. We recall that α, β, γ, δ range over ordinals, as they are defined in $\mathrm{PW}^- + \mathrm{GID}$ within the set-theoretic universes of §§ 22-23.

27.11. LEMMA

(i) *We can find a term RED such that* $\mathrm{PW}^- + \mathrm{GID}$ *proves*:

$\forall \alpha \forall x \forall y(\langle x,y\rangle \eta RED\alpha \leftrightarrow H(\langle x,y\rangle, RT(RED(<\alpha))))$,

(*where* $RED(<\alpha) := \{\langle x,y\rangle : \exists \beta < \alpha . \langle x,y\rangle \eta RED\beta\}$).

(ii) *Let* $\Rightarrow := \{\langle x,y\rangle : \exists \alpha . \langle x,y\rangle \eta RED\alpha\} := RC(LR)$ (*cf.* 23.7).
Then $\Rightarrow$ *extensionally coincides with* $I(LR)$, *the* $\subseteq$ *-least fixed point of* LR:

$$\forall x \forall y(x \Rightarrow y \leftrightarrow \langle x,y\rangle \eta I(LR)).$$

(iii) $\alpha \leq \beta \rightarrow RED\alpha \subseteq RED\beta$.

PROOF. As to (i) and (iii), choose $RED := Rec(LR)$ and apply 23.6 (i)-(ii),

27.10 (ii). Part (ii) is a consequence of 23.10. □

27.12. CONVENTION. We henceforth adopt the following notations:

$$\Rightarrow_\alpha := RED\alpha \ \text{ and } \ \rightarrow_\alpha := RT(RED(<\alpha));$$

therefore $\Rightarrow_\alpha = H(\rightarrow_\alpha)$, $(x \Rightarrow y) \leftrightarrow \exists\alpha.\, x \Rightarrow_\alpha y$.

The formal reconstruction of $\Rightarrow$ is indeed adequate to its informal counterpart of 27.6; this is perhaps better seen if we explicitly state the relevant introduction rules embodied in 27.11 (i) (from right to left):

27.13. LEMMA ($PW^- + GID$)

EXTN: $(x \rightarrow_\alpha y) \rightarrow (x \Rightarrow_\alpha y)$;

AT.1.1: $(Nx \rightarrow [Nx] \Rightarrow_\alpha \mathbb{I}) \wedge (x = y \rightarrow [x = y] \Rightarrow_\alpha \mathbb{I})$;

AT.1.2: $(\neg Nx \rightarrow [Nx] \Rightarrow_\alpha \mathbb{O}) \wedge (\neg x = y \rightarrow [x = y] \Rightarrow_\alpha \mathbb{O})$;

NEG.1: $(x \rightarrow_\alpha y) \rightarrow (\neg x \Rightarrow_\alpha \neg y)$;

NEG.2: $(x \rightarrow_\alpha \mathbb{I}) \rightarrow (\neg x \Rightarrow_\alpha \mathbb{O}). \wedge .(x \rightarrow_\alpha \mathbb{O}) \rightarrow (\neg x \Rightarrow_\alpha \mathbb{I})$;

CONJ.1: $Funcl(f) \wedge Funcl(g) \wedge \forall x(f(x) \rightarrow_\alpha g(x)) \rightarrow (\sqcap f \Rightarrow_\alpha \sqcap g)$;

CONJ.2: $Funcl(f) \wedge \forall x(f(x) \rightarrow_\alpha \mathbb{I}) \rightarrow (\sqcap f \Rightarrow_\alpha \mathbb{I})$;

CONJ.3: $Funcl(f) \wedge \exists x(f(x) \rightarrow_\alpha \mathbb{O}) \rightarrow (\sqcap f \Rightarrow_\alpha \mathbb{O})$;

DE.1: $(x_1 \rightarrow_\alpha y_1 \wedge x_2 \rightarrow_\alpha y_2) \rightarrow (x_1 \equiv x_2) \Rightarrow_\alpha (y_1 \equiv y_2)$;

DE.2: $(x \rightarrow_\alpha z \wedge y \rightarrow_\alpha z) \rightarrow (x \equiv y \Rightarrow_\alpha \mathbb{I})$;

DE.3: $(x \rightarrow_\alpha \mathbb{O} \wedge y \rightarrow_\alpha \mathbb{I}) \vee (x \rightarrow_\alpha \mathbb{I} \wedge y \rightarrow_\alpha \mathbb{O}) \rightarrow (x \equiv y \Rightarrow_\alpha \mathbb{O})$.

§28. The Church-Rosser theorem for $\Rightarrow$

In this section we are going to prove:

28.1. THEOREM. $PW_c + EA + GID \vdash CR(\Rightarrow)$.

By 27.8 and 27.11 it suffices to verify the Church-Rosser property for $\Rightarrow_\alpha$, where α is an arbitrary ordinal. As in the case of finitary reduction (see § 4), the theorem requires a standard inversion argument, which is here combined with double transfinite induction on ON(= ordinals; 23.3). The point is that if we know the form of x (e.g. x is a truth value $\mathbb{O}$ or $\mathbb{I}$, an infinitary conjunction, etc.), we can tell which form any given y, such that $x \Rightarrow_\alpha y$, will possibly have.

First, it is convenient to use a stage-by-stage characterization of transitive reflexive closure $RT(r)$ of a binary relation r: since $RT(r)$ is monotone in r, there is an operation $\lambda r \lambda y\, RT(r,y)$ (by 23.6-23.10), such that:

28.2. LEMMA. $\mathrm{PW}^- + \mathrm{GID}$ *proves:*

(i) $\forall x \forall y(\langle x,y\rangle \eta RT(r,\alpha) \leftrightarrow ((x=y) \vee (\langle x,y\rangle \eta r) \vee$
$\vee \exists z(\langle x,z\rangle \eta RT(r,<\alpha) \wedge \langle z,y\rangle \eta RT(r,<\alpha)))),$
where $RT(r,<\alpha) := \{\langle u,v\rangle : \exists \beta < \alpha . \langle u,v\rangle \eta RT(r,\beta)\};$

(ii) $\alpha \leq \beta \rightarrow RT(r,\alpha) \subseteq RT(r,\beta);$

(iii) $\forall x \forall y(\langle x,y\rangle \eta RT(r) \leftrightarrow \exists \alpha(\langle x,y\rangle \eta RT(r,\alpha))).$

28.3. LEMMA (*Inversion*). *The universal closure of each of the following sentences is provable in* $\mathrm{PW}^- + \mathrm{GID}$:

$\mathrm{SENT}^-(\alpha)$: $x \Rightarrow_\alpha y \wedge \neg PTER(x) \rightarrow x = y$ (*see* 27.3);

$\mathrm{TV}^-(\alpha)$: $x \Rightarrow_\alpha y \wedge x\eta\{\mathbb{O},\mathbb{I}\} \rightarrow x = y;$

$\mathrm{AT}^-(\alpha)$: *if* $At(x) := \exists u \exists v(x = [Nu] \vee x = [u=v]),$
$(x \Rightarrow_\alpha y. \wedge At(x)) \rightarrow (y = x) \vee (y = \mathbb{I} \wedge Tx\ ) \vee (y = \mathbb{O} \wedge T\neg x);$

$\mathrm{NEG}^-(\alpha)$: $(\neg x \Rightarrow_\alpha y) \rightarrow$
$\rightarrow [\exists z(y = \neg z \wedge x \rightarrow_\alpha z) \vee (y = \mathbb{I} \wedge x \rightarrow_\alpha \mathbb{O}) \vee (y = \mathbb{O} \wedge x \rightarrow_\alpha \mathbb{I})];$

$\mathrm{CONJ}^-(\alpha)$: $(\sqcap f \Rightarrow_\alpha y) \rightarrow$
$\rightarrow [\exists g(Funcl(g) \wedge y = \sqcap g \wedge \forall x(f(x) \rightarrow_\alpha g(x))) \vee$
$\vee (\forall x(f(x) \rightarrow_\alpha \mathbb{I}) \wedge y = \mathbb{I}\ ) \vee (\exists x(f(x) \rightarrow_\alpha \mathbb{O}) \wedge y = \mathbb{O})];$

$\mathrm{DE}^-(\alpha)$: $(x \equiv z \Rightarrow_\alpha y) \rightarrow$
$\rightarrow [\exists x_1 \exists z_1(y = (x_1 \equiv z_1) \wedge (x \rightarrow_\alpha x_1) \wedge (z \rightarrow_\alpha z_1)) \vee$
$\vee \exists w(x \rightarrow_\alpha w \wedge z \rightarrow_\alpha w \wedge y = \mathbb{I}) \vee$
$\vee (y = \mathbb{O} \wedge ((x \rightarrow_\alpha \mathbb{O} \wedge z \rightarrow_\alpha \mathbb{I}) \vee (x \rightarrow_\alpha \mathbb{I} \wedge z \rightarrow_\alpha \mathbb{O})))].$

PROOF. Consider the formulas:

$D_1(\alpha,x,y) := (\neg PTER(x) \wedge x = y);$

$D_2(\alpha,x,y) := (x\eta\{\mathbb{O},\mathbb{I}\} \wedge x = y);$

$D_3(\alpha,x,y) := At(x) \wedge ((y = x) \vee (y = \mathbb{I} \wedge Tx) \vee (y = \mathbb{O} \wedge T\neg x\ ));$

$$D_4(\alpha,x,y) := \exists v(x = \neg v \wedge (\exists z(y = \neg z \wedge v \rightarrow_\alpha z) \vee (y = \mathbb{I} \wedge v \rightarrow_\alpha \mathbb{O}) \vee$$
$$\vee (y = \mathbb{O} \wedge v \rightarrow_\alpha \mathbb{I})));$$

$$D_5(\alpha,x,y) := \exists f \exists g((Funcl(f) \wedge \sqcap f = x \wedge Funcl(g) \wedge y = \sqcap g \wedge$$
$$\wedge \forall x(f(x) \rightarrow_\alpha g(x))) \vee \forall x(f(x) \rightarrow_\alpha \mathbb{I} \wedge y = \mathbb{I}) \vee \exists x(f(x) \rightarrow_\alpha \mathbb{O} \wedge y = \mathbb{O}));$$

$$D_6(\alpha,x,y) := \exists u \exists v(x = (u \equiv v) \wedge (\exists x_1 \exists z_1(y = (x_1 \equiv z_1) \wedge (u \rightarrow_\alpha x_1) \wedge$$
$$\wedge (v \rightarrow_\alpha z_1)) \vee \exists w(u \rightarrow_\alpha w \wedge v \rightarrow_\alpha w \wedge y = \mathbb{I}) \vee$$
$$\vee (y = \mathbb{O} \wedge ((u \rightarrow_\alpha \mathbb{O} \wedge v \rightarrow_\alpha \mathbb{I}) \vee (u \rightarrow_\alpha \mathbb{I} \wedge v \rightarrow_\alpha \mathbb{O})))))).$$

Clearly the lemma is a consequence of $\forall \alpha C(\alpha)$, where $C(\alpha)$ is the formula

$$\forall x \forall y(x \Rightarrow_\alpha y \rightarrow M(\alpha,x,y)),$$

and $M(\alpha,x,y) := D_1(\alpha,x,y) \vee \ldots \vee D_6(\alpha,x,y)$. We verify $\forall \alpha C(\alpha)$ by transfinite induction on α (see 23.4 (iii)). Thus we assume $x \Rightarrow_\alpha y$ and

$$\forall \beta < \alpha . C(\beta), \tag{1}$$

as main induction hypothesis (MIH in short). By 27.11 (i), either we have used the clause EXTN, i.e. $x \rightarrow_\alpha y$ holds (this always happens if $\neg PTER(x)$ holds), or else $H_0(x,y,\rightarrow_\alpha)$ (see 27.9), and the required conclusion obtains. {As a typical instance, consider the case where $x \Rightarrow_\alpha y$ has the form $a \equiv b \Rightarrow_\alpha y$, but $x \Rightarrow_\alpha y$ does not follow by EXTN. Then we must have applied an inference among DE.1-DE.3: either $y = (x_1 \equiv z_1)$, $a \rightarrow_\alpha x_1$, $b \rightarrow_\alpha z_1$, for some x_1, z_1, or $a \rightarrow_\alpha w$, $b \rightarrow_\alpha w$, $y = \mathbb{I}$, for some w, or else $y = \mathbb{O}$, $a \rightarrow_\alpha \mathbb{O}$ ($\mathbb{I}$) and $b \rightarrow_\alpha \mathbb{I}$ ($\mathbb{O}$); then we are done. Observe that we implicitly rely upon the fact that $a \equiv b$ is provably different from every infinitary sentence built up by negation, $\sqcap$, and basic predicates ; cf. 27.1)}.
Therefore we are left with the verification of

$$\forall x \forall y((x \rightarrow_\alpha y) \rightarrow M(\alpha,x,y)). \tag{2}$$

Let us shorten $\langle a,b\rangle \eta RT(RED(<\alpha),\gamma) :=$ "$\langle a,b\rangle$ occurs at the γ-th stage in the reflexive transitive closure of $RED(<\alpha)$" with the simpler $(a \rightarrow_{\alpha_\gamma} b)$. By 28.2, (2) is equivalent to the statement

$$\forall \gamma \forall x \forall y((x \rightarrow_{\alpha_\gamma} y) \rightarrow M(\alpha,x,y)). \tag{3}$$

(3) is verified by induction on γ: so we assume $(x \rightarrow_{\alpha_\gamma} y)$ together with $\forall \delta < \gamma . \forall x \forall y((x \rightarrow_{\alpha_\delta} y) \rightarrow M(\alpha,x,y))$ as secondary induction hypothesis (SIH in short). If $(x \rightarrow_{\alpha_\gamma} y)$ holds by reflexivity, then $x = y$ and $M(\alpha,x,y)$ is trivial by SIH; on the other hand, if $\langle x,y\rangle \eta RED(<\alpha)$, we are done by MIH. Thus by 28.2 (i) we only check $M(\alpha,x,y)$, under the assumptions:

$$(x \rightarrow_{\alpha_\delta} z); \tag{3.1}$$

$$(z \rightarrow_{\alpha_\delta} y), \tag{3.2}$$

for some z, $\delta < \gamma$. We distinguish six main cases.

1: $\neg PTER(x)$ holds. By SIH applied to (3.1), $x = z$ and hence by (3.2) $(x \rightarrow_{\alpha_\delta} y)$, which implies $x = y$ again by SIH. Hence $D_1(\alpha, x, y)$.

2: $x = \mathbb{I}$ $(x = \mathbb{O})$. Then $z = \mathbb{I}$ $(z = \mathbb{O})$ by SIH applied to (3.1), whence $y = \mathbb{I}$ $(y = \mathbb{O})$ by SIH applied to (3.2). Hence $D_2(\alpha, x, y)$.

3: $At(x)$ holds. Apply SIH for (3.1). If $x = z$, the conclusion follows by SIH applied to (3.2). If Tx and $z = \mathbb{I}$ hold, then $y = \mathbb{I}$ by SIH for (3.2) and we are done. The remaining case is similar. In all cases, we have $D_3(\alpha, x, y)$.

4: $x = \neg a$. By SIH applied to (3.1), we must analyze three subcases.

4.1: $z = (\neg u)$, for some u such that $a \rightarrow_\alpha u$. Then we can apply SIH to (3.2) and we have $D_4(\alpha, z, y)$. If $y = \neg v$ and $u \rightarrow_\alpha v$, we have $a \rightarrow_\alpha v$; if $y = \mathbb{O}$ $(\mathbb{I})$ and $u \rightarrow_\alpha \mathbb{I}$ $(\mathbb{O})$, also $a \rightarrow_\alpha \mathbb{I}$ $(\mathbb{O})$. In all cases, we conclude to $D_4(\alpha, x, y)$; note that we need the fact that $\rightarrow_\alpha$ is a transitive closure.

4.2-4.3: $z = \mathbb{O}$ and $a \rightarrow_\alpha \mathbb{I}$ $(z = \mathbb{I}$ and $a \rightarrow_\alpha \mathbb{O})$. Then (3.2) and SIH imply $y = \mathbb{O}$ $(y = \mathbb{I})$, and we are done.

5: $x = \sqcap f$. By SIH for (3.1), we consider three subcases.

5.1. $z = \sqcap g$, where g is a function-class such that $f(c) \rightarrow_\alpha g(c)$, for every c. Then we can apply SIH to (3.2). If $y = \sqcap h$ and $g(c) \rightarrow_\alpha h(c)$, for every c and some function-class h, we get $f(c) \rightarrow_\alpha h(c)$ by transitivity for arbitrary c, i.e. $D_5(\alpha, x, y)$. If $z = \mathbb{I}$ and $f(c) \rightarrow_\alpha \mathbb{I}$, for every c, we have $y = \mathbb{I}$ by SIH for (3.2) and we are done; the case $z = \mathbb{O}$ is analogous.

5.2-5.3. If $z = \mathbb{I}$ and $f(c) \rightarrow_\alpha \mathbb{I}$, for every c (or $z = \mathbb{O}$ and $f(c) \rightarrow_\alpha \mathbb{O}$, for some c), we argue as in 5.1.

6. $x = (x_1 \equiv x_2)$. By SIH applied to (3.1) we consider three subcases.

6.1. $z = (z_1 \equiv z_2)$ and $(x_i \rightarrow_\alpha z_i)$ $(i = 1, 2)$. Hence we can apply SIH to (3.2). If $y = (y_1 \equiv y_2)$ and $(z_i \rightarrow_\alpha y_i)$, we get $(x_i \rightarrow_\alpha y_i)$ (for $i = 1, 2$) and hence $D_6(\alpha, x, y)$. The remaining cases are similar.

6.2-6.3. $z = \mathbb{I}$ and $x_1 \rightarrow_\alpha c$, $x_2 \rightarrow_\alpha c$, for some c, or $z = \mathbb{O}$ and $x_1 \rightarrow_\alpha \mathbb{I}$ $(\mathbb{O})$, $x_2 \rightarrow_\alpha \mathbb{O}$ $(\mathbb{I})$. Then we apply SIH to (3.2), in order to get $y = \mathbb{I}$ $(y = \mathbb{O})$. □

28.3.1. REMARK. $\text{SENT}^-(\alpha)$ implies:

$$\text{PW}^- + \text{GID} \vdash (a \Rightarrow b) \wedge PTER(b) \rightarrow PTER(a).$$

PROOF *of the main theorem* 28.1. We assume by IH that for every $\beta < \alpha$,

$CR(\Rightarrow_\beta)$ holds. We further observe that IH and 27.11 (iii) imply:

$$CR(RED(<\alpha)), \tag{*}$$

whence by 27.8:

$$CR(\rightarrow_\alpha). \tag{**}$$

Assume $a \Rightarrow_\alpha b$ and $a \Rightarrow_\alpha c$: we must produce d such that $b \Rightarrow_\alpha d$, $c \Rightarrow_\alpha d$. We distinguish six cases according to the form of a.

1. $\neg PTER(a)$ is assumed. Then by SENT$^-(\alpha)$, $a = b = c$ and we can choose $d = a$ ($x \Rightarrow_\alpha x$ holds by EXTN and closure of RT under reflexivity).

2. a is a truth value. Apply TV$^-(\alpha)$ and choose $d = a$.

3. $At(a)$ holds. We apply AT$^-(\alpha)$. If $a = b$, we choose $d = c$; if $b = \mathbb{I}$ ($\mathbb{O}$), we see that $d = \mathbb{I}$ ($\mathbb{O}$).

4. $a = \neg x$ for some x. Apply NEG$^-(\alpha)$.

4.1. Assume that $b = \neg y$, $c = \neg z$, for some y, z such that $x \rightarrow_\alpha y$, $x \rightarrow_\alpha z$. Then by (**), there exists w such that $y \rightarrow_\alpha w$, $z \rightarrow_\alpha w$, which yields (by 27.13) $b \Rightarrow_\alpha (\neg w)$, $c \Rightarrow_\alpha (\neg w)$: choose $d = \neg w$.

4.2. Assume $b = \neg y$, $c = \mathbb{I}$, where $x \rightarrow_\alpha y$ and $x \rightarrow_\alpha \mathbb{O}$. By (**), TV$^-(\alpha)$, 27.13 (NEG.2), we can choose $d = \mathbb{I}$; a similar argument yields $d = \mathbb{O}$ if $c = \mathbb{O}$ and $x \rightarrow_\alpha \mathbb{I}$.

4.3. Assume $b = \mathbb{I}$, $x \rightarrow_\alpha \mathbb{O}$. If $a \Rightarrow_\alpha c$ is introduced by NEG.1 (see 27.13), the argument is symmetric to 4.2. Else, notice that we cannot have $c = \mathbb{O}$ (this would imply the contradiction $\mathbb{O} = \mathbb{I}$ by (**), TV$^-(\alpha)$). Hence $c = \mathbb{I}$ and we can choose $d = \mathbb{I}$.

4.4. $b = \mathbb{O}$, $x \rightarrow_\alpha \mathbb{I}$: argue as in 4.3.

5. Assume $a = \sqcap f$. Apply CONJ$^-(\alpha)$. This is the most delicate step in the proof: it makes a(n essential ?) use of the selection theorem 20.4, and hence of EA, plus class induction on numbers.

5.1. Let $b = \sqcap g$, $c = \sqcap h$, where f, g, h are function-classes satisfying, for every u, the condition

$$f(u) \rightarrow_\alpha g(u) \text{ and } f(u) \rightarrow_\alpha h(u). \tag{1}$$

If $F(\alpha, g, h) = \{\langle u, w\rangle : (g(u) \rightarrow_\alpha w) \wedge (h(u) \rightarrow_\alpha w)\}$, (**) implies

$$\forall u \exists w (\langle u, w\rangle \eta F(\alpha, g, h)). \tag{2}$$

Hence by selection 20.4 we can find a function class $r = Sel(F(\alpha, g, h), V)$, such that, for every u, $g(u) \rightarrow_\alpha r(u)$ and $h(u) \rightarrow_\alpha r(u)$; hence we can choose $d = \sqcap r$ by CONJ.1 of 27.13.

5.2. Let $b = \mathbb{I}$, $f(u) \rightarrow_\alpha \mathbb{I}$, and $c = \sqcap h$, where $f(u) \rightarrow_\alpha h(u)$, for every u, and f, h are function classes. Then we can choose $d = \mathbb{I}$ by CONJ.2 (apply (**) and notice that $\mathrm{TV}^-(\alpha)$ implies $h(u) \rightarrow_\alpha \mathbb{I}$, for every u).
If we maintain the same hypothesis on c, but we assume $b = \mathbf{O}$ and $f(u) \rightarrow_\alpha \mathbf{O}$, for some u, we can choose $d = \mathbf{O}$ (by CONJ.3).

5.3. If both b and c are truth values, they must be identical and hence $d = b$; the remaining cases are symmetric to those already considered.

6. Let $a = x_1 \equiv x_2$. We apply $\mathrm{DE}^-(\alpha)$.

6.1. Assume that $b = \mathbb{I}$ and $x_1 \rightarrow_\alpha w$, $x_2 \rightarrow_\alpha w$. It is easily seen that $c \neq \mathbf{O}$ (by (**), $\mathrm{TV}^-(\alpha)$). Hence $d = \mathbb{I}$ if c is a truth value, and we can suppose that $c = y_1 \equiv y_2$ with $x_i \rightarrow_\alpha y_i$ $(i = 1, 2)$. Then by a first application of (**) we can find z_i such that $y_i \rightarrow_\alpha z_i$ $(i = 1, 2)$ and $w \rightarrow_\alpha z_1$, $w \rightarrow_\alpha z_2$; again with (**) we get z such that $z_i \rightarrow_\alpha z$ $(i = 1, 2)$; DE.2 grants that $d = \mathbb{I}$ is the right choice. A similar argument, together with DE.3, shows that $d = \mathbf{O}$ if $b = \mathbf{O}$.

6.2. Assume that $b = (y_1 \equiv y_2)$, $c = (z_1 \equiv z_2)$ with $x_i \rightarrow_\alpha y_i$, $x_i \rightarrow_\alpha z_i$ $(i = 1, 2)$. By (**) we can find d_i such that $y_i \rightarrow_\alpha d_i$, $z_i \rightarrow_\alpha d_i$, whence $x_i \rightarrow_\alpha d_i \, (i = 1, 2)$; an application of DE.1 yields $b \rightarrow_\alpha (d_1 \equiv d_2)$, $c \rightarrow_\alpha (d_1 \equiv d_2)$.

The remaining cases are symmetric to those already treated. □

§29. A model of type-free logic based on intensional equivalence

We first define a translation $^\varepsilon$ of $\mathcal{L}_B$ into $\mathcal{L}$; the basic idea is that $\equiv$ is mapped into the natural equivalence relation, generated by the confluent reduction $\Rightarrow$.

29.1.DEFINITION

(i) $\equiv_B := \{\langle x, y\rangle : \exists z(x \Rightarrow z \wedge y \Rightarrow z)\}$; we agree to write $x \equiv_B y$, for $\langle x, y\rangle \eta \equiv_B$.

(ii) Let $Gr(f) := \{u : \exists v(u = \langle v, fv\rangle)\}$;

$$Conj(x, y) := \{u : \exists v \exists w(u = \langle v, w\rangle \wedge ((v = \bar{0} \wedge w = x) \vee (\neg v = \bar{0} \wedge w = y)))\}.$$

$$ALL^* := \lambda f. \sqcap Gr(f); \quad AND^* := \lambda x \lambda y. \sqcap Conj(x, y).$$

(iii) The map $^\varepsilon$ is the identity map on variables and constants K, S, D, $PAIR$, $LEFT$, $RIGHT$, $\bar{0}$, SUC, $PRED$; it transforms the remaining constants of $\mathcal{L}_B$ according to the following stipulations:

$$(NAT\hat{\ })^{\mathcal{E}} := NAT; \qquad (ID\hat{\ })^{\mathcal{E}} := ID;$$
$$(NEG\hat{\ })^{\mathcal{E}} := NEG; \qquad (EQ\hat{\ })^{\mathcal{E}} := DE;$$
$$(ALL\hat{\ })^{\mathcal{E}} := ALL^*; \qquad (AND\hat{\ })^{\mathcal{E}} := AND^*;$$
$$(\mathbb{O}\hat{\ })^{\mathcal{E}} := \mathbb{O}\ ; \qquad (\mathbb{I}\hat{\ })^{\mathcal{E}} := \mathbb{I}.$$

{In (ii)-(iii) above, ID, NAT, NEG are the combinators of 7.1; DE, $\mathbb{O}$, $\mathbb{I}$, $\sqcap$ are defined in 27.1.}
The map $^{\mathcal{E}}$ is inductively extended to terms and formulas of $\mathcal{L}_B$:

$$(ts)^{\mathcal{E}} = t^{\mathcal{E}}s^{\mathcal{E}}; \quad (Nt)^{\mathcal{E}} = N(t^{\mathcal{E}}); \quad (t = s)^{\mathcal{E}} = (t^{\mathcal{E}} = s^{\mathcal{E}});$$
$$(t \equiv s)^{\mathcal{E}} = (t^{\mathcal{E}} \equiv_B s^{\mathcal{E}}); \quad (t \in s)^{\mathcal{E}} = (ts)^{\mathcal{E}} \equiv_B \mathbb{I};$$
$$(\neg A)^{\mathcal{E}} = \neg(A^{\mathcal{E}}); \quad (A \wedge B)^{\mathcal{E}} = A^{\mathcal{E}} \wedge B^{\mathcal{E}}; \quad (\forall x A)^{\mathcal{E}} = \forall x(A^{\mathcal{E}}).$$

The specific choices of 29.1 (ii) are motivated by the following:

29.2. LEMMA

(i) $\mathrm{MF}^- \vdash Funcl(Conj(a,b)) \wedge (AND^*ab)(\overline{0}) = a \wedge$
$$\wedge \forall x(\neg x = \overline{0} \rightarrow (AND^*ab)(x) = b);$$

(ii) $\mathrm{MF}^- \vdash Funcl(Gr(f)) \wedge \forall x((Gr(f))(x) = fx)$.

{Recall the definition 27.2 for the the notation $(\ldots)(x)$.}

29.3. THEOREM. *If* $\mathrm{BL}_c \vdash A$, *then* $\mathrm{PW}_c + \mathrm{GID} + \mathrm{EA} \vdash A^{\mathcal{E}}$.

Since $A^{\mathcal{E}} = A$, whenever A is in the language of OP without T, and $\mathrm{PW}_c + \mathrm{GID} + \mathrm{EA}$ is a conservative extension of OP (by 15.5 (i)), we can conclude:

29.4. COROLLARY. BL_c *is a conservative extension of* OP.

PROOF of 29.3. Let us prove the translation of class N-induction; we assume:

$$\forall x(bx \equiv_B \mathbb{I} \vee bx \equiv_B \mathbb{O}) \wedge \forall x(bx \equiv_B \mathbb{I} \rightarrow b(x+1) \equiv_B \mathbb{I}) \wedge b\overline{0} \equiv_B \mathbb{I}. \quad (1)$$

Then by the first conjunct of (1) and 28.3 (see $\mathrm{TV}^-(\alpha)$), we obtain, if we set $b_1 = \{x : bx \equiv_B \mathbb{I}\}$ and $b_2 = \{x : bx \equiv_B \mathbb{O}\}$:

$$\forall x(x\eta b_1 \vee x\eta b_2) \ \wedge \neg\exists x(x\eta b_1 \wedge x\eta b_2). \quad (2)$$

Hence by 16.11 and (2), there is a class c such that $c =_e b_1$ and $-c =_e b_2$; class N-induction yields $\forall x(Nx \rightarrow bx \equiv_B \mathbb{I})$.
It remains to check the provability of the $^{\mathcal{E}}$-translations of the $\equiv$-axioms of

groups 2-3 (p.164; 26.2). As to $(\mathrm{E}.1)^{\mathcal{E}}$, $\equiv_B$ is trivially symmetric, reflexive by the corresponding property of $\Rightarrow$ (27.6 (i), 27.13, EXTN), while it is transitive by the Church Rosser property of 28.1. $(\mathrm{E}.2.1)^{\mathcal{E}}$-$(\mathrm{E}.2.2)^{\mathcal{E}}$ are straightforward applications of 27.13 (see DE.1, CONJ.1, NEG.1) and 29.2. As a sample, consider $(\mathrm{E}.2.1)^{\mathcal{E}}$ and the case of $\wedge$-preservation. Assume $(a \equiv b)^{\mathcal{E}}$ and $(a' \equiv b')^{\mathcal{E}}$. By 27.11 (iii) and the closure properties of RT (see 27.6), there is an α such that $a \rightarrow_\alpha z$, $b \rightarrow_\alpha z$, $a' \rightarrow_\alpha w$, $b' \rightarrow_\alpha w$, for some z and w. By definition of AND^*, 29.2 and CONJ.1, we obtain:

$$AND^* aa' \Rightarrow_\alpha AND^* zw \text{ and } AND^* bb' \Rightarrow_\alpha AND^* zw,$$

whence $((a \wedge a') \equiv (b \wedge b'))^{\mathcal{E}}$.
As to $\equiv$-axioms of group 3, first observe that by $\mathrm{TV}^-(\alpha)$, $a \equiv_B \mathbb{I}$ ($\mathbb{O}$) is equivalent to $a \Rightarrow \mathbb{I}$ ($\mathbb{O}$). Then we argue as in the previous case, with introduction and inversion lemmata 28.3 and 27.13. Let us explicitly consider three instances.
Ad $(\mathrm{E}.3.1)^{\mathcal{E}}$. If $A = x \in y$, the translation reduces to an instance of the identity principle. Let $A = (x \equiv y)$: then we must check

$$x \equiv_B y \leftrightarrow (DExy) \equiv_B \mathbb{I} \leftrightarrow DExy \Rightarrow \mathbb{I}.$$

From left to right, we just apply closure under DE.2; in the opposite direction, we use $\mathrm{DE}^-(\alpha)$.
Ad $(\mathrm{E}.6)^{\mathcal{E}}$. Assume $(\forall a \equiv \mathbb{I}^{\wedge})^{\mathcal{E}} = (ALL^* a \equiv_B \mathbb{I})$: by 29.2 (ii) and $\mathrm{CONJ}^-(\alpha)$, we must have $a(x) \equiv_B \mathbb{I}$, for arbitrary x, i.e. $(\forall x.ax \equiv \mathbb{I}^{\wedge})^{\mathcal{E}}$. In the reverse direction, we apply CONJ.2.
Ad $(\mathrm{E}.7)^{\mathcal{E}}$. If $((a \equiv b) \equiv \mathbb{O}^{\wedge})^{\mathcal{E}}$, we get $(DEab) \Rightarrow \mathbb{O}$: by $\mathrm{DE}^-(\alpha)$, either $a \Rightarrow \mathbb{O}$ and $b \Rightarrow \mathbb{I}$, or else $a \Rightarrow \mathbb{I}$ and $b \Rightarrow \mathbb{O}$. In either case $(Da)^{\mathcal{E}}$,$(Db)^{\mathcal{E}}$ hold; were $(a \equiv b)^{\mathcal{E}}$, we could derive the contradiction $\mathbb{O} = \mathbb{I}$ by means of $CR(\Rightarrow)$ and $\mathrm{TV}^-(\alpha)$. The opposite direction is a consequence of DE.3. □

CHAPTER 7

ON THE GLOBAL STRUCTURE OF MODELS FOR REFLECTIVE TRUTH

§30. The lattice of fixed point models for the neutral minimal theory

§31. The sublattice of intrinsic fixed point models and the cardinality theorem

§32. Variations on the encoding technique: non-modularity and other oddities

§33. A model for an impredicative extension of reflective truth

§34. On Kripke's classification of self-referential sentences

§35. On the consistency of coinduction principles

Appendix: a variant to the basic operator Γ and the restriction axiom

According to chapter II, MF^- and its variants admit, as natural interpretations, the minimal fixed point models of a simple operator Γ, which can be described in the very language $\mathcal{L}$ of MF. This basic feature has been exploited at length, in order to establish the consistency of the approximation axioms and the generalized induction schema.

The starting point of this chapter is the observation that the fixed points of Γ are exactly the models of NMF^-, the subsystem of MF^-, which omits number-theoretic induction and consistency. This fact permits an extensive, natural application of lattice-theoretic techniques to the investigation of the global structure of NMF^--models. In fact, we concentrate upon the structure $FIX(\mathcal{M})$ of the fixed point models over *an arbitrary combinatory algebra* $\mathcal{M}$, and we prove that this structure is highly non-trivial and complex. An unexpected moral, which will emerge below, is that self-referential statements and conditions stand beforehand as flexible encoding tools for conveying information; their nonsensical character make them available to be filled with whatever sense we like. This idea is illustrated *ad nauseam* by the crucial results.

In particular, §30 first establishes a simple characterization theorem, implying that the collection of sentences *internally true in arbitrary models of the theory* MF^-(*extended with the restriction axiom of* §7) *is recursively axiomatizable.* Then we introduce a lattice-theoretic structure on $FIX(\mathcal{M})$.

It is shown that $FIX(\mathcal{M})$ is a complete involutive lattice, which is well-defined up to isomorphism of combinatory algebras. In a similar way, the least inductive model $\mathbb{O}(\mathcal{M})$ of MF^- only depends on the isomorphism type of $\mathcal{M}$.

§31 deals with the complete sublattice of the so-called intrinsic models (see Kripke 1975). Roughly speaking, a model is intrinsic if it is obtained from the minimal fixed point model $\mathbb{O}(\mathcal{M})$, by adding information which is compatible with that of any other model. The main result of §31 is an encoding method for producing plenty of intrinsic models; the cardinality theorem shows that there are $2^{card(M)}$ intrinsic models ($\mathcal{M}$ being the underlying combinatory algebra).

The subsequent section §32 establishes that $INT(\mathcal{M})$, the set of all intrinsic fixed points (and hence $FIX(\mathcal{M})$), is a non-distributive (indeed non-modular) lattice, which can be further "split" into disjoint intervals (in the sense of Whitman 1944). We finally produce infinite strictly descending sequences of models in $INT(\mathcal{M})$.

In §33 we use a self-referential property, in order to encode models of combinatory logic, embedded in second-order logic. The outcome is that we can consistently enlarge the framework of reflective truth with extensional objects, closed under impredicative comprehension. Again, this shows that the non-minimal models of MF^- can be quite complex.

§34 adapts to the present context Kripke's classification of self-referential sentences: the notions of grounded, intrinsic, paradoxical object are introduced and shown to be non-trivial.

The final section §35 is a complement to §12 of Ch. III. We concentrate on the pair $\langle\mathbb{O}(\mathcal{M}),\mathbb{I}(\mathcal{M})\rangle$, where $\mathbb{O}(\mathcal{M})$ and $\mathbb{I}(\mathcal{M})$ are the least and the largest fixed point of $FIX(\mathcal{M})$, respectively. We prove that $\mathbb{I}(\mathcal{M})$ satisfies a generalized coinduction principle, while $\mathbb{O}(\mathcal{M})$ verifies a generalized induction principle, which involves the complement of the largest fixed point of any given operator (for terminology, see 10.3).

To some extent, the results of this chapter are contained in Cantini (1989), (1993).

NB. We must warn the reader that we stick to the notations and simplifying conventions of chapters II-III, namely:

(i) Once $\mathcal{M}\models OP^-$ ($= OP$ without N-induction) is fixed, we assume that our basic languages are extended with distinct names for elements of M, the domain of $\mathcal{M}$; thus we deal with $\mathcal{L}_{op}(\mathcal{M})$, $\mathcal{L}(\mathcal{M})$ respectively. We do not typographically distinguish between elements of M and their names.

(ii) Syntax: $\neg$, $\wedge$, $\forall$ are also used for the proper terms NEG, AND, ALL of 7.1 (hence $\neg t$, $t \wedge s$, $\forall t$ are abbreviations for $NEGt$, $ANDts$, $ALLt$ (in the given order)); in addition, $\neg\neg t := \neg(\neg t)$ and $TA := T[A]$.

(iii) Semantics: if $\mathcal{M}$ is any given model of OP^- and b, $c \in M$, then bc, $\neg b$, $b \wedge c$, $\forall b$ shorten the proper $\mathcal{M}(Ap(b,c))$, $\mathcal{M}(NEGb)$, $\mathcal{M}(ANDbc)$, $\mathcal{M}(ALLb)$; $id(b,c)$, $tr(b)$, $nat(b)$ stand for $\mathcal{M}([a = b])$, $\mathcal{M}([Tb])$, $\mathcal{M}([Nb])$ (in the given order).

§30. The lattice of fixed point models for the neutral minimal theory

For the reader's sake, we repeat a few basic facts from §7. First of all, $PFOR(x)$ is the $\mathcal{L}_{op}$-formula, saying that x has one of the following forms: $\neg y$; $\forall y$; $[Ny]$; $[Ty]$; $[y = z]$; $y \wedge z$. The set M-$PFOR$ is

$$\{a : a \in M \text{ and } \mathcal{M} \models PFOR(a)\}.$$

30.1. LEMMA. *There exists an operator* $\Gamma : \mathcal{P}(M) \to \mathcal{P}(M)$ ((*see* 7.3 (v)) *such that*:

(i) *if* $P \subseteq M$, $\Gamma(P) \subseteq M$-$PFOR$.

(ii) $\Gamma : \mathcal{P}(M) \to \mathcal{P}(M)$ *is monotone*: $S \subseteq S' \Rightarrow \Gamma(S) \subseteq \Gamma(S')$.

(iii) *Assume* a, $b \in M$:
if $a \notin M$-$PFOR$, $(\neg a) \in \Gamma(S)$;
if A *is an e-atom or a negated e-atom*, $\mathcal{M}([A]) \in \Gamma(S)$ *iff* A *holds in* $\mathcal{M}$;
$(a \wedge b) \in \Gamma(S)$ *iff* $a \in S$ *and* $b \in S$;
$(\neg(a \wedge b)) \in \Gamma(S)$ *iff* $(\neg a) \in S$ *or* $(\neg b) \in S$;
$(\forall f) \in \Gamma(S)$ *iff* $\mathcal{M}(fa) \in S$, *for all* $a \in M$;
$(\neg(\forall f)) \in \Gamma(S)$ *iff* $(\neg(fa)) \in S$, *for some* $a \in M$;
$(\neg\, tr(a)) \in \Gamma(S)$ *iff* $(\neg a) \in S$;
$(tr(a)) \in \Gamma(S)$ *iff* $a \in S$;
$(\neg\neg a) \in \Gamma(S)$ *iff* $a \in S$.

30.2. DEFINITION

(i) $S \subseteq M$ is consistent (complete) iff for every $a \in M$, either $a \notin S$ or $(\neg a) \notin S$ ($a \in S$ or $(\neg a) \in S$).

$$CONS(\mathcal{M}) := \{S : S \subseteq M,\ S \text{ consistent}\};$$

$$COMP(\mathcal{M}) := \{S : S \subseteq M,\ S \text{ complete}\}.$$

(ii) $S \subseteq M$ is Γ-dense (Γ-closed) iff $S \subseteq \Gamma(S)$ ($\Gamma(S) \subseteq S$);

(iii) $FIX(\mathcal{M}) := FIX(\Gamma, \mathcal{M}) := \{S : S \subseteq M \text{ and } \Gamma(S) = S\}$;

$$FIX_{cs}(\mathcal{M}) := FIX(\mathcal{M}) \cap CONS(\mathcal{M});$$

$$FIX_{cp}(\mathcal{M}) := FIX(\mathcal{M}) \cap COMP(\mathcal{M}).$$

30.3. LEMMA. *If $S \subseteq M$ is Γ-dense and consistent (complete and Γ-closed), then $\Gamma(S)$ is Γ-dense and consistent (complete and Γ-closed).*

We recall from 7.10 that RES is the sentence:

$$\forall x((Tx \rightarrow PFOR(x)) \wedge (\neg PFOR(x) \rightarrow T\neg x));$$

NMF$^-$ is MF$^-$ without the consistency axiom (see 7.10, *T*.6). We now state a semantical *pendant* of proposition 7.12:

30.4. THEOREM (*Characterization*)

(i) $S \in FIX(\mathcal{M})$ *iff* $\langle \mathcal{M}, S\rangle \models$ NMF$^-$ + RES.

(ii) $S \in FIX_{cs}(\mathcal{M})$ *iff* $\langle \mathcal{M}, S\rangle \models$ NMF$^-$ + RES + CONS.

(iii) $S \in FIX_{cp}(\mathcal{M})$ *iff* $\langle \mathcal{M}, S\rangle \models$ NMF$^-$ + RES + COMP.

(iv) *If $\mathcal{M}$ is an ω-model (i.e. the interpretation of N is standard), we can omit the restriction* $^-$ *in* (i)-(iii) *above.*

PROOF. (ii)-(iv) are straightforward by (i).
(i): it is immediate from left to right, since the axioms simply establish that the interpretation of the truth-predicate is Γ-dense and Γ-closed. In the opposite direction, $\Gamma(S) \subseteq S$ is a consequence of the second part of RES, the appropriate axioms of NMF$^-$ and the inversion properties of lemma 30.1 (iii); the first part of RES with the truth axioms of NMF$^-$ yields $S \subseteq \Gamma(S)$. Alternatively, we could also observe that NMF$^-$ + RES is axiomatized by OP$^-$ + CONS + the fixed point axiom FPT of 7.12. □

If we regard the fixed points as *possible notions of truth*, we can immediately rephrase theorem 30.4 in the form of a completeness result, characterizing the "internal" and "external" logics of reflective truth.

30.5. DEFINITION. $SENT(\mathcal{L}) := \{A : A$ sentence of $\mathcal{L}\}$;

$\mathcal{ET} := \{A : A \in SENT(\mathcal{L})$ and $\langle \mathcal{M}, X\rangle \models A$, for every $\mathcal{M} \models$ OP$^-$ and every $X \in FIX_{cs}(\mathcal{M})\}$;

$\mathcal{IT} := \{A : A \in SENT(\mathcal{L})$ and $\mathcal{M}([A]) \in X$, for every $\mathcal{M} \models$ OP$^-$ and every $X \in FIX_{cs}(\mathcal{M})\}$.

$\mathcal{ET}(\mathcal{IT})$ might be called the *external* (*internal*) *logic of consistent reflective truth.*

30.6. COROLLARY. *$\mathcal{ET}$ and $\mathcal{IT}$ are axiomatizable (i.e. the sets of their Gödel numbers are recursively enumerable).*

PROOF. By the Gödel completeness theorem and 30.4. □

30.7. DEFINITION. $S^d := \{a : a \in M \text{ and } (\neg a) \notin S\}$ (= the dual of S).

30.8. LEMMA

(i) *If $S \in FIX(\mathcal{M})$, then $S^d \in FIX(\mathcal{M})$; moreover, if S is consistent (complete), S^d is complete (consistent).*

(ii) *If $S \in FIX(\mathcal{M})$, then $(S^d)^d = S$; $S \subseteq P$ implies $P^d \subseteq S^d$.*

PROOF. (ii) is trivial by definition of ^{d}, hypothesis, and lemma 30.1 (iii); as to (i), observe that $S^d \models \text{NMF}^-$ by theorem 8.11 and apply 30.4. □

We shall now investigate the closure properties of $FIX(\mathcal{M})$ and we directly introduce the appropriate lattice-theoretic operations.

30.9. DEFINITION. Let S be a subset of M. We define by transfinite recursion on ordinals ($ON =$ collection of ordinals):

(i) $UP(S,0) = S$; $UP(S,\alpha{+}1) = \Gamma(UP(S,\alpha))$;

$UP(S,\lambda) = \cup\{UP(S,\alpha) : \alpha < \lambda\}$, where λ is a limit;

(ii) $DOWN(S,0) = S$; $DOWN(S,\alpha{+}1) = \Gamma(DOWN(S,\alpha))$;

$DOWN(S,\lambda) = \cap\{DOWN(S,\alpha) : \alpha < \lambda\}$, where λ is a limit;

(iii) $UP(S) = \cup\{UP(S,\alpha) : \alpha \in ON\}$;

$DOWN(S) = \cap\{DOWN(S,\alpha) : \alpha \in ON\}$;

(iv) $\mathbb{O}(\mathcal{M}) = UP(\emptyset)$ and $\mathbb{I}(\mathcal{M}) = DOWN(M)$.

(v) If S is of the form $UP(S)$ ($DOWN(S)$), then $S(\alpha) := UP(S,\alpha)$ ($DOWN(S,\alpha)$ respectively) is called the *α-th approximation of S*.

30.10. LEMMA

(i) *If $X \subseteq Y$, then $UP(X) \subseteq UP(Y)$ and $DOWN(X) \subseteq DOWN(Y)$.*

(ii) *If S is Γ-dense, then $\alpha \leq \beta$ implies $UP(S,\alpha) \leq UP(S,\beta)$; if S is Γ-closed, then $\alpha \leq \beta$ implies $DOWN(S,\beta) \leq DOWN(S,\alpha)$.*

(iii) *If S is Γ-dense (and consistent), $UP(S)$ is the $\subseteq$-smallest (consistent) $S \in FIX(\mathcal{M})$ extending S; in particular, the minimum element $\mathbb{O}(\mathcal{M})$ of $FIX(\mathcal{M})$ is consistent.*

(iv) *If S is Γ-closed (and complete), $DOWN(S)$ is the $\subseteq$-largest (complete) $X \in FIX(\mathcal{M})$ contained in S. In particular, $\mathbb{I}(\mathcal{M})$ – the maximum of $FIX(\mathcal{M})$ – is complete and $\mathbb{I}(\mathcal{M})^d = \mathbb{O}(\mathcal{M})$.*

PROOF. Ad (i)-(ii): by induction on α, using Γ-monotonicity. In (ii), the assumption on S is essential for the case $\alpha = 0$.
Ad (iii). By (ii) and Cantor's theorem, the sequence $\{UP(S,\alpha) : \alpha \in ON\}$ is eventually constant. Hence, for some δ,

$$UP(S) = UP(S,\delta) = \Gamma(UP(S,\delta)) \in FIX(\mathcal{M}).$$

If X is Γ-closed with $S \subseteq X$, then $UP(S,\alpha) \subseteq X$ (easy ON-induction). Hence $UP(S)$ is the $\subseteq$-least Γ-closed $\supseteq S$. In addition, if S is consistent, each $UP(S,\alpha)$ is consistent by lemma 30.3 and (ii): hence $UP(S)$ is consistent.
Ad (iv). The dual claim for $DOWN(S)$ is left as exercise. As to $\mathbb{I}(\mathcal{M})^d = \mathbb{O}(\mathcal{M})$, apply 30.8. □

30.11. DEFINITION

(i) If $\mathfrak{C} \subseteq FIX(\mathcal{M})$, we set:
$\sqcup\mathfrak{C} := UP(\cup\mathfrak{C})(= \text{sup of the family } \mathfrak{C})$;
$\sqcap\mathfrak{C} := DOWN(\cap\mathfrak{C})\ (= \text{inf of the family } \mathfrak{C})$.
We adopt the standard abbreviation "X is a *poset*" for "X is a partially ordered set". As usual, $X \sqcup Y := \sqcup\{X, Y\}$ and $X \sqcap Y := \sqcap\{X,Y\}$. If X, $Y \in FIX(\mathcal{M})$, $X \leq Y$ iff $X \sqcap Y = X$ (iff $Y = X \sqcup Y$).

(ii) If $\mathfrak{P}$ is a poset with order $\leq$, $\mathfrak{C} \subseteq \mathfrak{P}$ is *coherent* iff for every pair a, b in $\mathfrak{C}$, there is $c \in \mathfrak{C}$, such that $a \leq c$ and $b \leq c$.

(iii) $\mathfrak{P}$ is a *complete coherent poset* (in short $\mathfrak{P}$ is c.c.p.o.) iff $\mathfrak{P}$ is a poset and every coherent subset of $\mathfrak{P}$ has a sup in $\mathfrak{P}$.

(iv) If $\mathfrak{P}$ is a poset with order $\leq$ and ^{d} is a map of $\mathfrak{P}$ into itself satisfying $(a^d)^d = a$ and $a \leq b \Rightarrow b^d \leq a^d$, then ^{d} is an *involution*. A poset $\mathfrak{P}$ with an involution ^{d} is called *involutive*.

30.12. LEMMA

(i) *If* $\mathfrak{C} \subseteq FIX(\mathcal{M})$, $\cup\mathfrak{C}$ *is* Γ*-dense and* $\cap\mathfrak{C}$ *is* Γ*-closed.*
(ii) *If* X, $Y \in FIX(\mathcal{M})$, *then* $X \leq Y$ *iff* $X \subseteq Y$.
(iii) $\sqcup\mathfrak{C}$ *is the sup of* $\mathfrak{C}$ *and* $\sqcap\mathfrak{C}$ *is the inf of* $\mathfrak{C}$ *(with respect to* $\leq$*).*

PROOF. (i): trivial with Γ-monotonicity and $\mathfrak{C} \subseteq FIX(\mathcal{M})$.
(ii): $X \subseteq Y$ implies $X \cup Y = Y$, whence

$$UP(X \cup Y) = X \sqcup Y = UP(Y) = Y, \text{ i.e. } X \leq Y.$$

Conversely, assume $X \leq Y$, i.e. $X \sqcup Y = Y$. Since $X \cup Y$ is Γ-dense (by (i)), we have $X \cup Y \subseteq X \sqcup Y = Y$, i.e. $X \subseteq Y$.
(iii) If $X \in \mathfrak{C}$, $X \subseteq \cup\mathfrak{C}$, whence $UP(X) = X \subseteq \sqcup\mathfrak{C}$ (by lemma 30.10 (i)), i.e. $X \leq \sqcup\mathfrak{C}$ by (ii) above. Let $Y \in FIX(\mathcal{M})$ with $X \leq Y$, for every $X \in \mathfrak{C}$;

let $S(\alpha) := UP(\cup \mathcal{C}, \alpha)$. Then we have, by induction on α:

$$S(\alpha) \subseteq Y, \text{ for every } \alpha. \qquad (+)$$

(If $\alpha = 0$, use the assumption and (ii); else, apply Γ-monotonicity and IH). But (+) implies $\sqcup \mathcal{C} \subseteq Y$ and hence $\sqcup \mathcal{C} \leq Y$ (again by (ii)). The argument for $\sqcap \mathcal{C}$ is dual. □

30.13. THEOREM

(i) $\langle FIX(\mathcal{M}), \sqcup, \sqcap, {}^d, \mathbb{O}(\mathcal{M}), \mathbb{I}(\mathcal{M})\rangle$ *is a complete lattice with minimum* $\mathbb{O}(\mathcal{M})$, *maximum* $\mathbb{I}(\mathcal{M})$, *and with an involution* ^{d}, *satisfying, for* $\mathcal{C} \subseteq FIX(\mathcal{M})$,

$$(\sqcup \mathcal{C})^d = \sqcap\{X^d : X \in \mathcal{C}\} \text{ and } (\sqcap \mathcal{C})^d = \sqcup\{X^d : X \in \mathcal{C}\} \qquad (*)$$

(ii) $\langle FIX_{cs}(\mathcal{M}), \leq\rangle$ *is a complete coherent poset with minimum* $\mathbb{O}(\mathcal{M})$, *which is closed under* $\sqcap$ *over non-empty subfamilies of consistent fixed points.*

(iii) ^{d} *is a dual isomorphism between* $\langle FIX_{cs}(\mathcal{M}), \leq\rangle$ *and* $\langle FIX_{cp}(\mathcal{M}), \leq\rangle$. *In other words:* ^{d} *is a bijection of* $FIX_{cs}(\mathcal{M})$ *onto* $FIX_{cp}(\mathcal{M})$ *satisfying* (*) *above (see* 30.2 *for* $FIX_{cs}(\mathcal{M})$, $FIX_{cp}(\mathcal{M})$).

PROOF. (i): straightforward by lemma 30.8 and 30.12. For instance, as to (*), observe that $X \leq \sqcup \mathcal{C}$ implies $(\sqcup \mathcal{C})^d \leq X^d$, for every $X \in \mathcal{C}$; hence $(\sqcup \mathcal{C})^d \subseteq \sqcap\{X^d : X \in \mathcal{C}\}$. Conversely, if $X \in \mathcal{C}$, then $\sqcap\{X^d : X \in \mathcal{C}\} \subseteq X^d$, whence $X \subseteq (\sqcap\{X^d : X \in \mathcal{C}\})^d$, which implies $\sqcup \mathcal{C} \subseteq (\sqcap\{X^d : X \in \mathcal{C}\})^d$ (definition of sup) and finally $\sqcap\{X^d : X \in \mathcal{C}\} \subseteq (\sqcup \mathcal{C})^d$.

(ii) Let $\mathcal{C} \subseteq FIX(\mathcal{M}) \cap CONS(\mathcal{M})$: if $\mathcal{C}$ is non-empty, $\cap \mathcal{C}$ is consistent; hence $\sqcap \mathcal{C}$ is the consistent inf of $\mathcal{C}$ (by 30.12 (i), 30.10 (iv)). If $\mathcal{C}$ is a coherent family of consistent fixed points, $\cup \mathcal{C}$ is Γ-dense and consistent (lemma 30.10). Hence by lemma 30.3 and induction on α, it follows that $UP(\cup \mathcal{C}, \alpha)$ is consistent; therefore $\sqcup \mathcal{C}$ is the consistent sup of $\mathcal{C}$ (lemma 30.12 (iii)).

(iii) the map ^{d} is onto because every complete fixed point X has the form Y^d for $Y \in FIX_{cs}(\mathcal{M})$: choose $Y = X^d$ (lemma 30.8). ^{d} is one-to-one: trivially $X^d = Y^d$ yields $(X^d)^d = (Y^d)^d$, which implies $X = Y$, again by lemma 30.8. As to the verification of (*), apply (i). □

30.13.1. REMARK. As to the map ^{d}, (i) above establishes that ^{d} is a dual automorphism of $FIX(\mathcal{M})$.

One may wonder to what extent the lattice structure of $FIX(\mathcal{M})$ depends on the underlying combinatory algebra $\mathcal{M}$; we prove that $FIX(\mathcal{M})$ is only determined by the isomorphism type of $\mathcal{M}$. Similarly, if $\mathcal{M}_1$ and $\mathcal{M}_2$ are isomorphic, then the expanded structures $\langle \mathcal{M}_1, \mathbb{O}(\mathcal{M}_1)\rangle$ and

$\langle \mathcal{M}_2, \mathbb{O}(\mathcal{M}_2)\rangle$ are isomorphic.
If t is a closed term of $\mathcal{L}_{op}(\mathcal{M})$, $\mathcal{M}(t)$ is the value of t in M. Below we make use of the following facts, well-known from introductory model theory:

30.14.1. *Let ϕ be an isomorphism between $\mathcal{M}_1$ and $\mathcal{M}_2$. Then*

$$\phi(\mathcal{M}_1(t(a_1,\ldots,a_n))) = \mathcal{M}_2(t(\phi(a_1),\ldots,\phi(a_n))),$$

for every term $t(x_1,\ldots,x_n)$ of $\mathcal{L}_{op}$ and arbitrary $a_1,\ldots,a_n \in M_1$.

30.14.2. *$\mathcal{M}_1$ and $\mathcal{M}_2$ are isomorphic modulo $\phi: M_1 \to M_2$, provided $\mathcal{M}_1 \models A(a_1,\ldots,a_n)$ iff $\mathcal{M}_2 \models A(\phi(a_1),\ldots,\phi(a_n))$, for every atomic formula A of $\mathcal{L}_{op}$, with free variables in the list $x_1,\ldots,x_n$, and for every $a_1,\ldots,a_n \in M_1$.* {Here we do not distinguish typographically between the $a_1,\ldots,a_n \in M_1$, $\phi(a_1),\ldots,\phi(a_n) \in M_2$ and their syntactical representatives in $\mathcal{L}(M_1)$ and $\mathcal{L}(M_2)$}.

Let $\Gamma_1 : \mathcal{P}(M_1) \to \mathcal{P}(M_1)$ and $\Gamma_2 : \mathcal{P}(M_2) \to \mathcal{P}(M_2)$, where each Γ_i is the Γ-operator of 30.1 and 7.3, relativized to $\mathcal{M}_i$, i = 1, 2. Similarly, let $UP_1 : \mathcal{P}(M_1) \to \mathcal{P}(M_1)$ $(UP_2 : \mathcal{P}(M_2) \to \mathcal{P}(M_2))$ denote the UP-operator induced by iterating Γ_1 (Γ_2 respectively) on subsets of M_1 (M_2).

30.15. LEMMA. *Let $\mathcal{M}_1 \models \mathrm{OP}^-$ and $\mathcal{M}_2 \models \mathrm{OP}^-$; let ϕ be an isomorphism between $\mathcal{M}_1$ and $\mathcal{M}_2$. Then:*

(i) *the map $\phi: \mathcal{P}(M_1) \to \mathcal{P}(M_2)$, defined by $\phi(X) = \{\phi(a) : a \in X\}$, commutes with the operator Γ, which characterizes the models of* NMF^-:

$$X \subseteq M_1 \Rightarrow \phi(\Gamma_1(X)) = \Gamma_2(\phi(X)). \qquad (*)$$

Hence, if $X \in FIX(\mathcal{M}_1)$, $\phi(X) \in FIX(\mathcal{M}_2)$.

(ii) *$\phi: \mathcal{P}(M_1) \to \mathcal{P}(M_2)$ commutes with the UP-operator:*

$$\textit{if } X \subseteq M_1, \textit{ then } \phi(UP_1(X)) = UP_2(\phi(X)). \qquad (**)$$

(iii) *ϕ preserves ^{d}, i.e for every $Y \subseteq M_1$,*

$$\phi(Y^d) = \phi(Y)^d.$$

PROOF. (i) Let ϕ be the given isomorphism between $\mathcal{M}_1$ and $\mathcal{M}_2$. First observe that 30.14.1 implies, for $a \in M_1$:

$$\mathcal{M}_1 \models PFOR(a) \text{ iff } \mathcal{M}_2 \models PFOR(\phi(a)). \qquad (1)$$

We now assume $X \in FIX(\mathcal{M}_1)$, and we check that $c \in \Gamma_1(X)$ implies $\phi(c) \in \Gamma_2(\phi(X))$. We freely apply 30.1 and 30.14.1-30.14.2 without explicit mention. We have to distinguish several cases.

Case 1. If $c = \mathcal{M}_1(\neg a)$ and $a \notin M_1\text{-}PFOR$, we can infer by (1) that $\phi(a) \notin M_2\text{-}PFOR$, whence $\mathcal{M}_2(\neg\phi(a)) = \phi(c) \in \Gamma_2(\phi(X))$.

Case 2. $c = \mathcal{M}_1([a = b])$: then $c \in \Gamma_1(X)$ implies $\mathcal{M}_1 \models (a = b)$; hence, by 30.14.2, $\mathcal{M}_2 \models \phi(a) = \phi(b)$, whence again:

$$\mathcal{M}_2([\phi(a) = \phi(b)]) = \mathcal{M}_2(\phi([a = b])) = \phi(c) \in \Gamma_2(\phi(X)).$$

Case 3. Let $c = \mathcal{M}_1(\forall a)$: then $c \in \Gamma_1(X)$ implies $\mathcal{M}_1(ad) \in X$, for every $d \in M_1$. Assume that u is an arbitrary element of M_2; since ϕ is onto, $u = \phi(v)$, for some $v \in M_1$ and hence by hypothesis $\mathcal{M}_1(av) \in X$, i.e. $\phi(\mathcal{M}_1(av)) = \mathcal{M}_2(\phi(a)u) \in \phi(X)$, which implies

$$\mathcal{M}_2(\forall(\phi(a))) = \mathcal{M}_2(\phi(\forall a)) \in \Gamma_2(\phi(X)).$$

Case 4. $c = \mathcal{M}_1(\neg\neg a)$: then $c \in \Gamma_1(X)$ implies $\mathcal{M}_1(a) \in X$, whence $\phi(\mathcal{M}_1(a)) = \mathcal{M}_2(\phi(a)) \in \phi(X)$, which yields

$$\mathcal{M}_2(\neg\neg\phi(a)) = \mathcal{M}_2(\phi(c)) \in \Gamma_2(\phi(X)).$$

The remaining cases are similar and left to the reader. The verification of $\Gamma_2(\phi(X)) \subseteq \phi(\Gamma_1(X))$ follows the same pattern, by repeated use of 30.14.1-30.14.2 and lemma 30.1. As to the final claim, if $X = \Gamma_1(X)$, $\phi(X) = \phi(\Gamma_1(X)) = \Gamma_2(\phi(X))$.

(ii) If $X \in M_1$, trivially

$$\phi(UP_1(X)) = \phi(\cup\{UP_1(X,\alpha) : \alpha \in ON\}) = \cup\{\phi(UP_1(X,\alpha)) : \alpha \in ON\}.$$

We check, by induction on α:

$$\phi(UP_1(X,\alpha)) = UP_2(\phi(X),\alpha). \qquad (***)$$

If $\alpha = 0$, $(***)$ holds by definition; if α is limit,

$$\phi(UP_1(X,\alpha)) = \phi(\cup\{UP_1(X,\beta) : \beta < \alpha\}) = \cup\{\phi(UP_1(X,\beta)) : \beta < \alpha\} =$$
$$= \cup\{UP_2(\phi(X),\beta) : \beta < \alpha\} = UP_2(\phi(X),\alpha),$$

where IH is applied in the third step. If $\alpha = \beta+1$, we get by definition and step (i):

$$\phi(UP_1(X,\alpha)) = \phi(\Gamma_1(UP_1(X,\beta))) = \Gamma_2(\phi(UP_1(X,\beta))),$$

whence by IH $\Gamma_2(\phi(UP_1(X,\beta))) = \Gamma_2(UP_2(\phi(X),\beta))) = UP_2(\phi(X),\alpha)$.

(iii) By definition of ^{d} and essential use of all the properties of ϕ. □

30.16. THEOREM

(i) *If ϕ is an isomorphism of the combinatory algebra $\mathcal{M}_1$ onto the combinatory algebra $\mathcal{M}_2$, then ϕ induces a complete lattice isomorphism between $FIX(\mathcal{M}_1)$ and $FIX(\mathcal{M}_2)$.*

(ii) *Under the same hypothesis of* (i), *the structures $\langle\mathcal{M}_1, \mathbb{O}(\mathcal{M}_1)\rangle$ and*

$\langle \mathcal{M}_2, \mathbb{O}(\mathcal{M}_2)\rangle$ *are isomorphic.*

PROOF. (i): define $\phi(X) = \{\phi(a) : a \in X\}$ $(X \in FIX(\mathcal{M}_1))$.
Obviously $\phi : FIX(\mathcal{M}_1) \to FIX(\mathcal{M}_2)$ by lemma 30.15 (i) and ϕ is bijective. Since ϕ^{-1} is an isomorphism between $\mathcal{M}_2$ and $\mathcal{M}_1$, we have, with the lemma above:

$$\Gamma_1(\phi^{-1}[Y]) = \phi^{-1}(\Gamma_2(Y)) = \phi^{-1}(Y) \text{ (for } Y \in FIX(\mathcal{M}_2)).$$

Therefore $\phi^{-1}(Y) \in FIX(\mathcal{M}_1)$ and ϕ is onto M_2. Since ϕ is a bijection between $FIX(\mathcal{M}_1)$ and $FIX(\mathcal{M}_2)$, ϕ induces a complete lattice isomorphism, as

$$\phi(\sqcup_1\{X_i : i \in I\}) = \sqcup_2\{\phi(X_i) : i \in I\}, \qquad (****)$$

where $\sqcup_1 = \sup$ in $FIX(\mathcal{M}_1)$; $\sqcup_2 = \sup$ in $FIX(\mathcal{M}_2)$. Indeed, using the lemma 30.15 (ii), we have:

$$\phi(\sqcup_1\{X_i : i \in I\}) = \phi(UP_1(\cup\{X_i : i \in I\})) = UP_2(\phi(\cup\{X_i : i \in I\})) =$$
$$= UP_2(\cup\{\phi(X_i) : i \in I\}) = \sqcup_2\{\phi(X_i) : i \in I\}.$$

(ii): straightforward corollary of lemma 30.15 and the recursive definition of $\mathbb{O}(\mathcal{M}_1)$. □

§31. The sublattice of intrinsic fixed point models and the cardinality theorem

Following Kripke (1975), we introduce an interesting subclass of the class of consistent fixed point models. Intuitively, an *intrinsic* model is a fixed point of Γ, which "smoothly and consistently" grows from $\mathbb{O}(\mathcal{M})$, in the sense that it is compatible with every possible fixed point. Below, $\mathcal{M}$ is a given model of OP^-.

31.1. DEFINITION

(i) Let $X \in FIX_{cs}(\mathcal{M})$: then X is *intrinsic* iff X is compatible with every $Y \in FIX_{cs}(\mathcal{M})$, i.e. for every $Y \in FIX_{cs}(\mathcal{M})$, there exists some $Z \in FIX_{cs}(\mathcal{M})$ with $X, Y \leq Z$.

$$INT(\mathcal{M}) := \{X \in FIX_{cs}(\mathcal{M}) : X \text{ is intrinsic}\}.$$

(ii) If $X \in CONS(\mathcal{M})$, X is intrinsic iff $X \subseteq Y$, for some intrinsic $Y \in FIX_{cs}(\mathcal{M})$.

(iii) $\mathbb{P} := \mathbb{P}(\mathcal{M}) := \sqcup INT(\mathcal{M})$ (= the maximum intrinsic fixed point, by 31.3 (iii) below).

(iv) A consistent fixed point X is *maximal* iff $X = Y$, whenever $X \leq Y$

and $Y \in FIX_{cs}(\mathcal{M})$;

$$MAX(\mathcal{M}) := \{X \in FIX_{cs}(\mathcal{M}) \colon X \text{ is maximal}\}.$$

That the previous notions make sense, is granted by the following lemmata.

31.2. LEMMA. *Every consistent fixed point can be extended to a maximal consistent fixed point.*

PROOF. Every $\leq$-chain in $FIX_{cs}(\mathcal{M})$ has an upper bound by 30.13; hence the existence of maximal elements of $FIX_{cs}(\mathcal{M})$ is a consequence of Zorn's lemma. □

31.3. LEMMA

(i) *If $X \subseteq M$ is consistent, intrinsic and Γ-dense (Γ-closed), then $UP(X)$ ($DOWN(X)$) is intrinsic.*

(ii) $X \in INT(\mathcal{M})$ *iff* $X \leq \sqcap MAX(\mathcal{M})$.

(iii) $\mathbb{P}(\mathcal{M}) = \sqcap MAX(\mathcal{M})$; *hence* $X \in INT(\mathcal{M})$ *iff* $X \leq \mathbb{P}(\mathcal{M})$.

PROOF. (i) By assumption and lemma 30.10.
(ii) $\Rightarrow$: by definition of maximality and $\sqcap$. $\Leftarrow$: let $X \leq \sqcap MAX(\mathcal{M})$ and pick an arbitrary $Z \in FIX_{cs}(\mathcal{M})$. By lemma 31.2, Z can be extended to a maximal element Z', whence $X \leq Z'$ and $Z \leq Z'$, i.e. $X \in INT(\mathcal{M})$.
(iii) $(\sqcup INT(\mathcal{M})) \leq \sqcap MAX(\mathcal{M})$: by (ii) and definition of $\sqcup, \sqcap$.
$\sqcap MAX(\mathcal{M}) \leq (\sqcup INT(\mathcal{M}))$: $MAX(\mathcal{M}) \subseteq FIX_{cs}(\mathcal{M})$ is non-empty by lemma 31.2; hence we have $\sqcap MAX \in FIX_{cs}(\mathcal{M})$ by 30.12 (ii) and $\sqcap MAX(\mathcal{M}) \in INT(\mathcal{M})$ by (ii) above, which implies

$$\sqcap MAX(\mathcal{M}) \leq \sqcup INT(\mathcal{M}).$$ □

By lemma 31.3, we immediately derive:

31.4. THEOREM

$INT(\mathcal{M}) := \langle INT(\mathcal{M}), \sqcup, \sqcap, \mathbb{O}(\mathcal{M}), \mathbb{P}(\mathcal{M})\rangle$ *is a complete sublattice of* $FIX(\mathcal{M})$.

We now show that $INT(\mathcal{M})$ is indeed very rich. To this aim, we encode arbitrary subsets of the ground combinatory structure by means of self-referential properties: as a by-product, we have a method for generating plenty of intrinsic fixed points. Recalling that $\lambda x.(x = x \wedge \neg(Jx \wedge \neg Jx))$ is an abbreviation for

$$\lambda x.AND(IDxx)(NEG(AND(Jx)(NEG(Jx)))),$$

we state:

31.5. LEMMA. *There exists a closed term J of $\mathcal{L}_{op}$ such that, if $\mathcal{M} \models \mathrm{OP}^{-}$, then:*

(i) $\mathcal{M} \models J = \lambda x.(x = x \wedge \neg(Jx \wedge \neg Jx))$;

$\mathcal{M} \models Jx = Jy \rightarrow x = y$;

(ii) *if* $X \in FIX_{cs}$, $X \models \forall x(x\eta J \leftrightarrow (x\eta J \vee x\overline{\eta} J)) \wedge \forall x(\neg x\overline{\eta} J)$;

(iii) $\mathbb{O}(\mathcal{M}) \models \forall x(\neg x\eta J)$.

PROOF.
(i): apply the paradoxical combinator to $\lambda y\lambda x(x = x \wedge \neg(yx \wedge \neg(yx)))$. As to the second claim, we apply the independence conditions of AND and ID (cf. 7.1.1): $Jx = Jy$ implies $[x = x] = [y = y]$, whence $x = y$.
(ii): apply type-free comprehension 9.2 and consistency of X.
(iii): pick α minimal such that $\mathbb{O}(\alpha) \models a\eta J$, for some $a \in M$, where $\mathbb{O}(\alpha)$ is the αth-approximation of $\mathbb{O}(\mathcal{M})$. By definition of $\mathbb{O}(\mathcal{M})$, $\alpha > 0$. Then by (i)-(ii), $\mathbb{O}(\alpha) \models T(a = a \wedge \neg(\neg Ja \wedge Ja))$, whence it follows, for some $\beta < \alpha$, either $\mathbb{O}(\mathcal{M}, \beta) \models a\eta J$ or $\mathbb{O}(\mathcal{M}, \beta) \models a\overline{\eta} J$: but the former is impossible by minimality of α, while the second implies that $\mathbb{O}(\mathcal{M})$ is inconsistent: contradiction! □

The reason for introducing J is that J plays the role of a generic property: if $P \subseteq M$, then there is an intrinsic fixed point, where J represents P. The redundancy $[x = x]$ in the choice of J is only needed to make J injective.

31.6. DEFINITION

(i) Let $P \subseteq M$: then

$$J(P) := \{\mathcal{M}(Ja) : a \in P\} \cup \{\mathcal{M}(\neg\neg Ja) : a \in P\} \cup$$
$$\cup \{\mathcal{M}(\neg(Ja \wedge \neg Ja)) : a \in P\} \cup \{\mathcal{M}(a = a) : a \in P\}.$$

(ii) $G(P) := UP(J(P))$.

If $a_1, \ldots, a_n \in M$, $G(a_1, \ldots, a_n)$ stands for $G(\{a_1, \ldots, a_n\})$.

We collect a few simple facts about G.

31.7. LEMMA

(i) *$J(P)$ is Γ-dense, consistent and $G(P) \in INT(\mathcal{M})$.*

(ii) *$G(P) \models a\eta J$ iff $a \in P$; $G(P) \models \neg Cl(J)$, whenever $P \subset M$.*

(iii) *$G(\emptyset) = \mathbb{O}(\mathcal{M})$ and $\mathbb{P}(\mathcal{M}) \models \forall x(x\eta J)$.*

(iv) *If* $P \subseteq M$ *is non-empty,*

$$\mathbb{O}(\mathcal{M}) < G(P) = G(P) \sqcap (G(P))^d < G(P) \sqcup (G(P))^d = G(P)^d < \mathbb{I}(\mathcal{M}).$$

PROOF. (i) As to Γ-density, if $\mathcal{M}(Ja) \in J(P)$, we have by definition: $\mathcal{M}(\neg\neg Ja) \in J(P)$ and $\mathcal{M}([a=a]) \in J(P)$; hence by lemma 30.1 and 31.5 $\mathcal{M}(a = a \wedge \neg(\neg Ja \wedge Ja)) = \mathcal{M}(Ja) \in \Gamma(J(P))$. If $\mathcal{M}(\neg\neg Ja) \in J(P)$, also $\mathcal{M}(Ja) \in J(P)$, whence $\mathcal{M}(\neg\neg Ja) \in \Gamma(J(P))$. On the other hand, if $\mathcal{M}([a=a]) \in J(P)$, $\mathcal{M}([a=a]) \in \Gamma(J(P))$ holds by definition of Γ.
If $\mathcal{M}(\neg(Ja \wedge \neg Ja)) \in J(P)$, then $\mathcal{M}(\neg\neg Ja) \in J(P)$, which implies, with 30.1, $\mathcal{M}(\neg(Ja \wedge \neg Ja)) \in \Gamma(J(P))$.
As to the consistency of $J(P)$, assume that there is an $a \in J(P)$, such that $(\neg a) \in J(P)$. Then we derive a contradiction, using the independence properties of ID, AND, NEG (see 7.1.1). There are sixteen cases to check, but we only consider two typical instances, because the verification is routine. Let $a = \mathcal{M}(\neg\neg Jb)$, $(\neg a) = \mathcal{M}(\neg\neg Jc)$, for some $b, c \in P$. Then we have $\mathcal{M}(\neg\neg\neg Jb) = \mathcal{M}(\neg\neg Jc)$, which implies $\mathcal{M}(\neg Jb) = \mathcal{M}(Jc)$, whence, by 31.5:

$$\mathcal{M}(\neg Jb) = \mathcal{M}(c = c \wedge \neg(Jc \wedge \neg Jc)). \tag{1}$$

But (1) yields the identity $\mathcal{M}(NEG) = \mathcal{M}(AND)$, against 7.1.1 !
Let $a = \mathcal{M}(Jb)$, $(\neg a) = \mathcal{M}(\neg(Jc \wedge \neg Jc))$, for some $b, c \in P$. Then we have $\mathcal{M}(Jb) = \mathcal{M}(Jc \wedge \neg Jc)$, which implies by 31.5:

$$\mathcal{M}(b = b \wedge \neg(Jb \wedge \neg Jb)) = \mathcal{M}(Jc \wedge \neg Jc),$$

whence, again with 31.5 and the properties of AND:

$$\mathcal{M}(b = b) = \mathcal{M}(c = c \wedge \neg(Jc \wedge \neg Jc)). \tag{2}$$

But (2) implies $\mathcal{M}(ID) = \mathcal{M}(AND)$, against 7.1.1.

We now prove that $G(P)$ is intrinsic. If $Y \in FIX_{cs}(\mathcal{M})$, then $Y_0 := Y \cup J(P)$ is Γ-dense, as both sets are Γ-dense (use Γ-monotonicity). Assume that Y_0 is inconsistent; since Y is consistent, as well as $J(P)$, there is an $a \in Y$ with $(\neg a) \in J(P)$ (or vice versa). If $(\neg a) = \mathcal{M}(b = b)$, then $\mathcal{M}(\neg b = b) \in Y$: contradiction! Let $(\neg a) = \mathcal{M}(Jb)$: as $(\neg\neg a) \in Y$, $Y \models b\overline{\eta}\, J$, against 31.5 (ii). If $\mathcal{M}(\neg a) = \mathcal{M}(\neg\neg Jb)$, we argue as in the preceding case. If $(\neg a) = \mathcal{M}(\neg(Jc \wedge \neg Jc))$, $a = \mathcal{M}(Jc \wedge \neg Jc) \in Y$, against the consistency of Y! If $a \in J(P)$ and $(\neg a) \in Y$, the argument is similar. It follows that Y_0 is consistent and $UP(Y_0) \in FIX_{cs}(\mathcal{M})$ (by lemma 30.10 (iii)); moreover, $J(P) \subseteq G(P) \leq UP(Y_0)$ and $Y \leq UP(Y_0)$, and hence $G(P) \in INT(\mathcal{M})$ by definition 31.1.

(ii) If $a \in P$, we have $G(P) \models a\eta J$ by definition of G. Conversely, assume that $G(P) \models a\eta J$. By consistency and definition of $G(P)$, we must have

$\mathcal{M}(Ja) \in J(P)$. As AND, ID, NEG denote pairwise distinct elements of $\mathcal{M}$, $\mathcal{M}(Ja) = \mathcal{M}(Jb)$, for some $b \in P$; but 31.5 (i) implies $a = b$ and $a \in P$. The second part is immediate by consistency of $G(P)$ and the preceding fact.
(iii) Trivially $J(\emptyset) = \emptyset \subseteq \mathbb{O}(\mathcal{M})$; but $G(\emptyset)$ is the least fixed point extending $J(\emptyset)$, hence $G(\emptyset) \subseteq \mathbb{O}(\mathcal{M})$. As to the second part, note that by (ii) $G(M) \models \forall x(x\eta J)$ and $G(M) \subseteq \mathbb{P}(\mathcal{M})$.
(iv) By consistency of $G(P)$, $G(P)$ is strictly contained in $G(P)^d$ and $G(P) < G(P)^d$ by lemma 30.12 (ii). Hence $G(P) \sqcup (G(P))^d = G(P)^d$, which also implies $G(P) \sqcap G(P)^d = G(P)$ (use 30.13 (i)).
$P \neq \emptyset$ implies $\mathbb{O}(\mathcal{M}) < G(P)$, whence $G(P)^d < \mathbb{I}(\mathcal{M})$, again by 30.13. □

31.7.1. REMARK. (iv) shows that $FIX(\mathcal{M})$ is not even an ortholattice with respect to ^{d} (see Birkhoff 1967, p. 52).

31.8. THEOREM (*Embedding*)

(i) *The map* $G: \mathcal{P}(M) \to INT(\mathcal{M})$ *is injective and*

$$G(\cup\{P_i : i \in I\}) = \sqcup\{G(P_i) : i \in I)\}.$$

(ii) *If* $X \in INT(\mathcal{M})$, *let* $J\hat{}(X) := \{a \in M : X \models a\eta J\}$.
Then $J\hat{}: INT(\mathcal{M}) \to \mathcal{P}(M)$ *is a complete lattice epimorphism, i.e.* $J\hat{}$ *is onto and*

$$J\hat{}(\sqcup \mathfrak{C}) = \cup\{J\hat{}(X) : X \in \mathfrak{C}\};\quad J\hat{}(\sqcap \mathfrak{C}) = \cap\{J\hat{}(X) : X \in \mathfrak{C}\}.$$

PROOF. (i) If $P \neq Q$ are distinct subsets of M, say $a \in P$ and $a \notin Q$, we have, by 31.7 (ii), $G(P) \models a\eta J$ and $G(Q) \models \neg a\eta J$, whence $G(P) \neq G(Q)$. On the other hand, $P \subseteq Q$ implies $J(P) \subseteq J(Q)$, whence $G(P) \leq G(Q)$ (by 30.10 (i), 30.12 (ii)), and $\sqcup\{G(P_i) : i \in I\} \subseteq G(\cup\{P_i : i \in I\})$. Conversely, we show by induction on α, with $Z := \cup\{P_i : i \in I\}$:

$$UP(J(Z), \alpha) \subseteq \sqcup\{G(P_i) : i \in I\}.$$

If $\alpha = 0$, $UP(J(Z), 0) = J(Z)$ and

$$\begin{aligned} J(Z) &\subseteq \cup\{J(P_i) : i \in I\} \\ &\subseteq \cup\{G(P_i) : i \in I\} \\ &\subseteq UP(\cup\{G(P_i) : i \in I\}). \end{aligned}$$

If α is a limit, we apply IH; if $\alpha = \beta{+}1$ and $a \in \Gamma(UP(J(Z), \beta))$, we easily obtain $a \in Y := \sqcup\{G(P_i) : i \in I\}$ by IH and Γ-closure of Y.
(ii) If $P \subseteq M$, $J\hat{}(G(P)) = P$ and hence $J\hat{}$ is onto. Trivially, $J\hat{}$ preserves $\leq$ and hence:

$$J\hat{}(\sqcap \mathfrak{C}) \subseteq \cap\{J\hat{}(X) : X \in \mathfrak{C}\};\ \cup\{J\hat{}(X) : X \in \mathfrak{C}\} \subseteq J\hat{}(\sqcup \mathfrak{C}). \qquad (*)$$

Assume $a \in J^{\frown}(X)$, for every $X \in \mathbb{C}$: then

$$J(a) := \{\mathcal{M}(Ja),\ \mathcal{M}(\neg\neg Ja),\ \mathcal{M}([a=a]),\ \mathcal{M}(\neg(Ja \wedge \neg Ja))\} \subseteq X,$$

for every $X \in \mathbb{C}$; thus $G(a) \leq \sqcap\mathbb{C}$, whence $\sqcap\mathbb{C} \models a\eta J$, i.e. $a \in J^{\frown}(\sqcap\mathbb{C})$. If $a \in J^{\frown}(\sqcup\mathbb{C})$, we have $UP(\cup\mathbb{C}) \models a\eta J$, which implies $\mathcal{M}(Ja) \in \cup\mathbb{C}$, i.e. $\mathcal{M}(Ja) \in X$, for some $X \in \mathbb{C}$, and finally $a \in \cup\{J^{\frown}(X) : X \in \mathbb{C}\}$. This completes the verification of the converse to $(*)$. □

The embedding theorem, together with 30.13 (iii), immediately yields:

31.9. COROLLARY

$card(FIX_{cs}(\mathcal{M})) = card(FIX_{cp}(\mathcal{M})) = card(FIX(\mathcal{M})) = 2^{card(\mathcal{M})}$, (*where* $card(P) =$ *the cardinal number of* P).

The "thickness" of $INT(\mathcal{M})$ is dramatically exemplified by a "local" version of the cardinality theorem: even if we consider an intrinsic fixed point "quite close" to $\mathbb{O}(\mathcal{M})$ (see $\mathbb{F}$ below), we still get an interval of the highest possible cardinality.

31.10. PROPOSITION. *Let* $\mathbb{F}(\mathcal{M}) := \sqcap\{G(P) : P \subseteq M$ *and* $P \neq \emptyset\}$. *Then*

$$card([\mathbb{O}(\mathcal{M}), \mathbb{F}(\mathcal{M})]) = 2^{card(M)}.$$

($[X,Y] := \{V \in FIX(\mathcal{M}) : X \leq V \leq Y\}$ = closed interval determined by X, Y).

PROOF. Choose, by fixed point for operations, a term E such that $E = \lambda x.(x = x \wedge \neg(\neg Ex \wedge \forall y \neg(Jy)))$; then, if $X \in FIX(\mathcal{M})$,

$$X \models \forall x(x\eta E \leftrightarrow (x\eta E \vee \exists y(y\eta J))).$$

If $P \subseteq M$, the set

$$H_0(P) := \{\mathcal{M}(Ea) : a \in P\} \cup \{\mathcal{M}(\neg\neg Ea) : a \in P\} \cup$$
$$\cup\{\mathcal{M}(a = a) : a \in P\} \cup \{\mathcal{M}(\neg(\neg Ea \wedge \forall y \neg(Jy))) : a \in P\}$$

is consistent (by 7.1.1), Γ-dense, and intrinsic: if $Q \neq \emptyset$, $G(Q) \models \exists y(y\eta J)$, hence $G(Q) \models a\eta E$, for every $a \in P$, yielding $H_0(P) \subseteq G(Q)$. Thus there exists the $\leq$-least $H(P) \in INT(\mathcal{M})$ with $H_0(P) \subseteq H(P)$ by lemma 31.3, and clearly $H(P) \subseteq G(Q)$. Since Q is an arbitrary non-empty subset of M, we actually obtain $H(P) \leq \mathbb{F}(\mathcal{M})$. But if $P \neq \emptyset$, then $\mathbb{O}(\mathcal{M}) \neq H(P)$ (recall the inductive definition of UP and lemma 31.7). In addition, $a \notin Q$ and $a \in P$ imply $H(Q) \models \neg a\eta E$ and $H(P) \models a\eta E$; in this step, we implicitly rely on the fact that, by choice of E, $\mathcal{M} \models Ea = Eb \rightarrow a = b$. Hence the map $P \mapsto H(P)$ is injective. □

§32. Variations on the encoding technique: non-modularity and other oddities

A lattice $\mathbb{L} = \langle L, \sqcap \sqcup \rangle$ is *modular* (see Birkhoff 1967, p.13) iff it satisfies the condition:

$$x \sqcup (y \sqcap z) = (x \sqcup y) \sqcap z \text{ , whenever } x \leq z \text{ and } x,\ y,\ z \in L.$$

32.1. THEOREM. *$INT(\mathcal{M})$ is non-modular.*

PROOF. Fix four distinct elements d, a, b, c of M; this can be done since $\mathcal{M} \models \mathrm{OP}^{-}$. Then there exists a term $V = V(a,b,c)$ such that, for every $X \in FIX_{cs}(\mathcal{M})$:

$$\mathcal{M} \models V = \lambda u.(\neg(\neg Vu \wedge (\neg(Ja \wedge Jb) \wedge \neg(Jb \wedge Jc)))); \qquad (*)$$

$$X \models \forall u(u\eta V \leftrightarrow (u\eta V \vee (a\eta J \wedge b\eta J) \vee (b\eta J \wedge c\eta J))); \qquad (**)$$

$$X \models \forall x(\neg x\overline{\eta} V). \qquad (***)$$

$(*)$-$(**)$ are clear; as to $(***)$, $X \models e\overline{\eta} V$ (e arbitrary) would imply either $X \models a\overline{\eta} J$, or $X \models b\overline{\eta} J$, or else $X \models c\overline{\eta} J$, contradicting lemma 31.5 (ii). Let us consider the consistent intrinsic fixed points N_0 and N_1, where

$$N_0 = G(b,c) \sqcap G(a,b);$$

$$N_1 = G(b) \sqcup (G(c) \sqcap G(a,b)).$$

We show:

$$\text{not } N_0 \subseteq N_1. \qquad (1)$$

Since we know by 31.8 that

$$G(b) \leq G(a,b) \text{ and } G(b,c) = G(b) \sqcup G(c), \qquad (2)$$

(1) will refute modularity of $INT(\mathcal{M})$. In order to check (1), we show

$$N_0 \models d\eta V\ (= V(a,b,c)); \qquad (3.1)$$

$$N_1 \models \neg d\eta V. \qquad (3.2)$$

Ad (3.1): if $N_0(\alpha) =$ the αth approximation of $G(b,c) \sqcap G(a,b)$ (see the inductive definition of $DOWN$ in 30.9), we check by transfinite induction:

$$\mathcal{M}(Vd) \in N_0(\alpha); \qquad (3.1.1)$$

$$\mathcal{M}(\neg\neg Vd) \in N_0(\alpha). \qquad (3.1.2)$$

$\alpha = 0$. Then $N_0(0) = G(b,c) \cap G(a,b)$; by 31.7 (ii), $\mathcal{M}(Jb) \in G(b) \subseteq G(b,c)$ and $\mathcal{M}(Jc) \in G(c) \subseteq G(b,c)$, whence $G(b,c) \models b\eta J \wedge c\eta J$, which implies by $(**)$ $\mathcal{M}(Vd) \in G(b,c)$. A similar argument yields $\mathcal{M}(Vd) \in G(a,b)$.

Moreover, $\mathcal{M}(\neg\neg Vd) \in G(b,c)$, since $G(b,c) \in FIX(\mathcal{M})$.
If α is a limit, the conclusion is trivial by IH.
$\alpha = \beta{+}1$. By IH we assume $\mathcal{M}(Vd) \in N_0(\beta)$ and $\mathcal{M}(\neg\neg Vd)\eta N_0(\beta)$. Then by definition of Γ, we obtain (in the given order): $\mathcal{M}(\neg\neg Vd) \in N_0(\alpha)$, $\mathcal{M}(Vd) \in N_0(\alpha)$.

Ad (3.2). We check by transfinite induction on α:

$$\mathcal{M}(Vd) \notin N_1(\alpha),$$

where $N_1(\alpha) =$ the αth approximation of N_1 (see 30.9). The limit case is trivial by IH. Let $\alpha = 0$. Since $N_1(0) = G(b) \cup (G(c) \sqcap G(a,b))$, we have to prove:

$$\mathcal{M}(Vd) \notin G(b); \tag{3.2.1}$$

$$\mathcal{M}(Vd) \notin G(c) \sqcap G(a,b). \tag{3.2.2}$$

Ad (3.2.1). By contradiction, let $\mathcal{M}(Vd) \in G(b)$ and choose α minimal such that $\mathcal{M}(Vd) \in G(b,\alpha)$; then $\alpha = \beta{+}1$, for some β, because $\mathcal{M}(Vd) \notin G(b,0)$ (we apply 7.1.1 to check that $Vd \not\equiv Je$, for all e). {Here $G(P,\alpha)$ is a temporary abbreviation for $UP(J(P),\alpha)$, the α-th approximation of $G(P)$}. Then by $(*)$

$$\mathcal{M}(\neg(\neg Vd \wedge (\neg(Ja \wedge Jb) \wedge \neg(Jb \wedge Jc)))) \in G(b,\beta{+}1).$$

By definition of $G(b,\beta{+}1)$ and 30.1, either $\mathcal{M}(\neg\neg Vd) \in G(b,\beta)$ or $\mathcal{M}(\neg(\neg(Ja \wedge Jb) \wedge \neg(Jb \wedge Jc))) \in G(b,\beta)$. In the first case, we get $\mathcal{M}(Vd) \in G(b,\gamma)$, for some $\gamma < \beta$, against the minimality of $\alpha = \beta{+}1$. In the second case, there exist ξ, $\zeta < \beta$ such that:

$$\text{either } \mathcal{M}(Ja),\ \mathcal{M}(Jb) \in G(b,\xi) \text{ or } \mathcal{M}(Jc),\ \mathcal{M}(Jb) \in G(b,\zeta).$$

Thus either $G(b) \models a\eta J$ or $G(b) \models c\eta J$, which imply either $a = b$ or $b = c$ (by 31.7): contradiction!

Ad (3.2.2): assume $\mathcal{M}(Vd) \in G(c) \sqcap G(a,b)$.
Then from $G(c) \sqcap G(a,b) \subseteq G(c) \cap G(a,b)$ we have $\mathcal{M}(Vd) \in G(c)$. If we repeat the argument of (3.2.1) by replacing in it b with c, we still reach a contradiction.
We conclude the verification of (3.2). Let $\alpha = \gamma{+}1$. Assume by IH $\mathcal{M}(Vd) \notin N_1(\gamma)$, and by contradiction:

$$\mathcal{M}(Vd) \in N_1(\gamma{+}1) = \Gamma(N_1(\gamma)). \tag{+}$$

Then we see that, for some $\delta < \gamma$, either $\mathcal{M}(Ja) \in N_1(\delta)$ or $\mathcal{M}(Jc) \in N_1(\delta)$. But $\mathcal{M}(Ja) \in N_1(\delta)$ implies $\delta = 0$, i.e. $\mathcal{M}(Ja) \in G(b) \cup (G(c) \sqcap G(a,b))$. Since $\mathcal{M}(Ja) \notin G(b)$ (by 31.7 with $a \neq b$), we must have $\mathcal{M}(Ja) \in G(c)$, i.e. $a = c$: contradiction. The case $\mathcal{M}(Jc) \in N_1(\delta)$ is similar. □

32.1.1. REMARK. Theorem 32.1 was stated by the author with a fake argument in a preliminary version of this work; fortunately, P.Minari (1987) found a nice proof of an analogous theorem for an infinitary propositional logic of truth and a trick of his could be adapted to the present framework; see also Visser (1984).

As a corollary, $FIX(\mathcal{M})$ is non-modular, hence non-distributive. We further indulge in self-referential constructions, in order to see that $INT(\mathcal{M})$ admits infinite strictly descending sequences (with respect to the lattice ordering).

32.2. LEMMA. *Let* $\mathcal{M}$ *be an* ω*-standard model of* OP *and let*

$$ADD(f) = FP(\lambda y \lambda x\,[yx \vee fx]).$$

Then we can find an operation $\lambda x.J(x)$ *such that, for every* n, a, c:

(i) $\mathcal{M} \models J(0) = J \wedge \forall n(J(n{+}1) = ADD(J(n)))$;

$\mathcal{M} \models \neg(J(n)a = J(n{+}1)c) \wedge \neg(J(n)a = \neg\neg J(n{+}1)c)$.

(ii) $\mathbb{O}(\mathcal{M}) \models \forall x(\neg x\eta J(n))$;

(iii) *if* $X \in FIX_{cs}(\mathcal{M})$, $X \models \forall x(\neg x\overline{\eta} J(n))$.

PROOF. (i) The first line is immediate by fixed point theorem for operations; the second part follows by induction on n, using the independence properties of AND, NEG (see 7.1.1).
(ii) By induction on n. If $n = 0$, $J(0) = J$ and we apply 31.5. Let $n = k{+}1$ and choose β minimal such that $\mathbb{O}(\mathcal{M}, \beta{+}1) \models a\eta J(k{+}1)$ ($\mathbb{O}(\mathcal{M}, \alpha) =$ the αth-approximation of $\mathbb{O}(\mathcal{M})$). Since (i) implies:

$$\mathcal{M} \models J(k{+}1)a = (ADD(J(k)))a = [(ADD(J(k)))a \vee (J(k))a],$$

then either for some $\delta < \beta$, $\mathbb{O}(\mathcal{M}, \delta) \models a\eta J(k{+}1)$ or $\mathbb{O}(\mathcal{M}, \delta) \models a\eta J(k)$. By minimality of β, $\mathbb{O}(\mathcal{M}, \delta) \models a\eta J(k)$, whence $\mathbb{O}(\mathcal{M}, \beta) \models a\eta J(k)$, against IH.
(iii) Easy induction on n, with lemma 31.5 in the case $n = 0$. □

32.3. DEFINITION (i) If $P \subseteq M$, let $J_0(P) = J(P)$ (see 31.6);
$J_{n+1}(P) = \{\mathcal{M}(J(n{+}1)a) : a \in P\} \cup \{\mathcal{M}(\neg\neg(J(n{+}1)a)) : a \in P\}$;

(ii) $G_0(P) = G(P)$ and $G_{n+1}(P) = UP(J_{n+1}(P))$.

32.4. THEOREM. *Let* $\mathcal{M}$ *be an* ω*-standard model of* OP. *Then, if* $P \subseteq M$ *is non-empty, we have, for every* $n \in \mathbb{N}$:

(i) $G_n(P) \in INT(\mathcal{M})$ *and* $\mathbb{O}(\mathcal{M}) < G_n(P)$;

(ii) $G_{n+1}(P) < G_n(P)$ (*here* $X < Y := X \leq Y$ *and* $Y \not\leq X$).

PROOF. (i) $J_n(P)$ is consistent by 7.1.1, and Γ-dense; hence $G_n(P)$ is a consistent fixed point $> \mathbb{O}(\mathcal{M})$, as $P \neq \emptyset$ and $G_n(P) \models a\eta J(n)$ for $a \in P$.
(ii) By construction, we have, for every $a \in P$, $G_n(P) \models a\eta J(n)$; hence, by 32.2, $G_n(P) \models a\eta J(n+1)$, which implies $J_{n+1}(P) \subseteq G_n(P)$. Since $G_{n+1}(P)$ is the least fixed point $\supseteq J_{n+1}(P)$, then $G_{n+1}(P) \leq G_n(P)$. On the other hand, if $a \in P$,

$$G_{n+1}(P) \models \neg\, a\eta J(n);$$

for $G_{n+1}(P) \models\, a\eta J(n)$ implies that $\mathcal{M}(J(n)a)$ is in $G_{n+1}(P)$ at stage 0: hence either $\mathcal{M}(J(n)a) = \mathcal{M}(J(n+1)c)$ or $\mathcal{M}(J(n)a) = \mathcal{M}(\neg\neg J(n+1)c)$, for some $c \in P$. This contradicts 32.2 (i).
Since $G_0(P)$ is intrinsic by 31.7 and $G_0(P) \geq G_n(P)$, $G_n(P) \in INT(\mathcal{M})$. □

A closer look into the structure of $FIX(\mathcal{M})$ can be gained through Whitman's notion of *splitting of a lattice* (see Whitman 1944).

32.5. DEFINITION. Let X, Y be elements of $FIX(\mathcal{M})$: $\langle X, Y\rangle$ is a *splitting pair* for $FIX(\mathcal{M})$ iff

(i) $[\mathbb{O}(\mathcal{M}), X] \cup [Y, \mathbb{I}(\mathcal{M})] = FIX(\mathcal{M})$;

(ii) $[\mathbb{O}(\mathcal{M}), X] \cap [Y, \mathbb{I}(\mathcal{M})] = \emptyset$;

(iii) X, Y are $\leq$-incomparable.

32.6. THEOREM. *There exist at least card(M)-many splitting pairs for $FIX(\mathcal{M})$.*

PROOF. Choose a $\in M$ and define

$$L(a) = \sqcup\{H \in FIX(\mathcal{M}) : H \models \neg\, a\eta J\}.$$

Then $L(a) \in FIX(\mathcal{M})$ by 31.4; we claim that $\langle L(a), G(a)\rangle$ is a splitting pair. First, observe that

$$L(a) \models \neg\, a\eta J. \tag{1}$$

If there were some $H \in FIX(\mathcal{M})$ such that $G(a) \leq H \leq L(a)$, we should have $L(a) \models a\eta J$, against (1). Hence condition (ii) of 32.5 holds for $\langle L(a), G(\mathrm{a})\rangle$ and also

$$G(a) \not\leq L(a). \tag{2}$$

Moreover $L(a) \not\leq G(a)$: indeed, assume by contradiction

$$L(a) \leq G(a). \tag{3}$$

Choose $c \neq a$ and a closed term S such that $S = \lambda x.\neg(\neg Sx \wedge \neg Jc)$ holds in $\mathcal{M}$. The set $W(c) = \{\mathcal{M}(Sc)\} \cup \{\mathcal{M}(\neg\neg Sc)\}$ is Γ-dense, consistent (with

7.1.1), and hence there exists a consistent fixed point $W^{\wedge}(c) = UP(W(c))$. Since $\mathcal{M}(Ja) \neq \mathcal{M}(Sc)$ and $\mathcal{M}(Ja) \neq \mathcal{M}(\neg\neg Sc)$, we have:

$$W^{\wedge}(c) \models \neg a\eta J.$$

Hence $W^{\wedge}(c) \leq L(a)$ and by (3), $W^{\wedge}(c) \leq G(a)$, i.e. $G(a) \models c\eta S$, which implies either $\mathcal{M}(Jc) \in G(a,0)$ or $\mathcal{M}(Sc) \in G(a,0)$ (as usual $G(a,\alpha) = \alpha$th approximation of $G(a)$). The first possibility is excluded by the choice of c and 31.7, while the second is impossible, as $\mathcal{M}(Sc) \notin J(a)$. Hence (3) is false and condition (iii) of 32.5 holds. Condition (i) of definition 32.5 is fulfilled, too: indeed, given any $X \in FIX(\mathcal{M})$, either $X \models a\eta J$ and hence $G(a) \leq X$, or $X \models \neg a\eta J$ and by definition $X \leq L(a)$.
As to the verification of the cardinality condition, it is enough to observe that, if $[\mathbb{O}(\mathcal{M}), L(a)] = [\mathbb{O}(\mathcal{M}), L(b)]$ and $[G(a), \mathbb{I}(\mathcal{M})] = [G(b), \mathbb{I}(\mathcal{M})]$, then $G(a) = G(b)$ and hence $a = b$ (by 31.7). □

32.6.1. REMARK. (i) The argument is readily extended to $INT(\mathcal{M})$.
(ii) A stronger form of 32.6 is proven by Minari (1987) for an infinitary propositional logic of self-referential truth.

We conclude with a useful modification of G. An essential feature of the map $P \mapsto G(P)$, is that it positively generates $G(P)$; but it also follows from 31.5 (ii) that J is in general never total, and hence P cannot be handled as a standard total object. This suggests a new map $P \mapsto D(P)$, such that P is represented by a class in $D(P)$. Moreover, the operation D yields additional cardinality information.

32.7. DEFINITION

(i) First of all, let EXT be the closed term satisfying

$$\mathcal{M} \models EXT = \lambda x.(x = x \wedge EXTx).$$

Clearly $\mathcal{M} \models EXTa = EXTb \rightarrow a = b$, for every $a, b \in M$.

(ii) If $P \subseteq M$, we put:

$$EXT(P) := \{\mathcal{M}(EXTa) : a \in P\} \cup \{\mathcal{M}(\neg EXTa) : a \notin P\} \cup \cup \{\mathcal{M}(a = a) : a \in P\}.$$

$$D(P) := UP(EXT(P)).$$

32.8. LEMMA

(i) $\mathbb{O}(\mathcal{M}) \models \forall x(\neg x\eta EXT \wedge \neg x\overline{\eta} EXT)$.

(ii) *If* $P \subseteq M$, $D(P)$ *is the least consistent non-intrinsic fixed point* $\supseteq EXT(P)$, *such that:*

$$D(P) \models Cl(EXT); \tag{$*$}$$

$$D(P) \models a\eta EXT \textit{ iff } a \in P, \textit{ for every } a \in M. \tag{$**$}$$

(iii) *The map $P \mapsto D(P)$ is injective.*

PROOF. (i) is a consequence of the inductive generation of $\mathbb{O}(\mathcal{M})$.
(ii) Clearly $EXT(P)$ is Γ-dense. As to consistency, assume by contradiction that $a \in EXT(P)$, $(\neg a) \in EXT(P)$ with $a = \mathcal{M}(EXTb)$ for some $b \in P$, $(\neg a) = \mathcal{M}(\neg EXTc)$, for some $c \notin P$. Then $\mathcal{M}(EXTb) = \mathcal{M}(EXTc)$ and hence $b = c$ by injectivity of EXT: contradiction! The remaining cases are disposed of with the independence properties of AND, ID, NEG. It follows that $D(P) \in FIX_{cs}(P)$. Moreover, $D(P)$ is non-intrinsic, since $D(P)$ is incompatible with $D(M - P)$.
($*$)-($**$) are immediate by construction (injectivity of EXT is required for checking $D(P) \models a\eta EXT \Rightarrow a \in P$).
(iii) If $a \in P$ and $a \notin Q$, $D(Q) \models a\overline{\eta} EXT$ and $D(P) \models a\eta EXT$, hence $D(P) \neq D(Q)$. □

32.9. THEOREM

(i) $card(MAX(\mathcal{M})) = card(FIX_{cs}(\mathcal{M}) - INT(\mathcal{M})) = 2^{card(M)}$.

(ii) *There are $2^{card(M)}$ $\leq$-incomparable elements of $FIX(\mathcal{M})$, which are neither complete nor consistent.*

PROOF. (i) Since every maximal consistent fixed point is non-intrinsic (by 31.3 (iii)), it suffices to check that there are $2^{card(M)}$-many elements in $MAX(\mathcal{M})$. Now by lemma 31.2, for each $D(P)$, there exists a consistent fixed point $D^*(P) \in MAX(\mathcal{M})$, which extends $D(P)$; furthermore, $P \neq Q$ implies $D^*(P) \neq D^*(Q)$.

(ii) Let P be a subset of M with at least two elements and define $I(P) := = \sqcup\{D(a) : a \in P\}$. Then $I(P)$ is inconsistent: if $a \neq b$ and a, $b \in P$, then:

$$D(b) \models a\overline{\eta} EXT \text{ and } D(a) \models a\eta EXT,$$

which implies $I(P) \models a\overline{\eta} EXT \wedge a\eta EXT$.
On the other hand, let L be such that $\mathcal{M} \models L = [\neg TL]$. Assume by contradiction that $I(P)$ is complete, and consider the case $\mathcal{M}(L) \in I(P)$. If $I(P,\alpha) = \alpha$-th approximation of $I(P)$, we should have $\mathcal{M}(L) \in I(P,\alpha{+}1)$, where α is minimal with respect to this property, and hence $\mathcal{M}(\neg L) \in I(P,\alpha)$, i.e. $\mathcal{M}(L) \in I(P,\gamma)$, for some $\gamma < \alpha$: contradiction !
A similar argument works for the extant case. It is also easy to check that $I(P) \neq I(Q)$ whenever $P \neq Q$. □

§33. A model for an impredicative extension of reflective truth

The systems we have been analysing so far, are basically intensional; also, the abstraction process is more or less reduced to an *inductive* valuation process, at least if we stick to the minimal model $\mathcal{O}(\mathcal{M})$. But it would be conceivable to have, besides properties, *sets (extensions), which can be circumscribed* by means of *impredicative* definitions.

We here assume that sets and properties coexist in the same universe; in particular, we identify sets with certain canonical *total* properties, whose specification is, so to speak, kept at a minimum of logical complexity. According to these desiderata, operations and truth predicates ought to apply to sets as well; moreover, extensional equality and the basic equality of the given universe (which is a model of combinatory logic) should coincide on sets.

33.1. DEFINITION

(i) We expand the language $\mathcal{L}$ of MF^- to the language $\mathcal{L}_\sigma$, in which there are new primitive individual constants $I\dot{D}$, $T\dot{R}$, $NA\dot{T}$, $NE\dot{G}$, $AN\dot{D}$, $AL\dot{L}$, $\forall_2$ and a denumerable list of set variables $X_1, X_2, \ldots$; X, Y, Z will be the corresponding syntactical variables.

(ii) We enlarge the notion of term by requiring that *set variables are individual terms.*

(iii) We admit quantification on set variables: if A is a formula of $\mathcal{L}_\sigma$, $\forall XA$ is a formula of $\mathcal{L}_\sigma$.

(iv) We define the map $A \mapsto [A]$ for arbitrary $\mathcal{L}_\sigma$-formulas by means of the standard clauses and the new constants of (i) (cf. Ch. II, §7; appendix III); in particular, we set $[\forall XA] = \forall_2(\lambda x.[A]))$.

Now the crucial question concerns set existence principles. The solution we choose comes from *a nominalistic interpretation of sets,* suggested, for instance, by Gilmore (1980, 1986). Under this interpretation, we must be careful in mixing up too freely occurrences, in which set variables are – so to speak – *mentioned* (as in the contexts $X\eta Z$ or $X = x$), with occurrences, in which X is properly *used* as an extension (see $Z\eta X$).

This point of view leads to conjecture the consistency of the basic theory of reflective truth with the principle stating the existence of a set corresponding to any given condition A, provided A does not contain *mixed* occurrences of set variables (*mixed* roughly means that no X can be used and mentioned at the same time). In order to give a precise version of this idea, we introduce analytical $\mathcal{L}_\sigma$-formulas, a sort of liberalized second order formulas.

33.2. DEFINITION

(i) The class EF of *η-free formulas* is the smallest class of formulas, which is closed under classical sentential connectives, quantification on either variable sort, and is generated from atoms of the form Nt, $t = s$, where t, s are *arbitrary* terms.

(ii) The class ST of *stratified formulas* is the smallest class, which is closed under classical sentential connectives, quantification on either variable sort, and is generated from atoms of the form Nt, $t = s$, $t\eta X$, provided *no set variable occurs free in* t, s (for related notions, see Feferman 1975, 1979; Cantini 1988).

(iii) A formula A of $\mathcal{L}_\sigma$ is *analytical* iff A is η-free or stratified.

(iv) MFS^- is the theory, which includes:

1. OP^- and (finitely many) axioms, establishing the independence of the primitive constants $\dot{ID}$, $\dot{TR}$, $\dot{NAT}$, $\dot{NEG}$, $\dot{AND}$, $\dot{ALL}$, $\forall_2$ (cf. Ch. II; axiom LOG of Appendix III);

2. the truth axioms T.1-T.6 of 7.10 with ID, TR, NAT, NEG, AND, and ALL replaced by $\dot{ID}$, $\dot{TR}$, $\dot{NAT}$, $\dot{NEG}$, $\dot{AND}$, $\dot{ALL}$ (respectively);

3. the natural axiom relating T and set quantifiers:

$T\forall_2$ $\quad (T\forall_2(f) \leftrightarrow \forall X T(fX)) \wedge (T\neg\forall_2(f) \leftrightarrow \exists X T\neg(fX))$;

4. the set principles SET.1-SET.3:

SET.1. $\quad Cl(X)$;

SET.2. $\quad X =_e Y \rightarrow X = Y \quad$ (extensionality);

SET.3. $\quad \exists Y(Y =_e \{x : A\})$,
provided A is analytical (as usual, $a =_e b$ stands for $\forall u(u\eta a \leftrightarrow u\eta b)$).

The logic underlying MFS^- is simply the two-sorted version of classical predicate logic. In particular, we include the generalization rule for introducing $\forall X$ and the schema $\forall X A \rightarrow A[X := Y]$ (Y free for X in A). MFS is MFS^- with the schema of N-induction for arbitrary $\mathcal{L}_\sigma$-formulas.

(v) $\aleph := \exists f(f : V \mapsto SET) =$ "there exists an operation establishing a bijection between the universe V and the collection of sets".

The operation f of $\aleph$ cannot be the identity operation (otherwise everything would be a class); in any case, there is at least a set X, which is left fixed by f. We show that $\text{MFS} + \aleph$ is interpretable in the theory of suitable models of the form $D(S)$ over arbitrary combinatory algebras, D being the operation of 32.8.

33.3. DEFINITION. $EXT(a) := \lambda x.EXT\langle a, x\rangle$, where EXT is the closed term of the previous section.

33.4. LEMMA. *Let* $\mathcal{M}\models\mathrm{OP}^-$. *Then:*

$$\mathcal{M}\models EXT(a) = EXT(b) \leftrightarrow a = b \ (a,\ b \text{ arbitrary elements of } M).$$

PROOF. Apply 32.7 and the pairing axioms. □

Let $S = \{S_a : a \in M\}$ be a M-indexed family of subsets of M. S can be encoded by a subset of M, also named S: define $S := \{\mathcal{M}(\langle a, b\rangle) : b \in S_a\}$ (where $\langle -, -\rangle$ is the term defining the pairing function).

33.5. LEMMA

(i) *If S is a subset of M, which encodes the family $\{S_a : a \in M\}$ and $\mathcal{M}\models\mathrm{OP}^-$, then*

$$D(S)\models \forall x Cl(EXT(x)) \text{ and } S_a = \{b \in M : D(S)\models b\eta EXT(a)\}.$$

(ii) *If $S = \{S_a : a \in M\}$ has no repetition (i.e. $S_a = S_b$ implies $a = b$), then*

$$D(S)\models EXT(a) =_e EXT(b) \rightarrow EXT(a) = EXT(b).$$

PROOF. By 32.8, definition of $EXT(a)$, assumption and 33.4. □

33.6. DEFINITION

(i) The language $\mathcal{L}_2{}^+$ is obtained from $\mathcal{L}_\sigma$ by omitting the predicate T. $\mathcal{L}_2{}^+$-terms are just the usual terms of OP (thus no set variable occurs in $\mathcal{L}_2{}^+$-terms). $\mathcal{L}_2{}^+$-atoms have the form $t = s$, Nt, Xt, where X is a set variable and t, s, are $\mathcal{L}_2{}^+$-terms; $\mathcal{L}_2{}^+$-formulas are inductively generated from atoms by means of $\neg$, $\wedge$ and quantification over individual variables and set variables.

(ii) OP_2 is the second-order extension of OP^-, which is based on classical predicate logic for the two-sorted language $\mathcal{L}_2{}^+$, and it includes the full second-order comprehension schema CA:

$$\exists X \forall u (Xu \leftrightarrow A(u)) \ (A \in \mathcal{L}_2{}^+,\ X \text{ not in } A).$$

(iii) If $\mathcal{M}\models\mathrm{OP}^-$, $\mathcal{M}_2 := \langle \mathcal{M}, \mathcal{P}(M)\rangle$, where $\mathcal{P}(M)$ is the power set of M; $\mathcal{L}_2{}^+(M)$ is $\mathcal{L}_2{}^+$, expanded with constants a, b, $c, \ldots$ for elements of M and with constants R, P, $S, \ldots$ for elements of $\mathcal{P}(M)$. Thus Ra is an atom of $\mathcal{L}_2{}^+(M)$, provided $a \in M$, $R \subseteq M$.

(iv) If $S \subseteq \mathcal{P}(M)$, A is a sentence of $\mathcal{L}_2{}^+(M)$, $\langle \mathcal{M}, S\rangle \models A$ is inductively defined according to the standard Tarskian clauses; in particular:

$$\langle \mathcal{M}, S\rangle \models \forall X A \ \text{ iff } \langle \mathcal{M}, S\rangle \models A[X := P], \text{ for } P \in S.$$

33.7. LEMMA

(i) *There exists a family without repetitions* $S = \{S_a : a \in M\}$ *of subsets of* M, *such that, if* $\mathcal{M}^+ = \langle \mathcal{M}, S \rangle$, *then for every* a, $b \in M$

$$\mathcal{M}_2 \models A(a, c, S_b) \text{ iff } \mathcal{M}^+ \models A(a, c, S_b),$$

for every $\mathcal{L}_2{}^+$*-formula* $A(u, v, X)$ *with the free variable shown.*

(ii) $\mathcal{M}^+ \models \mathrm{CA}$.

PROOF. This is a standard application of Skolem functions. Well-order $\mathcal{P}(M)$; then, if $A(u, X, Y) \in \mathcal{L}_2{}^+$, there exists a function f_A such that, if $a \in M$, $R \subseteq M$,

$$\mathcal{M}_2 \models \exists Y A(a, R, Y) \to A(a, R, F_A(a, R)),$$

where F_A is a new function symbol whose denotation is f_A. Let H be the Skolem hull of the empty set under the f_A' s, i.e.

$$H_0 = \emptyset,\ H_{n+1} = \{f_A(a, R) : R \in H_n,\ A \in \mathcal{L}_2{}^+,\ a \in M\};$$

$$H = \cup \{H_n : n \in \omega\}.$$

Since $card(H) = card(M)$, we pick a bijection $F : M \to H$ and we set $S_a = F(a)$, $S = \{S_a : a \in M\}$ and $\mathcal{M}^+ = \langle \mathcal{M}, S \rangle$. By construction, if $R \in S$ and $\mathcal{M}_2 \models \exists X A(a, R, X)$, then $\mathcal{M}_2 \models A(a, R, Q)$, for some $Q \in S$; by basic model theory, we derive (i) and hence (ii). □

We now embed $\mathcal{L}_\sigma$ into $\mathcal{L}$, essentially by interpreting sets as the denotations of terms having the form $EXT(x)$.

33.8. *Inductive definition of the translation* $^-$ *of* $\mathcal{L}_\sigma$ *into* $\mathcal{L}$.

(i) Translation of $\mathcal{L}_\sigma$-terms:

$(x_i)^- = x_{2i}$ and $(X_i)^- = EXT(x_{2i+1})$ (for $i \in \omega$);

if $c = \dot{ID}$ ($\dot{TR}$, $\dot{NAT}$, $\dot{NEG}$, $\dot{AND}$, $\dot{ALL}$), then $(c)^- = ID$ (TR, NAT, NEG, AND, ALL in the given order; ID, TR, NAT, NEG, AND, ALL being the terms of 7.11); else, $(c)^- = c$;

$(\forall_2)^- = \lambda f.ALL(\lambda x f(EXT(x)))$;

$(ts)^- = t^- s^-$;

(ii) Translation of $\mathcal{L}_\sigma$-formulas:

$(t = s)^- = (t^- = s^-)$; $(Nt)^- = N(t^-)$; $(Tt)^- = T(t^-)$;

$(\neg A)^- = \neg(A^-)$; $(A \wedge B)^- = (A)^- \wedge (B)^-$;

$(\forall x_i A)^- = \forall x_{2i} A^-$;

$$(\forall X_i A)^- = \forall x_{2i+1} A^-.$$

In practice, A^- and t^- are obtained 1) by relativizing set variables to the range of $\lambda x. EXT(x)$; 2) by replacing distinct variables of different sort by distinct general variables; 3) by renaming bound variables in order to avoid clash of free and bound variables. For instance, the term $\{x : \exists Y(Y = x)\}$ becomes $\{x : \exists y(EXT(y) = x)\}$.

33.9. THEOREM (*Interpretation*). *We can find* $S \subseteq M$ *such that, if* MFS $\vdash A(u, X)$, *then* $D(S) \models A^-(a, EXT(b))$, *for every* $a, b \in M$.

PROOF. Choose S by lemma 33.7 and define $D(S)$ as in §32. First of all, note that $(\aleph)^-$, $(\text{SET.1})^-$, $(\text{SET.2})^-$ hold in $D(S)$ by 33.4-33.5. As to $(T\forall_2)^-$, we have, by definition of $^-$-translation and renaming of bound variables:

$$\begin{aligned}(T\forall_2(f))^- &\leftrightarrow T\forall(\lambda x f(EXT(x)))\\ &\leftrightarrow \forall x T f(EXT(x)) \text{ (by } T.5.1)\\ &\leftrightarrow (\forall X T(fX))^-.\end{aligned}$$

The second part of $(T\forall_2)^-$ is similar. Thus it remains to check:

$$D(S) \models \exists y \forall x(x \eta EXT(y) \leftrightarrow A^-(x, c, EXT(b))),$$

$A(x, u, X)$ being analytical and $c, b \in M$.

Case 1. A is η-free. Then $\{a \in M : D(S) \models A^-(a, c, EXT(b))\} = S_a$, for some $a \in M$, as $A^-(u, c, EXT(b))$ is a formula of $\mathcal{L}_2{}^+(M)$ and S is closed under CA. Hence $S_a = \{d \in M : D(S) \models d\eta EXT(a)\}$ and we are done.

Case 2. If $A(x, Y)$ is stratified, we inductively define the transform $A_2(x, Y)$ of A, as follows:

$(t = s)_2 = t^- = s^-$; $(Nt)_2 = N(t^-)$; $(t\eta Y)_2 = Y(t^-)$;
the $_2$-transform commutes with $\neg$, $\wedge$, $\forall x$ and $\forall X$.

Then $A_2(x, Y)$ is a formula of $\mathcal{L}_2{}^+(M)$ and we can check, by induction on A, that for every $a, d \in M$:

$\mathcal{M}^+ \models A_2(a, S_d)$ iff $D(S) \models A^-(a, EXT(d))$ (we use the notations of 33.8).

But $\mathcal{M}^+ \models$ CA and hence, for some $c \in M$,

$$S_c = \{a \in M : D(S) \models a\eta EXT(c)\} = \{a \in M : \mathcal{M}^+ \models A_2(a, S_d)\},$$

which implies $D(S) \models \exists y(EXT(y) =_e \{x : A^-(a, S_d)\})$, which is $(\text{SET.3})^-$. □

33.10. COROLLARY. MFS + $\aleph$ *is consistent.*

We stress that MFS is a very experimental system, which proposes a blend of extensional and intensional elements, and we do not know whether MFS is stronger than full second order arithmetic. Of course, MFS has a few closure properties, which are typical of impredicative systems. They are quickly summarized below.

We say that $SET = \{u: \exists Y(Y = x)\}$ is closed under a given n-ary operation f, if $\mathrm{MFS} \vdash \exists Y(Y =_e fX_1 \ldots X_n)$. We then define:

(i) $pwX := \{u: \exists Y(u = Y \cap X)\}$ (= weak power set of X);

(ii) $igXR := \{u: \forall Z(Progr(X, R, Z) \rightarrow u\eta Z)\}$ (= inductive generation).

Of course: $Y \cap X := \{u: u\eta X \wedge u\eta Y\}$,

$$Progr(X, R, Z) := \forall x(x\eta X \wedge \forall y(yRx \rightarrow y\eta Z) \rightarrow x\eta Z),$$

where yRx is a shorthand for $\langle y, x\rangle\eta R$; note also that the defining conditions of pw, SET, ig are analytical.

33.11. PROPOSITION

(i) *SET contains the empty set and the universal set, is closed under boolean operations and analytical comprehension:*

$$\mathrm{MFS} \vdash \exists Y(Y =_e \{u: A(x, u, X)\} \wedge \forall v(v\eta Y \leftrightarrow A(v, u, X))),$$

where $A(x, u, Y)$ is analytical.

(ii) *SET is closed under weak power and inductive generation; in fact, pwX and $igXR$ are classes, provably in* MFS. *Moreover:*

$\exists Y(Y =_e SET)$ (= *there is the set of all sets*);

$\forall x(x\eta pwX \rightarrow \exists Y(Y \subseteq X \wedge Y =_e x) \wedge$

$\wedge \forall Y(Y \subseteq X \rightarrow \exists b(Cl(b) \wedge b =_e Y \wedge b\eta pwX)))$;

$Progr(X, R, Z) \rightarrow igXR \subseteq Z \wedge Progr(X, R, igXR)$.

Related principles have been investigated in the context of explicit mathematics by Feferman (1979). As to the relation between classes and sets, we remark that there are classes, which are provably not sets (e.g. $\{u: \exists Y(Y = u \wedge \neg u\eta Y)\}$).

§ 34. On Kripke's classification of self-referential sentences

In the seminal *Outline of a theory of truth* of 1975, Kripke shows how to use the fixed point semantics for classifying self-referential statements. We extend Kripke 's notions to the present framework.

34.1. DEFINITION. We fix $\mathcal{M}\models \mathrm{OP}^-$; the definitions are uniform in $\mathcal{M}$.

(i) If $X \in FIX(\mathcal{M})$, $X{\downarrow}a := X \models Ta \vee Fa$; $X{\uparrow}a :=$ not $(X{\downarrow}a)$. If $X{\downarrow}a$ $(X{\uparrow}a)$, we also say that X *converges* (*diverges*) on a, or X is (un-)defined on a. If $X \in FIX_{cs}(\mathcal{M})$, $\hat{X}$ is the partial function defined by the clauses: $\hat{X}(a) = 0$ if $X \models Ta$ and $\hat{X}(a) = 1$ if $X \models Fa$.

(ii) An element $a \in M$ is *grounded* (it is a proposition, according to chapter II), if $\mathbb{O}(\mathcal{M}){\downarrow}a$; $a \in M$ is ungrounded, otherwise.

(iii) An element $a \in M$ is *paradoxical* iff $X{\uparrow}a$, for every $X \in FIX_{cs}(\mathcal{M})$; a is unparadoxical, otherwise.

(iv) An element $a \in M$ is *intrinsic* (or has an intrinsic truth value) iff $X{\downarrow}a$, for some $X \in INT(\mathcal{M})$.

(v) An element $a \in M$ is *coherent* iff for every $X, Y \in FIX_{cs}(\mathcal{M})$ such that $X{\downarrow}a$ and $Y{\downarrow}a$, $\hat{X}(a) = \hat{Y}(a)$ (i.e. wherever a is defined, a always receives the same truth value). $a \in M$ is incoherent iff a is not coherent.

$$\mathrm{Gr}(\mathcal{M}) := \{a \in M : a \text{ grounded}\};$$

$$\mathrm{Intr}(\mathcal{M}) := \{a \in M : a \text{ intrinsic}\};$$

$$\mathrm{Coher}(\mathcal{M}) := \{a \in M : a \text{ coherent}\};$$

$$\mathrm{Paradox}(\mathcal{M}) := \{a \in M : a \text{ paradoxical}\}.$$

34.2. REMARK. It is immediate to see that:

1) if $a \notin M\text{-}PFOR$ (cf. 30.1), then $(\neg a)$ is grounded;
2) $\mathrm{Gr}(\mathcal{M}) \subseteq \mathrm{Intr}(\mathcal{M})$;
3) $\mathrm{Intr}(\mathcal{M}) \cup \mathrm{Paradox}(\mathcal{M}) \subseteq \mathrm{Coher}(\mathcal{M})$.

34.3. THEOREM (*Classification*)

(i) *There exist paradoxical elements;*
(ii) *there exist elements of M which are ungrounded, incoherent, unparadoxical;*
(iii) *there exist elements of M which are ungrounded, unparadoxical, coherent, non-intrinsic;*
(iv) *there exists ungrounded intrinsic elements of M.*

PROOF. (i): pick the "Liar" term L with $\mathcal{M}\models L = [\neg TL]$.
(ii) Pick the term s such that $\mathcal{M}\models s = [Ts]$, the so-called "truth-teller". s is ungrounded. Indeed, if $\mathbb{O}(\mathcal{M})\models Fs$, there would be α minimal such that $\mathcal{M}(\neg s) \in \mathbb{O}(\mathcal{M}, \alpha{+}1)$; but $\mathcal{M}(\neg s) = \mathcal{M}(\neg Ts)$, whence $\mathcal{M}(\neg s) \in \mathbb{O}(\mathcal{M}, \alpha)$: against minimality of α. The case $\mathbb{O}(\mathcal{M})\models Ts$ is similar. Consider $s^+ = \{\mathcal{M}(s)\}$ and $s^- = \{\mathcal{M}(\neg s)\}$; clearly either set is consistent and

Γ-dense. So there exist S^+, $S^- \in FIX_{cs}(\mathcal{M})$ such that $S^+ \models Ts$ and $S^- \models Fs$: hence s is incoherent (and also non-intrinsic) and unparadoxical.
(iii) Let $r = s \vee L$, where s, L are the closed terms of (i)-(ii) above. Then r is clearly ungrounded; also, using the above notation, $S^+ \models Ts$ implies $S^+ \models T(s \vee L)$, hence r is unparadoxical. r is coherent: if $X \models F(s \vee L)$, then $X \models Fs \wedge FL$, and therefore $X \models FL$: absurd! Hence, if $X \downarrow r$, $X(r) = 0$. r is non-intrinsic: if $X \models Tr$, then $X \models Ts$, hence $S^+ \leq X$. So a consistent fixed point extending both X and S^- would also extend S^+ and S^-, which is absurd.
(iv) We choose a closed term q such that $\mathcal{M} \models q = (\neg(q \wedge \neg q))$. q is obviously ungrounded (usual reasoning on the inductive definition of $\mathbb{O}(\mathcal{M})$). But q is unparadoxical: the set $\{\mathcal{M}(q), \mathcal{M}(\neg\neg q)\}$ is trivially Γ-dense, consistent (once more apply 7.1.1) and there exists $Z(q) \in FIX_{cs}(\mathcal{M})$ such that $Z(q) \models Tq$. On the other hand, if $X \in FIX_{cs}(\mathcal{M})$, the set $X_1 = X \cup \{\mathcal{M}(q), \mathcal{M}(\neg\neg q)\}$ is likewise Γ-dense and consistent (were X_1 inconsistent, $\mathcal{M}(\neg q) \in X$, whence $X \models Fq \wedge Tq$: against consistency of X). Thus $UP(X_1) \in FIX_{cs}(\mathcal{M})$, $UP(X_1) \geq X$, $Z(q)$, i.e. $Z(q) \in INT(\mathcal{M})$, and hence q is intrinsic. □

34.3.1. REMARK. If $X \in MAX(\mathcal{M})$ and a is intrinsic, then $X \downarrow a$. In fact, if a is intrinsic, $\mathbb{P}(\mathcal{M}) \downarrow a$, but $\mathbb{P}(\mathcal{M}) \leq \bigcap MAX(\mathcal{M})$ by 31.3 (iii). On the other hand, if a is a coherent, non-intrinsic element of M, there may be maximal consistent fixed points diverging on a: consider the fixed point S^- in the proof above. Then S^- has maximal consistent extension S_1^- (by 31.2) and clearly $S_1^- \models \neg T(s \vee L)$.

34.2 and 34.3 induce a corresponding classification on $\mathcal{L}$-sentences, uniformly in any given $\mathcal{M} \models \mathrm{OP}^-$. For instance, a sentence $A \in \mathcal{L}$ is intrinsic (paradoxical, etc.) iff $\mathcal{M}([A])$ is intrinsic (paradoxical, etc.).

The classification of self-referential notions produces highly complex sets (from the recursion-theoretic point of view). The moral is that "the truth is never simple" (Burgess 1986). Let us recall a couple of definitions from standard recursion theory.
A set $X \subseteq M$ is $\Sigma_1^1(M)$ (resp. $\Sigma_2^1(M)$) iff there exists a formula $B(x)$ of the language $\mathcal{L}_2{}^+(M)$ (see 33.6) such that $B(x) = \exists Y A(Y,x)$ $(\exists Y \forall Z A(Y,Z,x))$ and 1) no set quantifier occurs in A; 2) $X = \{a \in M : \langle \mathcal{M}, \mathcal{P}(M)\rangle \models B(a)\}$.
A set $X \subseteq M$ is $\Delta_1^1(M)$ iff both X and its complement are $\Sigma_1^1(M)$; $X \subseteq M$ is $\Pi_1^1(M)$ iff the complement of X is $\Sigma_1^1(M)$. Of course, similar classifications are readily extended to families $\mathcal{F} \subseteq \mathcal{P}(M)$ and it makes sense to speak of $\Sigma_1^1(M)$-families, etc.

34.4. LEMMA

(i) $FIX_{cs}(\mathcal{M})$, $FIX_{cp}(\mathcal{M})$, $FIX(\mathcal{M})$ *are* $\Delta_1^1(\mathcal{M})$ *families* $\subseteq \mathcal{P}(M)$;

(ii) *$MAX(\mathcal{M})$ and $INT(\mathcal{M})$ are $\Pi^1_1(\mathcal{M})$.*

PROOF. (i) By the characterization theorem 30.4, it suffices to show that the relation $\langle\mathcal{M}, X\rangle \models A$ is Δ^1_1, uniformly in X over $\mathcal{M}$, which is well-known (exercise or Moschovakis 1974).
(ii) $Y \in MAX(\mathcal{M}) \Leftrightarrow \forall X(X \in FIX_{cs}(\mathcal{M}) \wedge Y \subseteq X \Rightarrow Y = X)$, which is $\Pi^1_1(\mathcal{M})$ in a $\Delta^1_1(\mathcal{M})$-notion by (i), hence $\Pi^1_1(\mathcal{M})$ tout court. Define Compat(X,Y) iff $X \cup Y$ is consistent and Γ-dense. Clearly Compat(X, Y) is an elementary relation and hence $INT(\mathcal{M})$ is $\Pi^1_1(\mathcal{M})$, since we have:

$$Y \in INT(\mathcal{M}) \Leftrightarrow \forall X(X \in FIX_{cs}(\mathcal{M}) \Rightarrow \text{Compat}(X, Y)). \ \square$$

34.4 easily implies (with the notations of 34.1 and 13.3):

34.5. PROPOSITION

(i) $\text{Gr}(\mathcal{M}) \in IND(\mathcal{M})\text{–}HYP(\mathcal{M})$;

(ii) $\text{Paradox}(\mathcal{M}) \in \Pi^1_1(\mathcal{M})$;

(iii) $\text{Intrinsic}(\mathcal{M}) \in \Sigma^1_2(\mathcal{M})$.

PROOF. (i): $\text{Gr}(\mathcal{M}) \in IND(\mathcal{M})$, since $\mathbb{O}(\mathcal{M}) \in IND(\mathcal{M})$ (apply 13.4 and the trivial representability of $\mathbb{O}(\mathcal{M})$ by Tx). Were the complement $-\text{Gr}(\mathcal{M})$ of $\text{Gr}(\mathcal{M})$ inductive, there should be a closed term t (by 15.2) such that:

$$a \notin \text{Gr}(\mathcal{M}) \text{ iff } \mathbb{O}(\mathcal{M}) \models a\eta t.$$

It follows with consistency of $\mathbb{O}(\mathcal{M})$ that:

$$-\mathbb{O}(\mathcal{M}) = \{a \in M\colon\ \mathbb{O}(\mathcal{M}) \models a\eta\{x : x\eta t \vee T\neg x\}\},$$

whence, by 15.2, $-\mathbb{O}(\mathcal{M}) \in ENV(\mathbb{O}(\mathcal{M})) \subseteq IND(\mathcal{M})$, which contradicts theorem 13.4.
(ii): $a \in \text{Paradox}(\mathcal{M}) \Leftrightarrow \forall X(X \in FIX_{cs}(\mathcal{M}) \Rightarrow X{\uparrow}a)$.
(iii): $a \in \text{Intrinsic}(\mathcal{M}) \Leftrightarrow \exists Y(Y \in INT(\mathcal{M}) \wedge Y{\downarrow}a)$. $\square$

In the literature, sharp results are known about the logical complexity of the collections of paradoxical, intrinsic, classes, in the special case of models of reflective truth above the standard model of PA. The interested reader is sent to the paper of Burgess (1986).

34.6. REMARK (Cardinality theorems again). Consider the formula

$$C(\mathcal{M}) := \text{“if } \mathcal{M} \models \text{OP}^-,\ card(FIX_{cs}(\mathcal{M})) = 2^{card(\mathcal{M})}\text{”}.$$

Then $C(\mathcal{M})$ is $\Pi_1(\text{ZFC})$ and hence, by Levy's absoluteness lemma, in order to check that $C(\mathcal{M})$ is true for every $\mathcal{M}$, it is enough to verify $C(\mathcal{M})$ for *arbitrary countable* $\mathcal{M}$. Let $\mathcal{M}$ be countable: then by 34.4 and 13.4, $FIX_{cs}(\mathcal{M})$ is a $\Delta^1_1(\mathcal{M})$-family of subsets of M, containing an element which is not $\Delta^1_1(\mathcal{M})$. Thus we can apply the "Perfect Set Theorem" (see Moschovakis 1974), and $FIX_{cs}(\mathcal{M})$ is uncountable. {For the unexplained notions, the reader is sent to Barwise 1975}.

§ 35. On the consistency of coinduction principles

Assume that $A(x,v)$ is an operator of $\mathcal{L}$, in the sense of 10.3. As we know from chapter II, if we replace each occurrence of $t\eta v$ in $A(x,v)$ by means of the atom Pt (P fresh predicate symbol), we obtain a formula $A(x,P)$ of $\mathcal{L}(\mathcal{M},P) := \mathcal{L}(\mathcal{M}) \cup \{P\}$, which is positive in P. In chapter III we proved that there is a closed term $I(A) = FP(\lambda v.\{x\colon A(x,v)\})$, which represents in $\mathbb{O}(\mathcal{M})$ the least fixed point of the monotone operator:

$$\Gamma_A(S) = \{a \in M : \langle \mathcal{M}, \mathbb{O}(\mathcal{M}), S\rangle \models A(a,P)\} \text{ (here } P \text{ is interpreted by } S).$$

We now verify that $I(A)$ defines in $\mathbb{I}(\mathcal{M})$ (see 30.13) the largest fixed point of Γ_A. Formally, we are led to introduce a *generalized coinduction schema*, together with a generalized induction schema for the internal complement of $I(A)$.

35.1. DEFINITION

(i) Let $B(x)$ be a formula of $\mathcal{L}(\mathcal{M})$ and let $A(x,v)$ be an operator of $\mathcal{L}(\mathcal{M})$; we put:

$$Dense_A(B) := \forall x(B(x) \to A(x,B));$$

$$\mathrm{GID}\hat{} := Dense_A(B) \to \forall x(B(x) \to x\eta I(A));$$

$$^*\mathrm{GID}\hat{} := \forall x(B(x) \to FA(x,B)) \to \forall x(B(x) \to x\overline{\eta}I(A));$$

$$^*\mathrm{GID} := \forall x(FA(x,B) \to B(x)) \to \forall x(x\overline{\eta}I(A) \to B(x)).$$

GIDˆ is the *generalized coinduction schema*; for comparison, remind that the generalized induction schema has the form:

$$\mathrm{GID} := Clos_A(B) \to \forall x(x\eta I(A) \to B(x)),$$

where $Clos_A(B) := \forall x(A(x,B) \to B(x))$.

(ii) In order to give a dual version of 12.4, let MF_{cs} be the theory $\mathrm{NMF}^- + \mathrm{COMP}$, plus N-induction axiom for consistent properties:

$$Cs\text{-}N\text{-}IND := Cs(a) \wedge \overline{0}\eta a \wedge \forall x(x\eta a \to (x+1)\eta a) \to \forall x(Nx \to x\eta a),$$

where $Cs(a) := \forall x(\neg x\eta a \vee \neg x\overline{\eta}a)$.

(iii) We simultaneously introduce the transformations $^+$, $^-$ on $\mathcal{L}$-formulas (where $\exists$, $\vee$ are defined via $\forall$, $\wedge$, $\neg$):

$(A)^+ = A$, if A is an atom;

$(Tt)^- = T\neg t$; $\quad (A)^- = \neg A$, if $A = (t = s)$, Nt;

$(\neg B)^+ = (B)^-$; $\quad (\neg B)^- = (B)^+$;

$(\forall xB)^+ = \forall x(B)^+$; $\quad (\forall xB)^- = \exists x(B)^-$;

$(B \wedge C)^+ = (B)^+ \wedge (C)^+$; $\quad (B \wedge C)^- = (B)^- \vee (C)^-$.

As a preparatory step, we need two simple lemmata:

35.2. LEMMA

(i) *If* $X \subseteq M$ *and* X *is consistent,*

$$\langle \mathcal{M}, S \rangle \models (A^+ \rightarrow A) \wedge (A \rightarrow \neg A^-).$$

(ii) *If* $X \subseteq M$ *and* X *is complete,*

$$\langle \mathcal{M}, S \rangle \models (\neg A^- \rightarrow A) \wedge (A \rightarrow A^+).$$

(iii) *If* $X \in FIX(\mathcal{M})$, $X \models (A^+ \leftrightarrow TA) \wedge (A^- \leftrightarrow FA)$ *(A arbitrary $\mathfrak{L}(\mathcal{M})$-sentence).*

PROOF. (i)-(iii): induction on A, using consistency and completeness of X for $A = Tt$. □

35.3. LEMMA. *If A is an arbitrary sentence of $\mathfrak{L}(\mathcal{M})$,*

$\mathbb{O}(\mathcal{M}, \alpha+1) \models TA$ *implies* $\mathbb{O}(\mathcal{M}, \alpha) \models A^+$;

$\mathbb{O}(\mathcal{M}, \alpha+1) \models FA$ *implies* $\mathbb{O}(\mathcal{M}, \alpha) \models A^-$.

$\mathbb{I}(\mathcal{M}, \alpha) \models A^+$ *implies* $\mathbb{I}(\mathcal{M}, \alpha+1) \models TA$;

$\mathbb{I}(\mathcal{M}, \alpha) \models A^-$ *implies* $\mathbb{I}(\mathcal{M}, \alpha+1) \models FA$.

PROOF. This is a refinement of 12.1, which uses 30.1 and the fact that $\{\mathbb{O}(\mathcal{M},\alpha) : \alpha \in ON\}$ is a $\subseteq$-increasing sequence of consistent sets, while $\{\mathbb{I}(\mathcal{M},\alpha) : \alpha \in ON\}$ is a $\subseteq$-decreasing sequence of complete sets. □

35.4. THEOREM

(i) $\mathbb{I}(\mathcal{M}) \models \mathrm{GID}\hat{}$ *and* $\mathbb{I}(\mathcal{M}) \models {}^*\mathrm{GID}\hat{}$;

(ii) $\mathbb{O}(\mathcal{M}) \models \mathrm{GID}$ *and* $\mathbb{O}(\mathcal{M}) \models {}^*\mathrm{GID}$.

PROOF. (i) If $\mathbb{I}(\mathcal{M},\alpha) :=$ the αth-approximation of $\mathbb{I}(\mathcal{M})$, we must check by induction on α, assuming that $\mathbb{I}(\mathcal{M}) \models Dense_A(B)$:

$$\{a \in M : \mathbb{I}(\mathcal{M}) \models B(a)\} \subseteq \mathbb{I}(\mathcal{M}, \alpha).$$

We only consider the successor case. If $\mathbb{I}(\mathcal{M}, \alpha+1) \models \neg a\eta I(A)$, we have $\mathbb{I}(\mathcal{M}, \alpha) \models \neg A(a, I(A))$ by 35.2 (ii) and 35.3. But $\neg A(a, I(A))$ depends negatively on T and $I(A)$: since $\mathbb{I}(\mathcal{M}) \subseteq \mathbb{I}(\mathcal{M}, \alpha)$ we have $\mathbb{I}(\mathcal{M}) \models \neg A(a, B)$, whence $\mathbb{I}(\mathcal{M}) \models \neg B(a)$. The verification of *GID^ is similar.
(ii) First apply the induction theorem 12.4 for GID. As to *GID, let $-I(A) := \{x : x\overline{\eta} I(A)\}$ and assume:

$$\mathbb{O}(\mathcal{M}) \models \forall x(FA(x, B) \rightarrow B(x));$$

$$(-I(A))(\alpha) = \{a \in M : \mathbb{O}(\mathcal{M}, \alpha) \models a\overline{\eta} I(A)\} \subseteq \{a \in M : \mathbb{O}(\mathcal{M}) \models B(a)\}.$$

Then $\mathbb{O}(\mathcal{M}, \alpha+1) \models a\overline{\eta} I(A)$ and 35.3 imply $\mathbb{O}(\mathcal{M}, \alpha) \models A^{-}(a, I(A))$; since $A^{-}(a, I(A))$ depends positively on T and $(-I(A))$, we get with 35.2 (iii),

$$\mathbb{O}(\mathcal{M}) \models A^{-}(a, B), \text{ which implies } \mathbb{O}(\mathcal{M}) \models FA(a, B), \text{ i.e. } \mathbb{O}(\mathcal{M} \models B(a). \ \square$$

The theorem can be easily transformed into a strengthened conservation result:

35.5. COROLLARY

(i) PW_c + GID + *GID *is conservative over* OP;

(ii) MF_{cs} + GID^ + *GID^ *is a conservative extension of* OP.

Appendix: a variant to the basic operator Γ and the restriction axiom

Up till now, we avoided a critical evaluation of a strongly conventional assumption concerning the predicate T; indeed, the definition of Γ contains the clause "infer $T\neg x$ from $\neg PFOR(x)$", with the consequent truth of the restriction axiom RES:

$$(Tx \rightarrow PFOR(x)) \wedge (\neg PFOR \rightarrow T\neg x).$$

We briefly explore a natural alternative to RES.

First of all, why restriction axioms at all? The basic reason, already mentioned in Ch.II, 7.12, is semantically embodied in theorem 30.4: RES makes MF^{-} into a fixed point theory, and models of NMF^{-} + RES form a complete lattice with nice properties. A second point is that RES is self-dual with respect to the transformation of T into $\neg T\neg$ (see 8.11-8.12).

Can we imagine any reasonable variant to the operator Γ, with respect to the restriction axiom RES ? In particular, the condition RES.2 $\forall x(\neg PFOR(x) \rightarrow T\neg x)$ may appear unnatural. Therefore, we are going to consider a new operator Γ^{-} defined by the formula $\Gamma^{-}(x, P)$ which is obtained from $\Gamma(x, P)$ of §7 by omitting the clause corresponding to RES.2.

It turns out that the fixed points of Γ^- over a given model of OP^- are axiomatized by MF^- plus the restriction principle RES^-:

$$\forall x((Tx \rightarrow PFOR(x)) \wedge (T\neg x \rightarrow PFOR(x))).$$

More precisely, as in 7.12, we can derive:

1. PROPOSITION. *Let* FPT^- *be the sentence* $\forall x(Tx \leftrightarrow \Gamma^-(x,T))$, *where* $\Gamma^-(x,T)$ *is obtained from the formula* $\Gamma^-(x,P)$ *by replacing every subformula of* Pt *with* Tt. *Then* $\mathrm{MF}^- + \mathrm{RES}^-$ and $\mathrm{OP}^- + \mathrm{FPT}^-$ *have the same theorems.*

Of course, $\mathrm{MF}^- + \mathrm{RES}^-$ and $\mathrm{MF}^- + \mathrm{RES}$ are incompatible and we are left with a choice. In this respect, while we stress that both RES and RES^- are not required for relevant theoretic developments, we maintain a preference for the RES-axioms for the following reasons.

First, the alternative candidate theory $\mathrm{MF}^- + \mathrm{RES}^-$ does not enjoy the simple duality between T and $\neg T \neg$. On the semantic side, it turns out that the lattice-theoretic investigation of MF^--models is slightly simpler in presence of RES: with RES^- and Γ^- one has to relativize the lattice-theoretic operations to subsets of (the set defined) by $PFOR$.

Be that as it may, if we assume a neutral attitude, we can establish a few conservation relations between MF^-, $\mathrm{MF}^- + \mathrm{RES}$, $\mathrm{MF}^- + \mathrm{RES}^-$.

2. PROPOSITION

(i) $\mathrm{MF}^- + \mathrm{RES} \vdash A \Rightarrow \mathrm{MF}^- + \mathrm{RES}^- \vdash A$, *for any* T*-negative* A.

(ii) $\mathrm{MF}^- + \mathrm{RES}^- \vdash A \Rightarrow \mathrm{MF}^- \vdash A$, *for any* T*-positive* A.

(iii) MF^-, $\mathrm{MF}^- + \mathrm{RES}$, $\mathrm{MF}^- + \mathrm{RES}^-$ *have the same* T*-free theorems.*

PROOF. (iii) is a consequence of (i)-(ii).
(i): let A be T-negative and assume that A is not provable in $\mathrm{MF}^- + \mathrm{RES}^-$. Then for some structure $\langle \mathcal{M}, X\rangle$, $\langle \mathcal{M}, X\rangle \models \mathrm{MF}^- + \mathrm{RES}^- + \neg A$. Hence by the previous proposition, X is a fixed point of Γ^- and hence, since $\Gamma^-(X) \subseteq \Gamma(X)$, X is Γ-dense. Thus by lemma 30.10, there is a set $X\hat{}$ such that $X\hat{} = \Gamma(X\hat{})$ and $X \subseteq X\hat{}$. By the characterization theorem 30.4, since $\neg A$ is T-positive and hence upward persistent, we have

$$\langle \mathcal{M}, X\hat{}\rangle \models \mathrm{MF}^- + \mathrm{RES} + \neg A,$$

i.e. A is unprovable in $\mathrm{MF}^- + \mathrm{RES}$.

(ii): observe that if X is a model of MF^-, then X is Γ^--closed. Then argue as in (i), using downward persistence of T-negative formulas. □

Of course, there is room for exercises: we can change MF^- to systems including forms of N-induction, the approximation axioms, generalized induction. The point is that (iii) above remains true and it can be established by proof theoretic means; hence we conclude:

RES, RES^- *do not affect the proof theoretic strength of the theory involved.*

A possible addition to the basic MF^- is suggested by considering an essential feature of the usual truth predicates for formal languages: *they deal with inductively defined syntactical entities, i.e. sentences.* So we might replace the notion of pseudo-form $PFOR$ by an inductively defined subset, whose elements would play the role of traditional sentences. Then we could try to define a modification of T, which only acts on sentence-like objects (see Ch. II, for a related step).

PART D

LEVELS OF TRUTH AND PROOF THEORY

"*Für jede mathematische Disziplin ist es charakteristisch, dass 1) für sie ein derartiger Operationsbereich zugrunde liegt, wie wir ihn hier von Anfang an vorausgesetzt haben, dass diesem 2) stets die natürlrchen Zahlen samt der sie verknüpfenden Beziehung F assoziiert werden, und dass 3) über diesem kombinierten Operationsbereich durch den ev.sogar beliebig oft iterierten mathematischen Prozess ein Reich neuer idealer Gegenstände, von Mengen und funktionalen Zusammenhängen, aufgebaut wird.*"
(H.Weyl, 1918)

CHAPTER 8

LEVELS OF REFLECTIVE TRUTH

§36. A language and axioms for reflective truth with levels
§37. Simple consequences
§38. Universes and the Weyl extended iteration principle
§39. A recursion-theoretic model
§40. Levels of truth and predicatively reducible subsystems of second-order arithmetic
§41. Consistency of a reducibility principle for classes
§42. Levels of truth and impredicative subsystems of second-order arithmetic
Appendix: on projectibility and stronger reflection

We present a new formal framework TLR (= truth with levels and reflection), in which the theory of reflective truth is enriched by the notion of *level*. We will prove that TLR is able to internalize, to a certain extent, negation and quantification over classes. We will also verify that the resulting system is mathematically non-trivial and it yields a new characterization of predicative mathematics (this will follow from chapters IX and XI).

In the previous chapters, we have been pursuing a logical approach to abstraction, which is based on a self-referential truth predicate T. However, there appear severe limitations to the reflective power of T, even in presence of the approximation structure, investigated in chapters III-V. For instance, T cannot seriously think of itself to be consistent, without implying its completeness, and hence its very inconsistency by the Tarski-Russell arguments of 8.5-9.3. Nevertheless, it makes perfectly good sense to state that T is consistent, once, say, the inductive model $\mathbb{O}(\mathcal{M})$ of Ch. II (§7) is grasped as a whole. Therefore, we wonder whether we cannot design a new formal framework, which can better adjust *negative semantic information.*

A starting point towards a reasonable solution is the remark that the truth T is, after all, *a parametric notion*: it always depends on a set $\mathcal{K}_0$, involving complete information about given primitive predicates, which can also be regarded as *the context T is about.* As a consequence, T is really $T(\mathcal{K}_0)$, for some $\mathcal{K}_0$; and if we regard $T(\mathcal{K}_0)$ as grasped, *we are actually shifting from the context $\mathcal{K}_0$ to a new one $\mathcal{K}_1$*, which also includes a complete description of $T(\mathcal{K}_0)$ as primitive! This means that, if A is any sentence and $A \notin T(\mathcal{K}_0)$ ($A \in T(\mathcal{K}_0)$ respectively) holds, we must have $(\neg T_0 A) \in \mathcal{K}_1$ ($(T_0 A) \in \mathcal{K}_1$, T_0 being the formal counterpart of $T(\mathcal{K}_0)$). Of course, we must add $(\neg T_0 A)$ to $\mathcal{K}_1$ and not simply $(\neg A)$: $(\neg A)$ would dangerously conceal its context dependence and this would drive us immediately to contradiction.

These considerations naturally advice to make the parametric dependence of truth explicit by means of *levels*: the shift from $T(\mathcal{K}_0)$ to $T(\mathcal{K}_1)$ is regarded as a step to a higher stage of reflection, and, formally, from truth of ground level T_0 to truth of higher level T_1. Moreover, since the step from level 0 to level 1 can be understood as a general method to produce new truth predicates from given ones, we may identify levels simply with ordinals and devise a new formalism TLR, where T is accompanied by level dependent truth predicates T_i. Informally, we can summarize the basic tenets behind TLR as follows.

1) If i, j are levels and $i \prec j$ (where $\prec$ is the precedence order on the set of levels), T_i and T_j will be related in such a way that: (a) whatever is declared true by T_i, is declared true by T_j, i.e. $\forall x(T_i x \to T_j x)$; (b) T_i is decidable with respect to T_j, i.e. $T_j T_i A$ or $T_j \neg T_i A$ (A arbitrary; we neglect formalization details).

2) Each local truth predicate T_i satisfies the principles of the general theory MF^- of reflective truth (Ch. II, 7.11).

3) We still maintain a level-free truth predicate T with us, and we conceive it as the "limit" of the local truth predicates; in addition, we still assume that T itself has the same self-referential abilities of any T_i. To sum up, T is a model of MF^-, as well as each of its local approximations.

On the surface, we have restored a hierarchy of truth predicates, which is strictly reminiscent of *the Tarskian language/metalanguage hierarchy*; it might seem that we have destroyed the type-free style of the previous systems. As a matter of fact, the new framework is quite distant from the Tarskian one; in particular by 2) each T_i already encompasses the standard Tarskian predicates, as to closure properties and self-referential ability. Furthermore, the level structure greatly strengthens the deductive force, as it should appear from the summary of the results below.

In §36 we outline the new formal theory TLR. We mention that, in the axiomatic approach, the level ordering is not assumed to be total nor well-founded; it is also necessary to postulate an injection of levels into objects, in order to codify sentences involving levels. This is an important restriction for building models of TLR; it also requires non-trivial properties of admissible ordinals (projectibility). §37 states a few basic facts about

predicate abstraction relative to any given level i; in particular we can distinguish i-classes, i.e. predicates which are total relative to truth of level i. §38 investigates the influence of the local structure on the closure properties of level-free statements. It turns out that $CL := \{x : x \text{ class}\}$ splits into a directed family $\{CL_i : i \text{ level}\}$, where every $CL_i := \{x : x \text{ class of level } i\}$ is itself a class at any higher level $j \succ i$. As a consequence, classes are closed under an analogue of Weyl's *Iterationsprinzip* (see Weyl 1918), a transfinite recursion principle along CL-wellfounded linear orderings. We can also introduce, following Feferman (1982) and Jäger (1984), a satisfactory notion of *universe*. §39 describes a model $\mathfrak{C}_\iota$ for TLR, which is built-up by means of a suitable iterated inductive definition along the first recursively inaccessible ordinal. A byproduct of the effective nature of $\mathfrak{C}_\iota$ is that it validates a remarkable reducibility principle for classes RPC (§41): if an elementary condition with parameters in CL_i is satisfied by a class, then it is already satisfied by a j-class for some $j \succ i$. §40 and §42 establish a link between second-order arithmetic Z_2 and theories of reflective truth with levels. After a brief survey of Z_2 and its subsystems, we prove that TLR yields a model to Friedman's subsystem ATR_0, while MF_c and MF_p (see Ch.II) interpret suitable versions of hyperarithmetical analysis (namely Δ^1_1-CA_0 and Σ^1_1-DC_0 respectively). We shall later verify that TLR and ATR_0 have exactly the same arithmetical content (ATR_0 is a strong version of predicative analysis).
The final section shows that TLR plus RPC interprets impredicative subsystems of Z_2 (namely Π^1_1-CA_0 and Δ^1_2-CA_0 ; cf. §42). It is interesting to mention that RPC dispenses TLR with the primitive notion of natural number. We also stress that the interpretation results – coupled with well-known theorems of Friedman-Simpson's *reverse mathematics* – grant that the mathematical content of TLR and related systems is significant for mathematical practice. In the final appendix, we suggest the consistency of TLR with reflection principles, stemming from higher recursion theory and the related study of recursively large cardinals; this, however, is only a research to come.

As to the connection with the literature, the idea of iterating the abstraction procedure is implicitly involved in Ackermann's approach to type-free logic, in Lorenzen and Myhill (1959, pp.47-49), Schütte (1960) and Fitch (1964); it is then made explicit by Scott (1975), though in a very different context. More recently, we ought to mention Martin-Löf 's idea of adding "universes" (Martin-Löf 1984) and, in a classical set-theoretic context, the theories of iterated admissibility (Jäger-Pohlers 1982). For related ideas in the investigation of the so-called *logical frameworks* in Theoretical Computer Science, we send the reader to Aczel-Carlisle-Mendler (1991) and to the final chapter. Finally, ideas connected with truth and levels can be found in the philosophical papers of Burge (1979) and Gaifman (1983). A direct ancestor of TLR is outlined in Cantini (1987).

§ 36. A language and axioms for reflective truth with levels

The language $\mathcal{L}_V$ is the extension of the basic language $\mathcal{L}$ for reflective truth (§7), which includes:

(i) a new sort of variables for levels $i_0, i_1, \ldots$ (in short L-variables);
(ii) a new unary function symbol LT;
(iii) three new binary predicates $\preceq$, $=_l$ and V (for level ordering, level identity and local truth respectively).

The syntax of $\mathcal{L}_V$ requires the introduction of L(evel)-terms, besides terms in the usual sense.

36.1. DEFINITION. *L(evel)-terms, terms and formulas of* $\mathcal{L}_V$.

(i) L-terms are exactly the L-variables (i, j, k metavariables for L-terms);

(ii) the set of $\mathcal{L}_V$-terms is the least collection which is closed under the following clauses: individual variables and constants are terms; if j is an L-term, $LT(j)$ is a term; if t, s are terms, $Ap(t,s)$ is a term.

(iii) the set of $\mathcal{L}_V$-formulas is the least collection closed under the following clauses: if j and i are L-terms, $i \preceq j$ $i =_l j$ are atoms (and hence formulas); if t, s are terms and i is an L-term, Nt, $t = s$, Tt and $V(i,t)$ are atoms (and hence formulas); if A, B are formulas, $\neg A$, $A \wedge B$ are formulas; if A is a formula, x an individual variable and j is an L-variable, then $\forall x A$ and $\forall j A$ are formulas (where x, j occur bound).

36.2. NOTATION. We stick to conventions and notations of Ch. I, §1. In addition, we write $T_i t$, $t\eta_i s$, $t\overline{\eta}_i s$, $t\eta s$ and $Cl_i(t)$ as abbreviations for $V(i,t)$, $V(i,(st))$, $V(i, NEG(st))$, $T(st)$ and $\forall x(x\eta_i t \vee x\overline{\eta}_i t)$ (in the given order). If $Cl_i(t)$ is assumed, we say that t is *a class of level* i, or, simply, an i-class. If i, j are L-terms, $i = j$ stands for $i =_l j$; we also write $i \prec j$ for $(\neg i = j) \wedge (i \preceq j)$.

Before stating the axioms for the theory of truth, we adapt the abstraction operation of chapter II to the present context.

36.3. DEFINITION

(i) The set Δ^+ of acceptable formulas of $\mathcal{L}_V$ is the smallest collection, which includes the atoms Tt, $T_i t$, Nt, $t = s$ and is closed under negation, conjunction and universal quantification on object variables. Note that $\mathfrak{L}$-formulas are acceptable.

(ii) We assume the combinators ID, TR, NAT, NEG, ALL of §7. Then we inductively extend the map $A \mapsto [A]$ to arbitrary acceptable formulas: we only mention the new clause (which clarifies the use of the function symbol LT):

$$[T_i t] := \langle \overline{7}, \langle LT(i), t \rangle \rangle.$$

If A is acceptable, $T_i A := T_i[A]$ and $TA := T[A]$.

(iii) The standard definition of λ-abstraction (see Ch. I, 1.1) is extended by the new clause $\lambda x.LT(i) := K(LT(i))$ (where i is an arbitrary L-term). As a consequence, it makes sense to introduce the abstraction operator for arbitrary acceptable formulas by stipulating $\{x: A\} := \lambda x.[A]$.

36.4. The theory TL (= truth with levels) is based on two sorted classical predicate calculus (indeed, $\mathfrak{L}_V$ has L-variables and individual variables). The principles of TL are grouped into operational and number theoretic axioms, local truth axioms, level and connection axioms.

36.4.1. *Operational and number theoretic axioms.* They include the standard principles for basic combinators K and S, pairing, projections, zero, successor, predecessor, definition by cases on natural numbers (cf. COMB, PAIR, NAT.1-NAT.2 of Ch.I, 2.1), plus local number theoretic induction LIND, i.e. N-induction for i-classes:

LIND $\quad Cl_i(x) \wedge Clos_i(x) \rightarrow \forall u(Nu \rightarrow u\eta_i x)$,

where $Clos_i(x) := \overline{0}\eta_i x \wedge \forall v(v\eta_i x \rightarrow (v{+}1)\eta_i x)$. In addition, we require the projectibility axiom

PROJ $\quad \forall i \forall j(LT(i) = LT(j) \rightarrow i = j)$.

36.4.2. *Local truth axioms:*

4.2.1 $\quad T_i A \leftrightarrow A$, if $A = (x = y)$, Nx, $\neg x = y$, $\neg Nx$;

4.2.2 $\quad T_i x \rightarrow T_i T_i x$; $\qquad T_i \neg x \rightarrow T_i \neg T_i x$;

4.2.3. $\quad T_i \neg\neg x \leftrightarrow T_i x$;

4.2.4 $\quad T_i(x \wedge y) \leftrightarrow T_i x \wedge T_i y$; $\qquad T_i \neg(x \wedge y) \leftrightarrow T_i \neg x \vee T_i \neg y$;

4.2.5 $\quad T_i(\forall f) \leftrightarrow \forall x T_i(fx)$; $\qquad T_i \neg(\forall f) \leftrightarrow \exists x T_i \neg(fx)$;

4.2.6. $\quad \neg(T_i x \wedge T_i \neg x)$; $\qquad$ (Local consistency).

36.4.3. *Level axioms*: they include the standard equality axioms for level equality $=_l$, and state that $\preceq$ is a directed unbounded partial order:

4.3.1 $\forall i \forall j \forall k((i \preceq i) \wedge (i \preceq j \wedge j \preceq k \rightarrow i \preceq k) \wedge$

$\wedge (i \preceq j \wedge j \preceq i \rightarrow i = j))$;

4.3.2 $\forall i \forall j \exists k(i \prec k \wedge j \prec k)$.

36.4.4. *Connection axioms.* They are the crucial principles of the theory, relating truth predicates of different level.

4.4.1	$Tx \leftrightarrow \exists i T_i x$;	Limit
4.4.2	$i \preceq j \wedge T_i x \rightarrow T_j x$;	Persistence
4.4.3	$T_i Tx \leftrightarrow T_i x$; $T_i \neg Tx \leftrightarrow T_i \neg x$;	Localization
4.4.4	$i \prec j \rightarrow (T_j T_i x \vee T_j \neg T_i x)$;	Potential Completeness
4.4.5	$T_j T_i x \rightarrow i \preceq j \wedge T_i x$;	Positive Soundness
4.4.6	$T_j \neg T_i x \rightarrow (i = j \wedge T_i \neg x) \vee (i \prec j \wedge \neg T_i x)$	Negative Soundness

As usual, TL^- denotes TL without local N-induction. A word of comment. By the principles of groups 36.4.2, and 36.4.4.5, 36.4.4.6, local N-induction and soundness, it is obvious that the axioms of MF_c(= the system of reflective truth with class N-induction; see 10.7) are satisfied at every level. On the other side, potential completeness ensures that negative information about any level i becomes internal at higher levels, while limit and localization axioms grant that global truth statements always reduce to local truth statement (of sufficiently high level). Finally, by persistence and soundness, no information is lost at later levels, and later levels do not conflict with the earlier ones, even on negative information.

§37. Simple consequences

We begin with a few elementary consequences of TL.

37.1. DEFINITION

(i) Let i be any L-variable: the *i-transform* of $A \in \mathcal{L}$ is the $\mathcal{L}_V$-formula A_i, which results from A by substituting (each occurrence of atoms of the form) Tt by $T_i t$ (e.g. $(\forall x T(ax))_i = \forall x T_i(ax)$; $(TTt)_i = (T[Tt])_i = T_i Tt$).

(ii) An $\mathcal{L}_V$-formula A is T-positive iff A belongs to the least collection which contains expressions of the form $t = s$, $\neg t = s$, Nt, $\neg Nt$, $T_i t$, $\neg T_i t$,

Tt and is closed under conjunction, disjunction, and quantifiers (of either sort).

(iii) A is an i-formula iff A belongs to the least collection of formulas which is closed under $\wedge$, $\neg$, universal object quantification and contains only atoms of the form $t = s$, Nt, $T_i t$.

The first result allows to freely use i-transforms of MF_c-theorems within the system TL; it follows from local truth axioms and localization.

37.2. LEMMA. *If* $\mathrm{MF}_c \vdash A$, *then* $\mathrm{TL} \vdash A_i$.

37.2.1. APPLICATION. By the previous lemma and 9.6, we immmediately have the following useful facts:

(i) If $A(u,x)$ is an $\mathfrak{L}$-formula, quasi-elementary in x (cf. II, 9.5),

$$\mathrm{TL}^- \vdash Cl_i(x) \rightarrow (A_i(u,x) \leftrightarrow T_i A(u,x)).$$

(ii) If A is a T-positive $\mathfrak{L}$-formula,

$$\mathrm{TL}^- \vdash A_i \leftrightarrow T_i A.$$

(iii) If $A(u,x)$ is an $\mathfrak{L}$-formula, elementary in x,

$$\mathrm{TL}^- \vdash Cl_i(x) \rightarrow T_i A(u,x) \vee F_i A(u,x).$$

37.3. LEMMA. (i) TL^- *proves:*

$T.1$ $TA \leftrightarrow A$, if $A = (x = y)$, Nx, $\neg x = y$, $\neg Nx$;

$T.2$ $TTx \leftrightarrow Tx$; $\quad T\neg Tx \leftrightarrow T\neg x$;

$T.3$ $T\neg\neg x \leftrightarrow Tx$;

$T.4$ $T(x \wedge y) \leftrightarrow Tx \wedge Ty$; $\quad T\neg(x \wedge y) \leftrightarrow T\neg x \vee T\neg y$;

$T.5$ $T(\forall f) \rightarrow \forall x T(fx)$; $\quad \exists x T\neg(fx) \leftrightarrow T\neg(\forall f)$;

$T.6$ $\neg(Tx \wedge T\neg x)$.

(ii) Δ^+-*soundness:* $\quad \mathrm{TL}^- \vdash T_i A \rightarrow A \ (A \in \Delta^+)$.

(iii) $\mathrm{TL}^- \vdash T_i A \rightarrow A_i \ (A \in \Delta^+)$.

(iv) $\mathrm{TL}^- \vdash i \preceq j \wedge T_i A \rightarrow T_j A \ (A \in \Delta^+)$.

(v) *If A is a k-formula,*

$$\mathrm{TL}^- \vdash k \prec j \rightarrow (T_j A \vee T_j \neg A) \wedge (T_j A \leftrightarrow A);$$

$$\mathrm{TL}^- \vdash (TA \leftrightarrow A) \wedge (TA \vee T\neg A).$$

PROOF. (i): by limit, persistence, localization and local truth axioms, together with directedness of the level ordering. As a sample, let us check global consistency T.6. If Tx and $T\neg x$ are assumed, then by limit axiom $T_i x$ and $T_k \neg x$, for some i, k; hence, there exists some $j \succ i, k$, such that by persistence $T_j x$ and $T_j \neg x$, against local consistency.
(ii): induction on A. If A is an e-atom or has the form $T_j t$, $\neg T_j t$, we apply 36.4.2.1, positive and negative soundness and local consistency. Let $A := \neg Tt$ and assume $T_i \neg Tt$; then $T_i \neg t$ by localization, hence $T\neg t$ by limit and $\neg Tt$ by T.6 above. The remaining cases are straightforward by IH and local truth axioms.
(iii): exercise (induction on A).
(iv): apply persistence axiom.
(v): potential completeness and Δ^+-soundness imply the first statement, which yields the second by limit, persistence and unboundedness of the level ordering. □

We proceed by considering simple facts about abstraction, classes and i-classes. In particular, we see that the notion of i-class determines a class at any level $j \succ i$. Moreover, the Russell sentence relativized to level i becomes true at strictly higher levels. As usual, we let:

37.4. DEFINITION. $CL := \{x : Cl(x)\}$; $\quad CL_i := \{x : Cl_i(x)\}$;

$$R := \{x : \neg x\eta x\}; \qquad R(i) := \{x : \neg x\eta_i x\};$$

$$x \Leftrightarrow_i y := (T_i x \leftrightarrow T_i y) \wedge (T_i \neg x \leftrightarrow T_i \neg y);$$

$$x \Leftrightarrow y := (Tx \leftrightarrow Ty) \wedge (Fx \leftrightarrow Fy).$$

37.5. LEMMA

(i) *The extended abstraction schema for acceptable formulas: for every* $A \in \Delta^+$,

$$\mathrm{TL}^- \vdash \forall u(u\eta\{x : A(x)\} \Leftrightarrow A[x := u]).$$

(ii) *The local abstraction schema for acceptable formulas: if* $A \in \Delta^+$,

$$\mathrm{TL}^- \vdash \forall i \forall u(u\eta\{x : A\} \Leftrightarrow_i A[x := u]).$$

(iii) *If* $A(x)$ *is a* j*-formula,* TL^- *proves:*

$$j \prec i \rightarrow \forall u(u\eta_i\{x : A\} \leftrightarrow A[x := u]),$$

(*where* u *free for* x *in* A *in* (i)-(iii) *above*).

(iv) $\mathrm{TL}^- \vdash \forall i \neg Cl_i(R)$.

(v) $\mathrm{TL}^- \vdash \forall i(i \succ j \rightarrow Cl_i(R(j)) \wedge \neg Cl_j(R(j)) \wedge R(j)\eta_i R(j))$. *Hence:*

$$\mathrm{TL}^- \vdash \forall i Cl(R(i)).$$

(vi) $\mathrm{TL}^- \vdash i \prec k \rightarrow CL_i \eta_k CL_k \ \wedge \ CL_i \subset CL_k.$

PROOF. (i): trivial by β-conversion and $T.2$ of 37.3 (i).
(ii): immediate by β-conversion and localization axioms.
(iii): by (ii) and lemma 37.3 (v).
(iv): by lemma 37.2 and 9.3.
(v). Let $i \succ j$. As to the first conjunct, $R(j)$ is defined by a j-formula and hence we apply 37.3 (v) and local abstraction. The second conjunct is simply Russell rephrased for level j. The third conjunct is a consequence of the second one with (iii). The last statement follows by unboundedness of $\prec$, limit, persistence.
(vi). Assume $k \succ i$: then $CL_i \subset CL_k$ by persistence and (v). As to $CL_i \eta_k CL_k$, apply local abstraction and 37.3 (v). □

37.6. LEMMA. *Let $A(u,x)$ be an $\mathfrak{L}$-formula elementary in x (see definition 9.5 (i)). Then:*

(i) $\mathrm{TL}^- \vdash Cl_k(x) \rightarrow (A(u,x) \leftrightarrow A_k(u,x) \leftrightarrow T_k A(u,x)).$

(ii) *Closure of CL_k under elementary comprehension:*

$$\mathrm{TL}^- \vdash Cl_k(x) \rightarrow (Cl_k(\{u\colon A(u,x)\}) \wedge \forall v(v\eta\{u\colon A(u,x)\} \leftrightarrow A(v,x))).$$

(iii) *Closure of CL_k under join:*

$$\mathrm{TL}^- \vdash (Cl_k(x) \wedge f : x \rightarrow CL_k) \rightarrow Cl_k(\Sigma(x,f)) \wedge$$
$$\wedge \forall u(u\eta\Sigma(x,f) \leftrightarrow \exists v \exists w(u = \langle v,w\rangle \wedge v\eta x \wedge w\eta(fv))).$$

PROOF. (i) Assume that x is a k-class. The second equivalence holds by 37.2.1. We check the first equivalence by induction on A. If A is an atom different from $u\eta x$, we are done by 36.4.2.1. If $A(u,x) = u\eta x$, $u\eta_k x$ implies $u\eta x$ by limit axiom. In the opposite direction, we easily reach a contradiction from $u\eta x$ and $\neg u\eta_k x$ (apply $Cl_k(x)$, persistence, limit, unboundedness and local consistency). If $A(u,x)$ is a negation, a conjunction or a universal quantification, we simply apply IH.
(ii) If x is a k-class, so is $\{u\colon A(u,x)\}$ by 37.2.1 (iii). The second equivalence is an immediate consequence of (i) and 37.5 (iii).
(iii): by lemma 37.2 applied to the join principle 9.9 and (i) above. □

Up to this point, we have seen that the local approximations of T satisfy the basic axioms for self-referential truth, but it is not clear whether this happens for T itself, and hence whether there is *a harmony between global and local structure*. The time is ripe to introduce a simple reflection axiom which grants such correspondence.

37.7. DEFINITION

(i) *The Reflection principle* REF:

$$\forall i \forall y \forall z(\forall x \exists j(x\eta_i y \to x\eta_j z) \to \exists k \forall x \exists j(j \preceq k \wedge (x\eta_i y \to x\eta_j z))).$$

(ii) $\mathrm{TLR} := \mathrm{TL} + \mathrm{REF}$ and $\mathrm{TLR}^- :=$ TLR minus class N-induction.

The reflection principle implies that there are enough levels for T, in order to internalize universal statements about objects, and it is essential for showing that MF_c is a subtheory of TLR.

37.8. LEMMA

(i) *Positive reflection:*

$\mathrm{TLR}^- \vdash \forall x \exists i T_i(fx) \to \exists i \forall x T_i(fx)$;

(ii) $\mathrm{TLR}^- \vdash \forall x T(fx) \to T(\forall f)$;

(iii) *if A is acceptable and A is T-positive,*

$\mathrm{TLR}^- \vdash A \leftrightarrow \exists i.T_i A \leftrightarrow TA \leftrightarrow \exists i A_i$;

(iv) *if $A(x)$ is T-positive and acceptable,*

$\mathrm{TLR}^- \vdash \forall u(u\eta\{x{:}A\} \leftrightarrow A[x := u])$;

(v) $\mathrm{TLR}^- \vdash Cl(a) \leftrightarrow \exists i Cl_i(a)$;

(vi) *if $a \subseteq b := \forall x(x\eta a \to x\eta b)$,*

$\mathrm{TLR}^- \vdash a \subseteq b \leftrightarrow \forall i \exists k \forall x(x\eta_i a \to x\eta_k b)$;

(vii) *a class of classes is always an i-class, for some level i:*

$\mathrm{TLR}^- \vdash Cl(a) \wedge a \subseteq CL \to \exists i.\, a \subseteq CL_i$.

PROOF. (i): apply reflection with $y = \{u : u = u\}$ and persistence.
(ii): apply limit and positive reflection (i).
(iii) Let us consider the first equivalence. From right to left, it follows from Δ^+-soundness (lemma 37.3). As to the reverse direction, we argue by induction on A.
If $A = \neg T_j t$, choose $k \succ j$ by unboundedness of $\prec$: then $T_k \neg T_j t$ by potential completeness, positive and negative soundness. If $A := \forall x B$, we use IH, positive reflection and the local truth axiom for $\forall$. The other cases are easy and left as exercises. The second equivalence is just a restatement of the limit axiom. As to the third equivalence, $TA \to \exists i T_i A \to \exists i A_i$ (use lemma 37.3 (iii)). $A_i \to TA$ is inductively checked ((ii) above being used in the case $A := \forall x B$).

(iv): by (iii) above.
(v): apply the limit axiom from right to left. The reverse direction is a consequence of (iii), as the formula defining Cl is acceptable and T-positive.
(vi):$\Rightarrow$ by limit and reflection; the converse is trivial.
(vii): assume that a is a class of classes. Then by (v), a is an i-class and

$$a \subseteq CL \Rightarrow \forall j \exists k \forall x (x\eta_j a \rightarrow x\eta_k CL) \text{ by (vi);}$$
$$\Rightarrow \forall j \exists k \forall x (x\eta_j a \rightarrow Cl_k(x)), \text{ by localization and local abstraction;}$$
$$\Rightarrow \exists k \forall x (x\eta_i a \rightarrow Cl_k(x)) \text{ by logic;}$$
$$\Rightarrow a \subseteq CL_k \text{ for some } k,$$

as $x\eta a \leftrightarrow x\eta_i a$, by assumption on a and global consistency (37.3, T.6). □

37.9. THEOREM. *If* $\mathrm{MF_c} \vdash A$, *then* $\mathrm{TLR} \vdash A$ $(A \in \mathcal{L})$.

PROOF. By lemmata 37.3 and 37.8, it only remains to check:

$$\mathrm{TLR} \vdash Cl(x) \rightarrow (Clos(x) \rightarrow \mathbb{N} \subseteq x).$$

Since x is an i-class for some i, $Clos(x) \leftrightarrow Clos_i(x)$ by 37.6 (i) and hence we can apply local class N-induction. □

Of course, one may wonder *whether the set of provable level-free statements is really affected by the level structure*, that is: what more do we know about T within TL and TLR? Precisely this question is faced in the next section.

§38. Universes and the Weyl extended iteration principle

We present two significant level-free statements, whose verification essentially relies upon the level structure. The first principle roughly says that, as soon as we deal with logical constructions depending on *classes* as initial data, then we can always work within a nicely closed *universe*, which is itself a *class of classes* and to which the initial data belong. To make this idea precise, a few definitions are in order.

38.1. DEFINITION

(i) $y \models \mathrm{J} := \forall f \forall c ((c\eta y \wedge f : c \rightarrow y) \rightarrow (\Sigma(c,f)\eta y \wedge$
$\wedge \forall u (u\eta\Sigma(c,f) \leftrightarrow \exists v \exists w (u = \langle v, w\rangle \wedge v\eta c \wedge w\eta(fv)))));$

$y \models \mathrm{J}$ states that the join principle of 9.9 holds relativized to y.

(ii) We say that y is elementarily closed iff y contains the singleton $\{x\}$ for arbitrary x, the classes $\mathbb{APP} := \{\langle x,y,z\rangle : xy = z\}$, $\mathbb{ID} := \{\langle x,y\rangle : x = y\}$, $\mathbb{N} := \{x : Nx\}$ and is closed under intersection, complement and domain, plus the combinatorial operations of expansion, converse, cycle, transpose (see 9.11 and 9.13 in Ch.II).
In symbols, let $Elemclos(y)$ be the universal closure of the conjunction of the following two conditions:

EL.1 $\mathbb{APP}\eta y \wedge \mathbb{N}\eta y \wedge \mathbb{ID}\eta y \wedge \forall x(\{x\}\eta y)$;

EL.2 $\forall u \forall v(u\eta y \wedge v\eta y \rightarrow ((u \cap v)\eta y \wedge (-u)\eta y \wedge dom(u)\eta y \wedge$
$\wedge Exp(u)\eta y \wedge Cyc(u)\eta y \wedge Trans(u)\eta y \wedge Conv(u)\eta y)$.

(iii) y is *a universe of classes* iff y is an elementarily closed class of classes, which is also closed under join; in symbols:

$$Univ(y) := Cl(y) \wedge y \subseteq CL \wedge \ Elemclos(y) \wedge y \models \mathrm{J}.$$

38.2. THEOREM

(i) $\mathrm{TL}^- \vdash \forall k.Univ(CL_k)$.

(ii) $\mathrm{TLR}^- \vdash \forall y(Univ(y) \rightarrow \exists k(y \subseteq CL_k)$.

PROOF. (i) That CL_k is closed under join and elementary operations already follows from 37.6. $CL_k \subseteq CL$ and $CL_k\eta CL$ are consequences of persistence and lemma 37.5 (vi). (ii): immediate by lemma 37.8 (vii). □

38.3. COROLLARY. *Let* LIM := $\forall x(Cl(x) \rightarrow \exists y(Univ(y) \wedge x\eta y))$; *then*

$$\mathrm{TLR}^- \vdash \mathrm{LIM}.$$

PROOF. If x is a class, x is already a k-class (lemma 37.8), for some k and CL_k is a universe by the theorem. □

38.3.1. REMARK. (i) LIM is stated in Jäger(1984a) (but see also Feferman 1982) in the context of a theory of total classes; LIM is false in $\mathbb{O}(\mathcal{M})$.
(ii) Each CL_k is closed under the basic type constructors of Martin-Löf's type theory, and Martin-Löf's intuitionistic type theory with arbitrarily finite universes without W-types can be interpreted in the theory TLR.

The second principle we deal with, extends to the present non-set-theoretic framework the familiar transfinite recursion schema over wellorderings. However, it is a priori unclear how to render the notion of wellordering in the present context: shall we quantify over *classes* or arbitrary *partial predicates*? We anticipate that the two alternatives yield radically different notions and that the sharpest notion is obtained by quantifying over classes. This point will be clarified in the next chapter and with the help of proof-theoretic analysis.

38.4. DEFINITION

(i) Let w encode a binary relation, i.e. let w be a property of ordered pairs. We keep using the infix notation $x \prec_w y$ in place of $\langle x, y\rangle \eta w$.
$Field(\prec_w)$ is the term $\{x : \exists z(x \prec_w z \vee z \prec_w x)\}$ representing the field of $\prec_w$, while the x-segment of $\prec_w$ determined by x is defined by the term $\prec_w\lceil x := \{u : u \prec_w x\}$.

(ii) $LO(\prec_w) :=$

$$:= \forall x \forall y \forall z(\neg(x \prec_w x) \wedge (x \prec_w y \wedge y \prec_w z \rightarrow x \prec_w z) \wedge Conn(\prec_w)),$$

where $Conn(\prec_w)$ is the formula

$$\forall x \forall y(x \eta Field(\prec_w) \wedge y \eta Field(\prec_w) \rightarrow (x \prec_w y \vee x = y \vee y \prec_w x)).$$

Clearly $LO(\prec_w)$ states that $\prec_w$ is a linear ordering.

(iii) $Progr(\prec_w, b) := (\forall x \eta Field(\prec_w))(\forall y \prec_w x . y \eta b \rightarrow x \eta b)$.

$Progr(\prec_w, b)$ is to be read "b is progressive (relative to $\prec_w$)". We also define:

$$TI(\prec_w, b) := Progr(\prec_w, b) \rightarrow Field(\prec_w) \subseteq b.$$

(iv) A linear ordering $\prec_w$ is called a *pseudo-well-ordering* – in symbols $PWO(\prec_w)$, and, in short, $\prec_w$ is a pwo – iff $\forall b(Cl(b) \rightarrow TI(\prec_w, b))$.

(v) Let $A(u, x, y, z)$ be a formula with the free variables shown:

$$TR(y, A, \prec_w, z) := \forall x \forall u(x \eta Field(\prec_w) \rightarrow (u \eta y(x) \leftrightarrow A(u, x, y\lceil x, z))),$$

where $y(x) := \{v : \langle x, v\rangle \eta y\}$ and $y\lceil x := \{\langle u, v\rangle : u \prec_w x \wedge v \eta y(u)\}$.

38.4.1. REMARK. (i) The notion of pwo is deserved in the literature to HYP-wellfounded linear orderings of natural numbers (see Harrison 1968, Friedman 1976). We maintain this terminology for a more general situation, because of the analogy betwen hyperarithmetical sets and CL.
(ii) $TR(y, A, \prec_w, z)$ roughly says that y encodes a sequence of predicates $\{y_x\}$ indexed by elements in $\prec_w$-order; each y_x is recursively computed by application of the functional $a \mapsto \{u : A(u, x, a, z)\}$ to the collection (encoded by) $y\lceil x$ of previously defined predicates.

Now the main question is under which conditions there exists a class y

satisfying $TR(-, A, \prec_w, z)$. If the given z is a class, $\prec_w$ is a pwo and A is elementary extensional in the relevant parameters, the answer is positive and essentially requires the level structure of TL. For the reader's sake, we recall from 20.9 that a formula A of $\mathfrak{L}$ is elementary extensional in the list $x_1, \ldots, x_n$ iff A belongs to the least class of formulas inductively generated by means of $\wedge$, $\neg$, $\forall y$ (y distinct from $x_1, \ldots, x_n$), from atoms of the form $t = s$, Nt, $t\eta x_i$, provided $x_1, \ldots, x_n$ do not occur in t, s. Then we know:

38.5. FACT. *If $A(u, x, y)$ is elementary extensional in x, y and $=_e$ is extensional equality with respect to η* (see 9.11), *we can prove in pure logic:*

$$A(u, x, y) \wedge x =_e x' \wedge y =_e y' \rightarrow A(u, x', y').$$

38.6. THEOREM. *Let $A(u, x, y, z)$ be an $\mathfrak{L}$-formula elementary extensional in y, z with the free variables shown. Then we have, provably in* TLR^- :

(i) $Cl(\prec_w) \wedge PWO(\prec_w) \wedge Cl(z) \rightarrow \exists y(Cl(y) \wedge TR(y, A, \prec_w, z))$.

(ii) *Uniqueness: if y and y' are two classes satisfying $TR(-, A, \prec_w, z)$, then y and y' are pointwise extensionally equivalent, i.e.* TLR^- *proves*

$$[Cl(\prec_w) \wedge PWO(\prec_w) \wedge Cl(z) \wedge TR(y, A, \prec_w, z) \wedge TR(y', A, \prec_w, z) \wedge$$
$$Cl(y) \wedge Cl(y')] \rightarrow \forall x(x\eta Field(\prec_w) \rightarrow (y(x) =_e y'(x))).$$

PROOF. (i) Existence. Put $gxzy := \{u\colon A(u, x, y\lceil x, z)\}$. Then by the fixed point for operations we can find a term $RC[g, \prec_w]$ such that

$$RC[g, \prec_w]zx = gxz\Sigma(\prec_w\lceil x,\ \lambda u.RC[g, \prec_w]zu). \tag{1}$$

Also, if z and $\prec_w$ are classes, then z, $\prec_w$ and $Field(\prec_w)$ are k-classes for k large enough (by lemma 37.8 (v), 36.4.3.2, 37.5 (vi)).
Let us consider:

$$d := \{x : x\eta_k Field(\prec_w) \wedge Cl_k(RC[g, \prec_w]zx)\}.$$

If we choose $j \succ k$, d is a j-class (its defining condition being a k-formula, see 37.3 (v)). Hence d is a class and we can apply induction on $\prec_w$. Assume $x\eta Field(\prec_w)$ and $\forall y \prec_w x.\ y\eta d$: then by 37.5 (iii), $RC[g, \prec_w]zy$ is a k-class, for each $y \prec_w x$. Hence by closure of CL_k under join (37.6), the term $t := \Sigma(\prec_w\lceil x, \lambda u.RC[g, \prec_w]zu)$ is a k-class and so is $t\lceil x$. Since $A(u, x, y, z)$ is elementary in y and z and CL_k is closed under elementary comprehension, $gxzt = RC[g, \prec_w]zx$ is a k-class, which implies $x\eta_j d$, whence $x\eta d$. Therefore the class d is $\prec_w$-progressive and we can conclude that $RC[g, \prec_w]zx$ is a k-class for every x in the field of $\prec_w$, whence, again by join, $RC[A, \prec_w, z] := \Sigma(b, \lambda u. RC[g, \prec_w]zu)$ is a k-class (where b is $Field(\prec_w)$). If x is in the field of $\prec_w$, we have, with 37.6 and the extensionality property of A:

$$
\begin{aligned}
u\eta RC[A, \prec_w, z](x) &\leftrightarrow u\eta RC[g, \prec_w]zx \\
&\leftrightarrow u\eta gxz(\Sigma(\prec_w\lceil x, \lambda u.RC[g, \prec_w]zu)) \\
&\leftrightarrow A(u, x, \Sigma(\prec_w\lceil x, \lambda u.RC[g, \prec_w]zu)\lceil x, z) \\
&\leftrightarrow A(u, x, RC[A, \prec_w, z]\lceil x, z).
\end{aligned}
$$

{In the last step, we use the fact that if x is in $b := Field(\prec_w)$,

$\langle v, u\rangle\eta(\Sigma(\prec_w\lceil x, \lambda u.RC[g, \prec_w]zu))\lceil x \leftrightarrow \langle v, u\rangle\eta(\Sigma(b, \lambda u.RC[g, \prec_w]zu))\lceil x$}.

In conclusion, we proved that the term $RC[A, \prec_w, z]$ represents a class satisfying $TR(-, A, \prec_w, z)$.

(ii) The uniqueness (modulo extensional equivalence) follows by applying transfinite induction to $B(x) := x\eta Field(\prec_w) \rightarrow \forall u(u\eta y(x) \leftrightarrow u\eta y'(x))$. (Note that $\{u: B(u)\}$ is a class if y, y' are classes and $x\eta Field(\prec_w)$). □

38.6.1. REMARK. An axiom, which requires the existence of an ω-sequence of properties, obtained by iterating a given predicative operation (here the map $x \mapsto \{u: A(u, x, y\lceil x, z)\}$), is clearly stated in Weyl's *Das Kontinuum* (Weyl 1918, "Iterationsprinzip", p.27). Since the schema embodied in 38.6 can be naturally regarded as an extension of the Iterationsprinzip to pwos, we name it WP(= Weyl's principle).

38.6.2. APPLICATION (see 40.5). Interpret $\langle -, -\rangle$ as a number-theoretic pairing operation (i.e. an injection of $\mathbb{N} \times \mathbb{N}$ into $\mathbb{N}$) and assume that (1) the parameters $\prec_w$ and z in the statement of WP are subclasses of $\mathbb{N}$; (2) $\{u: A(u, x, b\lceil x, z)\} \subseteq \mathbb{N}$, whenever $x\eta Field(\prec_w)$, z, $y\lceil x \subseteq \mathbb{N}$ (A elementary extensional in b, z). Then we can find a subclass y of $\mathbb{N}$ such that $TR(y, A, \prec_w, z)$.
Verification. First observe that, under the closure assumption of $\mathbb{N}$ with respect to $\langle -, -\rangle$, the following version of join for numbers holds:

$$Cl(c) \wedge \forall u(u\eta c \rightarrow Cl(fu) \wedge fu \subseteq \mathbb{N}) \rightarrow Cl(\Sigma(c, f)) \wedge \Sigma(c, f) \subseteq \mathbb{N}. \qquad (*)$$

Then for $gxzb := \{u: A(u, x, b\lceil x, z)\}$:

$$\forall x(x\eta Field(\prec_w) \rightarrow RC[g, \prec_w]zx \subseteq \mathbb{N}). \qquad (**)$$

(**) readily follows by $\prec_w$-transfinite induction with (*) (apply the hypothesis on A, the fact that $\prec_w$ is a pwo, and $\mathbb{N}$, $RC[g, \prec_w]zx$ are classes if x is in the field of $\prec_w$). Again by (*) we conclude that $b := \Sigma(Field(\prec_w), \lambda u.RC[g, \prec_w]zu)$ is a subclass of $\mathbb{N}$, which satisfies $TR(-, A, \prec_w, z)$. □

§ 39. A recursion-theoretic model

We do not know yet whether our theory of self-referential truth with levels is consistent: we are going to describe a model of TLR within a suitable fragment of (powerless) set theory.

First of all, we fix a countable ω-model of the theory OP of operations, i.e. a model $\mathcal{M}$, where $N_{\mathcal{M}} := \{a : a \in |\mathcal{M}| \wedge \mathcal{M} \models Na\}$ is isomorphic with ω. It is essential – especially for the consistency theorem of the final section – to identify $\mathcal{M}$ with an *arithmetically definable model of* OP; to be definite, *we henceforth deal with* **CTM**, the closed term model of Ch. I, 4.10. This is because **CTM** can be identified with a simple arithmetical subset of ω; in addition, application, equality and the interpretation of the basic constants and predicates of OP in **CTM** are arithmetical by Ch. I, 4.13 and the appendix to Ch. I. Alternatively, we may choose the recursive graph model RE of §5. Then we inductively expand **CTM** with a suitable family $\mathcal{V}_\iota := \{V_\xi : \xi < \iota\}$ of truth predicates, indexed by the first recursively inaccessible ordinal ι; the level ordering is simply identified with the standard ordering relation on ordinals $< \iota$. The existence of ι derives from ordinal recursion theory and admissible set theory; the results we presuppose are covered by Richter and Aczel (1974), Barwise (1975), Hinman (1978). However, in order to make the present treatment reasonably self-contained, we define the essential notions and state the required results. Additional details are also available in the appendix.

The pure first-order set-theoretic language $\mathcal{L}_s^-$ is obtained from the set-theoretic language $\mathcal{L}_s$ of §22, by omitting all individual constants and predicates, except membership $\in$. If X is a predicate symbol $\neq \in$, $\mathcal{L}_s^-(X)$ is $\mathcal{L}_s^- \cup \{X\}$; $\mathcal{L}_s^-(X)$ contains new atoms of the form Xt; the intended meaning of X is that X is a class (in set-theoretic sense).

L_α is the collection of constructible sets up to the ordinal α, where $L_0 = \emptyset$, $L_\lambda = \cup\{L_\alpha : \alpha < \lambda\}$ (λ limit), and $L_{\beta+1}$ is the family of subsets of L_β first-order definable with parameters in the standard set-theoretic language over the structure $\langle L_\beta, \in \restriction L_\beta\rangle$. $L := \cup\{L_\alpha : \alpha \in ON\}$ is the constructible universe. We mention that a set theoretic formula A is Σ_1 iff A has the form $\exists z B$, for some bounded formula B; B is bounded if it only contains bounded set quantifiers (i.e. of the form $\forall y \in z$, $\exists y \in z$; see again §22). When we deal with semantical notions (e.g. definability over L), we tacitly assume that the set theoretic language is expanded with (distinct) constants for (distinct) parameters from a suitably large segment of L; but we use the same symbol for the object $a \in L$ and its name. Lower case Greek letters will range over the class ON of ordinal numbers.

39.1. DEFINITION

(i) An ordinal $\alpha > \omega$ is *admissible* iff α is a limit ordinal and L_α is closed under pairing, union, bounded separation and bounded collection (in other words, L_α is a model of Kripke-Platek set theory KP; cf. 22.4, for the axioms).

(ii) An admissible ordinal α is *recursively inaccessible* iff it is the limit of the admissible ordinals $< \alpha$.

(iii) $\iota :=$ the smallest recursively inaccessible ordinal.

(iv) If C is a class of ordinals and P is a (set-theoretic) class, a n-ary relation R is *uniformly* $\Sigma_1(L_\alpha)$ in P for $\alpha \in C$ iff there exists a Σ_1-formula $A(x_1, \dots, x_n, X)$ such that, if $\alpha \in C$,

$$R \cap L_\alpha = \{\langle c_1, \dots, c_n\rangle : c_1, \dots, c_n \in L_\alpha \text{ and } \langle L_\alpha, P \cap L_\alpha\rangle \models A(c_1, \dots, c_n, X)\},$$

where X is interpreted by $P \cap L_\alpha$.
R is uniformly $\Delta_1(L_\alpha)$ in P for $\alpha \in C$ iff R is uniformly $\Sigma_1(L_\alpha)$ in P for $\alpha \in C$, together with its complement. As a special case, a n-ary relation R on L_α is $\Sigma_1(L_\alpha)$ $(\Delta_1(L_\alpha))$ iff R is uniformly $\Sigma_1(L_\alpha)$ $(\Delta_1(L_\alpha))$ in $P = \emptyset$ for $\alpha \in C$, C being the singleton $\{\alpha\}$.

(v) A (possibly partial) function $F : L_\alpha \to L_\alpha$ is uniformly $\Sigma_1(L_\alpha)$ in P for $\alpha \in C$ iff its graph is uniformly $\Sigma_1(L_\alpha)$ in P for $\alpha \in C$.
The notions of $\Sigma_1(L_\alpha)$- and $\Delta_1(L_\alpha)$-function are obvious by (iv) and the preceding stipulation.

(vi) LEV_ι is the structure $\langle \iota, =, \leq \rangle$, where $=, \leq$ are equality and "less-than-equal" relation, restricted to ordinals $< \iota$ (respectively).

(vii) α is *projectible* iff there exists a $\Sigma_1(L_\alpha)$-injection from α into ω.

39.1.1. REMARK. A total $\Sigma_1(L_\alpha)$-function $F : L_\alpha \to L_\alpha$ is always $\Delta_1(L_\alpha)$. In general, every $\Sigma_1(L_\alpha)$-function $F : C \to L_\alpha$, whose domain $C \subseteq L_\alpha$ is $\Delta_1(L_\alpha)$, can be extended to a total $\Sigma_1(L_\alpha)$-function. As a consequence, the relation $R_F(a, x) := a \in F(x)$ is $\Delta_1(L_\alpha)$, provided F is $\Sigma_1(L_\alpha)$ and total, or defined on a $\Delta_1(L_\alpha)$-subset.

Obviously, if we let level variables range over ordinals below ι, while level identity and $\preceq$ are realized on ordinal theoretic $=$ and $\leq$ (in the given order), we have:

39.2. FACT. *LEV_ι is a model of the level axioms of* 36.4.3. *Indeed, the model satisfies linearity and wellfoundedness of $\prec$.*

For simplicity, it is convenient to identify closed terms of $\mathcal{L}_{op}$ with their respective number codes in the arithmetized version of **CTM** and hence to regard CTM as a subset of ω. Since every arithmetically definable subset becomes definable by a bounded formula on L_α, for $\alpha > \omega$, it follows, by inspection of the definition of the term models in 4.10:

39.3. LEMMA. *The sets* CTM, $N := \{t: t \in \text{CTM},\ t \geq \overline{n},$ *for some* $\overline{n}\}$ (*cf.* 4.1 for $\geq$, $\overline{n}$ *numeral*), *the application function* $*$: $\text{CTM} \times \text{CTM} \rightarrow \text{CTM}$, *the interpretations of the basic constants* 0, SUC, $PRED$, $PAIR$, $LEFT$, $RIGHT$, D *of* OP *are all elements of* L_α, *for every* $\alpha > \omega$.

In order to interpret the local truth predicates of TLR, a delicate step is the choice of a denotation for the function symbol LT; we must find an injection IN of ι into CTM, which does not spoil the self-referential abilities of the T_i's and is "reasonably" definable. Once IN is available, the model building only requires the well-known closure of admissible sets under Σ_1-recursion and Σ_1-inductive definitions (see Barwise 1975, pp. 24, 124, 208), plus the fact that ι is an admissible ordinal, which is limit of smaller admissibles.

39.4. LEMMA

(i) *The predicate* $Ad(\alpha) :=$ "α *is admissible*" *is uniformly* $\Delta_1(L_\beta)$ *for* β *limit* $> \omega$.

(ii) *The operation*

$$\beta \mapsto \beta^+ = \text{ the least admissible} > \beta$$

is uniformly $\Delta_1(L_\alpha)$, *for* α *limit of admissibles.*

(iii) *Let* $\tau_0 = \omega$ *and* $\tau_\alpha =$ *least admissible* $\gamma > \tau_\beta$, *for every* $\beta < \alpha$, *whenever* $\alpha > 0$. *Then the ordinal sequence* $\langle \tau_\alpha : \alpha < \delta \rangle$ *is uniformly* $\Delta_1(L_{\tau_\delta})$.

(iv) ι *is the least* α *such that* $\tau_\alpha = \alpha$. *In particular the restriction of* τ *to* ι *is* $\Sigma_1(L_\iota)$.

PROOF. (i): by standard techniques of formal set-theoretic semantics and the well-known uniform Δ_1-definability of the operation $\delta \mapsto L_\delta$ (see Barwise, cit.; Devlin 1977).
(ii): apply (i).
(iii): by (i)-(ii) and closure of admissible sets under Σ_1-recursion.
(iv): easy consequence of (iii). □

39.4.1. REMARK. If $\omega < \alpha < \iota$ and α is admissible, $\alpha = \beta^+$, for some $\beta < \alpha$.

We can now state the main fact needed for interpreting LT: each admissible β with $\omega < \beta \leq \iota$, is projectible into ω, uniformly via a Σ_1-map.

39.5. THEOREM. *There exists a function* IN, *uniformly* $\Sigma_1(L_\beta)$ *for* β *admissible* $> \omega$, *such that* $IN\lceil\tau_\alpha : \tau_\alpha \rightarrow \omega$ *is total and injective, for every* $0 < \alpha \leq \iota$ (*here* $IN\lceil\tau_\alpha$ *is the restriction of* IN *to* τ_α).

For a full proof, the reader is sent to Richter-Aczel (1974), while a sketch is given in the appendix.

39.5.1. REMARK. (i) In the statement 39.5 we can assume that the range of the projection IN is CTM. Indeed, it is enough to consider any Σ_1-bijection σ between ω and the set of numerals of CTM. Henceforth, we still maintain IN as a symbol for the projection of ι into CTM.
(ii) The uniformity of the function IN is not shared by all countable admissible projectible ordinals, since there exist non-projectible ordinals below projectible ones (see Barwise 1975; Hinman 1978, p.424).

Of course, if we extend the interpretation of 39.3 by realizing the function symbol LT on the map IN, we have:

39.6. COROLLARY. *The structure* $\mathrm{CTM}_\iota = \langle \mathbf{CTM}, LEV_\iota, IN\rangle$ *is a model of* OP *plus the level axioms and the projection axiom* PROJ of 36.4.1.

We now proceed to expand $\mathbf{CTM}_\iota$ with a family $\mathcal{V}$ of truth predicates, indexed by the ordinal ι, such that $\langle \mathrm{CTM}_\iota, \mathcal{V}\rangle$ is a model of TLR. To this aim, we assume that the language $\mathcal{L}_V$ of TLR is enlarged to a language $\mathcal{L}_V{}^+$ with constants for ordinals $< \iota$; lower case Greek letters represent both ordinals $< \iota$ and their names in $\mathcal{L}_V{}^+$. a, b, c are used as metavariables for *arbitrary elements of* CTM, while we keep i, j, k ranging over level variables.
If t is a closed term of $\mathcal{L}_V{}^+$, *possibly containing LT and ordinal constants,* $\mathrm{CTM}_\iota(t)$ is the value of t in **CTM**: $\mathrm{CTM}_\iota(t)$ is the closed term of $\mathcal{L}$, which results from t by replacing each subterm of the form $LT(\alpha)$ by $IN(\alpha)$ (which is identified with a closed term of CTM by 39.5.1; of course, the first occurrence of α stands for the name of α in the language).
We lift to the present context the simplified notations and conventions of §7 and 36.3. If a, b belong to CTM and $\alpha < \iota$, then ab, $\forall a$, $\neg a$, $a \wedge b$, $tr(\alpha, a)$, $id(a,b)$, $tr(a)$, $nat(a)$ denote the following elements of CTM (in the given order): $Ap(a,b)$, $ALLa$, $NEGa$, $ANDab$, $\mathrm{CTM}_\iota([T_\alpha a])$, $[a=b]$, $[Ta]$, $[Na]$.

Clearly by pairing, 39.5.1 and the remark 39.1.1 we have:

39.7. LEMMA
(i) $tr(\alpha,a) = \langle 7, \langle IN(\alpha), a\rangle\rangle$, *and the operation* $\langle\alpha, a\rangle \mapsto tr(\alpha,a)$ *is injective (in each coordinate separately)*;

(ii) *if* $\beta < \tau_\alpha$, *and* c, $a \in \mathrm{CTM}$, *the relation* $R(c,\beta,a) := (c = tr(\beta,a))$ *is uniformly* $\Delta_1(L_{\tau_\alpha})$ *for every* $\alpha \leq \iota$.

Hence if $\beta < \tau_\alpha$, the function $a \mapsto tr(\beta,a)$ is uniformly $\Delta_1(L_{\tau_\alpha})$, for every $\alpha \leq \iota$.

39.8. DEFINITION. If $S \subseteq \iota \times \mathrm{CTM}$, let, for $\alpha < \iota$:

$$S(\alpha) := \{a: a \in \text{CTM and } \langle \alpha, a\rangle \in S\}.$$

The structure $\mathfrak{C}_\iota = \langle \text{CTM}_\iota, S\rangle$ canonically determines a realization of $\mathcal{L}_V{}^+$, in which T_α is interpreted by $S(\alpha)$ (for $\alpha < \iota$) and T is assigned the set $\bigcup\{S(\alpha): \alpha < \iota\}$.

39.9. DEFINITION. If $\delta < \iota$, $S \subseteq \delta \times \text{CTM}$, $X \subseteq \text{CTM}$, let $\Gamma(\delta, S, X)$ be the subset of CTM, such that $a \in \Gamma(\delta, S, X)$ iff for some b, $c \in \text{CTM}$, one of the following cases holds:

1. $a = (\neg)tr(\beta, b)$ and $b \in S(\beta)(b \notin S(\beta))$, for some $\beta < \delta$;
2. $a = (\neg)id(b, c)$ and $\mathbf{CTM}\models(\neg)b = c$;
3. $a = (\neg)nat(b)$ and $\mathbf{CTM}\models(\neg)Nb$;
4. $a = (\neg)tr(b)$ and $b \in X$ $((\neg b) \in X)$;
5. $a = (\neg)tr(\delta, b)$ and $b \in X$ $((\neg b) \in X)$;
6. $a = \neg\neg b$ and $b \in X$;
7. $a = (\neg)(b \wedge c)$ and b, $c \in X$ (respectively $(\neg b) \in X$ or $(\neg c) \in X$);
8. $a = (\neg)\forall b$ and for every $d \in \text{CTM}$, $(bd) \in X$ (for some $d \in \text{CTM}$, $(\neg bd) \in X$).

39.10. LEMMA
(i) Γ *is monotone in the third variable: if* $\delta < \iota$, $S \subseteq \delta \times \text{CTM}$ *and* $X \subseteq Y \subseteq \text{CTM}$, *then* $\Gamma(\delta, S, X) \subseteq \Gamma(\delta, S, Y) \subseteq \text{CTM}$.

(ii) $\Gamma(\delta, S, X)$ *is uniformly* $\Delta_1(L_\alpha)$ *in* X, S, *for* α *admissible with* $\omega < \alpha < \iota$ *and* $\delta < \alpha$. *Hence* L_α *is closed under* Γ *in the following sense: if* $\delta < \alpha$, X, S *are* $\Delta_1(L_\alpha)$, *then* $\Gamma(\delta, S, X) \in \mathcal{P}(\text{CTM}) \cap L_\alpha$ ($\mathcal{P}$ = *the power set operation*).

PROOF. (i): its defining condition positively depends on X.
(ii): by inspection of definition 39.9, lemma 39.7 and remark 39.1.1, we see that $\Gamma(\delta, S, X)$ is uniformly $\Delta_1(L_\alpha)$ in X, S; so we can apply Δ_1-separation for L_α, since $\Gamma(\delta, S, X) \subseteq \text{CTM} \in L_\alpha$. □

39.11. LEMMA (*Inversion*). *Assume* $\delta < \iota$, $S \subseteq \delta \times \text{CTM}$, $X \subseteq \text{CTM}$ *and* a, $b \in \text{CTM}$. *Then, if* A *has the form* $a = b$, $\neg a = b$, Na, $\neg Na$,

1. $[A] \in \Gamma(\delta, S, X)$ *iff* $\mathbf{CTM}\models A$;
2. $tr(\beta, a) \in \Gamma(\delta, S, X)$ *iff either* $\beta = \delta$ *and* $a \in X$ *or* $\beta < \delta$ *and* $a \in S(\beta)$;
3. $(\neg tr(\beta, a)) \in \Gamma(\delta, S, X)$ *iff either* $\beta < \delta$ *and* $a \notin S(\beta)$ *or* $\beta = \delta$ *and* $(\neg a) \in X$;

4. $(a \wedge b) \in \Gamma(\delta, S, X)$ *iff* $a \in X$ *and* $b \in X$;

5. $(\neg(a \wedge b)) \in \Gamma(\delta, S, X)$ *iff* $(\neg a) \in X$ *or* $(\neg b) \in X$;

6. $(\forall a) \in \Gamma(\delta, S, X)$ *iff* $(ac) \in X$, *for all* $c \in \mathrm{CTM}$;

7. $(\neg(\forall a)) \in \Gamma(\delta, S, X)$ *iff* $(\neg(ac)) \in X$, *for some* $c \in \mathrm{CTM}$;

8. $tr(a) \in \Gamma(\delta, S, X)$ *iff* $a \in X$;

9. $(\neg tr(a)) \in \Gamma(\delta, S, X)$ *iff* $(\neg a) \in X$;

10. $(\neg\neg a) \in \Gamma(\delta, S, X)$ *iff* $a \in X$.

PROOF: similar to Ch. II, 7.4; in addition, we apply 39.7 above. □

39.12. DEFINITION. We introduce the β-th iteration $It(\Gamma, \delta, S, \beta)$ of Γ, for given $\delta < \iota$ and $S \subseteq \delta \times \mathrm{CTM}$, by recursion on β:

$$It(\Gamma, \delta, S, 0) = \emptyset; \quad It(\Gamma, \delta, S, \beta{+}1) = \Gamma(\delta, S, It(\Gamma, \delta, S, \beta));$$

$$\text{for } \lambda \text{ limit, } It(\Gamma, \delta, S, \lambda) = \cup\{It(\Gamma, \delta, S, \beta) : \beta < \lambda\}.$$

Clearly $\xi < \zeta$ implies $It(\Gamma, \delta, S, \xi) \subseteq It(\Gamma, \delta, S, \zeta)$ by monotonicity of Γ. Informally speaking, $It(\Gamma, \delta, S, \xi)$ exactly determines the ξ-th stage of truth of level δ, uniformly in a sequence $\{S(\gamma) : \gamma < \delta\}$ of possible candidates for truth predicates of lower level.

39.13. LEMMA. *Assume that* $S \subseteq \delta \times \mathrm{CTM}$ *and* $\delta < \iota$.

(i) *If* $a \in \mathrm{CTM}$, $\beta < \alpha$, $\delta < \alpha$, *then* $P(a, \delta, \beta, S) :=$ "$a \in It(\Gamma, \delta, S, \beta)$" *and the function* $\beta \mapsto It(\Gamma, \delta, S, \beta)$ *are uniformly* $\Delta_1(L_\alpha)$ *in* S, *for* α *admissible with* $\omega < \alpha < \iota$.
Hence, if S *is* $\Delta_1(L_\alpha)$, $It(\Gamma, \delta, S, -) : \alpha \rightarrow L_\alpha \cap \mathcal{P}(\mathrm{CTM})$.

(ii) *If* $\gamma = \alpha^+$, $\delta < \alpha$ *and* α *is admissible with* $\iota > \alpha > \omega$, *the set* $I := It(\Gamma, \delta, S, \alpha)$ *satisfies*:

$$I = \Gamma(\delta, S, I); \tag{$*$}$$

$$I \in L_\gamma \cap \mathcal{P}(\mathrm{CTM}). \tag{$**$}$$

PROOF. (i) $It(\Gamma,\delta,S,-)$ is recursively and uniformly defined by means of the operation Γ, which is uniformly $\Delta_1(L_\alpha)$ in S by 39.10 (ii).
(ii) If L_α is admissible, the least fixed point of any given positive operator which is $\Sigma_1(L_\alpha)$, is $\Sigma_1(L_\alpha)$ (this is Gandy's theorem, Barwise 1975, pp. 208-210). Hence $(*)$ is immediate by 39.10. As to $(**)$, $(*)$ implies that I is a $\Delta_1(L_\gamma)$-subset of $\mathrm{CTM} \in L_\gamma$. □

39.14. DEFINITION. If $\delta < \iota$ and It is the functional of definition 39.12, let

$$\mathcal{T}(\delta) = It(\Gamma, \delta, \mathcal{T}\rceil\delta, \tau_{\phi(\delta)}), \tag{$+$}$$

where $\mathcal{V}\rceil\delta = \{\langle\beta, a\rangle : \beta < \delta \text{ and } a \in \mathcal{V}(\beta)\}$ and $\phi(\delta) = \delta$ if δ is a limit; else $\phi(\delta) = \delta{+}1$. $\mathcal{V}$ is well-defined on ordinals $< \iota$, by Δ_1-recursion, 39.13 and 39.4.

In the following, $\mathcal{V}$ denotes the unique function satisfying (+) above.
We can now associate to the structure $\mathfrak{C}_\iota = \langle \mathrm{CTM}_\iota, \mathcal{V}\rangle$ the realization of $\mathcal{L}_V{}^+$, in which T_α is interpreted by $\mathcal{V}(\alpha)$ (for $\alpha < \iota$) and T is assigned the set $\bigcup\{\mathcal{V}(\alpha) : \alpha < \iota\}$.

39.15. LEMMA

(i) *The relation* $R(\delta, a) := a \in \mathcal{V}(\delta)$ *is uniformly* $\Delta_1(L_{\tau_{\phi(\delta)+1}})$ *for every* $\delta < \iota$. *Hence:*

$$\mathcal{V}(\delta) \in L_{\tau_{\phi(\delta)+1}} \quad \textit{and } \mathcal{V} : \iota \to L_\iota \cap \mathcal{P}(\mathrm{CTM}) \textit{ is } \Delta_1(L_\iota);$$

(ii) *if* $\delta < \iota$, $\mathcal{V}(\delta) = \Gamma(\delta, \mathcal{V}\rceil\delta, \mathcal{V}(\delta))$;

(iii) *if* $\delta < \iota$, *either* $a \notin \mathcal{V}(\delta)$ *or* $(\neg a) \notin \mathcal{V}(\delta)$, *for every* $a \in \mathrm{CTM}$;

(iv) *for every* $\gamma < \delta < \iota$, $a \in \mathrm{CTM}$, *either* $tr(\gamma, a) \in \mathcal{V}(\delta)$ *or* $(\neg(tr(\gamma, a)) \in \mathcal{V}(\delta)$;

(v) *if* $\gamma < \delta < \iota$, $\mathcal{V}(\gamma)$ *is a proper subset of* $\mathcal{V}(\delta)$.

PROOF. (i) and (ii) follow from 39.4, 39.13 and closure of admissible sets under Δ_1-recursion.
(iii): by main transfinite induction on $\delta < \iota$, and secondary induction on $\tau_{\phi(\delta)}$, using (i) and inversion lemma 39.11 at the successor step (see 7.4).
(iv) If $\gamma < \delta$ and $a \in \mathcal{V}(\gamma)$, then by definition of Γ, monotonicity and (ii):

$$tr(\gamma, a) \in \Gamma(\delta, \mathcal{V}\rceil\delta, \emptyset) \subseteq \Gamma(\delta, \mathcal{V}\rceil\delta, \mathcal{V}(\delta)) \subseteq \mathcal{V}(\delta).$$

For $a \notin \mathcal{V}(\gamma)$, the argument is similar.
(v). If $\gamma < \delta$ and $\mathcal{V}(\gamma, \xi) := It(\Gamma, \gamma, \mathcal{V}\rceil\gamma, \xi)$, it is enough to verify by induction on $\xi < \tau_{\phi(\gamma)}$:

$$\mathcal{V}(\gamma, \xi) \subseteq \mathcal{V}(\delta). \tag{+}$$

If $\xi = 0$ or ξ is a limit ordinal, the proof is trivial. Assume (+) by IH and $a \in \Gamma(\gamma, \mathcal{V}\rceil\gamma, \mathcal{V}(\gamma, \xi))$: we show $a \in \mathcal{V}(\delta)$ as a consequence of the inversion lemma and the property (ii) above. We distinguish several cases according to the form of a. Let $a = (\neg(tr(\nu, b))$ for some ν: then by inversion either $\nu < \gamma$ and $b \notin \mathcal{V}(\nu)$ or $\nu = \gamma$ and $(\neg b) \in \mathcal{V}(\gamma, \xi)$. In the first case, since $\nu < \delta$, $a \in \Gamma(\delta, \mathcal{V}\rceil\delta, \emptyset) \subseteq \mathcal{V}(\delta)$ by definition of Γ and (ii). In the second case, $(\neg b) \in \mathcal{V}(\gamma)$ by definition and hence $b \notin \mathcal{V}(\gamma)$ by consistency (see (iii) above). Since $\gamma < \delta$, $a \in \Gamma(\delta, \mathcal{V}\rceil\delta, \emptyset) \subseteq \mathcal{V}(\delta)$. Let $a = (b \wedge c)$; by assumption and inversion $b \in \mathcal{V}(\gamma, \xi)$ and $c \in \mathcal{V}(\gamma, \xi)$, whence b, $c \in \mathcal{V}(\delta)$ by IH. By definition of Γ, $a \in \Gamma(\delta, \mathcal{V}\rceil\delta, \mathcal{V}(\delta))$ and $a \in \mathcal{V}(\delta)$ by (ii). The extant cases are

easily checked as exercise. As to proper inclusion, consider the terms $R(\gamma) = \{x : \neg x\eta_\gamma x\}$ and $c = R(\gamma)R(\gamma) = [\neg R(\gamma)\eta_\gamma R(\gamma)]$.
Then observe that (ii) and (iv) imply $c \in \mathcal{V}(\gamma+1) - \mathcal{V}(\gamma)$ (see 37.5 (v)). □

39.16. THEOREM. $\mathfrak{C}_\iota = \langle \mathrm{CTM}_\iota, \mathcal{V} \rangle \models \mathrm{TLR}$.

PROOF. LIND holds in $\mathfrak{C}_\iota$, since **CTM** is an ω-model, while corollary 39.6 takes care of the level axioms and LT-injectivity. Local truth axioms and connection axioms (cf. 36.4) are straightforward consequences of the definition of Γ, inversion, definition 39.8 and the previous lemma. As to the reflection principle, assume, for a, $b \in \mathrm{CTM}$ and $\gamma < \iota$:

$$\mathfrak{C}_\iota \models \forall x \exists j (x\eta_\gamma a \rightarrow x\eta_j b). \tag{$*$}$$

By $\Delta_1(L_\iota)$-definability of $R(\alpha, a) := a \in \mathcal{V}(\alpha)$ (see 39.15 (i)), condition ($*$) is equivalent, by well-known absoluteness of Δ_1-conditions, to:

$$L_\iota \models (\forall x \in \mathrm{CTM})(\exists \beta) A(x, \gamma, \beta, a, b), \tag{$**$}$$

for a suitable Δ_1-formula $A(x, y, z, u, v)$; hence by Σ_1-reflection, for some $\xi < \iota$, we have $L_\iota \models (\forall x \in \mathrm{CTM})(\exists \beta < \xi)\, A(x, \gamma, \beta, a, b)$, which yields by ($*$) and ($**$), the required conclusion

$$\mathfrak{C}_\iota \models \exists k \forall x \exists j (j \preceq k \wedge (x\eta_\gamma a \rightarrow x\eta_j b)). \qquad \square$$

39.17. REMARK

(i) $\mathfrak{C}_\iota$ satisfies EA, the full schemata of N-induction and $\prec$-induction. Note also that the proof step from ($*$) to ($**$) still works if we replace $T_i(ax)$ by any formula B, which belongs to the least collection $\mathrm{E}\Delta^+$ of formulas containing atoms of the form Nt, $t = s$, Tt, $T_i t$, closed under bounded level quantifiers, object quantifiers and logical connectives (i.e. if $B \in \mathrm{E}\Delta^+$, also $\forall j \prec k.B \in \mathrm{E}\Delta^+ := \forall j(j \prec k \rightarrow B)$). Hence $\mathfrak{C}_\iota$ makes true the schema REFL^+:

$$\forall x \exists i B(x, i) \rightarrow \exists k \forall x \exists i \prec k.B(x, i) \ (B \in \mathrm{E}\Delta^+).$$

(ii) We might pursue the recursion-theoretic analysis of the subsets of CTM, which are extensions of classes in $\mathfrak{C}_\iota$. It should not be difficult to show that the sets $\subseteq \omega$, which are representable by classes, are exactly the sets $\subseteq \omega$, which are recursive in the Tugué functional E_1(or hyperjump), where $E_1(F) = 0$, if F codes a well-founded tree of number sequences, and $E_1(F) = 1$, else, for F: $\omega \rightarrow \omega$ (see Hinman 1978).

(iii) If we replace ι by ω in the model construction of the theorem, we obtain a model CTM_∞ of TL, the theory without reflection.

§40. Levels of truth and predicatively reducible subsystems of second-order arithmetic

Is there any standard mathematical system naturally related to the theory TLR? It turns out that, although TLR is based *on the logical notions of truth and iteration of the reflection process*, TLR is strictly connected with an important subsystem ATR_0 of second-order arithmetic Z_2; it is well-known that ATR_0 has a non trivial mathematical content and is actually a strong version of predicative mathematics. In order to clarify this point, we are forced to digress into the classical realm of Z_2.

As Hilbert and Bernays already showed in Supplement IV of the *Grundlagen der Mathematik*, the language of Z_2 is remarkably expressive, and fundamental theorems of ordinary mathematics (including non trivial parts of the theory of countable sets and ordinals) can be already derived in Z_2. On the other hand, proof-theoretic investigations of the sixties produced a systematic classification of "natural" subsystems of Z_2, which were mainly suggested by definability criteria and limitation of logical complexity (see Kreisel 1968 for a vivid picture of the intertwined technical and conceptual motivations). In the seventies, this research thread progressively shifted toward investigations of formal theories related to mathematical practice (Feferman 1977, Takeuti 1978).

In particular, a new twist in the investigation of Z_2 and its foundational significance was impressed by Friedman (1976), and subsequently by Simpson. Friedman discovered that, in order to formalize ordinary non-set-theoretic mathematics, only five set existence principles formalizable in Z_2 are actually needed; he further noticed that in a number of relevant cases, the set existence axiom, applied to prove a given paradigmatic theorem, is actually implied by the theorem itself over a weak basis theory (say, a form of recursive analysis), whence the name of *reverse mathematics* to the systematic development of the program. Since then, this phenomenon has been widely investigated and Friedman's subsystems – accompanied by sophisticated model-theoretic and recursion-theoretic techniques – have become an interesting tool to calibrate chapters of ordinary mathematics (from calculus to countable algebra, logic included; see the announced monograph by Simpson and several research papers).

We compare TLR with exactly the fourth level of Friedman-Simpson reverse mathematics, namely the above mentioned ATR_0. We also apply a theorem of Ch.V about choice principles, to give inner models of weaker subsystems of Z_2, based on corresponding choice schemata. In the final section, we shall interpret the fifth level $\Pi^1_1\text{-}CA_0$ of reverse mathematics in a natural extension of TLR.

40.1. DEFINITION. The language $\mathcal{L}_2$ of second-order arithmetic contains:

(i) a denumerable list of number variables $x_1, x_2, x_3, \ldots$;

(ii) a denumerable list of set variables $X_0, X_1, X_2, \ldots$;

(iii) the individual constant $\bar{0}$, the function symbols $'$ (successor, 1-ary), $+$ (addition, 2-ary), $\cdot$ (product, 2-ary);

(iv) the binary predicates $<$ (ordering on ω) and ϵ (membership);

(v) classical logical operations (say $\neg$, $\forall$, $\wedge$) and $=$.

Terms are inductively generated from number variables and the constant $\bar{0}$ by application of the function symbols $'$, $\cdot$, $+$; thus, if t, s are terms, so are t', $t+s$, $t\cdot s$. The atoms of $\mathcal{L}_2$ have the form $t=s$, $t\epsilon X$, $t<s$, where t, s are terms and X is a set variable. Formulas are inductively generated by means of the following clauses: atoms are formulas; if A, B are formulas, $\neg A$, $A \wedge B$, $\forall x A$, $\forall X A$ are formulas.

An $\mathcal{L}_2$-formula is *arithmetical* if no set variable occurs bound in A.

$\Sigma^1_0 = \Pi^1_0 =$ the collection of arithmetical formulas. If A is Π^1_0, then $\exists Y A$ ($\forall Y A$) is Σ^1_1 (Π^1_1). If A is Σ^1_1, then $\forall Y A$ ($\exists Y A$) is Π^1_2 (Σ^1_2). An arithmetical formula A is *bounded* (Δ_0) if it contains only bounded number quantifiers, i.e. quantifiers of the form $\forall x(x<t \rightarrow \ldots)$, $\exists x(x<t \wedge \ldots)$ (x not occurring in t). A is a Σ^0_1- (or Π^0_1-) formula iff A has the form $\exists x B$ ($\forall x B$) with B bounded.

40.2. *Second-order arithmetic (in short Analysis) and its subsystems*

40.2.1. Z_2 is the theory in the language $\mathcal{L}_2$, which contains:

(i) classical predicate calculus with identity for $\mathcal{L}_2$;

(ii) $\forall x\, (\neg x' = \bar{0}) \wedge \forall x \forall y (x' = y' \rightarrow x = y)$;

(iii) $\forall x (\neg x < \bar{0}) \wedge \forall x \forall y (x < y \leftrightarrow \exists z (z' + x = y))$;

(iv) $\forall x (x + \bar{0} = x) \wedge \forall x (x \cdot \bar{0} = \bar{0}) \wedge$

$\wedge \forall x \forall y (x + y' = (x+y)' \wedge x \cdot y' = x \cdot y + x)$.

(v) Induction axiom Ax-IND: $\bar{0} \epsilon X \wedge \forall x (x \epsilon X \rightarrow x' \epsilon X) \rightarrow \forall x (x \epsilon X)$.

(vi) Full comprehension schema CA: $\exists X \forall u (u \epsilon X \leftrightarrow A(u,Y))$, where $A(u,Y)$ is an arbitrary $\mathcal{L}_2$-formula, X does not occur in A.

40.2.2. If $\mathcal{F}$ is a collection of $\mathcal{L}_2$-formulas, $\mathcal{F}$-CA is the schema CA restricted to formulas A of $\mathcal{F}$; $\mathcal{F}$-IND is the schema of induction for every $A \in \mathcal{F}$ (that is, we replace $u \epsilon X$ in Ax-IND with any $A(u) \in \mathcal{F}$).

40.2.3. If $\mathcal{F}$ is a collection of $\mathcal{L}_2$-formulas, $\Delta(\mathcal{F})$-CA is the schema:

$$\forall x (A(x) \leftrightarrow \neg B(x)) \rightarrow \exists Y \forall x (x \epsilon Y \leftrightarrow A(x)),$$

where A, $B \in \mathfrak{F}$ and Y does not occur in A, B.

40.2.4. Let $\mathfrak{P}(\omega)$ be the power set of ω, the set of natural numbers. The standard set theoretic model of Z_2, also named $\mathfrak{P}(\omega)$, is given by letting the set variables range over arbitrary subsets of ω, while the individual variables range over ω and $\overline{0}$, $'$, $+$, $\cdot$, $<$, ϵ are assigned the intended meaning of zero, successor, addition and product, natural ordering and membership $\in$.
If $P_1,\ldots,P_k$ are sets $\subseteq \omega$, $m_1,\ldots,m_n \in N$, and $B(x_1,\ldots,x_n,Y_1,\ldots,Y_k)$ is a formula of $\mathcal{L}_2$ (with free variables occurring in the given list), $\mathfrak{P}(\omega) \models A[m_1,\ldots,m_n,P_1,\ldots,P_k]$ stands for "A is satisfied in the standard model of Z_2, whenever $m_1,\ldots,m_n,P_1,\ldots,P_k$, are assigned to $x_1,\ldots,x_n$, $Y_1,\ldots,Y_k$" (respectively).

40.2.5. If $P \subseteq \omega$, we say that $R \subseteq \omega$ is *arithmetical* (Σ^1_1, Π^1_1 respectively) in P iff there exists an arithmetical (Σ^1_1, Π^1_1) formula $A(u,X)$ such that $R = \{n: \mathfrak{P}(\omega) \models A[n,P]\}$.

40.3. For the sake of comparison with type-free systems, we explicitly introduce the following subtheories of Z_2.

40.3.1 ACA_0 is the subsystem of Z_2 which only contains Π^1_0-CA; ACA_0 is known in the literature as *arithmetical analysis* (with induction axiom).

40.3.2. Let Δ^1_1-CA be the schema $\Delta(\Sigma^1_1)$-CA (see 40.2.3): Δ^1_1-CA_0 also labels the subsystem of Z_2, in which full CA is replaced by Δ^1_1-comprehension; this subsystem is known in the literature as *hyperarithmetical analysis* (with induction axiom).

40.3.3. Σ^1_1-DC is the schema of dependent choice:

$$\forall x \forall X \exists Y A(x,X,Y) \rightarrow \forall X \exists Z (Z_0 = X \wedge \forall x A(x, Z_x, Z_{x+1})).$$

Here A is any Σ^1_1-formula, $u \epsilon Z_x$ stands for $\langle u,x\rangle \epsilon Z$ and $\langle u,x\rangle$ denotes any fixed $\mathcal{L}_2$-definable injective pairing function of $\omega \times \omega$ into ω. Σ^1_1-DC_0 is the theory ACA_0 plus the schema Σ^1_1-DC.

40.3.4 We now introduce the subsystem ATR_0 of *arithmetical transfinite recursion.* Let $WO(<_X)$ be the Π^1_1-formula, stating that X encodes a linear ordering of ω such that $\forall Y(\forall x(\forall y(y <_X x \rightarrow y \epsilon Y) \rightarrow x \epsilon Y) \rightarrow \forall x (x \epsilon Y))$ {here $y <_X x := \langle y,x\rangle \epsilon X$}.
ATR is the schema:

$$\forall X \forall Z \exists Y (WO(<_X) \rightarrow \forall y \forall u (y \epsilon Y_u \leftrightarrow A(y,u,Y\lceil u, Z))),$$

where A is an arithmetical formula and $Y\lceil u$ is contextually defined by

$$\langle v, y\rangle \epsilon Y\lceil u := v <_X u \wedge \langle v,y\rangle \epsilon Y.$$

ATR_0 is the theory $ACA_0 + ATR$. ATR is a consequence of Π^1_0-CA, plus the classical schema of bar induction, namely:

40.3.5. $BI := WO(<_X) \to TI(<_X, B)$, where B is arbitrary.

40.3.6. We finally define the strongest subsystems we are going to consider. Δ^1_2-CA is the schema $\Delta(\Sigma^1_2)$-CA; Δ^1_2-CA_0, Π^1_1-CA_0 are the subsystems of Z_2, where CA is replaced by Δ^1_2-CA and Π^1_1-CA (respectively).

The weakest subsystem of reverse mathematics is recursive analysis RCA_0, i.e. the subsystem obtained from Z_2, by replacing Ax-IND with the induction schema for Σ^0_1-formulas, and CA with $\Delta^0_1\text{-CA} := \Delta(\Sigma^0_1)$-CA, the schema of recursive comprehension.

Given two theories $\mathcal{T}_1$ and $\mathcal{T}_2$ in the language $\mathcal{L}_2$, we set $\mathcal{T}_1 \geq \mathcal{T}_2$ iff $PA \vdash Cons(\mathcal{T}_1) \to Cons(\mathcal{T}_2)$, where PA is Peano arithmetic (cf. appendix to Ch. I) and $Cons(\mathcal{T})$ is a standard formalization of the metamathematical statement "$\mathcal{T}$ is consistent" in the arithmetical language.
$\mathcal{T}_1$ and $\mathcal{T}_2$ are *proof-theoretically equivalent* (in short, $\mathcal{T}_1 \equiv \mathcal{T}_2$) iff $\mathcal{T}_1 \geq \mathcal{T}_2$ and $\mathcal{T}_2 \geq \mathcal{T}_1$; $\mathcal{T}_1 > \mathcal{T}_2$ iff $\mathcal{T}_1 \geq \mathcal{T}_2$ and not $\mathcal{T}_1 \equiv \mathcal{T}_2$. If $\mathcal{T}_1 > \mathcal{T}_2$, we say that $\mathcal{T}_1$ *is proof-theoretically stronger than* $\mathcal{T}_2$.

We now state without proof the following known results:

40.4. THEOREM

(i) $\Delta^1_2\text{-}CA_0 \equiv \Pi^1_1\text{-}CA_0$ *and* $\Pi^1_1\text{-}CA_0 > ATR_0$.

(ii) $ATR_0 \equiv$ *Predicative Analysis (in the sense of Schütte* 1977, *Feferman* 1964, cf.Ch. XI). *Moreover* $ATR_0 > \Sigma^1_1\text{-}DC_0$.

(iii) $\Sigma^1_1\text{-}DC_0 > \Delta^1_1\text{-}CA_0$.

(iv) $\Delta^1_1\text{-}CA_0 \equiv ACA_0 \equiv PA$.

(v) $RCA_0 \equiv PRA$.

(ii)-(iv) above can be obtained as a corollary of the proof-theoretical analysis of Ch. XI (but see Feferman-Sieg 1981, for (i), (iii)-(iv), Friedman, Simpson and Mc Aloon 1981, for (i) and Simpson (199?) for 40.4 (v)). It is to be mentioned that each instance of Δ^1_1-CA is derivable in ATR_0; however Σ^1_1-DC is unprovable in ATR_0 (actually $ATR_0 + \Sigma^1_1$-DC is strictly stronger than ATR_0, by theorems of Friedman).

We are now ready to state the promised interpretation result. We recall that PW_p is obtained from the system PW_c of 16.1 by replacing number-theoretic induction for classes with number-theoretic induction for properties.

40.5. THEOREM

(i) ATR_0 *is interpretable in* TLR.

(ii) Σ^1_1-DC_0 *is interpretable in* $\mathrm{PW}_p+\mathrm{FA}$.

(iii) Δ^1_1-CA_0 *is interpretable in* PW_c.

PROOF. The proof is easy, as the essential work was already carried out in earlier sections.
(i) We first define a translation * of $\mathcal{L}_2$ into the level-free part of the language of TLR. Informally speaking, we simply verify that N, plus the subclasses of N, is a model of ATR in TLR. More formally, we choose combinators 0, $\cdot$, $+$, $'$, in order to interpret the basic function symbols of $\mathcal{L}_2$ (we adopt the same notation). Hence we can inductively assign to each $\mathcal{L}_2$-term t a term t^* in the language $\mathcal{L}_{op}$ (= the operational fragment of $\mathcal{L}_V$), with the same free variables. Moreover, if $t=s$, $t\epsilon X$, $t<s$ are atoms of $\mathcal{L}_2$, we put $(t=s)^*=(t^*=s^*)$; $(t\epsilon X)^*=(t^*\eta x)$ (x fresh variable); $(t<s)^*=(t^*<s^*)$ (the second occurrence of $<$ being a canonically chosen $\mathcal{L}_{op}$-definition of $<$; see 3.6). We then extend * to arbitrary formulas of $\mathcal{L}_2$ by stipulating that * commutes with $\neg$, $\wedge$ and

$$(\forall XA)^*=\forall x(Cl_N(x)\rightarrow A^*),\ (\forall xA)^*=\forall x(Nx\rightarrow A^*)=\forall nA^*,$$

where $Cl_N(x):=Cl(x)\wedge\forall u(u\eta x\rightarrow Nx)$. It is clear that * is a well-defined translation of $\mathcal{L}_2$ into $\mathcal{L}$. Let A be an $\mathcal{L}_2$-formula with free variables in the list $X=X_0,\dots,X_n$, $y=y_0,\dots,y_k$: then we check by induction on the definition of ATR_0-provability:

$$\text{if } \mathrm{ATR}_0\vdash A(y,X),\text{ then } \mathrm{TLR}\vdash Ny\wedge Cl_N(x)\rightarrow A^*(y,x). \tag{1}$$

It suffices to see that the *-translations of Ax-IND, Π^1_0-CA and ATR are provable in TLR. Now (Ax-IND)*and $(\Pi^1_0\text{-CA})^*$ become instances of class N-induction and elementary comprehension and hence are provable in TLR by theorem 37.9. Note also that, if $(Cl_N(X)\wedge WO(<_X))^*$ is assumed, then $<_X$ encodes a subclass of N which is a pwo. Hence if z is any subclass of N and $A(u,x,Y,Z)$ is arithmetical, $u\eta\mathbb{N}\wedge A^*(u,x,y,z)$ is elementary extensional in y, z (y, z fresh variables). Now the hypothesis of 38.6.2 are trivially met, and there exists a subclass of N satisfying the *-translation of the ATR-consequent.
(ii): apply the elementary dependent choice schema EDC of 20.10 (ii).
(iii): we simply apply Δ-comprehension 16.7 of Ch. IV to the translation of hyperarithmetical comprehension. □

To conclude, it is time to reconsider the opening problem of the section, concerning the theoretical relevance of TLR and its ability to represent significant parts of mathematical knowledge. The answer is implicit in the interpretation result above. Here we freely rely on results of Feferman, Takeuti, Friedman and Simpson (op. cit.).

First of all, significant parts of ordinary mathematics, like elementary calculus and countable algebra, can be already developed in conservative extensions of Peano Arithmetic (Takeuti, Feferman) and actually in fragments of arithmetical analysis ACA_0, which are not proof-theoretically stronger than primitive recursive arithmetic (hence a fortiori in fragments of MF_c).

A typical example thereof is the Cauchy-Peano theorem CP, asserting the existence of solutions for ordinary differential equations; CP is indeed equivalent to König's lemma for *binary* trees WKL (modulo RCA_0) by a theorem of Simpson (1984). Furthermore, the very principle of arithmetical comprehension is equivalent in RCA_0 to the statement that every Cauchy sequence of reals converges to a limit in $\mathbb{R}$ and also to the existence of maximal ideals for countable abelian rings or even to the König lemma for *finitary* trees.

On the other hand, ATR_0 has a good theory of countable ordinals and it proves non-trivial classical results of descriptive set theory. In particular, as Friedman and Simpson observed, ATR_0 is mathematically much more effective then the subsystems of hyperarithmetical analysis and even predicative analysis in the sense of Feferman-Schütte: there are important consequences of ATR_0, which are false in the model of hyperarithmetical sets and hence independent of Predicative Analysis. Here is a sample of significant results. ATR is equivalent (modulo RCA_0) to:

(i) comparability of well-orderings, i.e. countable well-orderings are comparable;

(ii) the Lusin-Sierpinski theorem: every analytic set in the Baire space ${}^{\omega}\omega$ of unary functions $\omega \longrightarrow \omega$ is either countable or has a perfect subset);

(iii) the Gale-Stewart theorem: every open game $\subseteq {}^{\omega}\omega$ is determined;

(iv) the Ulm structure theorem for countable reduced abelian p-groups (Friedman, Simpson and Smith 1984).

At the same time, the strength estimate of 40.4 (ii) assigns precise limits to ATR_0 and hence, by the equivalence theorem of Ch. XI, to the theory TLR of truth with levels. For instance, there is no way to prove in ATR_0 that every arithmetical set of Dedekind reals has a least upper bound, nor ATR_0 proves the classical Cantor-Bendixson theorem (every closed subset of the Baire space is the union of a countable sets of reals plus a perfect set); by contrast these two theorems are derivable in Π^1_1-CA_0 (and hence in the extension of TLR of §42).

§41. Consistency of a reducibility principle for classes

We wish to have a closer look to the recursion-theoretic model $\mathfrak{C}_\iota$, in order to investigate quantification on classes. In the usual inductive models, like $\mathbb{O}(\mathbf{CTM})$ (see §7), $CL = \{x : Cl(x)\}$ is generally not closed under quantification of classes: the best we can afford is a sort of Δ-comprehension (see corollary 16.7). However, we shall verify that, *if an elementary predicate of classes, possibly depending on additional class parameters, is non-empty, then the same predicate is already satisfied by some class of any level* $\succ i$ (e.g. a solution is to be found in CL_{i+1}, if i has a successor level $i{+}1$), provided i is an upper bound on the level of the given class parameters. Hence, at least for elementary predicates, *quantification on arbitrary classes is reducible to quantification on classes of a fixed level.* But we know that quantification on classes of fixed levels does not push outside the realm of classes.
The result we hint at above, is in essence a consequence of the effective nature of **CTM** and generalizes to the present framework the classical Kleene basis theorem (Kleene 1959). Formally, we consider a reducibility schema for classes RPC:

41.1. $i \prec k \wedge Cl_i(x) \wedge \exists y(Cl(y) \wedge A(u,x,y)) \rightarrow \exists y(Cl_k(y) \wedge A(u,x,y))$,

for every $\mathfrak{L}$-formula $A(u,x,y)$ with the free variables shown, which is elementary extensional in x, y.

The rest of the section is devoted to convince the reader that RPC holds in $\mathfrak{C}_\iota$. The proof combines the classical analysis of Π^1_1-sets by means of recursive trees with a straightforward transfer argument from the standard model of Z_2 to $\mathfrak{C}_\iota$, coupled with representability of inductive predicates in our language.

41.2. First, we recall the relevant recursion-theoretic notions and results. As usual, n, m, k range over natural numbers; e, f, g denote indexes of partial recursive functions. *Par abus de langage*, we keep using $\langle \dots \rangle$ for a fixed number-theoretic primitive recursive function, which injectively maps finite sequences of natural numbers into numbers; on the same par, $(n)_i$ will denote the corresponding primitive recursive projection, such that, if n encodes $\langle n_0, \dots, n_k \rangle$ and $i \leq k$, then $(n)_i = n_i$. $lh(s)$ is the primitive recursive function which computes the length of the sequence code s. $Seq(x)$ is the predicate "to be a (number which encodes a) finite number sequence", while $\langle\,\rangle := 0$ is the code of the empty sequence. If s, r, range over elements of Seq, the concatenation of r with s is $r{*}s := \langle r_0, \dots, r_n, s_0, \dots, s_m \rangle$, where $r = \langle r_0, \dots, r_n \rangle$, $s = \langle s_0, \dots, s_m \rangle$. t is a subsequence of s (in symbols $t \subseteq s$) iff $s = t{*}r$, for some r in Seq; if $r \neq \langle\,\rangle$, we write $t \subset s$. If $F : \omega \rightarrow \omega$, we inductively define $\overline{F(0)} = \langle\,\rangle$ and $\overline{F(n{+}1)} = \overline{F(n)}{*}\langle F(n) \rangle$. The expression

$\{e\}^P(n) \simeq m$ means that the e-th partial function, which is recursive in the set $P \subseteq \omega$, converges on n with value m; clearly the Kleene bracket relation can be defined in the standard model by a Σ^0_1-formula, which is also denoted by $\{e\}^X(m) \simeq n$ and contains a second-order parameter X; $Tot^X(e) := \forall n \exists m(\{e\}^X(n) \simeq m)$ means that the function with index e is total on natural numbers. $g\lceil s$ is the index for the partial recursive function defined by $\{g\lceil s\}^X(s') := \{g\}^X\,(s{*}s')$ (s, s' in Seq). If $s = \langle n \rangle$, we simply write $g\lceil n$ instead of $g\lceil\langle n\rangle$. In order to define recursive trees, we consider the arithmetical formula

41.2.1 $W(e, X, Y) := Tot^X(e) \wedge (\forall s(\neg\{e\}^X(s) \simeq 0) \vee \forall n((e\lceil n)\epsilon Y)))$.

Note that W is positive in Y, while X possibly occurs in negated atoms of the form $\neg t\epsilon X$. Then we can define:

41.2.2. $$e\epsilon W(X) := \forall Y(Clos(W, X, Y) \rightarrow e\epsilon Y),$$

where $Clos(W, X, Y) := \forall e(W(e, X, Y) \rightarrow e\epsilon Y)$.

By standard arguments:

41.3. LEMMA

W.1 $\mathcal{P}(\omega) \models \forall X \forall e(W(e, X, W(X)) \rightarrow e\epsilon W(X))$;

W.2 $\mathcal{P}(\omega) \models \forall X(\forall e(W(e, X, B) \rightarrow B(e)) \rightarrow \forall e(e\epsilon W(X) \rightarrow B(e)))$,

where $B(x)$ is an arbitrary $\mathcal{L}_2$-formula and $W(e, X, B)$ results from $W(e, X, Y)$ by replacing each atom of the form $t\epsilon Y$ by $B[x := t]$.

If P is a fixed set $\subseteq \omega$, $\mathcal{W}(P)$ denotes the Π^1_1-set defined by the formula $W(X)$ in $\mathcal{P}(\omega)$, when X is assigned P as value. In general, to any index e of a total recursive function and any $P \subseteq \omega$, we can associate the tree

$$T^P_e = \{s : \forall s' \subseteq s.\, \{e\}^P(s') \simeq 0\},$$

which is closed under the subsequence relation $\subset$. $\mathcal{W}(P)$ encodes the set of well-founded trees which are recursive in P:

41.4. LEMMA. *If $P \subseteq \omega$, then $e \in \mathcal{W}(P)$ iff T^P_e is well-founded with respect to the converse of $\subset$, i.e. there is no function $F : \omega \rightarrow \omega$ such that for every $n \in \omega$, $\overline{F(n)} \in T^P_e$.*

Let ${}^\omega\omega = \{F : F$ is a unary function $\omega \rightarrow \omega\}$; then Π^1_1-predicates enjoy a simple, but essential property:

41.5. LEMMA (*Normal form*). *If $R \subseteq \omega$ is Π^1_1 in a given $P \subseteq \omega$, then we can find a primitive recursive predicate S_R such that, if F_P is the characteristic function of P, then:*

$$R = \{n \in \omega : (\forall G \in {}^{\omega}\omega)(\exists m \in \omega)(\langle n, \overline{F_P(m)}, \overline{G(m)}\rangle \in S_R)\}.$$

41.6. THEOREM (*Tree theorem for Π^1_1-sets*). *If $R \subseteq \omega$ is Π^1_1 in $P \subseteq \omega$, we can find a primitive recursive function $TREE_R$ such that*

$$n \in R \textit{ iff } TREE_R(n) \in \mathcal{W}(P).$$

PROOF. Let $TREE_R(n)$ denote the primitive recursive index of the characteristic function of the set

$$T_R(n) = \{s : Seq(s) \wedge \forall s' \subset s.\langle n, \overline{F_P(lh(s'))}, s'\rangle \notin S_R\}$$

(we apply the normal form lemma and the related notations). Then $T_R(n)$ is a tree, which is well-founded with respect to the converse of $\subset$ iff $TREE_R(n) \in \mathcal{W}(P)$ iff $n \in R$ (use 41.4 for the second equivalence). □

41.7. THEOREM (*The Kleene basis theorem*). *Let F be the characteristic function of $P \subseteq \omega$, fix $n \in \omega$ and let S be primitive recursive. Assume that for some function $G \in {}^{\omega}\omega$ and for every m, $\langle n, \overline{F_P(m)}, \overline{G(m)}\rangle \in S$. Then there exists $H \in {}^{\omega}\omega$ which is recursive in $\mathcal{W}(P)$, such that, for every m, $\langle n, \overline{F_P(m)}, \overline{H(m)}\rangle \in S$.*

PROOF (after Shoenfield 1967). By the so-called leftmost branch selection, one recursively defines:
$H(p)$ = the least number k such that, for some $G \in {}^{\omega}\omega$ and for every m, $\langle n, \overline{F_P(m)}, \overline{H(p)}{*}k{*}\overline{G(m)}\rangle \in S$. By inspection H is recursive in a set, which is Σ^1_1 in P; hence, by the tree theorem, it is recursive in $\mathcal{W}(P)$. □

41.8. COROLLARY. *Let $A(u, X, Y)$ be an arithmetical formula with the free variables shown and fix $n \in \omega$, $P \subseteq \omega$. If $\mathcal{P}(\omega) \models A[n, P, Q]$, for some Q, then there exists R recursive in $\mathcal{W}(P)$, such that $\mathcal{P}(\omega) \models A[n, P, R]$.*
PROOF : by 41.5-41.7.□

We finally apply the given machinery to $\mathfrak{C}_\iota$. We adopt the notations of §39 and assume that the language is expanded with names for ordinals $< \iota$ (for which we use lower case Greek letters). We fix a Gödel numbering $\ulcorner - \urcorner$ of CTM; by 10.11 of Ch. I, there is a term $Val(x)$ such that if $b \in$ CTM, $Val(\ulcorner b \urcorner) = b$. $CT(x)$ is an abbreviation for "$x = \ulcorner b \urcorner$, for some $b \in$ CTM".

41.9. DEFINITION. If $b \in$ CTM, we set:

$b^{ct} := \{x : Val(x)\eta b \wedge CT(x)\}$; $b^{ct}_\iota := \{\ulcorner a \urcorner : \mathfrak{C}_\iota \models Val(\ulcorner a \urcorner)\eta b$ and $a \in$ CTM$\}$;

$b(\alpha) := \{x : x\eta_\alpha b\}$; $\quad CL_\iota(\alpha) := \{b \in$ CTM: $\mathfrak{C}_\iota \models Cl_\alpha(b)\}$;

$b^\omega := \{x : x\eta b \wedge Nx\}$; $\quad b^\omega_\iota := \{n \in \omega : \mathfrak{C}_\iota \models \overline{n}\eta b\}$.

If $P \subseteq \omega$, $P^* := \{\overline{n} : n \in P\}$; so $P^* \subseteq$ CTM.

Clearly b_ι^{ct}, b_ι^ω are subsets of ω, represented in $\mathfrak{C}_\iota$ by b^{ct}, b^ω (respectively).

41.10. LEMMA (*Translation*)

(i) *Let $A(u,x,y)$ be an $\mathfrak{L}$-formula with the free variables shown, which is elementary extensional in x, y: then we can find an arithmetical formula $A_2(u,X,Y)$ of $\mathfrak{L}_2$ such that, if a, b, $c \in$ CTM, then*

$$\mathfrak{C}_\iota \models A[c,a,b] \text{ iff } \mathfrak{P}(\omega) \models A_2[\ulcorner c \urcorner, a_\iota^{ct}, b_\iota^{ct}]. \qquad (*)$$

(see §40, for $\mathfrak{P}(\omega) \models A$).

(ii) *Let $A(u,X,Y)$ be an arithmetical $\mathfrak{L}_2$-formula with the free variables shown: then we can find an $\mathfrak{L}$-formula $A^*(u,x,y)$, which is elementary extensional in the fresh variables x, y and satisfies, for $a,b \in$ CTM, $n \in \omega$:*

$$\mathfrak{P}(\omega) \models A[n, a_\iota^\omega, b_\iota^\omega] \text{ iff } \mathfrak{C}_\iota \models A^*[n, a^\omega, b^\omega]. \qquad (**)$$

(iii) *Under the same assumption of (ii), if $A^*(u,X,Y)$ is the formula of $\mathfrak{L} \cup \{X,Y\}$, which is obtained from $A^*(u,x,y)$ by replacing all atoms of the form $t\eta x, s\eta y$ by $t^* \in X$, $s^* \in Y$ (respectively), then*

$$\mathfrak{P}(\omega) \models A[n,P,Q] \text{ iff } \mathbf{CTM} \models A^*(\overline{n}, P^*, Q^*) \qquad (***)$$

*for $n \in \omega$, $P,Q \subseteq \omega$. {The right member of (∗∗∗) means that $A^*(u,X,Y)$ is satisfied in **CTM** by the assignement $\langle \overline{n}, P^*, Q^* \rangle$}.*

PROOF. (i) Here is the recipe to obtain $A_2(u,X,Y)$ from $A(u,x,y)$:
1) we replace each subformula of the form $t\eta a$, $t\eta b$ in A respectively by $\ulcorner t \urcorner \epsilon a_\iota^{ct}, \ulcorner t \urcorner \epsilon b_\iota^{ct}$;
2) we replace the subformulas $t = s$, Nt by the corresponding arithmetical formulas of $\mathfrak{L}_2$, which define $=$ and N in the term model **CTM**;
3) we relativize the the universal quantifiers of A to CTM.
By assumption on A, we can inductively check that A_2 is arithmetical in the sets a_ι^{ct}, b_ι^{ct} and that (∗) holds.
(ii)-(iii): apply the *-translation of 40.5 with a simple inductive argument (using $\mathbf{CTM} \models t^* = s^* \Rightarrow \mathfrak{P}(\omega) \models t = s$ and definition of P^*). □

If $a \in CL_\iota(\alpha)$, the set $\mathcal{W}(a_\iota^\omega)$ of wellfounded trees, recursive in a_ι^ω, is definable in $\mathfrak{C}_\iota$ by a term of $\mathfrak{L}_V$, which is a β-class in $\mathfrak{C}_\iota$, whenever $\beta > \alpha$.

41.11. LEMMA. *We can find a term $I(W,x,i)$ such that if $a \in CL_\iota(\alpha)$ (where $a \in$ CTM, $\alpha < \iota$), then*

$$\mathcal{W}(a_\iota^\omega) = \{n \in \omega : \mathfrak{C}_\iota \models \overline{n} \eta_\alpha I(W, a^\omega, \alpha)\}; \qquad (*)$$

$$\mathfrak{C}_\iota \models \forall k (k \succ \alpha \to Cl_k(\{u : u\eta_\alpha I(W, a^\omega, \alpha)\})). \qquad (**)$$

PROOF. Consider the formula $W^*(e,x,y)$, obtained from $W(e,X,Y)$ of 41.2.1 by means of 40.10 (ii); then W^* is elementary extensional in x, y, and operative in y (see Ch. II, 9.5 and 10.3). If $W_i(e,x,y)$ is the i-transform of $W^*(e,x,y)$ (37.1), we have, for every $e \in \omega$, $a \in CL_\iota(\alpha)$, $b \in$ CTM:

$$\mathfrak{P}(\omega) \models W[e, a^\omega_\iota, b(\alpha)^\omega_\iota] \text{ iff } \mathfrak{C}_\iota \models W^*(\bar{e}, a^\omega, b(\alpha)^\omega) \text{ iff } \mathfrak{C}_\iota \models W_\alpha(\bar{e}, a^\omega, b^\omega). \tag{1}$$

Choose $FP(\lambda y.\{e : W_i(e,x,y)\}) := I(W,x,i)$: since $\mathfrak{C}_\iota \models$TLR, we apply 37.5, 37.2.1 under assumption that $a \in CL_\iota(\alpha)$, and we obtain, for every $e \in \omega$:

$$\mathfrak{C}_\iota \models \bar{e} \eta_\alpha I(W, a^\omega, \alpha) \leftrightarrow W_\alpha(\bar{e}, a^\omega, I(W, a^\omega, \alpha)). \tag{2}$$

Now set $I := (I(W, a^\omega, \alpha)(\alpha))^\omega_\iota$ and $F(Y) := \{e \in \omega : \mathfrak{P}(\omega) \models W[e, a^\omega_\iota, Y]\}$. Then I is F-closed by (2) and 40.10 (ii). On the other hand, I is the least F-closed subset of ω: apply the inductive generation of $\mathcal{V}_\alpha$, (2), 40.10 (iii) and adapt the proof of 13.4. As to (**), we use lemma 37.3 (v). □

41.12. THEOREM. *The schema* RPC *is true in* $\mathfrak{C}_\iota$.

PROOF. Let $a \in CL_\iota(\alpha)$, for some $\alpha < \iota$, and $\mathfrak{C}_\iota \models \exists y(Cl(y) \wedge A(c,a,y))$, where $c \in$ CTM: then by lemma 41.10 (i) and 41.8, the set

$$\{P : P \subseteq \text{CTM and } \mathfrak{P}(\omega) \models A_2[\ulcorner c \urcorner, a^{ct}_\iota, P]\} := G(c,a)$$

contains an element P, which is recursive in the set $\mathcal{W}(a^{ct}_\iota) = \mathcal{W}((a^{ct})^\omega_\iota)$. By 41.11, P is arithmetical in some d^{ct}_ι, for $d \in CL_\iota(\beta)$, $\beta > \alpha$, and by closure of β-classes under elementary comprehension, there is a β-class e such that:

$$e^{ct}_\iota = P \text{ and } \mathfrak{P}(\omega) \models A_2[\ulcorner c \urcorner, a^{ct}_\iota, e^{ct}_\iota];$$

again by lemma 41.10 (i), $\mathfrak{C}_\iota \models \exists y(Cl_\beta(y) \wedge A(c,a,y))$. □

41.13. COROLLARY (Bar Induction)
If $\mathfrak{C}_\iota \models Cl(r) \wedge PWO(r)$, *then* $\{\langle a,b\rangle : a,b \in \text{CTM}$ *and* $\mathfrak{C}_\iota \models \langle a,b\rangle \eta r\}$ *is a well-ordering.*

41.14. REMARK. If we replace ι by ω in the model construction of the previous section, we obtain a model CTM_∞ of TL, which makes RPC true, once we replace $Cl(x)$ by $Cl_\infty(x) := \exists i Cl_i(x)$.

§42. Levels of truth and impredicative subsystems of second-order arithmetic

We are now ready to derive an analogue of Π^1_1-comprehension for classes from the reducibility principle for classes; in this section TLR^- stands for TLR without LIND, the local number theoretic induction axiom.

42.1. THEOREM. *Let* $A(u,x,y)$ *be an* $\mathfrak{L}$*-formula, which is elementary extensional in* x, y. *Then:*

$$\text{TLR}^- + \text{RPC} \vdash Cl(x) \rightarrow \exists z(Cl(z) \wedge \forall u(u \eta z \leftrightarrow \forall y(Cl(y) \rightarrow A(u,x,y)))).$$

PROOF. Let i such that $Cl_i(x)$ and let $i \prec k$. Then by RPC, definition of k-transform and lemma 37.6 (i):

$$\forall y(Cl_k(y) \to A_k(u,x,y)) \leftrightarrow \forall y(Cl(y) \to A(u,x,y)). \qquad (+)$$

If we choose $z := \{u : \forall y(Cl_k(y) \to A_k(u,x,y))\}$, then by lemma 37.3 (v), z is a j-class for any $j \succ k$ and hence a class such that, for every u,

$$u\eta z \leftrightarrow u\eta_j z \leftrightarrow \forall y(Cl_k(y) \to A_k(u,x,y)) \leftrightarrow \forall y(Cl(y) \to A(u,x,y));$$

(apply 37.5 (iii) for the second equivalence and (+) for the last one). □

42.1.1. REMARK. (i) After remark 41.14, it is easy to see that the theorem holds if we replace TLR + RPC by TL + RPC and we let $Cl_\infty(\dots)$ occur in place of Cl.

Numbers can be explained away in the classical style of Dedekind by means of the above comprehension principle. Consider the sublanguage $\mathcal{L}_V^-$, which omits the individual constants $\overline{0}$, SUC, $PRED$, D, $PAIR$, $LEFT$, $RIGHT$ and the predicate N. Let LTL = "Logicistic truth theory with levels" be the restriction of TLR + RPC to $\mathcal{L}_V^-$: thus LTL omits LIND and the OP-axioms PAIR, NAT.1-NAT.2. Then we already know how to define zero, successor and the other individual constants listed above in combinatory logic (see Ch. I, 3.8). By 42.1 there esists a class c such that, if $Clos(y)$ stands for $\overline{0}\eta y \wedge \forall u(u\eta y \to (u+1)\eta y)$, then:

$$\forall u(u\eta c \leftrightarrow \forall y(Cl(y) \wedge Clos(y) \to u\eta y))$$

and we can interpret Nx as $x\eta c$, so that class N-induction for classes and NAT.1-2 are derivable in LTL. Therefore we have:

42.2. THEOREM. TLR + RPC *is interpretable in* LTL.

We conclude by exploring two additional principles:

(i) we assume that *levels are objects*, and hence that the projection function LT collapses to identity;

(ii) we apply reflective truth to expressions containing bounded level quantification.

42.3. DEFINITION

(i) $\mathcal{L}_V^r$ is $\mathcal{L}_V$ without LT. The definition of term is modified by stipulating that level terms are terms *tout court*; all the rest is unchanged. We introduce the abbreviations:

$$\forall i \prec k.A := \forall i(i \prec k \to A); \quad \exists j \prec k.A := \exists j(j \prec k \wedge A).$$

(ii) The class $\mathrm{E}\Delta^+$of extended acceptable formulas of $\mathcal{L}_V$ is the smallest class which contains all atoms of the form $t = s$, Nt, $T_i t$ and is closed under bounded level quantifiers $\forall j \prec k$, negation, conjunction and universal object quantifiers.

(iii) We modify the map $A \mapsto [A]$ for arbitrary $A \in \mathrm{E}\Delta^+$ as follows. First we let $\forall j \prec i.f := \langle \overline{11}, \langle i, f \rangle\rangle$ (this makes sense now, since the level variable i is a term). Then we inductively add the new clause

$$[\forall j \prec i.A] := (\forall j \prec i.(\lambda x.[A[j := x]])) := \langle \overline{11}, \langle i, \lambda x[A] \rangle\rangle.$$

It is easy to check that $[A]$ is a term of $\mathcal{L}_V^r$, if $A \in \mathrm{E}\Delta^+$.

(iv) The bounded level quantifier axioms BLQ are given by the formulas:

$$T_k(\forall j \prec i.f) \leftrightarrow i \prec k \wedge \forall j \prec i.T_k(fj);$$

$$T_k\neg(\forall j \prec i.f) \leftrightarrow i \prec k \wedge \exists j \prec i.T_k(\neg(fj)).$$

(v) The ontological axiom ONT is $\forall i \exists x(i = x)$.

(vi) $\mathrm{TLR}^* := \mathrm{TL} + \mathrm{BLQ} + \mathrm{REFL}^+ + \mathrm{RPC} + \mathrm{ONT}$;

(for REFL^+, cf. 39.17), where the axioms containing LT are of course omitted.

We leave to the reader the verification that the elementary facts of §37 still hold for the extended notion of acceptable formula and with TLR^* in place of TLR. The definition of $\Gamma(\delta, S, X)$ in §39.9 is easily adapted for interpreting the new system: the idea is that level variables now range on the set $\{IN(\alpha) : \alpha < \iota\}$. The typical clauses which concern bounded level quantification in the definitions of the modified operator (compare with 39.9) have the form:

1) if for some $\gamma < \delta$ with $IN(\gamma) = i$ we have

$$\forall \beta < \gamma.\forall a \in \mathrm{CTM}.(IN(\beta) = a \Rightarrow fa \in X),$$

then $(\forall j \prec i.f) \in \Gamma(\delta, S, X)$ (where $\delta < \iota$, $S \subseteq \delta \times \mathrm{CTM}$, $X \subseteq \mathrm{CTM}$, i, b, $f \in \mathrm{CTM}$);

2) if for some $\gamma < \delta$ with $IN(\gamma) = i$, there exist $\beta < \gamma$, $a \in \mathrm{CTM}$ with $IN(\beta) = a$ and $(\neg fa) \in X$, then $(\neg\forall j \prec i.f) \in \Gamma(\delta, S, X)$ (where $\delta < \iota$, $S \subseteq \delta \times \mathrm{CTM}$, $X \subseteq \mathrm{CTM}$, $f, i, b \in \mathrm{CTM}$).

Since the relevant lemmas 39.10-15 still hold for the modified Γ, we obtain:

42.4. THEOREM. *There is a set* $\mathcal{V}^* \subseteq \mathrm{CTM}$ *such that:*

$$\mathfrak{C}_\iota^* := \langle \mathbf{CTM}, \mathcal{V}_\iota^* \rangle \models \mathrm{TLR}^*.$$

At this point one can proceed to strengthen the interpretation results. $\Delta^1_2(A,B)$ is the formula:

$$\forall u(\forall w(Cl(w)\rightarrow\exists z(Cl(z)\wedge B(w,z,u)))\leftrightarrow\exists x(Cl(x)\wedge\forall y(Cl(y)\rightarrow A(x,y,u))))$$

42.5. THEOREM. *Let $A(x,y,u)$ and $B(w,z,u)$ be elementary extensional in x,y and w,z respectively; then we can derive in* TLR* *without local number-induction:*

$$\Delta^1_2(A,B)\rightarrow\exists v(Cl(v)\wedge\forall u(u\eta v\leftrightarrow\forall w(Cl(w)\rightarrow\exists z(Cl(z)\wedge B(w,z,u))))).$$

PROOF. Assume $\Delta^1_2(A,B)$; then by 37.8 (v) we have, for arbitrary u:

$$\forall j\forall w(Cl_j(w)\rightarrow\exists k\exists z(Cl_k(y)\wedge B(w,z,u)))\leftrightarrow$$
$$\leftrightarrow\exists j\exists x(Cl_j(x)\wedge\forall k\forall y(Cl_k(y)\rightarrow A(x,y,u))). \tag{1}$$

By logic, we have, for every u,

$$\exists j\exists w\forall k(Cl_j(w)\wedge\forall z(Cl_k(z)\rightarrow\neg B(w,z,u)))\ \vee \tag{2}$$
$$\vee\ \exists j\exists x\forall k(Cl_j(x)\wedge\forall y(Cl_k(y)\rightarrow A(x,y,u))).$$

Hence:

$$\exists j\forall k\exists w(Cl_j(w)\wedge\forall z(Cl_k(z)\rightarrow\neg B(w,z,u)))\ \vee \tag{3}$$
$$\vee\ \exists j\forall k\exists x(Cl_j(x)\wedge\forall y(Cl_k(y)\rightarrow A(x,y,u))).$$

Again by logic:

$$\forall u\exists j\forall k\{\exists w(Cl_j(w)\wedge\forall z(Cl_k(z)\rightarrow\neg B(w,z,u)))\vee \tag{4}$$
$$\vee\ \exists x(Cl_j(x)\wedge\forall y(Cl_k(y)\rightarrow A(x,y,u)))\}.$$

By (4), $\forall j\exists k(j\prec k)$ and persistence, we get:

$$\forall u\exists k.\exists j\prec k.\{\exists w(Cl_j(w)\wedge\forall z(Cl_k(z)\rightarrow\neg B(w,z,u)))\vee \tag{5}$$
$$\vee\ \exists x(Cl_j(x)\wedge\forall y(Cl_k(y)\rightarrow A(x,y,u)))\}.$$

But (5) has the form $\forall u\exists k\, C$ where $C\in \mathrm{E}\Delta^+$; hence by REFL$^+$ there is a level p such that:

$$\forall u.\exists k\prec p.\exists j\prec k.\{\exists w(Cl_j(w)\wedge\forall z(Cl_k(z)\rightarrow\neg B(w,z,u)))\vee \tag{6}$$
$$\vee\ \exists x(Cl_j(x)\wedge\forall y(Cl_k(y)\rightarrow A(x,y,u)))\}.$$

Choose $c:=\{u\ :\ \exists k\prec p.\exists j\prec k.\exists x(Cl_j(x)\wedge\forall y(Cl_k(y)\rightarrow A(x,y,u)))\}$. Then c is a q-class if $p\prec q$. Indeed, assume $\neg u\eta_q c$, i.e.

$$\neg T_q\exists k\prec p.\exists j\prec k.\ \exists x(Cl_j(x)\wedge\forall y(Cl_k(y)\rightarrow A(x,y,u))). \tag{7}$$

By the *axiom* BLQ of 42.3, we have:

$$\forall k \prec p.\forall j \prec k.\ \neg T_q(\exists x(Cl_j(x) \wedge \forall y(Cl_k(y) \rightarrow A(x,y,u)))). \tag{8}$$

Given $k \prec p$ and $j \prec k$, we have $j \prec q$ and hence that CL_j, CL_k are q-classes (37.5 (vi)), whence we can erase T_q in (8):

$$\forall k \prec p.\forall j \prec k.\forall x(Cl_j(x) \rightarrow \exists y(Cl_k(y) \wedge \neg A(x,y,u))). \tag{9}$$

In order to check that c is a q-class , we must prove:

$$F_q(\exists k \prec p.\exists j \prec k.\exists x.(Cl_j(x) \wedge \forall y(Cl_k(y) \rightarrow A(x,y,u)))), \tag{10}$$

i.e., by BLQ, $\forall k \prec p \forall j \prec k.F_q(\exists x(Cl_j(x) \wedge \forall y(Cl_k(y) \rightarrow A(x,y,u))))$, which is easily seen to be equivalent to (9). We now claim:

$$\forall u(u\eta c \leftrightarrow \exists x(Cl(x) \wedge \forall y(Cl(y) \rightarrow A(x,\ y,u)))). \tag{11}$$

$\Rightarrow$: by assumption, for some $k \prec p$, $j \prec k$ and some j-class x, we get:

$$\forall y(Cl_k(y) \rightarrow A(x,y,u)). \tag{12}$$

Hence by RPC, we also have:

$$\forall y(Cl(y) \rightarrow A(x,y,u)). \tag{13}$$

$\Leftarrow$: were $u\eta c$ false, we should have:

$$\forall k \prec p.\forall j \prec k.\forall x(Cl_j(x) \rightarrow \exists y(Cl_k(y) \wedge \neg A(x,y,u))). \tag{14}$$

Hence by (6), we can find $k \prec p$, $j \prec k$ and w in CL_j such that

$$\forall z(Cl_k(z) \rightarrow \neg B(w,z,u)). \tag{15}$$

By RPC, we get

$$\forall z(Cl(z) \rightarrow \neg B(w,z,u)), \tag{16}$$

whence

$$\exists w(Cl(w) \wedge \forall z(Cl(z) \rightarrow \neg B(w,z,u))). \tag{17}$$

But (17) implies, by assumption on A, B,

$$\forall x(Cl(x) \rightarrow \exists y(Cl(y) \wedge \neg A(x,y,u))). \ \square$$

By theorem 42.1 and 42.5, with application of the *-translation of 40.5, it is straightforward to conclude:

42.6. COROLLARY

(i) Π^1_1-CA_0 *is interpretable in* TLR + RPC (*actually* LTL *of theorem* 42.2 *is enough*).

(ii) Δ^1_2-CA_0 *is interpretable in* TLR*.

42.7. REMARK. (i) The corollary holds true even if we add to the

subsystems based on Π^1_1-CA and Δ^1_2-CA the bar induction schema of 40.3.5, and at the same time we strengthen TLR* with the schema $Cl(r) \wedge PWO(r) \rightarrow TI(r, B)$ (B arbitrary). The new extension of TLR* is obviously consistent by corollary 41.13.
(ii) Let $Progr(r, c)$ be the formula $\forall x(\forall y(\langle y, x\rangle \eta r \rightarrow y\eta c) \rightarrow x\eta c)$. Then TLR$^-$ + RPC proves:

$$Cl(r) \wedge Cl(a) \rightarrow \exists c(Cl(c) \wedge$$

$$\wedge\, Progr(r, c) \wedge \forall d(Cl(d) \wedge Progr(r, d) \rightarrow c \subseteq d)).$$

Since classes are closed under join and elementary comprehension in TLR$^-$, one can immediately verify that Feferman's system $EM_0\restriction + J +$ the non-uniform version of the so-called inductive generation axiom is interpretable in TLR$^-$ + RPC (see Feferman 1979).

Appendix: on projectibility and stronger reflection

A (relatively) quick strategy for proving 39.5 (and actually far-reaching generalizations of it) is to devise an arithmetical notation system N_ι for ι; this leads to reconsider work of the early seventies about non-monotone inductive definitions.

The most comprehensive treatment we are aware of, is the long paper of Richter-Aczel (1974), which is our reference text below for all details not included here. There exists also a category-theoretic approach to generalized recursion (e.g. Girard-Vauzeilles 1981, Ressayre 1982).

Coming to the point, we adopt the terminology and conventions of §40 as far as the language of second-order arithmetic is concerned.

1. DEFINITION. If $\Gamma : \mathcal{P}(\omega) \rightarrow \mathcal{P}(\omega)$, where Γ is possibly non-monotone, we recursively define for $\alpha \in ON$:

(i) $I(\Gamma, \alpha) = \cup\{\Gamma(I(\Gamma, \beta)) : \beta < \alpha\}$;

$I(\Gamma) = \cup\{I(\Gamma, \alpha) : \alpha \in ON\}$;

if $n \in I(\Gamma)$, $|n| =$ least α such that $n \in I(\Gamma, \alpha+1)$.

(ii) Since the sequence $\langle I(\Gamma, \alpha) : \alpha \in ON\rangle$ is non-decreasing with respect to inclusion, it makes sense to define the closure ordinal $|\Gamma|$ of Γ

$$|\Gamma| := \text{least } \alpha \text{ such } I(\Gamma, \alpha) = I(\Gamma, \alpha+1).$$

(iii) Let $A(u, X)$ be an arbitrary formula of second-order arithmetic with the only free variables shown; then A defines the operator:

$$\Gamma(P) = \{m \in \omega : \mathcal{P}(\omega) \models A[m, P]\}.$$

If $\mathcal{F}$ is a class of formulas of Z_2 (e.g. $\mathcal{F}$ is the class of arithmetical formulas), an operator Γ is said to be $\mathcal{F}$, if Γ can be defined by a formula of $\mathcal{F}$.

2. LEMMA. *If Γ is arithmetical, the sequence $\langle \mathrm{I}(\Gamma, \beta) : \beta < \alpha \rangle$ is uniformly $\Sigma_1(L_\alpha)$ for α admissible $> \omega$. In particular,* $\mathrm{I}(\Gamma, \beta) \in L_\alpha$, *for each $\beta < \alpha$.*

PROOF: arithmetical formulas become Δ_1 in L_α, if $\alpha > \omega$; then we can apply Δ_1-recursion. □

3. THEOREM

(i) *There exists an arithmetical operator Γ whose closure ordinal is ι, i.e. $|\Gamma| = \iota$.*

(ii) *Indeed, Γ can be chosen so that, for every $n \in \omega$, $P \subseteq \omega$,*

$$n \in \Gamma(P) \text{ iff } n \in \Gamma_0(P) \text{ or } (\Gamma_0(P) \subseteq P \text{ and } n \in \Gamma_1(P)),$$

where Γ_1 is a Π^0_1-operator and Γ_0 is a bounded operator.

The proof can be found after lemmas 7.8 and 9.5 of Richter-Aczel (1974).

PROOF of 39.5: if Γ is the operator given by the previous theorem, $\mathrm{I}(\Gamma, \alpha)$ is a proper subset of $\mathrm{I}(\Gamma, \alpha+1)$ for every $\alpha < \iota$. Hence the function:

$$IN(\alpha) = \text{least } n \in \omega \text{ with } n \in \mathrm{I}(\Gamma, \alpha+1) \text{ and } n \notin \mathrm{I}(\Gamma, \alpha)$$

is always defined on ι and is trivially injective; lemma 2 grants that IN satisfies the required definability conditions. □

This is not the end of the story, however; indeed theorem 3 is only the starting point of a bold generalization, including ordinals, which immensely overcome recursively inaccessibles. As a sample, we state a theorem of Richter (1971).

4. DEFINITION

(i) A set-theoretic formula A is Π_2 iff A is logically equivalent to a formula of the form $\forall x_1 \ldots \forall x_n \exists y_1 \ldots \exists y_k A$, where A is bounded.

(ii) If P is a class of ordinals, α reflects the sentence A on P iff

$$L_\alpha \models A \text{ implies } L_\beta \models A, \text{ for some } \beta \in P.$$

(iii) α is Π_2-reflecting on P iff α reflects every Π_2-sentence on P.

(iv) α is recursively Mahlo iff α is Π_2-reflecting on Ad, the class of admissible ordinals.

4.1. REMARK. α is Π_2-reflecting on ON, or Π_2-reflecting tout court iff α is admissible $> \omega$. It is also easy to see that α is recursively Mahlo iff α is Π_2-reflecting on the class of recursively inaccessible ordinals.

5. THEOREM (Richter 1971)

(i) *There exists an arithmetical operator* Γ, *whose closure ordinal is the least recursively Mahlo ordinal* μ.

(ii) *Moreover, there exist* Π^0_1-*operators* Γ_1, Γ_0, *such that for every* $n \in \omega$, $P \subseteq \omega$, *then* $n \in \Gamma(P)$ *iff* $n \in \Gamma_0(P)$ *or* $(\Gamma_0(P) \subseteq P$ *and* $n \in \Gamma_1(P))$.

6. COROLLARY. *The projectibility theorem of* 39.5 *holds if we replace* ι *with* μ.

Below, we stick to the notations of 42.3 for bounded level quantifiers and extended acceptable formulas of $\mathcal{L}_V$. We now formalize the notions of admissibility and inaccessibility in the language of truth with levels.

7. DEFINITION

(i) $Adm(i)$ is the conjunction of the following two formulas of $\mathcal{L}_V$

$$Lim(i) := \forall j \prec i.\forall k \prec i.\exists p \prec i.(j \prec p \wedge k \prec p);$$

$$\Pi_2\text{-}refl(i) := \forall f \forall g[\forall j \prec i.\forall x(T_j(fx) \to \exists p \prec i.T_p(gx)) \to$$
$$\to \forall q \prec i.\exists k(q \prec k \prec i \wedge \forall j \prec i.\forall x(T_j(fx) \to T_k(gx)))].$$

(ii) $Inac(i)$ is the conjunction of $Lim(i)$ with

$$T\text{-}Lim(i) := \forall x(T_i x \to \exists k \prec i.T_k x) \text{ and}$$
$$\forall f \forall g[\forall x(T_i(fx) \to T_i(gx)) \to$$
$$\to \forall j \prec i.\exists k(Adm(k) \wedge j \prec k \prec i \wedge \forall x(T_k(fx) \to T_k(gx)))].$$

(iii) TMA, the Mahlo principle for truth, is the schema

$$\forall f \forall g[\forall x(T(fx) \to T(gx)) \to \forall j \exists k(j \prec k \wedge Inac(k) \wedge \forall x(T_k(fx) \to T_k(gx)))].$$

7.1. REMARK. The definitions are equivalent to the usual ones in the intended model of §39 (see the remark 4.1 above; if one chooses $fx = [x = x]$ in the second condition of $Adm(i)$, one readily has that, if $Adm(i)$ holds, then $\forall x \exists p \prec i.T_p(gx) \to \exists k \prec i.\forall x T_k(gx)$, i.e. positive reflection relativized to i; cf. 37.8).

8. THEOREM. *The theory* TL + RPC + TMA *is consistent.*

PROOF. We modify the construction of theorem 39.16 as follows; if $\alpha < \mu$, the least recursively Mahlo, we recursively define a new operation $\mathcal{V}_1$ such that, for $\delta < \mu$:

$$\mathcal{V}_1(\delta) = \cup\{\mathcal{V}_1(\beta) : \beta < \delta\} \text{ if } \delta \text{ is recursively inaccessible;}$$

$$\text{else, } \mathcal{V}_1(\delta) = It(\Gamma, \delta, \mathcal{V}_1 \lceil \delta, \tau_{\phi(\delta)}).$$

Of course Γ, It, τ are the operations of §39. Observe that the notion of recursively inaccessible ordinal is uniformly $\Delta_1(L_\alpha)$ for α admissible $> \omega$; thus the definition by cases used in $\mathcal{V}_1$ does not lead out of the class of the Δ_1-operations. It is straightforward to check that, mutatis mutandis, all conditions of 39.15 hold for $\mathcal{V}_1$; hence we can define $\mathfrak{C}_\mu = \langle \mathbf{CTM}, \mathcal{V}_1 \rangle$ and extend the interpretation of 39.16. Then TMA holds as immediate application of Π_2-reflection of μ over the class of recursively inaccessible ordinals. □

On the axiomatic side, we propose a few easy consequences of TMA:

9. PROPOSITION. $TL^- + TMA$ *proves*:

(i) $\forall i \exists k (i \prec k \wedge Inac(k))$ (choose $f = g = \lambda x.x$ in TMA);

(ii) $\forall x \exists i T_i(fx) \rightarrow \forall j \exists k (j \prec k \wedge Inac(k) \wedge \forall x T_k(fx))$;

(iii) $f : CL \rightarrow CL. \rightarrow \exists k (Adm(k) \wedge f : CL_k \rightarrow CL_k)$.

10. PROBLEMS

1) Study the relation with recursion in the superjump functional sJ (see Hinman 1978 for the relevant definition); for instance, is every set $X \subseteq \omega$ recursive in sJ, definable by a class in $\mathfrak{C}_\mu$? Are the sets $\subseteq \omega$ recursively enumerable in sJ (which are known to coincide with Π^1_2-sets), definable by closed terms of $\mathfrak{C}_\mu$?

2) Charachterize the least ordinal for which the $\mathcal{V}$-construction is no more possible. There exist similar ordinals already below the first stable ordinal σ_0 (see Barwise 1975, Hinman 1978). For σ_0 the lemma 39.5 fails badly: σ_0 is projectible into ω, but there is an $\alpha < \sigma_0$ which is not projectible into ω.

3) What is the proof theoretic strength of $TL^- + TMA$? The work of Rathjen (1991) is probably relevant here.

CHAPTER 9

LEVELS OF TRUTH AND PREDICATIVE WELL-ORDERINGS

§ 43. On well-orderings
§ 44. Ramified hierarchies
§ 45. Predicative well-orderings I
§ 46. Predicative well-orderings II

In the previous chapter we announced that the system TLR of reflective truth with levels is equivalent, as to its arithmetical content, to Feferman-Schütte predicative analysis (henceforth FS). In order to prove such a claim, we are going to develop a proof-theoretic analysis of TLR. The first step is to describe the standard primitive recursive well-ordering of type Γ_0, the Feferman-Schütte ordinal, within the context of TLR. Indeed, we shall work in a fragment MFR(p) of TLR, which includes: 1) the ground system MF_c of Ch. II with number-theoretic induction for classes; 2) axioms stating the existence of the ramified hierarchy, *generated by conditions of a fixed logical complexity* p, along suitable explicitly presented pseudo-well-orderings (in the sense of 38.4).

In § 43 we discuss two different notions in MF: *pseudo-well-orderings* (orderings, which are well-founded with respect to *classes*, in short *pwos*) and *quasi-well-orderings* (orderings which are well-founded with respect to *properties*, in short *qwos*). We shall prove that MF_p derives the analogue of the Weyl principle for qwos and related transfinite recursion schemata. In § 44 we construct a formalized version of the second-order ramified hierarchy $\mathfrak{R}$, the classical model-theoretic counterpart of predicativity. §§ 45-46 contain an elementary presentation of the so-called predicative standard well-ordering of type Γ_0 and a well-ordering proof within the fragment MFR(p) of TLR. More precisely, we verify that for each $\alpha < \Gamma_0$, the segment of type α of the standard well-ordering is a pwo, provably in MFR(p); in the special case where $\alpha = \epsilon_0$ (respectively $\phi\omega 0$, the first ω-critical ordinal) MFR(p) can be replaced by MF_c (MF_p). The results are optimal by the upper bound theorems of Ch. XI; there, we will establish a constructive consistency proof of the theory TLR (MF_c, MF_p) within Peano arithmetic extended by transfinite induction up to Γ_0 (ϵ_0, $\phi\omega 0$).

For the applicative-minded reader, we mention that, since a few years,

Γ_0, as well as more powerful proof-theoretic ordinals, have found non-trivial applications in the study of term rewrite systems.

§ 43. On well-orderings

It is well-known that the notion of well-ordering (and more generally of well-foundedness) is essentially second-order and it depends on the extension of the universe $\mathfrak{U}$ of second-order objects, be they sets, predicates or functions over the ground level. In the familiar arithmetical case, this dependence shows up in *the non-absoluteness of the well-ordering notion, with respect to the standard predicative interpretation* (that is, suitable segments of the hyperarithmetical hierarchy); there are straightforward examples of primitive recursive linear orderings on ω, which are HYP-well-founded, i.e. well-founded with respect to hyperarithmetical sets of numbers, and yet not truly well-founded (see Rogers 1967, Harrison 1968). On the other hand, the whole power set of ω is not necessary to test well-foundedness. By the Kleene Basis Theorem of 41.7, if a linear ordering $\prec_w$ of ω is well-founded within any collection $\mathfrak{U}$ of sets $\subseteq \omega$, containing $\prec_w$ and closed under the hyperjump operation $X \mapsto \mathcal{W}(X)$ (for $\mathcal{W}(X)$, see 41.2.2), then $\prec_w$ is a well-ordering "in the real world" (compare with the bar induction corollary of 41.13). {Such $\mathfrak{U}$'s are properly included by Friedman 1969 in the well-known collection of Mostowski's β-models, the ω-models of second-order arithmetic Z_2, which are absolute with respect to Π^1_1-conditions}.
It is then natural to see how far we can proceed – within the theories of reflective truth – in dealing with countable well-orderings. As we mentioned in § 38, there are at least two possible versions of the well-ordering notion; and we already know that pseudo-well-orderings are pleasantly closed under forms of transfinite recursion, provably in TLR.

Below we argue informally; nevertheless, it should be always clear how to work out the results in the indicated axiomatic systems.

43.1. DEFINITION (we repeat 38.4).

(i) Let us first remind that if w defines a binary relation, i.e. w is a property of ordered pairs, we keep using the infix notation $x \prec_w y$ in place of $\langle x,y\rangle \eta w$. Also, $Field(\prec_w)$ is the term $\{x : \exists z(x \prec_w z \vee z \prec_w x)\}$ representing the field of $\prec_w$ and $LO(\prec_w)$ means that $\prec_w$ is a linear ordering. If $B(x)$ is a formula with the free variable shown,

$$Progr(\prec_w, B) := (\forall x \eta Field(\prec_w))(\forall y \prec_w x . B(y) \rightarrow B(x)),$$

where $Progr(\prec_w, B)$ is to be read "B is progressive" (relative to $\prec_w$). If $B(x) = x\eta b$, we write $Progr(\prec_w, b)$.
As usual, $TI(\prec_w, B) := Progr(\prec_w, B) \to \forall x(x\eta Field(\prec_w) \to B(x))$, while $TI(\prec_w, b)$ stands for $Progr(\prec_w, b) \to Field(\prec_w) \subseteq b$.

(ii) We recall that $\prec_w$ is called a *pseudo-well-ordering* – in symbols $PWO(\prec_w)$, and, in short, $\prec_w$ is a pwo – iff $\forall b(Cl(b) \to TI(\prec_w, b))$.

(iii) A linear ordering $\prec_w$ is called a *quasi-well-ordering – in symbols $QWO(\prec_w)$, and, in short, $\prec_w$ is a qwo – iff $\forall b TI(\prec_w, b)$).*

(iv) A qwo (pwo) $\prec_w$ is *acceptable* iff $\prec_w$ is a class; $\prec_w$ is *unbounded* (on its field) iff $\forall x(x\eta Field(\prec_w) \to \exists y(y\eta Field(\prec_w) \wedge x \prec_w y))$.
Given an acceptable unbounded pwo or qwo, we introduce the standard notions of "zero" ($\prec_w$-least element), "successor" and "limit":

$0_{\prec_w}$:= the $\prec_w$-least element of $Field(\prec_w)$;

$S_{\prec_w}(x)$:= the $\prec_w$-least element of $\{y : y\eta Field(\prec_w) \wedge x \prec_w y\}$;

$Lim(x)$ iff "x is nor $0_{\prec_w}$ neither a successor".

(v) A qwo (pwo) $\prec_w$ is *locally decidable via f and h*, if $h : a \to a$ and $f : a \to \{\overline{0}, \overline{1}, \overline{2}\}$, (where $a = Field(\prec_w)$), and for every x in a, we have

$$fx = \begin{cases} \overline{0} & \text{if } x = 0_{\prec_w} \\ \overline{1} & \text{if } x \text{ is a } \prec_w\text{-sccessor;} \\ \overline{2} & \text{if } x \text{ is a limit;} \end{cases} \qquad (*)$$

$$hx = \text{the predecessor of } x, \text{ whenever } fx = 1. \qquad (**)$$

A pwo (qwo) is locally decidable iff it is locally decidable via some f and h. If a pwo (qwo) $\prec_w$ is acceptable, its field and every initial segment of $\prec_w$ (of the form $\{x : x \prec_w y\}$) are classes.

43.2.REMARK. The expression "quasi-well-ordering" is used by Crossley (1969) for linear well-orderings of ω, which are well-founded with respect to recursive sets.

We now show that a strong transfinite recursion principle holds for acceptable locally decidable qwos.

43.3. DEFINITION (cf. 9.14, for $\otimes$ below).

$f : CL \to CL := \forall x(Cl(x) \to Cl(fx))$;

$g : Field(\prec_w) \otimes CL \otimes CL \to CL :=$

$:= \forall x \forall u \forall v(x\eta Field(\prec_w) \wedge Cl(u) \wedge Cl(v) \to Cl(gxuv))$;

$$\prec_w\lceil x := \{u : u \prec_w x\}.$$

43.4. THEOREM (*Special transfinite recursion along qwos*). *We can find – provably in* MF^- *– a closed term* $\lambda h \lambda f \lambda w \lambda z \lambda x.Rc[h,f,\prec_w]zx$ *such that, if* $h : CL \to CL$, $f : CL \to CL$ *and* $\prec_w$ *is an acceptable qwo, which is locally decidable via* g_1, g_2, *then for every* x *in* $Field(\prec_w)$,

$$g_1 x = \overline{0} \to Rc[h,f,\prec_w]zx = hz; \tag{1}$$

$$g_1 x = \overline{1} \to Rc[h,f,\prec_w]zx = f(Rc[h,f,\prec_w]z(g_2 x));$$

$$g_1 x = \overline{2} \to Rc[h,f,\prec_w]zx = \{u : \exists v(v \prec_w x \wedge u\eta(Rc[h,f,\prec_w]zv)\};$$

$$\Sigma(Field(\prec_w), \lambda x Rc[h,f,\prec_w]zx) \text{ is a class, whenever } z \text{ is a class.} \tag{2}$$

If $hz \subseteq CL$ *and* $fx \subseteq CL$, *whenever* z, x *are classes, then*

$$Rc[h,f,\prec_w]zx \subseteq CL. \tag{3}$$

{NB: *for simplicity, we omit the uniform dependence of Rc from* g_1, g_2}.

PROOF. (1). Observe that the operation g_1 makes possible to define by cases over the field of $\prec_w$. Then the fixed point for operations implies the existence of $Rc[h,f \prec_w]$ satisfying the given equations.

(2) It is enough to check that, if z is a class, then

$$\forall x(x\eta Field(\prec_w) \to Cl(H(x,z))), \tag{*}$$

where $H(x,z) = Rc[h,f,\prec_w]zx$; then we can apply the join principle of 9.9. But (*) is straightforward by $\prec_w$-induction applied to $\{x : Cl(H(x,z))\}$ with the assumption on h, f and $\prec_w$.

(3) If $G(z) = \{x : Rc[h,f,\prec_w]zx \subseteq CL\}$, (*) implies

$$x\eta G(z) \leftrightarrow Rc[h,f,\prec_w]zx \subseteq CL. \tag{**}$$

Hence we can proceed by $\prec_w$-induction to show
$\forall x(x\eta Field(\prec_w) \to x\eta G(z))$, under the additional assumption of (3). □

We also mention that a version of the Weyl principle §38 holds for qwos. With the notations of 38.4, $TR(y,A,\prec_w,z)$ is a shortening for the formula $\forall u \forall x \forall u(x\eta Field(\prec_w) \to (u\eta y(x) \leftrightarrow A(u,x,y\lceil x,z)))$.
Then we have, by an easy adaptation of 38.6:

43.5. THEOREM (WP *for qwos*). *Let* $A(u,x,y,z)$ *be a formula, which is elementary extensional in* y, z *with the free variables shown. Then we prove in* MF^-:

(i) *if* $\prec_w$ *is an acceptable qwo,* z *is a class, then there exists a class* y *such that* $TR(y,A,\prec_w,z)$ *holds.*

(ii) *Under the same hypothesis, if* y, y' *are two classes satisfying* $TR(-, A, \prec_w, z)$, *then for every* x *in* $Field(\prec_w)$, *we have*

$$\forall u(u\eta y(x) \leftrightarrow u\eta y'(x)).$$

In general, the hypothesis of 43.5 (i) above cannot be weakened to $PWO(\prec_w)$, unless we substantially enrich MF^-, e.g. to TL^-. Indeed, by a theorem of Spector and Gandy, there is an elementary condition A such that $\prec_w$ is a well-founded recursive linear ordering of ω iff a solution to $TR(-, A, \prec_w)$ exists in HYP (= the collection of hyperarithmetical sets $\subseteq \omega$). Thus no HYP-solution (in our case, no solution in CL) to $TR(-, A, \prec_w)$ exists in general, whenever $\prec_w$ is only HYP-well-founded; furthermore, there are recursive pwos on ω such that no solution at all exists for $TR(-, A, \prec_w)$ (Friedman 1976). In positive form: *postulating closure of* CL *under elementary transfinite recursion along arbitrary pwos is to require that* CL *is really much richer than in the simple inductive models of* MF^-.

§44. Ramified hierarchies

In this section we apply the special transfinite recursion theorem 43.4 to prove the existence of the second-order ramified hierarchy $\mathfrak{R}$ of classes of natural numbers along any given acceptable unbounded locally decidable qwo. The result can be extended to pwos in the formal setting of TL.

Let us first recall an informal definition of $\mathfrak{R}$. By $\mathfrak{L}_2$ we understand the language of second-order arithmetic, introduced in the previous chapter (40.1). Thus $\mathfrak{L}_2$-formulas are also closed under quantifications $\forall X$ and $\exists Y$ on sets of numbers, and atoms have the form $t \in X$ and $t = s$ (t, s terms built up from variables by means of $+$, $\cdot$ and successor function symbols).

44.1. DEFINITION. Let $\mathcal{S}$ be a family of subsets of ω ($\omega =$ the set of natural numbers):

(i) if A is a formula of $\mathfrak{L}_2$ with free variables in the lists $x_1, \dots, x_k, X_1, \dots, X_n$, $\mathcal{S} \models A[n_1, \dots, n_k, P_1, \dots, P_n]$ stands for the usual satisfaction relation: it is understood that $n_1, \dots, n_k \in \omega$ are assigned in the given order to $x_1, \dots, x_k$ and number quantifiers range over ω; $P_1, \dots, P_n \in \mathcal{S}$ are assigned to $X_1, \dots, X_k$ and set quantifiers range over the family $\mathcal{S}$.

$P \subseteq \omega$ is $\mathcal{S}$-*definable* (with set parameters) iff

$$P = \{n \in \omega : \langle \omega, \mathcal{S} \rangle \models A[n, Q_1, \ldots, Q_n]\},$$

for some $\mathcal{L}_2$-formula $A(x, X_1, \ldots, X_n)$ and a (possibly empty) list $Q_1, \ldots, Q_n \in \mathcal{S}$ of set parameters.

(i) $Def(\mathcal{S}) = \{P \subseteq \omega : P \text{ is } \mathcal{S}\text{-definable}\}$.

(ii) We define by recursion on countable ordinals an operation $\mathcal{R}$ such that

$$\mathcal{R}_0 = Def(\{\omega\}); \qquad \mathcal{R}_{\alpha+1} = Def\ (\mathcal{R}_\alpha) \text{ and } \mathcal{R}_\lambda = \bigcup_{\beta < \lambda} \mathcal{R}_\beta$$
(where λ is a limit).

44.1.1. REMARK. (i) $Def(\{\omega\})$ extensionally coincides with the family of the subsets $\subseteq \omega$, which are definable by $\mathcal{L}_2$-formulas with no set quantifier and no set parameter. If $P \in \mathcal{R}_0$, P is called *arithmetical.*

(ii) Once one has accepted ω and a segment λ of ordinals, the definition of $\mathcal{R}$ is predicative in the traditional sense: set quantifiers range over countable collections, which are already built up.

(iii) We mention a few basic facts about $\mathcal{R}$ (for definitions and proofs, see Apt-Marek 1974, Moschovakis 1974, Boyd-Hensel-Putnam 1969, Kleene 1959, Jockusch-Simpson 1975).

1. $\mathcal{R}$ is a hierarchy, i.e. $\alpha < \beta$ implies $\mathcal{R}_\alpha \subseteq \mathcal{R}_\beta$.
2. There exists a countable ordinal β_0 such that $\mathcal{R}_{\beta_0} = \mathcal{R}_{\beta_0+1}$ and $\mathcal{R}_{\beta_0}$ is the smallest β-model of second-order arithmetic Z_2 (Gandy-Putnam).
3. $\beta_0 > \omega_1^{ck}$ (= the first non recursive ordinal) and β_0 is Δ^1_2-definable;
4. Every set of $\mathcal{R}_{\beta_0}$ is Δ^1_2 and so is $\mathcal{R}_{\beta_0}$.
5. $\mathcal{R}_{\omega_1^{ck}} = HYP =$ the collection of hyperarithmetical sets $\subseteq \omega$.

44.2. DEFINITION. Recall that $a =_e b := \forall x(x\eta a \leftrightarrow x\eta b)$; if we put

$$Embed(g, \mathcal{S}_1, \mathcal{S}_2) := \forall a(a\eta\mathcal{S}_1 \rightarrow (ga\eta\mathcal{S}_2 \wedge a =_e ga));$$

then we define:

(i) $\mathcal{S}_1 \subseteq^+_e \mathcal{S}_2 := \exists g Embed(g, \mathcal{S}_1, \mathcal{S}_2)$;

(ii) $\mathcal{S}_1 =^+_e \mathcal{S}_2 := \mathcal{S}_1 \subseteq^+_e \mathcal{S}_2 \wedge \mathcal{S}_2 \subseteq^+_e \mathcal{S}_1$.

Any pair g, h such that $Embed(g, \mathcal{S}_1, \mathcal{S}_2)$ and $Embed(h, \mathcal{S}_2, \mathcal{S}_1)$ is said *to witness* $\mathcal{S}_1 =^+_e \mathcal{S}_2$. Of course, $\subseteq^+_e$ is reflexive and transitive and, by definition of *Embed* with $=_e$, it follows that, if g and h witness $\mathcal{S}_1 =^+_e \mathcal{S}_2$,

g is 1-1 onto with respect to $=_e$. Indeed, we have $\forall a \forall b(ga =_e gb \to a =_e b)$; furthermore

$$Embed(g, \mathcal{I}_1, \mathcal{I}_2) \wedge Embed(h, \mathcal{I}_2, \mathcal{I}_1) \to$$
$$\to \forall a(a\eta\mathcal{I}_1 \to h(ga) =_e a) \wedge \forall b(b\eta\mathcal{I}_2 \to g(hb) =_e b).$$

44.2.1. FACT. MF$^-$ *proves:*

$$Cl(\mathcal{I}_1) \wedge Cl(\mathcal{I}_2) \wedge \mathcal{I}_1 \subseteq CL \wedge \mathcal{I}_2 \subseteq CL\,. \to T(\mathcal{I}_2 \subseteq_e^+ \mathcal{I}_1) \vee F(\mathcal{I}_1 \subseteq_e^+ \mathcal{I}_2).$$

(The same holds with $=_e^+$ in place of $\subseteq_e^+$).

Of course, the relations $=_e^+$, $\subseteq_e^+$ are natural strengthenings of extensional equality and inclusion; we need them to capture the extensional features of the ramified hierarchy in our non-extensional framework. We also remind that $MF_p := MF^- + \text{P-IND}$, and that property induction P-IND has the form:

$$Clos(a) \to \mathbb{N} \subseteq a;$$

here $Clos(a) := \overline{0}\eta a \wedge \forall x(x\eta a \to (x+1)\eta a)$; $\mathbb{N}$ stands for the class $\{x: Nx\}$ of MF$^-$.

44.3. THEOREM (MF$_p$)

We can find – uniformly in any given acceptable unbounded qwo $\prec_w$, *locally decidable via* g_1, g_2 *– an operation* $\lambda x.\mathcal{R}_x$ *such that, if* x *is in the field of* $\prec_w$, *then*

(i) $g_1 x = \overline{0} \to \mathcal{R}_x = Def(\{\mathbb{N}\})$ (*i.e. if* x *is the* $\prec_w$*-first element*);

$g_1 x = \overline{1} \to \mathcal{R}_x = Def(\mathcal{R}_{g_2 x}) \cup \mathcal{R}_{g_2 x}$ (*if* x *is the* $\prec_w$*-successor of* $g_2 x$);

$g_1 x = \overline{2} \to \mathcal{R}_x = \{u: \exists v(v \prec_w x \wedge u\eta\mathcal{R}_v)\}$, (*if* x *is a* $\prec_w$*-limit*);

(ii) $\mathcal{R}_x$ *is a class of classes and* $\mathcal{R}_x \subseteq_e^+ \mathcal{R}_y$, *whenever* $x \prec_w y$.

PROOF. The argument requires a number of separate steps and definitions; the essential point is to introduce the operation Def in our language.
First of all, we associate to each primitive function symbol f of $\mathcal{L}_2$ a closed λ-term f^*, which formally represents it (see Ch. I, 3.6).
We then define a satisfaction predicate $SAT[\mathcal{I}, f_1, f_2]$ for $\mathcal{L}_2$-formulas, uniformly in any fixed class $\mathcal{I}$ of classes $\subseteq \mathbb{N}$, and in any pair of operations $f_1 : \mathbb{N} \to \mathbb{N}$, $f_2 : \mathbb{N} \to \mathcal{I}$.
To this aim, we fix a Gödel numbering GD of $\mathcal{L}_2$, in such a way that all the syntactical notions (term, formula, occurrence of a free variable, etc.) define classes (provably in MF$_c$); $GD(E)$ ambiguously denotes the Gödel number of the expression E and the term, formally representing it. $For_2(x)$

is the predicate "x is the Gödel number of an $\mathcal{L}_2$-formula"; $Ter(x)$ is the predicate "x is the Gödel number of an $\mathcal{L}_2$-term." Put

$$var_2(i) := GD(X_i); \quad ins(GD(t); \quad var_2(i)) := GD(t \in X_i);$$

$$Iden(GD(t), GD(s)) := GD(t = s).$$

If $f : \mathsf{N} \to \mathfrak{S}$ ($\mathfrak{S} = \mathsf{N}$ or $\mathcal{I}$) and b is in $\mathfrak{S}$, then $f(^i_b) : \mathsf{N} \to \mathfrak{S}$ is the operation defined by: $f(^i_b)n = fn$, if $i \neq n$, $f(^i_b)i = b$ (here $f(^i_b)$ is well-defined with definition by cases on N). Moreover $val(u, f_1)$ is (the formal presentation of) the operation, which associates to the term encoded by u its numerical value under the assignment f_1.

44.3.1. Let $W(\mathcal{I}, f_1, f_2, b, k)$ stand for (the formalization of) the following condition (as to logical complexity, cf. Introduction, 5.5):

(i) every element of b has the form $\langle GD(A), f_1, f_2 \rangle$, for some $\mathcal{L}_2$-formula A of logical complexity $\leq k$ (k being a natural number), where $f_1 : \mathsf{N} \to \mathsf{N}$, $f_2 : \mathsf{N} \to \mathcal{I}$;

(ii) $\langle GD(A), f_1, f_2 \rangle \eta b$ iff either:

1.1. $A = (t = s)$ and $val(GD(t), f_1) = val(GD(s), f_1)$; or

1.2. $A = t \in X_i$ and $val(GD(t), f_1)\eta(f_2 i)$; or

2. $A = \neg B$ and not $\langle GD(B), f_1, f_2 \rangle \eta b$; or

3. $A = B_1 \wedge B_2$ and $\langle GD(B_1), f_1, f_2 \rangle \eta b$, $\langle GD(B_2), f_1, f_2 \rangle \eta b$; or

4. $A = \forall x_i B$ and $\forall k \eta \mathsf{N}. \langle GD(B), f_1(^i_k), f_2 \rangle \eta b$; or

5. $A = \forall X_i B$ and $\forall P \eta \mathcal{I}. \langle GD(B), f_1, f_2(^i_P) \rangle \eta b$.

Then we prove:

44.3.2. There is a term $F = F(\mathcal{I}, f_1, f_2)$ such that

(i) $\mathrm{MF}_p \vdash (Cl(\mathcal{I}) \wedge \mathcal{I} \subseteq CL \wedge f_1 : \mathsf{N} \to \mathsf{N} \wedge f_2 : \mathsf{N} \to \mathcal{I}) \to$

$\to \forall n \eta \mathsf{N}. (Cl(Fn) \wedge W(\mathcal{I}, f_1, f_2, Fn, n))$;

(ii) for each $n \in \omega$,

$\mathrm{MF}_c \vdash (Cl(\mathcal{I}) \wedge \mathcal{I} \subseteq CL \wedge f_1 : \mathsf{N} \to \mathsf{N} \wedge f_2 : \mathsf{N} \to \mathcal{I}) \to$

$\to (Cl(F\bar{n}) \wedge W(\mathcal{I}, f_1, f_2, F\bar{n}, \bar{n}))$.

Verification of 44.3.2 (i). Let $C_0(f_1, f_2)$, $C_1(f_1, f_2)$ be the terms (in the given order):

$\{x : \exists u \exists v (x = \langle iden(u, v), f_1, f_2 \rangle \wedge Ter(u) \wedge Ter(v) \wedge val(u, f_1) = val(v, f_1))\}$;

$\{x: \exists u \exists i(N(i) \wedge Ter(u) \wedge x = \langle ins(u, var_2(i)), f_1, f_2\rangle \wedge val(u, f_1)\eta(f_2 i))\}$.

By recursion on $\mathbb{N}$, we define, uniformly in $\mathcal{S}$, f_1, f_2:

$F0 = C_0(f_1, f_2) \cup C_1(f_1, f_2)$;

$F(n+1) = Fn \cup \{\langle GD(A), f_1, f_2\rangle : A$ has logical complexity exactly $n+1$ and

1) $A = \neg B$ and $\neg\langle GD(B), f_1, f_2\rangle\eta(Fn)$; or

2) $A = B_1 \wedge B_2$ and $\langle GD(B_1), f_1, f_2\rangle\eta(Fn)$ and $\langle GD(B_2), f_1, f_2\rangle\eta(Fn)$; or

3) $A = \forall x_i B$ and $\forall k\eta\mathbb{N}. \langle GD(B), f_1(^i_k), f_2\rangle\eta(Fn)$; or

4) $A = \forall X_i B$ and $\forall P\eta\mathcal{S}. \langle GD(B), f_1, f_2(^i_P)\rangle\eta(Fn)$.

Then we easily have, by hypothesis on $\mathcal{S}$, f_1, f_2:

$$Cl(F\bar{0}) \wedge W(\mathcal{S}, f_1, f_2, F\bar{0}, \bar{0}).$$

On the other hand, since the formula which defines $F(n+1)$ is elementary extensional in $\mathcal{S}$ and Fn, then $F(n+1)$ is a class and satisfies $W(\mathcal{S}, f_1, f_2, -, n+1)$ (argue by N-induction on properties and elementary comprehension).
As to the second part of 44.3.2, one applies metamathematical induction on natural numbers instead of property N-induction. □

Let $Kom: \mathbb{N} \to \mathbb{N}$ be the (primitive recursive) operation, which assigns to each element of For_2 its logical complexity. Then we define:

44.3.3

(i) $SAT^n[\mathcal{S}, f_1, f_2] = \{x: For_2(x) \wedge \langle x, f_1, f_2\rangle\eta Fn\}$;

$SAT[\mathcal{S}, f_1, f_2] = \{x: For_2(x) \wedge \langle x, f_1, f_2\rangle\eta F(Kom(x))\}$.

(ii) By induction on A (A arbitrary $\mathfrak{L}_2$-formula):

$(t \in X_i)[\mathcal{S}, f_1, f_2] = val(t, f_1)\eta(f_2 i)$;

$(t = s)[\mathcal{S}, f_1, f_2] = (val(t, f_1) = val(s, f_1))$;

$(\neg B)[\mathcal{S}, f_1, f_2] = \neg((B)[\mathcal{S}, f_1, f_2])$;

$(B \wedge C)[\mathcal{S}, f_1, f_2] = (B[\mathcal{S}, f_1, f_2] \wedge C[\mathcal{S}, f_1, f_2])$;

$(\forall x_i B)[\mathcal{S}, f_1, f_2] = \forall k\eta\mathbb{N}. (B[\mathcal{S}, f_1(^i_k), f_2])$;

$(\forall X_i B)[\mathcal{S}, f_1, f_2] = \forall P(P\eta\mathcal{S} \to B[\mathcal{S}, f_1, f_2(^i_P)])$.

Then we obtain:

44.3.4

(i) $\mathrm{MF}_p \vdash (Cl(\mathcal{S}) \wedge \mathcal{S} \subseteq CL \wedge f_1 : \mathbb{N} \to \mathbb{N} \wedge f_2 : \mathbb{N} \to \mathcal{S}) \to$

$\to \forall n \eta \mathbb{N}. Cl(SAT^n[\mathcal{S}, f_1, f_2]) \wedge Cl(SAT[\mathcal{S}, f_1, f_2])$.

Moreover $\mathrm{MF}_c \vdash (Cl(\mathcal{S}) \wedge \mathcal{S} \subseteq CL \wedge f_1 : \mathbb{N} \to \mathbb{N} \wedge f_2 : \mathbb{N} \to \mathcal{S}) \to$

$\to Cl(SAT^n[\mathcal{S}, f_1, f_2])$, for each $n \in \omega$.

(ii) For every $A \in \mathcal{L}_2$ of complexity n (n fixed),

$\mathrm{MF}_c \vdash (Cl(\mathcal{S}) \wedge \mathcal{S} \subseteq CL \wedge f_1 : \mathbb{N} \to \mathbb{N} \wedge f_2 : \mathbb{N} \to \mathcal{S}) \to$

$\to (GD(A) \eta SAT^n[\mathcal{S}, f_1, f_2] \leftrightarrow A[\mathcal{S}, f_1, f_2])$;

(iii) $\mathrm{MF}_p \vdash [Cl(\mathcal{S}) \wedge \mathcal{S} \subseteq CL \wedge f_1 : \mathbb{N} \to \mathbb{N} \wedge f_2 : \mathbb{N} \to \mathcal{S} \wedge f_3 : \mathbb{N} \to \mathcal{S} \wedge$
$\wedge \forall k \eta \mathbb{N}. (f_2 k =_e f_3 k)] \to \forall x (x \eta SAT[\mathcal{S}, f_1, f_2] \leftrightarrow x \eta SAT[\mathcal{S}, f_1, f_3])$.

The previous statement 44.3.4 (iii) can be proved in MF_c if we replace SAT by SAT^n for fixed $n \in \omega$.

Verification of 44.3.4. (i) holds by 44.3.2 (i), while (ii) follows by outer induction on the complexity of A. As to (iii), proceed by property induction on $Kom(x)$, while the second part applies the second part of 44.3.4 (i). □

Of course, we can define the operations:

$$ext^{\mathcal{S}}_n[x, i, f_1, f_2] = \{k : k \eta \mathbb{N} \wedge x \eta SAT^n[\mathcal{S}, f_1(^i_k), f_2]\}.$$

$$Def^n(\mathcal{S}) =$$

$$= \{b : (\exists x)(\exists f_1 : \mathbb{N} \to \mathbb{N})(\exists f_2 : \mathbb{N} \to \mathcal{S})(\exists i)(For_2(x) \wedge b = ext^{\mathcal{S}}_n[x, i, f_1, f_2])\}.$$

$$Def(\mathcal{S}) = \{b : \exists n (b \eta Def^n(\mathcal{S}))\}.$$

44.3.5. Then we have in MF_p:

(i) if $\mathcal{S} \subseteq CL$ and $Cl(\mathcal{S})$, then $Cl(Def(\mathcal{S}))$ and $Def(\mathcal{S}) \subseteq CL$;

(ii) if $\mathcal{S} =^+_e \mathcal{U}$ and $\mathcal{S}, \mathcal{U} \subseteq CL$ are classes, then $Def(\mathcal{S}) =^+_e Def(\mathcal{U})$;

(iii) If Def^n, for n fixed, replaces Def in (i)-(ii) above, then the corresponding statements are provable in MF_c.

Verification of 44.3.5.
(i) follows by 44.3.4 (i). As to the subsequent point (ii), let $g : \mathcal{S} \to \mathcal{U}$, $h : \mathcal{U} \to \mathcal{S}$ be a pair of maps witnessing $\mathcal{S} =^+_e \mathcal{U}$. Then we can introduce two operations $g\hat{} : Def(\mathcal{S}) \to Def(\mathcal{U})$ and $h\hat{} : Def(\mathcal{U}) \to Def(\mathcal{S})$, such that

$$g\hat{}b = ext^{\mathcal{U}}_n[u, i, f_1, gf_2] \text{ and } h\hat{}c = ext^{\mathcal{S}}_n[v, i, f_1, hf_3]. \qquad (*)$$

In (*), $b = ext^{\mathcal{S}}_n[u, i, f_1, f_2] \eta Def(\mathcal{S})$, $c \eta Def(\mathcal{U})$, u, v are in For_2, i is the index of the individual variable free in u, v, $f_1 : \mathbb{N} \to \mathbb{N}$, $f_2 : \mathbb{N} \to \mathfrak{S}$, $f_3 : \mathbb{N} \to \mathfrak{U}$ and $(gf_2)j = g(f_2 j)$, $(hf_3)j = h(f_3 j)$ (where j is in $\mathbb{N}$). That $g\hat{}$, $h\hat{}$ witness $Def(\mathfrak{S}) =_e^+ Def(\mathfrak{U})$ is checked by class N-induction on the complexity of the formula (encoded by) u (or v), using 44.3.4 (iii). As to (iii), the proof is similar. □

By 44.3.5 and the assumption on $\prec_w$, we are under the hypothesis of the special transfinite recursion theorem of 43.4: hence there exists an operation $\lambda x.\mathfrak{R}_x$ satisfying the equations of 44.3 (i). It only remains to check:

$$\mathfrak{R}_x \subseteq_e^+ \mathfrak{R}_y, \text{ for } x \prec_w y.$$

Let

$$I_{xy} b = \{k : k\eta\mathbb{N} \wedge GD(v_0 \in X_0)\eta SAT[\mathfrak{R}_x, f_1(\begin{smallmatrix}0\\k\end{smallmatrix}), f_2(\begin{smallmatrix}0\\b\end{smallmatrix})]\}.$$

Clearly I_{xy} embeds $\mathfrak{R}_x$ in $\mathfrak{R}_y$, provided y is the $\prec_w$-successor of $x\eta Field(\prec_w)$. Then we recursively define:

$$\text{if } gy = \overline{0} \text{ or } gy = \overline{2},\ \phi_{xy} b = b;\ \text{ else, if } gy = \overline{1},\ \phi_{xy} b = I_{(hy)y}(\phi_{x(hy)} b).$$

By $\prec_w$-induction for classes, we obtain

$$(\forall y \eta Field(\prec_w))(\forall x \prec_w y) Embed(\phi_{xy}, \mathfrak{R}_x, \mathfrak{R}_y). \ \square$$

We next consider a refinement of 44.3, where we deal with a hierarchy whose defining conditions have an *a priori* bounded logical complexity.

44.4. COROLLARY (*Existence of the bounded ramified hierarchy, provably in* MF_c). *Let k be a natural number. Then:*

(i) *for every acceptable unbounded locally decidable qwo $\prec_w$, there is an operation $\lambda x.\mathfrak{R}_x^k$ such that:*

1. *if x is in the field of $\prec_w$, $\mathfrak{R}_x^k$ is a class of classes $\subseteq \mathbb{N}$;*

1.1 $\mathfrak{R}_x^k = Def^k(\{\mathbb{N}\})$, *whenever x is the first element of $\prec_w$;*

1.2 $\mathfrak{R}_x^k = Def^k(\mathfrak{R}_y^k) \cup \mathfrak{R}_y^k$, *if x is the $\prec_w$-successor of y;*

1.3 $\mathfrak{R}_x^k = \{v : \exists y \prec_w x.\, v\eta\mathfrak{R}_y^k\}$, *if x is a $\prec_w$-limit;*

2. $\mathfrak{R}_x^k \subseteq_e^+ \mathfrak{R}_y^k$, *whenever $x \prec_w y$.*

(ii) *The same holds, provably in* TLR, *if we replace qwo with pwo in* (i) *above.*

PROOF. (i) Simply replace Def by Def^k in the definition of $\mathfrak{R}$ and observe that by 44.3.4 – 44.3.5 (iii) we can proceed by N-induction on classes, once we have a fixed bounded logical complexity.
(ii): similar to the proof of WP in § 38. Fix a natural number k and choose

a level j such that $Field(\prec_w)$ is a class of level j. By 37.8 (vi), there exists a level i, such that $CL_j \eta CL_i$, $i \succ j$ ($\succ$ is the level ordering !) and hence

$$d = \{x : x\eta_j Field(\prec_w) \wedge Cl_j(\mathcal{R}_x^k)\}$$

is an i-class. By 44.3.5 and lemma 37.2, CL_j is provably closed under Def^k, i.e. TLR proves

$$\mathcal{I} \subseteq CL_j \wedge Cl_j(\mathcal{I}) \rightarrow Def^k(\mathcal{I}) \subseteq CL_j \wedge Cl_j(Def^k(\mathcal{I})).$$

Hence d is $\prec_w$-progressive and, being an i-class, is also a class. So the conclusion follows by class induction on $\prec_w$. □

44.5. *Concluding Remarks*

(i) *Relativization.* The construction of $\mathcal{R}$ is easily generalized to classes other than $\mathbb{N}$. Given any fixed subclass P of $\mathbb{N}$, we can introduce the hierarchy $\lambda x.\mathcal{R}_x(P)$. The definition is analogous, except that $\mathcal{L}_2$ is expanded with a new predicate, whose interpretation is P itself, and we put $\mathcal{R}_x(P) = Def(\{P\})$, if x is the first element of the given qwo. It is also true that if $P =_e Q$, then $\mathcal{R}_x(P) =_e^+ \mathcal{R}_x(Q)$.

(ii) *Closure of limit segments under comprehension.*
Let a be in $\mathcal{R}_y$ for $y \prec_w x$, where x is a limit of $Field(\prec_w)$. Assume that A is a $\mathcal{L}_2$-formula with the free variables X_0, v_0 and let

$$A(k, a) = A[\mathcal{R}_y, f_1(\tbinom{0}{k}), f_2(\tbinom{0}{a})] \text{ (see 44.3.3)},$$

where $f_1 : \mathbb{N} \rightarrow \mathbb{N}$, $f_2 : \mathbb{N} \rightarrow \mathcal{R}_y$. Then we can find a class $b =_e \{k : A(k, a)\}$ such that $b\eta\mathcal{R}_x$ (apply 44.3.5 (ii)); in such a case we simply say that $\{k : A(k, a)\}$ is in $\mathcal{R}_x$ (this is not misleading, if we are working in contexts, which extensionally depend on $\{k : A(k, a)\}$). Of course, the closure property above also holds for limit segments of the bounded ramified hierarchy, as soon as we deal with A's of the appropriate logical complexity.

(iii) $\mathcal{R}$ *only depends on the isomorphism type of the given qwo.*
Indeed, let $\prec_1$, $\prec_2$ be qwos, which are acceptable and locally decidable. $\prec_1$ is isomorphic to $\prec_2$ (in short $\prec_1 \simeq \prec_2$) iff there are operations $f_1 : Field(\prec_1) \rightarrow Field(\prec_2)$, $f_2 : Field(\prec_2) \rightarrow Field(\prec_1)$, such that:

1. $f_2(f_1 x) = x$ and $f_1(f_2 y) = y$, whenever x belongs to $Field(\prec_2)$ and y belongs to $Field(\prec_2)$;
2. $u \prec_1 v$ implies $f_1 u \prec_2 f_1 v$ and $x \prec_2 y$ implies $f_2 x \prec_1 f_2 y$, for every u, v in $Field(\prec_1)$ and x, y in $Field(\prec_2)$.

Then we can easily check, by induction and using 44.3.5 above:
if $\prec_1 \simeq \prec_2$ via g, h, $\mathcal{R}(1)_x =_e^+ \mathcal{R}(2)_{gx}$ and $\mathcal{R}(2)_y =_e^+ \mathcal{R}(1)_{hy}$, where $\mathcal{R}(i)$ is the ramified hierarchy along $\prec_i$ $(i = 1, 2)$ and $x\eta Field(\prec_1)$, $y\eta Field(\prec_2)$.

§ 45. Predicative well-orderings I

The aim of §§ 45-46 is to introduce a specific primitive recursive ordering $\prec_w$ of $\mathbb{N}$ and to show that $\prec_w$ is isomorphic to a certain ordinal, which is known in the literature as Γ_0. Γ_0 is intimately connected with the theory TLR of § 36, and it is essential for the proof-theoretic investigation of the formal systems, outlined in this book.

In the foundational literature there are classical and exhaustive expositions of this matter (in particular Schütte 1977); so we shall only state the basic facts, needed for describing a notation system for Γ_0. Proofs are mostly omitted or simply outlined. However, § 46 will contain a detailed well-ordering proof for Γ_0 in a fragment of TLR.

45 A. *Preliminaries: Veblen hierarchies* (*Veblen* 1908)

In this subsection α, β, γ, δ, ξ range over elements of Ω, the set of countable ordinals; $<$ is the natural ordering of Ω.

45.1. DEFINITION

(i) $f:\Omega\rightarrow\Omega$ is *increasing* iff $\alpha<\beta$ implies $f(\alpha)<f(\beta)$, for every α, β. In particular, if f is increasing, $f(\alpha)\geq\alpha$ holds for arbitrary α.

(ii) $f:\Omega\rightarrow\Omega$ is *continuous* iff for every limit ordinal λ, $f(\lambda)=\sup_{\beta<\lambda} f(\beta)$. If f is increasing and continuous, f is called *normal.*

(iii) If $Y\subseteq X\subseteq\Omega$, Y is *bounded* in X iff there is a $\gamma\in X$ such that, for all $\beta\in Y$, $\beta<\gamma$.

(iv) $X\subseteq\Omega$ is *closed* iff $(\sup Y)\in X$, for every $Y\subseteq X$, which is bounded in X.

(v) $X\subseteq\Omega$ is *unbounded* iff for every α there is a $\beta\in X$ such that $\alpha<\beta$. If $X\subseteq\Omega$ is closed and unbounded, we say for short that X is a *club.*

(vi) Let $X\subseteq\Omega$: by transfinite recursion we may define

$$E_X(\alpha) = \text{the least } \delta\in X \text{ such that for every } \beta<\alpha,\ E_X(\beta)<\delta.$$

E_X is defined on a segment of Ω and is called *the canonical enumeration* of X. The following properties hold:

P.1. $X\subseteq\Omega$ is a club iff E_X is normal. Conversely, if f is increasing, f is normal iff $f[\Omega]$ is a club.

P.2. Let $f:\Omega\rightarrow\Omega$ be normal and consider $fix(f)=\{\alpha : f(\alpha)=\alpha\}$. Then $fix(f)$ is a club. As usual, we write f' for $E_{fix(f)}$ and f' is called *the derivative of f*. f' can be given a recursive definition, which displays the

iterative nature of f'.
Set $f^0(\alpha)=\alpha$; $f^{k+1}(\alpha)=f(f^k(\alpha))$. By transfinite recursion, we define:

$$f'(0)=\sup_{n\in\omega} f^n(f(0)+1);$$
$$f'(\alpha+1)=\sup_{n\in\omega} f^n(f'(\alpha)+1);$$
$$f'(\lambda)=\sup_{\beta<\lambda} f'(\beta).$$

P.3. Let $\{X_\alpha\}_{\alpha<\delta}$ be a family of clubs such that, if $\beta<\gamma<\delta$, $X_\gamma\subseteq X_\beta$. Then $\bigcap_{\alpha<\delta} X_\alpha$ is a club.

P.1 – P.3 justify the introduction, again by transfinite recursion of a hierarchy of normal functions, uniformly in any given normal F.

45.2. *The Veblen hierarchy starting with F.*

$$V_F^0 = F;$$
$$V_F^{(\beta+1)} = (V_F^\beta)';$$
if λ is a limit, $V_F^\lambda = E_X$, where $X=\bigcap_{\beta<\lambda}(V_F^\beta)[\Omega]$.

Here $V_F^\beta(\alpha)$ stands for the value of V_F^β applied to α. By construction, if $\gamma<\delta$, we have

45.2.1. $$V_F^\gamma(V_F^\delta(\alpha))=V_F^\delta(\alpha).$$

We write $g(\beta,\alpha)$ for $V_F^\beta(\alpha)$, whenever F is clear from the context; in the special case, where $F=\lambda\alpha.\omega^\alpha$, we use, following Schütte, $\phi\beta\alpha$ for $g(\beta,\alpha)$. $\phi 1\alpha$ is sometimes replaced by the more familiar notation ϵ_α; so, as a special case, we have $\epsilon_0=\phi 10=$ the least solution of $\omega^\alpha=\alpha$.

P.4. For all α_1, β_1, α_2, β_2 and $R := <, =$, we have

$g(\alpha_1,\beta_1)\ R\ g(\alpha_2,\beta_2)$ iff

$\alpha_1<\alpha_2$ and $\beta_1 R\ g(\alpha_2,\beta_2)$, or

$\alpha_1=\alpha_2$ and $\beta_1 R\ \beta_2$, or

$\alpha_2<\alpha_1$ and $g(\alpha_1,\beta_1)\ R\ \beta_2$.

N.B. P.4 does not depend on the initial normal function F.

Besides that, we have:

P.5. The function $g(\alpha)=g(\alpha,0)$ is normal.

45.2.2. DEFINITION: Γ_α is the α-th solution of $\phi\beta 0=\beta$.

From P.4, P.5, using the definition of $f'(0)$ for $f = \lambda\beta.\phi\beta 0$, it follows that Γ_0 is the least ordinal > 0, which is closed under $\lambda\alpha\lambda\beta.\phi\alpha\beta$.

45 B. *Normal forms. Notations*

The problem of constructing a notation system for a given countable ordinal α is essentially algebraic: one has to find out a system of finitary operations, which unambiguously represent the ordinals $< \alpha$, starting with an effective set of generators. In this section, *we face the problem for the predicative ordinals from an elementary point of view* (for the category-theoretic methods of Π^1_2-logic, we send the reader to Girard 1982).

We first state the Cantor normal form theorem (for a proof, Schütte 1977):

45.3. THEOREM. *For every ordinal* $\gamma \neq 0$, *there are unique* $\gamma_1 \geq \ldots \geq \gamma_n$, $n \geq 1$, *such that* $\gamma = \omega^{\gamma_1} + \ldots + \omega^{\gamma_n}$.

Then we apply 45.A and extend theorem 45.3.

45.4. LEMMA. *For every* $\alpha \in (0, \Gamma_0)$ *of the form* ω^δ, *there are unique* α_1, α_2 *such that* $\alpha = \phi\alpha_1\alpha_2$ *and* $\alpha > \alpha_1, \alpha_2$.

PROOF. The set $X = \{\gamma : \gamma \leq \alpha \wedge \alpha < \phi\gamma\alpha\}$ is non-empty ($\alpha < \phi\alpha\alpha$ by P.4 and P.5). Hence we can pick out its minimum α_1. If $\alpha_1 = 0$, choose $\alpha_2 = \delta$; since $\alpha < \phi 0\alpha$, $\alpha_2 = \delta < \phi 0\delta$. If $\alpha_1 \neq 0$, we get: $(\forall\gamma < \alpha_1)(\phi\gamma\alpha = \alpha)$, i.e. $\alpha = \phi\alpha_1\xi$, for some ξ. By choice of α_1, $\phi\alpha_1\xi < \phi\alpha_1(\phi\alpha_1\xi)$, i.e. $\xi < \phi\alpha_1\xi$ (P.4). On the other hand, since $\alpha_1 \leq \alpha < \Gamma_0$, $\phi\alpha_1\xi \geq \phi\alpha_1 0 > \alpha_1$.
As to the uniqueness, assume $\alpha = \phi\alpha_1\alpha_2 = \phi\beta_1\beta_2$ and $\phi\alpha_1\alpha_2 > \alpha_2, \alpha_1$, $\phi\beta_1\beta_2 > \beta_1, \beta_2$. By P.4, we must have $\alpha_1 = \beta_1$ and hence $\beta_2 = \alpha_2$. □

45.5. COROLLARY (*Extended normal form for ordinals* $< \Gamma_0$). *For every ordinal* $\alpha \in (0, \Gamma_0)$, *there exist unique* $\alpha_1, \beta_1, \ldots, \alpha_n, \beta_n$, $n \geq 1$, *such that*

(i) $\phi\alpha_i\beta_i > \beta_i, \alpha_i$ $(i \in [1, n])$;

(ii) $\phi\alpha_1\beta_1 \geq \ldots \geq \phi\alpha_n\beta_n$;

(iii) $\alpha = \phi\alpha_1\beta_1 + \ldots + \phi\alpha_n\beta_n$.

PROOF: apply 45.4 to the Cantor normal form. □

Below, we temporarily adopt the symbol $=$ also for literal identity ($t = s$ means that t and s coincide).

45.6. *A system of ordinal notations.* We consider a term-language which contains:

(i) a new constant $\boldsymbol{c}$.

(ii) a concatenation operation $*$ and a binary function symbol $\boldsymbol{f}$.

The set TER of terms is inductively generated from $\boldsymbol{c}$ by closing off under the clause:
if $t_1, s_1, \ldots, t_n, s_n$ $(n \geq 1)$ are terms, so is $\boldsymbol{f}(t_1, s_1) * \ldots * \boldsymbol{f}(t_n, s_n)$.

The degree of $t \in TER$ is recursively assigned:

$$gr(\boldsymbol{c}) = 0;$$
$$gr(\boldsymbol{f}(t_1, s_1)) = gr(t_1) + gr(s_1) + 1 \text{ and}$$
$$gr(t_1 * \ldots * t_n) = gr(t_1) + \ \ldots \ + gr(t_n) \ (n > 1).$$

We recursively define the character $cr(t) \in TER$, for each $t \in TER$:

$$cr(\boldsymbol{c}) = \boldsymbol{c}; \ \text{ if } t = \boldsymbol{f}(t_1, s_1) * \ldots * \boldsymbol{f}(t_n, s_n), \ \text{ let } cr(t) = \begin{cases} \boldsymbol{c} \text{ if } n > 1; \\ t_1 \text{ if } n = 1. \end{cases}$$

We now introduce a structure $\mathfrak{CN} = \langle CN, \dot{<}, \boldsymbol{c} \rangle$, where $CN \subseteq TER$, $\dot{<} \subseteq ON \times ON$. We write $t \dot{\leq} s$ for $t \dot{<} s$ or $t = s$, and $t \dot{\geq} s$ for $s \dot{\leq} t$.

45.6.1. *Simultaneous recursive definition of* CN, $\dot{<}$:

(i) $t \in CN$ iff either $t = \boldsymbol{c}$, or
$t = \boldsymbol{f}(t_1, s_1) * \ldots * \boldsymbol{f}(t_n, s_n)$, for some $t_1, s_1, \ldots, t_n, s_n \in CN$,
such that $\boldsymbol{f}(t_1, s_1) \dot{\geq} \ldots \dot{\geq} \boldsymbol{f}(t_n, s_n)$ and $cr(s_i) \dot{\leq} t_i$, for $i = 1, \ldots, n$.

(ii) $t \dot{<} s$ holds iff $s \neq \boldsymbol{c}$ and one of the following cases holds:

1. $t = \boldsymbol{c}$, or
2. $t = t_1 * \ldots * t_n$, $s = s_1 * \ldots * s_m$, $t, s \in CN$, $n{+}m > 2$ and

2.1. $n < m$ and $t_i = s_i$, for all $i \in [1, n]$, or

2.2. there is a $j \leq n, m$, such that $t_j \dot{<} s_j$ and $t_i = s_i$, for all $1 \leq i < j$, or

3. $t = \boldsymbol{f}(t_1, s_1)$, $s = \boldsymbol{f}(t_2, s_2)$, $t, s \in CN$ and

3.1. $t_1 = t_2$ and $s_1 \dot{<} s_2$, or

3.2. $t_2 \dot{<} t_1$ and $t \dot{<} s_2$, or

3.3. $t_1 \dot{<} t_2$ and $s_1 \dot{<} s$.

Clearly 45.6.1 can be reduced to a standard course-of-value recursion, and it is straightforward to check:

45.7. LEMMA. *CN, $\dot{<}$ are primitive recursive.*

Then we have:

45.8. LEMMA. (i) *CN is linearly ordered by* $\dot{<}$.
(ii) *if* $t, s \in CN$, $cr(s) \dot{\leq} t$ *iff* $s \dot{<} f(t,s)$;
(iii) *if* $f(t,s) \in CN$, *then* $s \dot{<} f(t,s)$ *and* $t \dot{\leq} f(t,s)$;
(iv) *if* $t, s \in CN$ *and* $t \dot{<} cr(s)$, *then* $t \dot{<} s$.

PROOF (Hint). The proof of (i) entirely relies on the recursive definitions of 45.6.1 and is carried out by appropriate induction on the (sum of the) degrees of the terms involved (details can easily be supplied by the reader or adapted from Schütte 1977, Feferman 1968). (iii) easily follows from (i)-(ii).
Verification of (ii). $\Rightarrow$: by induction on $gr(s)$. If $s = c$, the claim is trivial. Let $s = f(t_1, s_1) \in CN$ and $cr(s) = t_1 \dot{\leq} t$. As $s \in CN$, we have $t_1, s_1 \in CN$, $cr(s_1) \dot{\leq} t_1$ and hence also $cr(s_1) \dot{\leq} t$. So, by IH, since $gr(s_1) < gr(s)$,

$$s_1 \dot{<} f(t_1, s_1); \tag{1a}$$

$$s_1 \dot{<} f(t, s_1). \tag{1b}$$

If $t = t_1$, (1a) suffices for $s \dot{<} f(t,s)$. Let $t_1 \dot{<} t$: then (1a)-(1b) and $\dot{<}$-transitivity yield $s_1 \dot{<} f(t,s)$, whence $s \dot{<} f(t,s)$.
If $s = f(t_1, s_1) * r \in CN$, then $f(t_1, s_1) \in CN$; so $t_1, s_1 \in CN$ and $cr(s_1) \dot{\leq} t_1$. As in the previous case, (1a) holds by IH. We then apply $\dot{<}$-linearity. If $t_1 = t$, we have $s_1 \dot{<} f(t_1, s_1) \dot{<} f(t_1, s_1) * r$, whence $s \dot{<} f(t,s)$. If $t_1 \dot{<} t$, also (1b) above holds and we can repeat the same argument. If $t \dot{<} t_1$, we have by definition of $\dot{<}$, $f(t_1, s_1) \dot{<} f(t_1, s_1) * r$, and hence $s \dot{<} f(t,s)$.
$\Leftarrow$: if $s = c$ or $s = t_1 * \ldots * t_n$ with $n > 1$, the claim is trivial. Assume by contradiction $s = f(t_1, s_1)$, $s \dot{<} f(t,s)$ and $t \dot{<} cr(s) = t_1$. Then by 45.6.1 (ii) $s \dot{<} s$: against (i).
Verification of (iv). $t \dot{<} cr(s)$ implies $s = f(t_1, s_1)$; so, by definition of $cr(s)$ and (iii), $t \dot{<} t_1 \dot{<} f(t_1, s_1) = s$. □

The proof of 45.8 is formalizable in a a weak fragment of OP.

45.9. DEFINITION. By recursion on $gr(t)$ we define a map $F: CN \rightarrow \Gamma_0$, such that:

$$F(c) = 0;$$

$$F(f(t_1, s_1)) * \ldots * f(t_n, s_n)) = \phi F(t_1) F(s_n) + \ldots + \phi F(t_n) F(s_n).$$

45.10. LEMMA. *If* t, $s \in CN$, *then* $t \dot{<} s$ *implies* $F(t) < F(s)$.

PROOF: by induction on the definition of $\dot{<}$ with P.4. □

45.11. COROLLARY. $\dot{<}$ *is well-founded on* CN.

PROOF: any strictly descending sequence in CN $t_0 \dot{>} t_1 \dot{>} t_2 \ldots$ becomes (via F) a strictly descending sequence in Γ_0. □

45.12. THEOREM. *F is an isomorphism between* $\mathfrak{CN}$ *and* $\Gamma = \langle \Gamma_0, <, 0 \rangle$.

PROOF: F is already $\dot{<}$-preserving by 45.10. If t, $s \in CN$ and $t \neq s$, we have $t \dot{<} s$ or $s \dot{<} t$ (45.8); hence, by 45.10, either $F(t) < F(s)$ or $F(s) < F(t)$, i.e. F is injective. But F is surjective, i.e., for every $\alpha < \Gamma_0$ there is a term $t \in CN$ such that $F(t) = \alpha$. We argue by induction on α. If $\alpha = 0$, put $t = \boldsymbol{c}$. If $\alpha \neq 0$, let by lemma 45.4, α_1, $\beta_1, \ldots,$ α_n, β_n such that $\phi\alpha_i\beta_i > \beta_i$ $(1 \leq i \leq n)$, $\alpha > \phi\alpha_1\beta_1 \geq \ldots \geq \phi\alpha_n\beta_n$ and

$$\alpha = \phi\alpha_1\beta_1 + \ldots + \phi\alpha_n\beta_n.$$

By IH, there are t_1, $s_1, \ldots,$ t_n, s_n in CN with $F(t_i) = \alpha_i$, $F(s_i) = \beta_i$ for every $1 \leq i \leq n$. Since F is $<$-preserving and $\dot{<}$ is linear, $\boldsymbol{f}(t_i, s_i) \dot{>} s_i$ for every $1 \leq i \leq n$ and $\boldsymbol{f}(t_1, s_1) \dot{\geq} \ldots \dot{\geq} \boldsymbol{f}(t_n, s_n)$. Hence by 45.8 (ii) we have $cr(s_i) \dot{\leq} t_i$ for $i = 1, \ldots, n$ and we can choose

$$t = \boldsymbol{f}(t_1, s_1) * \ldots * \boldsymbol{f}(t_n, s_n). \qquad \square$$

We do not develop ordinal arithmetic in $\mathfrak{CN}$. It is possible to represent the standard operations of ordinal sum, product, exponentiation and the ϕ-function, as primitive recursive operations on the terms of CN. As a by-product, the elementary properties, which are usually proven by transfinite induction, become provable by N-induction. As samples, we present the definitions of ordinal sum, successor, ϕ-function, ω-exponential.

45.13. DEFINITION. Let $t, s \in CN$.

(i) $t + \boldsymbol{c} = t = \boldsymbol{c} + t$.

Let $t = t_1 * \ldots * t_m$, $s = s_1 * \ldots * s_n$ $(m + n \geq 2)$, where $t_1, \ldots, t_m$, $s_1, \ldots, s_n$ are principal terms, i.e. of the form $\boldsymbol{f}(a, b)$; then

$$t + s = \begin{cases} s & \text{if } t_1 \dot{<} s_1; \\ t_1 * \ldots * t_j * s_1 * \ldots * s_n, & \text{for } j = \text{the largest } j \text{ such that } t_j \dot{\geq} s_1. \end{cases}$$

(ii) The successor of t: $t + 1 = t + \boldsymbol{f}(\boldsymbol{c}, \boldsymbol{c})$.

(iii) The internal ϕ-function $\dot{\boldsymbol{f}} : CN \times CN \to CN$:

$$\dot{\boldsymbol{f}}(t, s) = \begin{cases} \boldsymbol{f}(t, s), & \text{if } cr(s) \dot{\leq} t; \\ s & \text{otherwise}; \end{cases}$$

(iv) $\omega^t = \dot{\boldsymbol{f}}(c, t)$;

(v) $\omega^t \cdot 0 = \boldsymbol{c}$; $\omega^t(n+1) = \omega^t \cdot n + \omega^t$.

45.14. LEMMA. *If* $t,\ s \in CN$, *then* $t+s,\ t+1,\ \dot{f}(t,s),\ \omega^t,\ \omega^t \cdot n$ *are in* CN.

If we apply 45.8 (iii)-(iv) and the basic definitions, we readily see that $\dot{f}$ behaves like ϕ, and has a useful syntactical property:

45.15. LEMMA. *Let* $t,\ s,\ t_i,\ s_i \in CN\ (i = 1,2)$.

(i) $s \leq \dot{f}(t,s)$ *and* $t < \dot{f}(t,s)$;

(ii) $s < r \Rightarrow \dot{f}(t,s) < \dot{f}(t,r)$;

(iii) *if* R *is* $=$ *or* $\dot{<}$, $\dot{f}$ *and* R *satisfy the analogue of* P.4:

$\dot{f}(t_1,s_1)R\dot{f}(t_2,s_2)$ *iff* $t_1 = t_2$ *and* s_1Rs_2, *or* $t_2 \dot{<} t_1$ *and* tRs_2, *or* $t_1 \dot{<} t_2$ *and* s_1Rs.

(iv) $\dot{f}(t,r)+1 \dot{<} s \dot{<} \dot{f}(t,r+1) \Rightarrow gr(\dot{f}(t,r)+1) \dot{<} gr(s)$.

45.15 implies that the Cantor normal form theorem and its extended version (cf. 45.3-45.5) have natural counterparts for notations. So, in particular, by definition of CN and ω-exponential, we get:

45.16. LEMMA. *For every* $t \in CN$, *if* $t \neq \mathbf{c}$, *there are unique canonical terms* $t_1 \geq \ldots \geq t_n$, $n \geq 1$, *such that* $t = \omega^{t_1} + \ldots + \omega^{t_n}$.

45.17. DEFINITION. If $t, s \in CN$,

$$\omega^t \cdot s = \begin{cases} \mathbf{c}, \text{ if } s = \mathbf{c}; \\ \omega^{t+s_1} + \ldots + \omega^{t+s_n}, \text{ provided } s = \omega^{s_1} + \ldots + \omega^{s_n}, \\ \text{and } s_1 \dot{\geq} \ldots \dot{\geq} s_n. \end{cases}$$

Of course we have:

45.17.1. *If* $t,\ s \in CN$, *then* $\omega^t \cdot s \in CN$.

45 C. *Enriching the ordinal structure of* CN: *fundamental sequences*

From a constructive point of view, a (countable) limit ordinal λ is given only if we have an effective sequence $\langle \lambda[n] : n \in \omega \rangle$ such that

(i) $\lambda[n] < \lambda[n+1] < \lambda$, for each $n \in \omega$;

(ii) if $\beta < \lambda$, we can find – effectively in β and λ – a number k such that $\beta < \lambda\,[k]$.

We now assign to each limit term a primitive recursive sequence $\langle t[n] : n \in \omega \rangle$ of CN-terms, which converges to t. First, we need a few preliminary notions.

45.18. DEFINITION

(i) If $t \in CN$, t is a successor iff t has the form $s+1$, for $s \in CN$.

(ii) If $t \in CN$, $t \in Lim$ iff $t \not\equiv \mathbf{c}$ and t is not a successor.

(iii) We also define, by induction on $n \in \omega$:

$$\dot{\mathbf{f}}(0)(s,r) = r; \quad \dot{\mathbf{f}}(i+1)(s,r) = \dot{\mathbf{f}}(s, \dot{\mathbf{f}}(i)(s,r)).$$

45.19. DEFINITION OF $\langle t[n] : n \in \omega \rangle$ $(t \in Lim \cap CN)$. Such definition is carried out by main induction on $gr(t)$ and secondary induction on $n \in \omega$:

(i) $t = \mathbf{f}(\mathbf{c}, s+1)$: then $t[0] = \omega^s + 1$ and $t[n+1] = t[n] + \omega^s$;

(ii) $t = \mathbf{f}(s+1, r+1)$: $t[n] = \dot{\mathbf{f}}(n+1)(s, \dot{\mathbf{f}}(s+1, r)+1)$;

(iii) $t = \mathbf{f}(s, r+1)$ and $s \in Lim$: then $t[n] = \dot{\mathbf{f}}(s[n], \dot{\mathbf{f}}(s,r)+1)$;

(iv) $t = \mathbf{f}(s+1, \mathbf{c})$: $t[n] = \dot{\mathbf{f}}(n+1)(s, \mathbf{c})$;

(v) $t = \mathbf{f}(s, \mathbf{c})$, $s \in Lim$: $t[n] = \mathbf{f}(s[n], \mathbf{c})$;

(vi) $t = \mathbf{f}(s,r)$ and $r \in Lim$: $t[n] = \dot{\mathbf{f}}(s, r[n])$;

(vii) $t = s * r$ and $r = \mathbf{f}(r_1, r_2)$: $t[n] = s * r[n]$.

For technical reasons (cf. the proof of 45.20 (ii), 3), we set $(t+1)[n] = t$.

45.20. THEOREM

(i) *If $t \in CN \cap Lim$, then for every $n \in \omega$, $t[n] \dot{<} t[n+1] \dot{<} t$ and $t[n] \in CN$.*

(ii) *If $t \in CN \cap Lim$, then*
1. $gr(t) \leq gr(t[0]) + 1$;
2. $s \in CN,\ n \in \omega,\ t[n] \dot{<} s \dot{<} t \Rightarrow gr(t[n]) < gr(s)$;
3. $s \in CN,\ n \in \omega,\ t[n] \dot{<} s \dot{<} t \Rightarrow t[n] \leq s[0]$;
4. $s \in CN,\ s \dot{<} t \Rightarrow s \leq t[gr(s)]$.

PROOF (Hint). (i): by main induction on $gr(t)$ and secondary induction on $n \in \omega$.
(ii): 1 is easily checked by induction on $gr(t)$.
The verification of 2 runs by induction on $gr(s) + gr(t)$. The essential step is to check the claim whenever $t = \mathbf{f}(t_1, s_1)$ and $s = \mathbf{f}(t_2, s_2)$. There are three main cases to distinguish, according to $t_1 = t_2$, $t_1 \dot{<} t_2$ or $t_2 \dot{<} t_1$, together with several subcases, depending on the form of t and s; one repeatedly applies lemma 45.15. 1-2 imply 3-4, as noticed by Buchholz-Cichon-Weiermann (1993). Indeed, let $s \in CN$, $n \in \omega$, $t[n] \dot{<} s \dot{<} t$ and assume $s[0] \dot{<} t[n]$. Then, by 1-2, $gr(t[n]) < gr(s) \leq gr(s[0]) + 1 \leq gr(t[n])$:

contradiction. So we have 3. As to 4, 2 implies

$$gr(t[0]) < gr(t[1]) < gr(t[2])\ldots,$$

whence $n \leq gr(t[n])$ for all $n \in \omega$. In particular, $gr(s) \leq gr(t[gr(s)])$, which together with 2 yields $(s \dot{<} t \Rightarrow s \dot{<} t[gr(s)])$. □

45.20.1. REMARK. For a uniform approach to fundamental sequences and their relations with hierarchies of number-theoretic functions, the reader should consult the cited paper of Buchholz-Cichon-Weiermann. According to their terminology, 45.20 (ii) 1-2 states that the fundamental sequences for Γ_0 form a *normed Bachmann system*; by 3-4 the system is *regular* and has the so-called *nesting (or Bachmann) property*.

45.21. ENCODING. Since we aim at a constructive well-ordering proof of $\mathcal{CN}$ in the next section, it is convenient to fix an arithmetical copy of $\mathcal{CN}$ by standard Gödel numbering. So there are – within a fragment of OP – closed λ-terms, f_1, f_2 representing the number-theoretic characteristic functions of $\dot{<}$ and of CN^*, the set of Gödel numbers of CN-terms. To simplify the matter, we also assume that CN^* coincides with $\mathbb{N}$ (fix a primitive recursive bijection from $\mathbb{N}$ onto CN^*). It is immediate to prove:

45.22. LEMMA (MF_c). *$\dot{<}$ is acceptable (i.e. $\{\langle n,m\rangle : n \dot{<} m\}$ is a class) and locally decidable.*

We can find an operation distinguishing zero, limits and successors, because these notions correspond to syntactic properties of CN-terms. Similarly a predecessor operation on CN-terms can be produced.

45.23. CONVENTION. Henceforth we systematically identify Γ_0 with its arithmetical copy CN^*. In particular we use lower case Greek letters, ordinal predicates and operations within the language of MF_c, instead of the corresponding codified notions and notations. So, in particular, $\phi\alpha\beta$ will henceforth stand for $\dot{f}(t,s)$ where $t = \alpha$ and $s = \beta$; $\lambda \in LIM$ will formalize the property of being a limit term, etc.

§46. Predicative well-orderings II

While the elementary properties of the natural well-ordering of type Γ_0 can be dealt with in MF_c, the verification of well-foundedness for arbitrary segments of Γ_0 is non-elementary and essentially requires a transfinite iteration of elementary abstraction. To this aim, we single out a subsystem of TL, which is tailored for carrying out the well-ordering proof.

46.1. DEFINITION

(i) Fix a natural number p; we introduce the following abbreviations:

$$CL_N := \{x: x \subseteq \mathbb{N} \wedge Cl(x)\};$$

$$Good(\mathfrak{R}^p_\alpha(z)) := \forall\beta(\beta < \alpha \to Cl(\mathfrak{R}^p_\beta(z)) \wedge \mathfrak{R}^p_\beta(z) \subseteq CL_N);$$

$$TI(\beta, B) := Progr(<, B) \to \forall\alpha < \beta. B(\alpha);$$

$$I^U(\beta) := \forall b(b\eta U \to TI(\beta, b)).$$

$$\alpha_0 = \omega; \quad \alpha_{n+1} = \phi\alpha_n 0.$$

$\mathfrak{R}^p_\alpha(z)$ is the term constructed by corollary 44.4 and 44.5 (i); if $U = CL$, we write $PWO(\alpha)$ instead of $I^{CL}(\alpha)$.

(ii) $RAM(p, \alpha) := PWO(\alpha) \wedge Cl(z) \wedge z \subseteq \mathbb{N} \to Good(\mathfrak{R}^p_\alpha(z))$;

MFR(p, α) is the theory MF$_c$ + $RAM(p, \alpha)$;

MFR$(p) := \cup\{\text{MFR}(p, \alpha): \alpha < \Gamma_0\}$

We stick to convention 45.23; thus $\alpha < \beta$ is a shortening for the formula that says: α, β are codes for CN-terms ordered by $\dot{<}$, the term ordering of type Γ_0.

46.1.1. FACT. *For each* p, MFR$(p) \subseteq$ TLR.

PROOF: immediate by 45.22 and the corollary 44.4 (ii). □

Let $|\mathcal{T}|$ denote ϵ_0 ($\phi\omega 0$, Γ_0), if $\mathcal{T}$ is MF$_c$ (MF$_p$, MFR(p) with $p \geq 10$, respectively); we are going to prove:

46.2. THEOREM. *For each* $\alpha < |\mathcal{T}|$, $\mathcal{T} \vdash PWO(\alpha)$.

The core of the proof for $\mathcal{T} = \text{MFR}(p)$ hinges upon an idea of Feferman (1982) and is based on the fact that, if the ramified hierarchy is well-defined up to $\omega^{\alpha+1}$, then $I^U(\phi\alpha 0)$ holds for any class U of classes $\subseteq \mathbb{N}$, extending $\mathfrak{R}_{\omega^{\alpha+1}}(X)$, where X is an arbitrary class $\subseteq \mathbb{N}$. Actually, we obtain a sharper result: we show that only the bounded ramified hierarchy is required. We begin with two simple observations:

46.2.1 MF$_c \vdash I^U(0) \wedge (I^U(\alpha) \to (\forall\beta < \alpha) I^U(\beta))$;

MF$_c \vdash \lambda \in LIM \to (\forall n I^U(\lambda[n]) \to I^U(\lambda))$ (apply 45.20 (ii), item 4).

A basic step in the proof of the main theorem is the introduction, due to Schütte, of the *-operation, where

$$A^*(\alpha) = \forall\beta(\forall\delta < \beta. A(\delta) \to \forall\delta < \beta + \omega^\alpha. A(\delta)).$$

If $A(\alpha) = \alpha\eta a$, $a^* = \{x\colon A^*(x)\}$. Then we observe:

46.2.2. LEMMA
(i) $\mathrm{MF}_c \vdash Cl(a) \to Cl(a^*)$;
(ii) $Progr(<, A) \to Progr(<, A^*)$ *is provable in any subtheory of* MF, *which derives* N*-induction on formulas positive in* A *(hence* MF_c *proves* 46.2.2 *if* $\{x\colon A(x)\}$ *is a class).*

(i) follows by elementary comprehension. The verification of (ii) is immediate if $\alpha = 0$ or α is a limit. If $\alpha = \gamma + 1$, one applies the hypothesis and N-induction to $(\forall\delta < \beta + \omega^\gamma \cdot k)A(\delta)$ (k in $\mathbb{N}$).
If we define $Clos(U;*) := \forall x(x\eta U \to \exists y(y\eta U \wedge y =_e x^*)$ ($= U$ is $*$-closed), we have:

46.2.3. LEMMA. *The statements* (i)-(iii) *below are provable in* MF_c:

(i) $Cl(a) \to (TI(\alpha, a^*) \to TI(\omega^\alpha, a))$;

(ii) $PWO(\alpha) \to PWO(\omega^\alpha)$;

(iii) $(Cl(U) \wedge U \subseteq CL \wedge Clos(U;*)) \to ((I^U(\alpha) \to I^U(\omega^\alpha)) \wedge I^U(\epsilon_0))$;

(iv) $PWO(\omega_k)$, *where* $\omega_0 = 1$ *and* $\omega_{n+1} = \omega^{\omega_n}$;

(v) $\mathrm{MF}_p \vdash QWO(\omega^k)$, *for each* $k > 0$.

PROOF. (i) Assume $Progr(<, a)$, $TI(\alpha, a^*)$ and $a\eta CL$. By 46.2.2, a^* is a progressive class, whence $\forall\delta < \alpha.\delta\eta a^*$ (with $TI(\alpha, a^*)$) and $\alpha\eta a^*$ (again by $Progr(<, a^*)$. If we choose $\beta = 0$, we obtain $\forall\delta < \omega^\alpha.\delta\eta a$.
(ii): immediate by (i) and 46.2.2.
(iii) The first part is immediate by (i) and assumptions on U. As to $I^U(\epsilon_0)$, apply class N-induction to $\forall X\eta U.TI(\omega_n, X) = B(n)$; this is possible since U is a class of classes.
(iv): metamathematical induction on k with (ii) at the successor step and class-N-induction for $k = 0$.
(v): metamathematical induction on k. We apply $QWO(\omega)$ and the fact that if $\omega^k m \leq u < \omega^k(m+1)$, $u = \omega^k m + y$, for some $y < \omega^k$. □

It is convenient to adopt a stronger notion of progressiveness:

$$Pr^+(A) := A(0) \wedge \forall\beta(A(\beta) \to A(\beta+1)) \wedge \forall\lambda(Lim(\lambda) \wedge \forall n A(\lambda[n]). \to A(\lambda)).$$

$Pr^+(A) :=$ "A is strongly progressive". Clearly $Pr^+(A)$ implies $Pr(A)$; the reverse implication holds whenever $\forall\alpha(A(\alpha) \to (\forall\beta < \alpha)A(\beta))$ holds (e.g. if $A(\alpha) = TI(\alpha, B)$).

46.2.4. LEMMA. *Let* U *be a class of classes* $\subseteq \mathbb{N}$ *closed under* *. *Then:*

(1) $A^U(\alpha) := \forall\beta(I^U(\beta) \to I^U(\phi\alpha\beta))$ *implies that*

$S^U(\alpha) := \{\beta : I^U(\phi(\alpha+1)\beta)\}$ *is strongly progressive, provably in* MF_c.
{*Formally:* $\mathrm{MF}_c \vdash Cl(U) \wedge U \subseteq CL \wedge Clos(U;*) \rightarrow (A^U(\alpha) \rightarrow Pr^+(S^U(\alpha)))$}.

(2) *If* $\forall n.A^U(\lambda[n])$, *where* λ *is a limit, then* $L^U(\lambda) := \{\beta : I^U(\phi\lambda\beta)\}$ *is strongly progressive, provably in* MF_c.

PROOF. Ad (1).
1.1: by hypothesis on U, we can apply class-N-induction to check $\forall n I^U(h[n])$ (this suffices by 46.2.1), where $h[0] = \phi\alpha 0$, $h[n+1] = \phi\alpha h[n]$ and $\phi(\alpha+1)0 = lim\ h[n]$. The case $n = 0$ is trivial. If $I^U(h[n])$ holds by IH and we choose $\beta = h[n]$ in $A^U(\alpha)$, we obtain $I^U(h[n+1])$.
1.2 Assume

$$I^U(\phi(\alpha+1)\beta). \tag{$*$}$$

We check (again by N-induction for classes): $I^U(f[n])$, for every n in $\mathbb{N}$, where $f[0] = \phi\alpha(\phi(\alpha+1)\beta+1)$, $f[n+1] = \phi\alpha f[n]$. If $n = 0$, we are done by $(*)$, 46.2.3 (iii) and 46.2.1; in the successor case, apply $A^U(\alpha)$ and IH.
1.3 If λ is a limit and we assume $\forall n. I^U((\phi(\alpha+1)\lambda)[n])$, we get $\forall n I^U(\phi(\alpha+1)\lambda[n])$ by definition of the fundamental sequences and hence $I^U(\phi(\alpha+1)\lambda)$ by 46.2.1.

Ad(2): as in (1), we distinguish three subcases and we make use of the fundamental sequences for $\phi\lambda\beta$. □

If $p \in \omega$ and $p \geq 10$, we define – uniformly in X, γ – the formula

$$A_\gamma^X(\alpha) := (\forall\delta > 0)\ (\omega^{\alpha+1}\cdot\delta < \omega^{\gamma+1} \rightarrow \forall\beta(I^U(\beta) \rightarrow I^U(\phi\alpha\beta))),$$

where $U = U(\alpha, \delta, X) = \mathfrak{R}^p_{\omega^{\alpha+1}\cdot\delta}(X)$.

46.2.5. LEMMA (*provably in* MF_c). *Let* $p \geq 10$. *If* $X \subseteq \mathbb{N}$ *is a class and* $\mathfrak{R}^p_{\omega^{\gamma+1}}(X)$ *is good (see* 46.1), *then* $\{\alpha : A_\gamma^X(\alpha)\}$ *is a strongly progressive class.*

PROOF. By hypothesis $\{\alpha : A_\gamma^X(\alpha)\}$ is a class. We need a simple fact of ordinal arithmetic (provable in OP; see Schütte 1977, p.93), which explains the choice of the ordinal terms:

$$\text{if } \nu < \omega^\alpha\cdot\delta,\ \xi < \alpha, \text{ then we can find a } \delta' \text{ with } \nu < \omega^\xi\cdot\delta' < \omega^\alpha\cdot\delta. \tag{1}$$

Let $\alpha = 0$: once we note that $\mathfrak{R}^p_{\omega\cdot\delta}(X)$ is closed under * (by remark 44.5(ii) and since $p \geq 10$), we get $A_\gamma^X(0)$ as a consequence of 46.2.3 (iii) and the fact that $\mathfrak{R}^p_{\omega^{\gamma+1}}(X)$ is good. Next, we show:

$$A_\gamma^X(\alpha) \rightarrow A_\gamma^X(\alpha+1)\ (X \subseteq \mathbb{N} \text{ class}). \tag{2}$$

Fix $\delta > 0$ and assume

$$k = \omega^{\alpha+2} \cdot \delta < \omega^{\gamma+1}; \tag{3}$$

$$A_\gamma^X(\alpha); \tag{4}$$

$$I^W(\beta), \text{ where } W = \mathcal{R}_k^p(X). \tag{5}$$

If c is in $\mathcal{R}_k^p(X)$ and is progressive, we want

$$(\forall u < \phi(\alpha+1)\beta).u\eta c. \tag{$*$}$$

By (1), choose $\delta' > 0$ such that $c\eta\mathcal{R}_\nu^p(X)$, where $\nu = \omega^{\alpha+1} \cdot \delta' < k$. Obviously $\mathcal{R}_\nu^p(X) \subseteq CL_N$ is a class, being a segment of $\mathcal{R}_k^p(X)$, and it is closed under the map * (ν being a limit). If $U = \mathcal{R}_\nu^p(X)$ in 46.2.4, we have with (4):

$$\mathrm{S}^U(\alpha) \text{ is strongly progressive and hence progressive.} \tag{6}$$

But $S^U(\alpha)$ is in $\mathcal{R}_k^p(X)$ because $\nu < k$, ν is a limit and $I^U(\phi(\alpha+1)\beta)$ has logical complexity $\leq 10 \leq p$; hence by (5),

$$\forall u < \beta.\, u\eta S^U(\alpha). \tag{7}$$

By progressiveness, $\beta\eta S^U(\alpha)$, i.e. $I^U(\phi(\alpha+1)\beta)$, which implies ($*$), as c is progressive and in U. As to the limit case, we may assume, for $X \subseteq \mathbb{N}$, X class:

$$\forall n.\ A_\gamma^X(\lambda[n])\ (\lambda \text{ limit}); \tag{8}$$

$$W = \mathcal{R}^p_{\omega^{\lambda+1} \cdot \delta}(X) \text{ and } \kappa = \omega^{\lambda+1} \cdot \delta < \omega^{\gamma+1} \text{ (where } \delta > 0); \tag{9}$$

$$I^W(\beta). \tag{10}$$

As in the successor case, let c be a progressive class of W, i.e. c is in $U = \mathcal{R}_\nu^p(X)$, where $\nu = \omega^\lambda \cdot \delta' < \kappa$, for some $\delta' > 0$. In order to apply the second part of 46.2.4, we show:

$$\forall n \forall \beta (I^U(\beta) \rightarrow I^U(\phi\lambda[n]\beta)). \tag{11}$$

Indeed, fix n arbitrary, assume $I^U(\beta)$ and pick a progressive class a of U.

Then $a\eta\mathcal{R}_\xi^p(X)$, for some $\xi < \nu$ and by (1) we can find a δ_n such that

$$\xi < \omega^{\lambda[n]+1} \cdot \delta_n < \nu, \text{ i.e. } a\eta\mathcal{R}^p_{\omega^{\lambda[n]+1} \cdot \delta_n}(X) = U_n$$

Hence, since $U = \mathcal{R}_\nu^p(X) \supseteq U_n$, we have, by downward persistence, $I^{U_n}(\beta)$. But (8) and (9) imply:

$$I^{U_n}(\beta) \rightarrow I^{U_n}(\phi\lambda[n]\beta), \text{ for every } \beta. \tag{12}$$

Therefore $I^{U_n}(\phi\lambda[n]\beta)$ holds and by progressiveness of a, we conclude that $\forall u < \phi\lambda[n]\beta.\, u\eta a$; this completes the verification of (11), because U is the union over the U_n's. But (11) and 46.2.4 yield that $L^U(\lambda)$ is strongly progressive. Since $\nu < \kappa$, $L^U(\lambda)$ is an element of W (see (9)), whence by (10)

$$\forall u < \beta . u \eta L^U(\lambda). \tag{13}$$

But $L^U(\lambda)$ is progressive, hencc $\beta \eta L^U(\lambda)$, i.e. $I^U(\phi\lambda\beta)$ which yields $(*)$ since c is progressive. □

46.2.6. *First part of theorem* 46.2:

$$\mathrm{MF}_p \vdash PWO(\alpha), \textit{ for each } \alpha < \phi\omega 0.$$

$$\mathrm{MF}_p \vdash Cl(X) \wedge X \subseteq \mathbb{N} \rightarrow Good(\mathfrak{R}_\omega m(X)), \text{ for each } m \in \omega.$$

PROOF. Let $\alpha < \phi\omega 0$, and choose m such that $\alpha < \phi m 0$. Since ω^{m+2} is a qwo (46.2.3 (v)), $\mathfrak{R}^p_{\omega^{m+2}}(X)$ is good by 44.3 (X is assumed to be in CL_N). Then 46.2.5 implies $A^X_{m+1}(m)$; for $\delta = 1$, we get $\forall \beta (I^U(\beta) \rightarrow I^U(\phi m \beta))$, where $U = \mathfrak{R}^p_{\omega^{m+1}}(X)$. If we choose $\beta = 0$ and we remark that X belongs to U, we get $TI(\phi m 0, X)$; but X is an arbitrary class, whence the conclusion follows by 46.2.1. □

Proof of the theorem 46.2 (*conclusion*). By 46.2.3 and 46.2.6, it remains to check the case of $\mathcal{T} = \mathrm{MFR}(p)$ with $p \geq 10$. By the axiom schema $RAM(p)$, we have for arbitrary $\alpha < \Gamma_0$:

$$\mathrm{MFR}(p) \vdash (PWO(\alpha) \wedge X \eta CL \wedge X \subseteq \mathbb{N}) \rightarrow Good(\mathfrak{R}^p_\alpha(X)). \tag{1}$$

We prove (i)-(ii) below by metamathematical induction on n, where $\alpha_0 = \omega$ $\alpha_{n+1} = \phi\alpha_n 0$:

(i) $$\mathrm{MFR}(p) \vdash \forall X (X \eta CL_N \rightarrow Good(\mathfrak{R}^p_{\alpha_n}(X)));$$

(ii) $$\mathrm{MFR}(p) \vdash PWO(\alpha_n).$$

If $n = 0$, (ii) simply reduces to class N-induction, while (i) follows from (1). Let $n = m+1$. By IH we may assume $PWO(\alpha_m)$, whence $PWO(\alpha_m + 2)$, which implies by 46.2.3 – provably in MF_c – $PWO(\omega^{\alpha_m+2})$. Hence, if X is any subclass of $\mathbb{N}$ and $\kappa = \omega^{\alpha_m+2}$, we have by (1) that $\mathfrak{R}^p_\kappa(X)$ is good. But we can apply lemma 46.2.5 and we get that $A^X_{\alpha_m+1}(\alpha)$ is progressive, and hence with $PWO(\omega^{\alpha_m+2})$, $A^X_{\alpha_m+1}(\alpha_m)$, i.e. $PWO(\phi\alpha_m 0) = PWO(\alpha_{m+1})$. A final application of (1) yields (i) for $n = m+1$. □

46.3. REMARK. Γ_0 *and the applications.* H.Friedman showed that the well-foundedness of the standard well-ordering $\dot{<}$ for Γ_0 follows from a theorem of Kruskal about well-quasi-orderings (see Gallier 1991); hence Kruskal's theorem is unprovable in Predicative Analysis. On the other hand,

Kruskal's theorem is a powerful tool for investigating term rewrite systems used in computer science. Thus Friedman's result suggests that there may be connections between (segments of) the standard well-ordering $\dot{<}$ of type Γ_0 and the term orderings involved in termination proofs of term rewrite systems. Indeed, Dershowitz and Okada established interesting relations with proof-theoretic ordinals; for instance, it can be shown that *the order-type of the so-called multiset path ordering on the terms of an alphabet whose precedence ordering is* ω, *is exactly* $\phi\omega 0$. For a survey on term rewrite systems, the reader can consult Dershowitz and Jouannaud (1990).
The relevance of Γ_0 for combinatorics and computer science is discussed by Gallier (1991), where the results of Dershowitz and Okada are also reviewed.

CHAPTER 10

REDUCING REFLECTIVE TRUTH WITH LEVELS TO FINITELY ITERATED REFLECTIVE TRUTH

§ 47. A sequent calculus STLR for a theory of reflective truth with levels
§ 48. Basic properties of STLR
§ 49A. Elimination of the full level induction schema
§ 49B. Elimination of unbounded level quantifiers
§ 50. The infinitary sequent calculus IT_n^∞ of n-iterated reflective truth
§ 51. Embedding STLR_n into IT_n^∞

In semantic form, the main theorem we are going to establish sounds as follows: *the first recursively inaccessible ordinal can be replaced by* ω *in the construction of the recursion-theoretic model* of § 39, insofar as we deal with TLR-consequences of the form

$$\forall i \exists j \forall x(Cl_i(x) \rightarrow \exists y(Cl_j(y) \wedge A(x,y))) \ (A \text{ elementary extensional in } x,\ y).$$

Indeed, something stronger will be true: as a consequence of proof-theoretic analysis, we shall prove that the theory TLR of reflective truth, with variable levels and *full transfinite induction schema on level ordering*, can be constructively reduced to a family $\{\mathrm{IT}_n^\infty : n \in \omega\}$ of theories of arbitrary finitely iterated truth predicates. In each system IT_n^∞, level variables and quantifiers are explained away in favour of a sequence $\{T_k : k \leq n\}$ of self-referential truth predicates of increasing logical complexity. The reduction is carried out in four steps that we summarize below.

1) $\mathrm{TLR} + TI(lev) \mapsto \mathrm{STLR}$: we give a sequent style presentation STLR (= *sequent calculus of reflective truth with levels and reflection*) of a system, which contains TLR and *the full transfinite induction schema on levels* (see §§ 47-48).

2) $\mathrm{STLR} \mapsto \mathrm{STLR}^\infty$: STLR is embedded into an infinitary system STLR^∞, where $TI(lev)$ is eliminated in favour of an ω-rule, which forces the level variables to range over *finite standard ordinals*. Since STLR^∞ contains a reflection principle for levels, STLR^∞ cannot have ω-standard models; yet, *due to the weak number-theoretic induction*, STLR^∞ is consistent. STLR^∞ enjoys a crucial quasi-normalization property (§ 49A): cut-rule can be

restricted to formulas, which contain only unbounded universal or existential level quantifiers.

3) $STLR^{\infty} \mapsto \{STLR_n : n \in \omega\}$: this is the central step of the constructive interpretation. First, we define a sequence of finitary approximations $STLR_n$ to $STLR^{\infty}$, in which *only bounded level quantifiers are allowed and where we can explicitly refer only to the first n levels.* The main theorem 49.18 ensures that $STLR^{\infty}$-theorems can be suitably interpreted in the $STLR_n$'s. The result is based on an *asymmetric treatment of unbounded universal and existential level quantifiers*; the informal idea is to reinterpret unbounded quantifiers on levels according to a "potentialistic" point of view, so that $\forall j$ only refers to arbitrary *finite segments* of the level ordering. As a consequence, the meaning of $\exists j$ depends on the given initial segments, and this dependence is expressed by majorizing functions, whose complexity depends upon the *transfinite ordinal height* of the given quasi-normalized $STLR^{\infty}$-derivations.

4) $STLR_n \mapsto IT_n^{\infty}$. In §§50-51 we carry out a complete elimination of level quantification and level structure: each $STLR_n$-system is embedded into a level-free infinitary system IT_n^{∞}, where the number-theoretic induction schema is replaced by an infinitary rule for N.

§ 47. A sequent calculus STLR for a theory of reflective truth with levels

We describe a sequent calculus STLR in the style of Tait (1968), which strengthens TLR.

47.1. *The syntax of* STLR. The language $\mathcal{L}_V^+$ of STLR is, in essence, $\mathcal{L}_V$. For technical reasons, it is convenient to adopt a different set of logical constants and new predicate symbols for falsehood. Here is the list of primitive symbols:

(i) individual variables $x_0, x_1, x_2 \ldots (x, y, z$ are metavariables);
(ii) individual constants $\underline{0}$, SUC, $PRED$, $PAIR$, $LEFT$, $RIGHT$, D;
(iii) function symbols Ap (binary) and LT (unary);
(iv) predicate symbols $=$, Vr, Fl, $\preceq$, $=_l$ (binary); T, F, N (unary);
(v) level variables $i_0, i_1, \ldots$;
(vi) countably many individual constants $\{\boldsymbol{m} : m \in \omega\}$ for levels;
(vii) the logical constants $\vee$, $\wedge$, $\neg$, $\forall$, $\exists$.

The L-*terms* are exactly the L(evel)-variables and the L-constants; i, j, k will ambiguously range over L-terms. The *terms* of $\mathcal{L}_V^+$ form the least collection which is closed under the following clauses: individual variables and constants are terms; if j is an L-term, $LT(j)$ is a term; if t, s are terms, $Ap(t,s)$ is a term.

Atoms and formulas of $\mathcal{L}_V^+$. If t, s are terms, $t = s$, Nt and their negations are e-atoms (e = elementary); if i, j are L-terms, $i \preceq j$, $i =_l j$, and their negations are L-atoms. If t is a term and j is an L-term, then Tt, Ft, $Vr(i,t)$, $Fl(i,t)$ and their negations are T-atoms. A is an *atom* iff either A is an e-atom or A is an L-atom, or else A is a T-atom. An atom A is *positive* if $\neg$ does not occur in A; an atom A is *negative* iff A has the form $\neg B$, where B is a positive atom.
The collection of $\mathcal{L}_V^+$-formulas is inductively generated from atoms by closing off under $\wedge$, $\vee$ and quantification over either variable sort.

NB: in the previous chapters "atom" was used for what is here called "positive atom"; this change of terminology should not cause any trouble, since it only concerns the languages for sequent calculi.

47.2. *Preliminary definitions and conventions*

47.2.1 We assume the conventions and notations of Chapters I, § 1 and VIII, 36.2; $T_i t$, $F_i t$, $t\eta_i s$, $t\bar{\eta}_i s$, $t\eta s$, $t\bar{\eta} s$ and $Cl_i(t)$ abbreviate $Vr(i,t)$, $Fl(i,t)$, $Vr(i, Ap(s,t))$, $Fl(i, Ap(s,t))$, $T(st)$, $F(st)$ and $\forall x(x\eta_i t \vee x\bar{\eta}_i t)$ (in the given order). If i, j are L-terms, $i = j$ stands for $i =_l j$; we also write $i \prec j$ for $(\neg i = j) \wedge (i \preceq j)$. We set:
$\forall i \preceq j.\, A := \forall i(i \preceq j \rightarrow A)$ and $\exists j \preceq i.\, A := \exists j(j \preceq i \wedge A)$; $\forall i \preceq j$ and $\exists j \preceq i$ are called *bounded level quantifiers*.

47.2.2. Negation is inductively extended to arbitrary $\mathcal{L}_V^+$-formulas, according to the following clauses:

$$\neg\neg A = A, \text{ if } A \text{ is a positive atom; } \neg(A \overset{\vee}{\underset{\wedge}{}} B) = (\neg A) \overset{\wedge}{\underset{\vee}{}} (\neg B);$$

$$\neg(\overset{\forall}{\underset{\exists}{}} v.\, A) = (\overset{\exists}{\underset{\forall}{}} v.\, \neg A) \text{ (provided } v \text{ is an individual or level variable).}$$

The other connectives $\rightarrow$, $\leftrightarrow$ are introduced according to their classical definitions. Observe that, if A is arbitrary, $\neg\neg A$ and A coincide.

47.2.3. Σ is the smallest collection of $\mathcal{L}_V^+$-formulas, which includes e-atoms, L-atoms, T-atoms of the form Ft, Tt, $\neg T_i t$, $T_i t$ and is closed under conjunction, disjunction, existential quantification on level variables, bounded universal level quantification, universal and existential quantification on individual variables.

We also define: $\Pi := \{\neg A : A \in \Sigma\}$ and $\Delta_0 := \Pi \cap \Sigma$.

Following 36.3, *the collection* Δ^+ of *acceptable formulas of* $\mathcal{L}_V^+$ is the smallest collection which includes e-atoms, T-atoms and is closed under conjunction, disjunction and universal and existential quantification on individual variables.

47.2.4. Once the combinators ID, TR, NAT, NEG, AND, OR, ALL, $EXIST$ of Ch. II, §7 are given, we inductively extend the map $A \mapsto [A]$ to arbitrary formulas of Δ^+ (see §36; we recall that $[T_i t] := \langle \bar{7}, \langle LT(i), t \rangle \rangle$) and in addition we stipulate: $[F_i t] := [T_i \neg t]$ and $[Ft] := [T \neg t]$.
If $A \in \Delta^+$, $T_i A := T_i[A]$ and $TA := T[A]$. λ-abstraction is extended to $\mathcal{L}_V^+$-terms by adding $\lambda x.LT(i) := K(LT(i))$ (where i is an arbitrary L-term). Hence, if $A \in \Delta^+$, it makes sense to set $\{x : A\} := \lambda x.[A]$.

47.2.5. As usual in proof-theoretic investigations, we partition variables of either sort into *bound* and *free variables*: for instance, we may use variables of even index as bound variables, while variables of odd index are only used as free variables, or, in short, *parameters*.

We say that a formula of $\mathcal{L}_V^+$ has the variable separation property (= VSP) iff every free variable of A has odd index, while every bound variable of A has even index. Henceforth, we stick to the following convention VSP (which is not restrictive, by trivial logical considerations):

(i) by $\mathcal{L}_V^+$-formula we always mean a formula which satisfies the variable separation property;

(ii) by $\mathcal{L}_V^+$-term we always understand a term whose free variables are actually parameters, i.e. have odd index.

If Γ is a set of $\mathcal{L}_V^+$-formulas, $FV(\Gamma)$ is the set of parameters occurring in the formulas of Γ; of course, $FV(A) := FV(\{A\})$.
In the special case of individual variables, x, y, z range over bound variables, while a, b, c stand for parameters.

47.2.6. *Inductive definition of $Lc(A)$ (A arbitrary $\mathcal{L}_V^+$-formula)*:

$Lc(A) = 0$, for A atom; $Lc(A \circ B) = max(Lc(A), Lc(B))+1$, if $\circ = \wedge, \vee$; $Lc(QxA) = Lc(A)+1$, if $Q = \forall, \exists$.

$Lc(A)$ is the logical complexity of A. Clearly $Lc(A) = Lc(\neg A)$.

47.2.7. *Inductive definition of $rk(A)$ (A arbitrary $\mathcal{L}_V^+$-formula).*

$rk(A) = 0$ if $A \in \Sigma \cup \Pi$; else

$rk(A \circ B) = max(rk(A), rk(B))+1$, if $\circ = \wedge, \vee$;

$rk(QxA) = rk(A)+1$, if $Q = \forall, \exists$.

$rk(A)$ is called the *rank* of A.

47.2.8. For a compact presentation of STLR-axioms, it is convenient to introduce the following abbreviations:

(i) $T_i\text{-}Clause(t) := \exists x\exists y((t = [x = y] \wedge x = y) \vee (t = [Nx] \wedge Nx) \vee$
$\vee ((t = \lfloor Tx \rfloor \vee t = \lfloor T_i x \rfloor) \wedge T_i x) \vee \exists j(j \prec i \wedge t = [T_j x] \wedge T_j x) \vee$
$\vee (t = (\neg x) \wedge F_i x) \vee (t = (x \wedge y) \wedge T_i x \wedge T_i y) \vee (t = (\forall x) \wedge \forall v. T_i(xv))).$

(ii) $F_i\text{-}Clause(t) := \exists x\exists y((t = [x = y] \wedge \neg x = y) \vee$
$\vee (t = [Nx] \wedge \neg Nx) \vee ((t = [Tx] \vee t = [T_i x]) \wedge F_i x) \vee$
$\vee \exists j(j \prec i \wedge t = [T_j x] \wedge \neg T_j x) \vee (t = (\neg x) \ \wedge T_i x\ ) \vee$
$\vee (t = (x \wedge y) \ \wedge \ (F_i x \vee F_i y)) \vee (\ t = (\forall x) \wedge \exists v. F_i(xv))).$

47.2.9. REMARK. T_i-*Clause*(t) and F_i-*Clause*(t) formalize the conditions, which are necessary and sufficient for t to fall under T_i and F_i respectively. The choice of T_i-*Clause* and F_i-*Clause* is motivated by the definition of the operator which generates the recursion-theoretic model for TLR (39.9).

47.3. *The system* STLR

The language of STLR is $\mathcal{L}_V^+$. We present STLR as a Tait-style sequent calculus, where sequents are derived instead of formulas. Sequents are finite sets of $\mathcal{L}_V^+$-formulas, denoted by capital Greek letters Γ, Δ,.... The intended meaning of a sequent $\Gamma = \{A_1, \ldots, A_n\}$ is the finite disjunction $A_1 \vee \ldots \vee A_n$. The expression "Γ, Δ" stands for the set-theoretic union of Γ with Δ. If p is a parameter of either sort and t is a term of the same sort of p, $\Gamma[p := t]$ denotes the sequent obtained by substituting each occurrence of the parameter p by t in Γ; $\Gamma[p := t]$ is called a *substitution instance* of Γ.
A set $\mathcal{S}$ of sequents is closed under substitution whenever

$$\Gamma \in \mathcal{S} \text{ implies } \Gamma[p := t] \in \mathcal{S}, \text{ for arbitrary } p(\text{arameter}),\ t(\text{erm}).$$

The substitution closure of $\mathcal{S}$ is the smallest set of sequents, which contains $\mathcal{S}$ and is closed under substitution.

47.3.1. *Axioms of* STLR: they form the substitution closure of the following sets of sequents.

A.1. *Logical axioms.* Let p, q be parameters of the same sort and let $=$ denote the corresponding equality predicate. Then we postulate:

(i) $p = p$;
(ii) $\neg p = q$, $\neg A(p)$, $A(q)$, where $rk(A) = 0$;
(iii) $\neg A$, A , where $rk(A) = 0$.

A.2. *Operational axioms*

(i) $Kab = a$ and $Sabc = ac(bc)$;

(ii) $\neg C = C'$ (C, C' distinct individual constants);
(iii) $(\langle a_1, a_2 \rangle)_i = a_i$ $(i = 1, 2)$;
(iv) $\neg Na, \neg Nb, \neg a = b, Dabcd = c$;
(v) $\neg Na, \neg Nb, a = b, Dabcd = d$.

A.3. *Peano axioms*

(i) $N\overline{0}$ and $\neg Na, N(a+1)$;
(ii) $\neg Na, \neg a+1 = \overline{0}$;
(iii) $\neg Na, PRED(a+1) = a$.

A.4. *Level axioms*

(i) $i \preceq i$;
(ii) $\neg i \preceq j, \neg j \preceq k, i \preceq k$;
(iii) $\exists k(i \prec k \wedge j \prec k)$;
(iv) $\neg i \preceq j, \neg\, j \preceq i, i = j$;
(v) $\neg LT(i) = LT(j), i = j$.

A.5. *Persistence* : $\neg i \preceq j, \neg T_i a, T_j a$;

A.6. *Consistency*: $\neg T_i a, \neg F_i a$;

A.7. *Limit axioms*: $\neg Ta, \exists i. T_i a$; $\neg T_i a, Ta$; $\neg Fa, \exists i. F_i a$; $\neg F_i a, Fa$.

A.8. *Fixed point axioms*

(i) $\neg T_i\text{-}Clause(t), T_i t$ and $\neg T_i t, T_i\text{-}Clause(t)$;
(ii) $\neg F_i\text{-}Clause(t), F_i t$ and $\neg F_i t, F_i\text{-}Clause(t)$.

A.9. Δ_0-*N-induction*: $\neg A(\overline{0}), \neg\forall x(A(x) \rightarrow A(x+1)), \neg Nb, A(b)$ (A arbitrary Δ_0-formula).

A.10. Δ_0-*Reflection*: $\neg\forall x \exists i. A(x,i), \exists j. \forall x. \exists i \preceq j. A(x,i)$ (A arbitrary Δ_0-formula).

A.11. *Level Induction*: $\neg Progr(\prec, A), \forall i A(i)$, where A is arbitrary and $Progr(\prec, A)$ abbreviates $\forall i(\forall j(j \prec i \rightarrow A(j)) \rightarrow A(i))$.

47.3.1.1. REMARK. All formulas occurring in the axioms, except for A.11, are at most Σ or Π (see A.4 (iii), A.7, A.10).

47.3.2. *Rules of* STLR. STLR contains the following finitary rules:

$$(\wedge)\ \frac{\Gamma, A \quad \Gamma, B}{\Gamma, A \wedge B}; \qquad (\vee)\ \frac{\Gamma, A}{\Gamma, A \vee B} \text{ and } \frac{\Gamma, B}{\Gamma, A \vee B};$$

$$(\forall x)\ \frac{\Gamma, A(a)}{\Gamma, \forall x A} \text{ with } a \notin FV(\Gamma, A); \quad (\exists x)\ \frac{\Gamma, A(t)}{\Gamma, \exists x A} \ (t \text{ individual term});$$

$(\forall j)\ \frac{\Gamma,\, A(k)}{\Gamma,\, \forall j A}$ with $k \notin FV(\Gamma, A)$; $(\exists j)\ \frac{\Gamma,\, A(i)}{\Gamma,\, \exists j A}$ (i L-term);

$(\mathrm{Cut})\ \frac{\Gamma,\, A \quad \Gamma,\, \neg A}{\Gamma}$.

Terminology. $(\wedge)$, $(\vee)$, $(\forall x)$, $(\forall j)$, $(\exists x)$, $(\exists j)$ are called *logical rules.* In a given rule $\mathfrak{R}$, the elements of Γ are called *side formulas.* A formula which occurs in the premises (in the conclusion) of $\mathfrak{R}$, but is not a side formula, is called *minor* (*active*) formula of the inference. The minor formulas of (Cut) are called *cut formulas*; observe that they have the same rank. A sequent Γ is said to be *initial* iff $\Gamma \supseteq \Delta$, for some axiom Δ.

§ 48. Basic properties of STLR

We introduce a notion of STLR-derivability.

48.1. *Inductive definition of the derivability relation* $\mathrm{STLR} \vdash^m_n \Gamma$ (for m, $n \in \omega$).

48.1.1. If Γ is an initial sequent, $\mathrm{STLR} \vdash^m_n \Gamma$, for every m, $n \in \omega$.

48.1.2. Assume that (i) Γ is the conclusion from the premises Γ_i of a logical rule, or of a cut rule of rank $< n$; (ii) $\mathrm{STLR} \vdash^{m_i}_n \Gamma_i$ $(i \leq 2)$ and $m_i < m$. Then $\mathrm{STLR} \vdash^m_n \Gamma$.

$\mathrm{STLR} \vdash^m_n \Gamma$ is read as "Γ is STLR-derivable with length $\leq m$ and cut rank $< n$". We immediately have from the definition:

48.1.3. FACT. *If* $\mathrm{STLR} \vdash^m_n \Gamma$ *and* $m \leq k$, $n \leq p$, *then* $\mathrm{STLR} \vdash^k_p \Gamma$.

By 48.1.3, it is not restrictive to assume that the premises of a rule are derivable with the same length.

48.1.4. REMARK. $\mathrm{STLR} \vdash^m_n \Gamma$ iff there exists a finitary tree $\mathfrak{T}$ of sequents with root Γ, such that (i) every top sequent of $\mathfrak{T}$ is initial; (ii) every other sequent S occurring in $\mathfrak{T}$ is obtained by means of a rule of STLR from sequents standing immediately above S; (iii) the height of $\mathfrak{T}$ is $\leq m$ and every cut formula occurring in the tree has rank $< n$. $\mathfrak{T}$ is usually called *derivation* of the given sequent.

We now state a few elementary properties of the derivability relation.

48.2. LEMMA (*Substitution*)

(i) *If* STLR $\vdash_n^m \Gamma(a)$, *then* STLR $\vdash_n^m \Gamma[a := t]$ (t individual term);

(ii) *If* STLR $\vdash_n^m \Gamma(i)$ (i *L-parameter*), *then* STLR $\vdash_n^m \Gamma[i := j]$ (j *L-term*).

48.3. LEMMA (*Weakening*) *If* STLR $\vdash_n^m \Gamma$, *then* STLR $\vdash_n^m \Gamma, A$.

48.4. LEMMA

(i) *Tautology*: STLR $\vdash_0^m \Gamma, A, \neg A$ (A arbitrary), *for every* $m \geq 2 \cdot rk(A)$.

(ii) *Substitutivity*: STLR $\vdash_0^m \Gamma, \neg t = s, \neg A[x := t], A[x := s]$ (*A arbitrary*), *for every* $m \geq 2 \cdot rk(A)$.

We show that STLR is not weaker than the axiomatic system TLR of Ch. VIII.

48.5. NOTATION. STLR $\vdash A$ stand for STLR $\vdash \{A\}$; we also adopt the abbreviations: STLR $\vdash_n \Gamma :=$ STLR $\vdash_n^m \Gamma$, for some m, and we write STLR $\vdash \Gamma :=$ STLR $\vdash_n \Gamma$, for some n.

48.6. LEMMA (*Independence*)

(i) *Let* C, D *be distinct elements of the set*

$$\{[a = b],\ [Na],\ [Ta],\ [T_j a],\ a \wedge b,\ \neg a,\ \forall a\}.$$

Then STLR $\vdash \neg(C = D)$.

(ii) STLR $\vdash [T_j t] = [T_i s] \rightarrow i = j \wedge t = s$.

PROOF. Apply A.2 (iii), A.4 (v) and STLR $\vdash \neg(\overline{n} = \overline{m})$, (which follows from A.3, if $\overline{n} = \overline{m}$ are distinct numerals). □

48.7. LEMMA. *Each substitution instance of the following sequents is derivable in* STLR:

(i) $\neg T_i$*-Clause*$([A]), A$ *and* $\neg F_i$*-Clause*$([A]), \neg A$, *if* A *is* $a = b$, Na;

(ii) $\neg T_i$*-Clause*$([Ta]), T_i a$;

(iii) $\neg T_i$*-Clause*$([T_j a]), j \preceq i \wedge T_j a$;

(iv) $\neg F_i$*-Clause*$([T_j a]), i = j \wedge F_j a, j \prec i \wedge \neg T_j a$;

(v) $\neg T_i$*-Clause*$(\neg a), F_i a$ *and* $\neg F_i$*-Clause*$(\neg a), T_i a$;

(vi) $\neg T_i$*-Clause*$(a \wedge b), T_i a \wedge T_i b$; $\neg F_i$*-Clause*$(a \wedge b), F_i a, F_i b$;

(vii) $\neg T_i\text{-}Clause(\forall a), \forall x T_i(ax); \quad \neg F_i\text{-}Clause(\forall a), \exists u. F_i(au)$.

The straightforward verification makes use of the previous independence lemma and is left to the reader.

Now observe that every formula A of the language of TLR has a canonical equivalent $NF(A)$, written in the language of STLR; $NF(A)$ is the so-called negation normal form of A, i.e. the formula obtained from A by pushing $\neg$ in front of atoms with the help of De Morgan's laws and standard equivalences between $\neg\forall$ and $\exists\neg$, $\neg\exists$ and $\forall\neg$, and by deleting double negations. Consider the schema:

$$TI(lev) := \forall i(\forall j(j \prec i \rightarrow A(j)) \rightarrow A(i)) \rightarrow \forall i A(i) \ (A \text{ arbitrary}).$$

Then we can state:

48.8. THEOREM. *If* TLR + $TI(lev) \vdash A$, *then* STLR $\vdash NF(A)$ (*A arbitrary formula of* $\mathcal{L}_V$).

PROOF: it suffices to prove that the non-logical axioms of TLR are derivable in STLR. First of all, observe that the local N-induction axiom LIND of 36.4.1 and the reflection axiom of 37.7 are special cases of A.9 and A.10 (respectively). The operational and number-theoretic axioms (36.4.1), the level axioms 36.4.3, local consistency, limit and persistence (i.e. 36.4.2.6, 36.4.4.1, 36.4.4.2) are disposed of by means of A.2, A.3, A.4, A.6, A.7, A.5 (respectively). As to the other axioms of TLR, which concern T and $T\neg$, they follow from lemma 48.7 and the fixed point axioms A.8. To this aim, observe that, by use of logic only, we can "invert" the sequents of 48.7, e.g. STLR $\vdash \neg T_i a, T_i\text{-}Clause([Ta])$, etc. The schema of level induction is derivable with A.11. □

Henceforth, we generally disregard the fact that TLR and STLR have different logical primitives, and we simply identify A and $NF(A)$.

§49 A. Elimination of the full level induction schema

The idea is to let the level variable range over *finite ordinals.* Formally, this leads to the following infinitary variant of STLR.

49.1. (i) The language $\mathcal{L}_V^*$ of STLR$^\infty$: it is obtained from $\mathcal{L}_V^+$ *by omitting free level variables.*

(ii) L-terms are exactly the level constants $\boldsymbol{m}$, for each $m \in \omega$.

(iii) The notion of formula is inductively defined in the standard way, as well the classes of Σ-, Π-, Δ_0-formulas; the notions of logical complexity and rank are lifted without change to $\mathcal{L}_V^*$.

49.2. *Axioms of* STLR$^\infty$: they are obtained from the axioms of STLR by means of the following changes:

(i) all level parameters occurring in the axioms of § 47 are replaced by level constants;

(ii) level induction A.11 is omitted; the logical axioms A.1 (ii)-(iii) are restricted to atoms of $\mathcal{L}_V^*$;

(iii) the level axioms of A.4 are replaced by

A$'$.4.1: $\neg LT(\boldsymbol{i}) = LT(\boldsymbol{j})$ if $i \neq j$ $(i,\ j \in \omega)$;

A$'$.4.2: if A is a true L-atom of $\mathcal{L}_V^*$, then A is an axiom;

("true" refers to the structure with support $\{k\colon\ k \in \omega\}$, where level identity and level ordering are interpreted by number-theoretic identity and natural ordering respectively; of course, the axioms of A.4 become provable from A$'$.4).

(iv) The persistence axiom A.5 takes the form:

A$'$.5. $\neg T_{\boldsymbol{i}} t,\ T_{\boldsymbol{k}} t,$ if $i < k$.

49.2.1. REMARK. The formulas occurring in the axioms of STLR$^\infty$ have rank 0.

49.3. *Rules of* STLR$^\infty$: STLR$^\infty$ has the same rules as STLR, except that $(\forall i)$, $(\exists i)$ are now replaced by

$$(\forall\omega)\ \frac{\ldots\Gamma,\ A(\boldsymbol{i})\ldots \text{ for each } i \in \omega}{\Gamma,\ \forall j.A}\ ;\ (\exists\omega)\ \frac{\Gamma,\ A(\boldsymbol{i}),\text{ for some } i \in \omega}{\Gamma,\ \exists j.A}\ .$$

Below lower case Greek letters $\alpha, \beta, \ldots$ range over arbitrary ordinals $< \epsilon_0 = \phi 10$ (cf. Ch. IX).

49.4. *Inductive definition of the derivability relation*

$$\text{STLR}^\infty \vdash^{\alpha}_{k} \Gamma\ (k \in \omega,\ \alpha < \epsilon_0).$$

1.1. If Γ contains (as a set) an axiom of STLR$^\infty$, STLR$^\infty \vdash^{\alpha}_{k} \Gamma$ for every α and every $k \in \omega$;

1.2. Assume that Γ is the conclusion from the premises Γ_i of a finitary logical rule, or of $(\forall\omega)$, $(\exists\omega)$, or else of a cut rule applied to formulas of rank $< k \in \omega$; assume also that STLR$^\infty \vdash^{\alpha_i}_{k} \Gamma_i$ $(i \leq \omega)$ and $\alpha_i < \alpha$. Then STLR$^\infty \vdash^{\alpha}_{k} \Gamma$.

49.5. NOTATIONS

(i) STLR$^\infty \vdash^{\alpha}_{k} \Gamma$:= Γ is STLR$^\infty$-derivable with length $\leq \alpha$ and cut rank $< k$.

(ii) $\mathrm{STLR}^\infty \vdash^{<\omega} \Gamma := \mathrm{STLR}^\infty \vdash^{m}_{k} \Gamma$, for some finite m, k.

(iii) $\mathrm{STLR}^\infty \vdash \Gamma := \mathrm{STLR}^\infty \vdash^{\alpha}_{k} \Gamma$, for some α and $k \in \omega$.

There are two main reasons for introducing STLR^∞: the system satisfies a weak cut elimination property; STLR is embeddable into STLR^∞.

49.6. LEMMA

(i) *If* $\mathrm{STLR}^\infty \vdash^{\alpha}_{n} \Gamma(a)$, *then* $\mathrm{STLR} \vdash^{\alpha}_{n} \Gamma[a := t]$ (t *individual term*);

(ii) $\mathrm{STLR}^\infty \vdash^{\alpha}_{n} \Gamma, A_1 \wedge A_2$ *implies* $\mathrm{STLR}^\infty \vdash^{\alpha}_{n} \Gamma, A_i$ ($i = 1,\ 2$), *provided* $rk(A_1 \wedge A_2) > 0$;

(iii) $\mathrm{STLR}^\infty \vdash^{\alpha}_{n} \Gamma,\ A_1 \vee A_2$ *implies* $\mathrm{STLR}^\infty \vdash^{\alpha}_{n} \Gamma, A_1, A_2$, *provided* $rk(A_1 \vee A_2) > 0$;

(iv) $\mathrm{STLR}^\infty \vdash^{\alpha}_{n} \Gamma,\ \forall x A$ *implies* $\mathrm{STLR}^\infty \vdash^{\alpha}_{n} \Gamma,\ A[x := t]$ (t *individual term*), *provided* $rk(\forall x A) > 0$;

(v) $\mathrm{STLR}^\infty \vdash^{\alpha}_{n} \Gamma, \forall j A$ *implies* $\mathrm{STLR}^\infty \vdash^{\alpha}_{n} \Gamma,\ A[j := \mathbf{i}]$ ($i \in \omega$), *provided* $rk(\forall j A) > 0$.

(vi) *If* $\mathrm{STLR}^\infty \vdash^{\alpha}_{n} \Gamma$, *then* $\mathrm{STLR}^\infty \vdash^{\alpha}_{n} \Gamma,\ A$.

49.7. LEMMA

(i) *Tautology*: $\mathrm{STLR}^\infty \vdash^{\alpha}_{0} \Gamma,\ A,\ \neg A$ (A *arbitrary*), *for every* $\alpha \geq 2 \cdot rk(A)$.

(ii) *Substitutivity*: $\mathrm{STLR}^\infty \vdash^{\alpha}_{0} \Gamma,\ \neg t = s,\ \neg A[x := t],\ A[x := s]$ (A *arbitrary*), *for every* $\alpha \geq 2 \cdot rk(A)$.

Proofs of the lemmata can be easily carried out by appropriate inductive arguments; the inversion properties need the proviso, essentially because we allow formulas of logical complexity > 0 in the axioms (e.g. logical axioms, A.9). If we apply the usual cut elimination procedure to STLR^∞, we see that it is not restrictive to replace *full* cut rule by cut on Σ- or Π-formulas.

49.8. THEOREM (*Weak Cut Elimination*)

If $\mathrm{STLR}^\infty \vdash \Gamma$, *then* $\mathrm{STLR}^\infty \vdash^{\alpha}_{1} \Gamma$, *for some* $\alpha < \epsilon_0$.

The theorem is a consequence of the so-called reduction and elimination lemmata:

49.8.1. LEMMA (*Reduction*). *If* $rk(A) = n{+}1$, $\mathrm{STLR}^\infty \vdash^{\alpha}_{n+1} \Gamma,\ A$ *and* $\mathrm{STLR}^\infty \vdash^{\beta}_{n+1} \Gamma, \neg A$, *then* $\mathrm{STLR}^\infty \vdash^{\alpha \# \beta}_{n+1} \Gamma$

($\#$ is the natural ordinal sum; see below § 53).

49.8.2. LEMMA (*Elimination*)

If $n > 0$ *and* $\mathrm{STLR}^\infty \vdash^{\alpha}_{n+1} \Gamma$, *then* $\mathrm{STLR}^\infty \vdash^{\omega^\alpha}_{n} \Gamma$.

To avoid repetitions, proofs will be given for the infinitary systems of the next chapter.

49.9. LEMMA

(i) *If* A *is an arbitrary* $\mathcal{L}^*_V$*-formula,*

$$\mathrm{STLR}^\infty \vdash^{\omega}_{0} \neg Progr(\prec, A), \forall i A(i);$$

$\{Progr(\prec, A)$ *abbreviates* $\forall i(\forall j(j \prec i \rightarrow A(j)) \rightarrow A(i))\}$.

(ii) $\mathrm{STLR}^\infty \vdash^{<\omega}_{1} \Gamma$, *provided* Γ *is a sequent of the following form:*

$$\{\neg \boldsymbol{i} \preceq \boldsymbol{k}, \neg T_{\boldsymbol{i}} t, T_{\boldsymbol{k}} t\};$$

$$\{\boldsymbol{i} \preceq \boldsymbol{i}\}; \ \{\neg \boldsymbol{i} \preceq \boldsymbol{j}, \neg \boldsymbol{j} \preceq \boldsymbol{k}, \boldsymbol{i} \preceq \boldsymbol{k}\}; \ \{\exists k(\boldsymbol{i} \prec k \wedge \boldsymbol{j} \prec k)\};$$

$$\{\neg \boldsymbol{i} \preceq \boldsymbol{j}, \neg \boldsymbol{j} \preceq \boldsymbol{i}, \boldsymbol{i} = \boldsymbol{j}\}; \ \{\neg LT(\boldsymbol{i}) = LT(\boldsymbol{j}), \boldsymbol{i} = \boldsymbol{j}\};$$

(i, j, $k \in \omega$ *arbitrary*).

PROOF. (i): let $\Delta = \{\neg Progr(\prec, A)\}$. Then it is enough to check by induction on $i \in \omega$:

$$\mathrm{STLR}^\infty \vdash^{<\omega}_{0} \Delta, A(\boldsymbol{i}). \tag{1}$$

If $i = 0$, (1) follows with the derivability of $\{\neg \boldsymbol{n} \prec \mathbf{0}\}(= \{\neg \boldsymbol{n} \preceq \mathbf{0} \vee n = \mathbf{0}\})$. In the induction step, we get by IH and tautology (respectively):

$\mathrm{STLR}^\infty \vdash^{<\omega}_{0} \Delta, \forall j \prec \boldsymbol{i{+}1}. A(j)$ and $\mathrm{STLR}^\infty \vdash^{<\omega}_{0} \Delta, \neg A(\boldsymbol{i{+}1}), A(\boldsymbol{i{+}1})$.

From this we get $\mathrm{STLR}^\infty \vdash^{<\omega}_{0} \Delta, \neg A(\boldsymbol{i{+}1}) \wedge \forall j \prec \boldsymbol{i{+}1}. A(j), A(\boldsymbol{i{+}1})$ by

($\wedge$). Now $\mathrm{STLR}^\infty \vdash^{<\omega}_{0} \Delta, A(\boldsymbol{i{+}1})$ follows by ($\exists\omega$).

(ii): by A′.4.1-4.2, A′.5, logical axioms and ($\exists\omega$). □

49.10. DEFINITION. Let Γ be a set of $\mathcal{L}^+_V$-formulas: Γ' is a $[0, m]$-instance of Γ, iff Γ' is obtained from Γ by replacing the free level variables occurring in Γ with level constants of value $\leq m$; clearly, once Γ' is a $[0, m]$-instance of Γ, Γ' is a set of $\mathcal{L}^*_V$-formulas.

49.11. THEOREM (*Embedding*). *If* $\mathrm{STLR} \vdash \Gamma$, *there exist* $\alpha < \omega^2$ *and* $k < \omega$, *such that, for each* $m \in \omega$ *and each* $[0, m]$*-instance* Γ' *of* Γ, *then*

$$\mathrm{STLR}^\infty \vdash^{\alpha}_{k} \Gamma'.$$

PROOF: straightforward by application of 49.9. □

§ 49 B. Elimination of unbounded level quantifiers

We define a sequence $\{\mathrm{STLR}_n : n \in \omega\}$ of subsystems of STLR^∞, such that STLR^∞ is locally embeddable into $\cup\{\mathrm{STLR}_n : n \in \omega\}$, in a sense to be made precise below.

49.12. *Syntax of* STLR_n $(n \in \omega)$

49.12.1. The language $\mathcal{L}_n^*$ of STLR_n is the fragment of $\mathcal{L}_V^*$, which omits the unary predicates T and F, and only contains the first n level constants $\{\boldsymbol{m} : m \leq n\}$. Thus L-terms of $\mathcal{L}_n^*$ coincide with the level constants $\boldsymbol{m}$, for each $m \leq n \in \omega$, and Tt, Ft are no more atoms of $\mathcal{L}_n^*$. We also define

$$\mathcal{L}_\omega := \cup\{\mathcal{L}_n^* : n \in \omega\}.$$

49.13. The axioms of STLR_n are obtained from the axioms of STLR^∞ by means of the following changes:

(i) all level constants occurring in the axioms of § 47 must be level constants of value $\leq n \in \omega$;

(ii) the limit axiom A.7 and the Δ_0-Reflection axioms A.10 are omitted; the logical axioms A.1 (ii)-(iii) are restricted to atoms of $\mathcal{L}_n^*$.

STLR_n has the same rules as STLR^∞, except that $(\forall\omega)$, $(\exists\omega)$ are now replaced by

$$(\forall)^b\ \frac{\ldots\Gamma,\ A(\boldsymbol{i})\ldots \text{ for each } i \leq k \leq n}{\Gamma,\ \forall j \preceq \boldsymbol{k}.\,A}\ ; \qquad (\exists)^b\ \frac{\Gamma,\ A(\boldsymbol{i}), \text{ for some } i \leq k \leq n}{\Gamma, \exists j \preceq \boldsymbol{k}.\,A}\,.$$

We assume that the (finitary) notion of STLR_k-derivability is made precise by rephrasing it in the style of the definition 48.1. To save space, we also assume that the obvious analogues of substitution, weakening and tautology lemmata 48.2, 48.3, 48.4 (i) have been stated and checked for STLR_n.

49.14. $\mathrm{STLR}_k \vdash^m_n \Gamma :=$ "Γ is STLR_k-derivable with length $\leq m$ and cut rank $< n$";

$$\mathrm{STRL}_\omega \vdash \Gamma := \text{“ } \Gamma \text{ is } \mathrm{STLR}_k\text{-derivable for some } k \in \omega\text{”}.$$

We now proceed to a systematic translation of the language with unbounded level quantifiers into the language of STLR_n, which can only deal with quantification on levels $\leq n$. The result is that we can associate to each provable statement of STLR^∞ a family of "approximations", each provable in some STLR_k, for k big enough.

49.15

(i) *Inductive definition of* $A[m,n]$, *for each* $\mathcal{L}_V^*$*-formula* A (m, $n \in \omega$).

1. $(Tt)[m,n] = \exists i \preceq \boldsymbol{n}.\, T_i t$ and $(\neg Tt)[m,n] = \forall i \preceq \boldsymbol{m}.\, (\neg T_i t)$;

   $(Ft)[m,n] = \exists i \preceq \boldsymbol{n}.\, F_i t$ and $(\neg Ft)[m,n] = \forall i \preceq \boldsymbol{m}.\, (\neg F_i t)$;

2. $A[m,n] = A$, for every other atom of $\mathcal{L}_V^*$;

3. $[m,n]$ commutes with $\wedge$, $\vee$, $\forall x$, $\exists x$ and bounded level quantifiers;

4. $(\forall i A)[m,n] = \forall i \preceq \boldsymbol{m}.(A[m,n])$ and $(\exists i A)[m,n] = \exists i \preceq \boldsymbol{n}.(A[m,n])$.

NB: an occurrence of Tt, Ft within a term of the form $[A]$ is not affected by the $[m,n]$-transform; for instance:

$$(F_i Tt)[m,n] = F_i Tt \text{ and } (TTt)[m,n] = \exists i \preceq \boldsymbol{n}.\, T_i Tt.$$

(ii) If A is a $\mathcal{L}_V^*$-formula and $\boldsymbol{k}$ is a level constant, then A^k is the expression, which is obtained from A by replacing each unbounded quantifier $\mathrm{Q}j$ of A by $\mathrm{Q}j \preceq \boldsymbol{k}$ ($\mathrm{Q} = \forall$, $\exists$).

49.15.1. FACT. (i) If $A \in \mathcal{L}_V^*$ and all the L-constants of A have value $\leq k$, then $A[m,n]$ is a formula of $\mathcal{L}_n^*$, provided k, $m \leq n$.

(ii) Let $A \in \mathcal{L}_V^*$: then $A[m,n] = A^m$, if $A \in \Pi$; $A[m,n] = A^n$, if $A \in \Sigma$; $A[m,n] = A$, if $A \in \Delta_0$.

If $\Gamma = \{A_1, \ldots, A_k\}$ is a set of $\mathcal{L}_V^*$-formulas,

$$\Gamma[m,n] = \{A_1[m,n], \ldots, A_k[m,n]\}.$$

We state a simple property, which motivates the $[m,n]$-transform.

49.16. LEMMA (*Persistence*)

(i) *Let A be an $\mathcal{L}_V^*$-formula with L-terms of values $\leq k$, and assume $m' \leq m \leq n \leq n' \leq k$. Then* $\mathrm{STLR}_k \vdash \neg(A[m,n]),\ A[m',n']$.

(ii) *Let Γ be a finite set $\mathcal{L}_V^*$-formulas with L-terms of value $\leq k$, and assume $m' \leq m \leq n \leq n' \leq k$. Then*

$\mathrm{STLR}_k \vdash \Gamma[m,n]$ *implies* $\mathrm{STRL}_k \vdash \Gamma[m',n']$.

PROOF. (i) Induction on A. If A is an atom with $A[m,n] = A$, we are done by tautology lemma. If $A = Tt$ (or Ft), then we must check $\mathrm{STLR}_k \vdash \forall i \preceq \boldsymbol{n}.\neg T_i t,\ \exists i \preceq \boldsymbol{n}'.\, T_i t$. But for each $i \leq n$, $\mathrm{STLR}_k \vdash \neg T_i t,\ T_i t$ (logical axiom), and the conclusion follows with $(\forall)^b$ and $(\exists)^b$. If $A = B \wedge C$, $B \vee C$, $\exists x B$ or $\forall x B$, we simply apply IH and the rules corresponding to the principal connective of A; if $A = \forall j B$ or $\exists j B$, we apply IH and the bounded level quantifier rules.

(ii) follows by repeated application of (i) and cut rule. □

49.17. *Preliminaries.* We fix the initial segment $\prec$ of order type ϵ_0 of the primitive recursive well-ordering of type Γ_0, which was defined in §45. Lower case Greek letters range over elements of *Field*($\prec$); $\alpha < \beta$ is a shortening for "α, $\beta \in$ *Field*($\prec$) and $\alpha \prec \beta$". We also remind that

$$\omega_0 = 1,\ \omega_{m+1} = \omega^{\omega_m}.$$

To each derivability statement $\mathrm{STLR}^\infty \vdash^{\alpha}_{k} \Gamma$, we can naturally associate a construction tree $\mathfrak{D}$, labelled by formulas and locally correct with respect to the rules of STLR^∞, to be regarded as a *derivation of* Γ. By standard proof theory, it is not restrictive to assume that infinitary derivations of STLR^∞ are represented by *primitive recursive trees* and hence encoded by numbers. The exact choice for encoding derivation trees is largely unessential; it suffices that we can primitive recursively read off from $\mathfrak{D}$ all relevant data (final sequent, information on final inference and immediate subderivations of $\mathfrak{D}$, ordinal length and cut rank). Details are given in the appendix of the next chapter for the ramified system RS_n, and they can easily be adapted to the simpler STLR^∞. As a temporary notation, $\mathfrak{D} \vdash^{\alpha}_{k} \Gamma$ will stand for "$\mathfrak{D}$ is a derivation of Γ in STLR^∞ with cut rank k and length $\leq \alpha$". For the sake of simplicity, we do not distinguish the derivation $\mathfrak{D}$ from its code. If Γ is a sequent of STLR^∞ or STLR_ω, define:

49.17.1. $LevPar(\Gamma) := \{k\colon \boldsymbol{k}$ occurs as a level constant in $\Gamma\}$;

$$|\,\Gamma\,| := max\ LevPar(\Gamma).$$

49.18. THEOREM (*Asymmetric interpretation of* STLR^∞ *into* STLR_ω).

We can find a partial recursive function F such that if $\mathfrak{D} \vdash^{\alpha}_{1} \Gamma$, then for every m, $F(\mathfrak{D}, m)$ is defined, $m \leq F(\mathfrak{D}, m)$ and for every $n \geq F(\mathfrak{D}, m)$,

$$|\,\Gamma[m,n]\,| \leq n \text{ and } \mathrm{STLR}_n \vdash \Gamma[m,n].$$

PROOF. We define $F(\mathfrak{D}, m)$ by induction on the length α of $\mathfrak{D}$.

Case 1: Γ is an axiom. We choose $F(\mathfrak{D}, m) = max(m, |\,\Gamma\,|)$. If $F(\mathfrak{D}, m) \leq n$, then trivially $n \geq |\,\Gamma[m,n]\,| = max\{m, n, |\,\Gamma\,|\}$. It suffices to check that the $[m,n]$-transforms of STLR^∞-axioms are derivable in STLR_n, for $n \geq F(\mathfrak{D}, m)$ (we apply persistence and weakening, if necessary).
If Γ is an instance of A.1 (i), A.2, A.3, A′.4, A′.5, A.6, A.8, A.9, the verification is trivial, because $\Gamma[m,n] = \Gamma$ is an axiom of STLR_n. If Γ is an instance of A.1 (ii)-(iii), the conclusion follows by tautology, substitution and a simple persistence argument:

if $rk(B) = 0$ and $B \in \Sigma$, $(\neg B)[m,n] = \neg B^m$ and $B[m,n] = B^n$; but

$\mathrm{STLR}_n \vdash \neg B^m,\ B^n.$

The $[m,n]$-transforms of instances of A.7 have the form $\neg T_{\boldsymbol{i}}t,\ \exists i \preceq \boldsymbol{n}. T_i t$ and $\forall i \preceq \boldsymbol{m}. \neg T_i t,\ \exists i \preceq \boldsymbol{n}. T_i t$, for $m \leq n$; they are easily derivable in STLR_n by use of logical axioms $\neg T_{\boldsymbol{i}} a, T_{\boldsymbol{i}} a$ and $\neg F_{\boldsymbol{i}} a, F_{\boldsymbol{i}} a$ with bounded level quantifier rules. As to the first sequent, we need that the parameter $\boldsymbol{i}$ has value $\leq F(\mathfrak{D}, m)$ (this is ensured by the definition of F at the outset). If we have an instance $\{\neg \forall x \exists i A(x,i),\ \exists j \forall x \exists i \preceq j. A(x,i)\}$ of A.10 (A is Δ_0), its asymmetric transform becomes

$$\{\neg \forall x \exists i \preceq \boldsymbol{m}. A(x,i),\ \exists j \preceq \boldsymbol{n}. \forall x \exists i \preceq j. A(x,i)\},$$

which is trivially STLR_n-derivable as $m \leq n$.

Case 2: (Cut). Then for some $\beta < \alpha$, we can assume (by the analogue of 48.1.3) that $\mathfrak{D}_1 \vdash^{\beta}_{1} \Gamma, B$ and $\mathfrak{D}_2 \vdash^{\beta}_{1} \Gamma, \neg B$, where $\mathfrak{D}_1$, $\mathfrak{D}_2$ are the immediate subderivations of $\mathfrak{D}$. Since $rk(B) = 0$, we may assume without restriction that $B \in \Sigma$ and hence $(\neg B) \in \Pi$.
Fix $m,\ n \geq F(\mathfrak{D}, m) = F(\mathfrak{D}_2, F(\mathfrak{D}_1, m))$ and choose $p_1 = F(\mathfrak{D}_1, m)$. By IH we obtain:

$$F(\mathfrak{D}, m) \geq F(\mathfrak{D}_1, m) \geq m;$$

$$n \geq p_1 \geq |\,\Gamma[m, p_1], B^{p_1}\,| \text{ and } n \geq |\,\Gamma[p_1, n], \neg B^{p_1}\,|;$$

$$\mathrm{STLR}_n \vdash \Gamma[p_1,\ n],\ \neg B^{p_1};$$

$$\mathrm{STLR}_n \vdash \Gamma[m, p_1],\ B^{p_1}.$$

Since $m \leq p_1 \leq n$, we conclude, by downward and upward persistence (respectively),

$$\mathrm{STLR}_n \vdash \Gamma[m,n],\ \neg B^{p_1} \text{ and } \mathrm{STLR}_n \vdash \Gamma[m,n], B^{p_1},$$

whence $\mathrm{STLR}_n \vdash \Gamma[m,n]$ by (Cut).

Case 3. ($\exists\omega$): then $\mathfrak{D}_1 \vdash^{\beta}_{1} \Gamma, A(\boldsymbol{i})$, for some $\beta < \alpha$, $i \in \omega$, $\Gamma = \Delta, \exists j A$ and $\mathfrak{D}_1$ immediate subderivation of $\mathfrak{D}$. Fix m and $n \geq F(\mathfrak{D}, m) = F(\mathfrak{D}_1, m)$. In fact, by IH,

$$F(\mathfrak{D}, m) \geq m \text{ and } n \geq |\,\Gamma[m,n], A[m,n](\boldsymbol{i})\,| \geq i; \tag{1}$$

$$\mathrm{STLR}_n \vdash \Gamma[m,n],\ A[m,n](\boldsymbol{i}). \tag{2}$$

By the second part of (1), we conclude with $(\exists)^b$, applied to (2),

$$\mathrm{STLR}_n \vdash \Delta[m,n],\ \exists j \preceq \boldsymbol{n}. A[m,n](j).$$

Case 4. ($\forall\omega$): we have, for each $k \in \omega$, if $\Gamma = \Delta,\ \forall i A(i)$

$$\mathfrak{D}_k \vdash^{\alpha_k}_{1} \Gamma, A(\boldsymbol{k}).$$

Fix m and $n \geq F(\mathfrak{D}, m) = max\{F(\mathfrak{D}_0, m), \ldots, F(\mathfrak{D}_m, m)\}$; then we have by IH $F(\mathfrak{D}, m) \geq m$ and for each $k \leq m$, $n \geq |\Gamma[m,n], A[m,n](\boldsymbol{k})|$ and

$$\mathrm{STLR}_n \vdash \Gamma[m,n], A[m,n](\boldsymbol{k}).$$

The conclusion follows by the rule of bounded universal level quantification. The other cases are disposed of with similar arguments (persistence and the appropriate logical inference). Finally, if we consider derivations as primitive recursively encoded objects, the defining conditions of F can be regarded as effective equations, having thereby a partial recursive solution by the second recursion theorem. □

49.19. COROLLARY

(i) *If* STLR $\vdash \Gamma$ *and* Γ' *is the* $[0,0]$*-instance of* Γ, *we can effectively find a natural number* k *such that* $\mathrm{STLR}_n \vdash \Gamma'[0,n]$, *for all* $n \geq k$.

(ii) *If* $\mathrm{TLR} + TI(lev) \vdash A$ *and* $A \in \mathfrak{L}_{op}$, *then* $\mathrm{STLR}_\omega \vdash A$.

PROOF. (i) Apply 49.11, 49.8 and 49.18.
(ii) Apply 48.8 and (i). □

49.20. FINAL REMARKS

(i) Let STLR(−) be STLR without transfinite induction on levels A.11. Then STLR(−) still contains TLR and enjoys weak cut elimination (i.e. if $\mathrm{STLR}(-) \vdash^k_q \Gamma$, then also $\mathrm{STLR}(-) \vdash^p_1 \Gamma$ for some p and every Γ). Thus we can avoid the detour through STLR^∞ and we obtain a simpler version of 49.18: if $\mathrm{STLR}(-) \vdash^k_1 \Gamma$, then $\mathrm{STLR}_n \vdash \Gamma'[m,n]$, for all m, $n \geq m+2^k$, and each $[0,m]$-instance Γ' of Γ.

(ii) So as it stands, the index for F in the proof of 49.18 is not primitive recursive. Indeed, if $\mathfrak{D}$ ends with $(\forall\omega)$, the definition of F involves a non-primitive recursive enumeration function U_1 for primitive recursive 1-ary functions: for F satisfies

$$F(\mathfrak{D}, m) = max\{F(U_1(h(\mathfrak{D}), i), m) : i \leq m\},$$

where $h(\mathfrak{D}) = e$ is a primitive recursive index, primitive recursively read off via h from $\mathfrak{D}$, and $U_1(e,n)$ encodes $\mathfrak{D}_n$. Of course, this does not exclude the existence of sharper upper bounds on the complexity of F; we leave this as a problem: is there any primitive recursive function F satisfying 49.18?

(iii) As a refinement, observe that the partial recursive function F of theorem 49.18 can be made total and indeed "ω_k-recursive" (for $k \in \omega$ sufficiently big). In fact, if we adapt to STLR^∞ the notions of derivation code and label from the appendix of Ch. XI , we can directly introduce F by combining definition by cases on primitive recursive clauses, course-of-value recursion and recursion on the (possibly) infinitary tree $\mathfrak{D}$ of height α such that $\mathfrak{D} \vdash^{\alpha}_{1} \Gamma$. For the reader's sake, we define ω_k-recursion (after Schwichtenberg 1977):

49.20.1. $\mathfrak{R}(\omega_k)$ – *the class of* ω_k*-recursive number-theoretic functions* – is the least class which contains the primitive recursive functions and is closed under composition and the schema of (nested) ω_k-recursion (for $k > 0$):

$$G(0, \vec{a}) = H(\vec{a});$$

$$G(\beta, \vec{a}) = t(\beta, \vec{a}, G\lceil\beta), \text{ for } 0 < \beta < \omega_k;$$

here $\vec{a}$ stands for a finite list of natural numbers, $t(y, \vec{x}, \Phi)$ is a term built up from the number variables $y, \vec{x}$, function symbols for elements of $\mathfrak{R}(\omega_k)$ already introduced, and the function variables Φ (of suitable arity);

$$\text{further } (G\lceil\beta)(\delta, \vec{r}) = \begin{cases} G(\delta, \vec{r}) \text{ if } \delta < \beta \\ 0 \text{ , otherwise.} \end{cases}$$

By 46.2.3 we know:

49.20.2. For each $n \in \omega$ and $A \in \mathcal{L}_{op}$, OP proves

$$TI(\omega_n, A) := \forall\alpha(\forall\beta(\beta < \alpha \to A(\beta)) \to A(\alpha)) \to \forall\alpha(\alpha{<}\omega_n \to A(\alpha)).$$

As a consequence of 49.20.2 and proof theory, we can also show:

49.20.3. A function F is ω_k-recursive iff F is provably recursive in OP.
{If F is a m-ary number-theoretic function ($m > 0$), F is *provably recursive in* OP iff there is an $\exists(+)$-formula $A(x_1, \ldots, x_m, y)$ of OP (see Ch. I, 4.13), such that:

$$F(n_1, \ldots, n_m) = p \text{ implies } \mathrm{OP} \vdash A(\overline{n}_1, \ldots, \overline{n}_m, \overline{p});$$

$$\mathrm{OP} \vdash \forall x_1 \ldots \forall x_m(Nx_1 \wedge \ldots \wedge Nx_m \to \exists! y(Ny \wedge A(x_1, \ldots, x_m, y)))\}.$$

Now 49.18 can be formalized in $\mathrm{OP} + TI(\omega_k)$, for each given $k > 0$ and derivations of length $<\omega_k$. Therefore the partial recursive function F of the asymmetric interpretation theorem should be at most ω_k-recursive. These remarks are useful for the conservation theorems of Ch. XI.

We conclude with a second problem, concerning a requirement of greater uniformity on F: is it possible to define a majorizing function F, which satisfies the conditions of the asymmetric interpretation theorem 49.18 and *only depends on the height* α *of* $\mathfrak{D}$ *and* m?

§50. The infinitary sequent calculus IT_n^∞ of n-iterated reflective truth

We devise a new system IT_n^∞, in which level variables, bounded level quantifiers and level atoms are omitted.

50.1. *The syntax of* IT_n^∞

50.1.1. The language $\mathcal{L}_n$ of IT_n^∞ is $\mathcal{L}_n^*$ without: (i) bound level variables and bounded level quantifiers; (ii) the function symbol LT; (iii) the predicate symbols $=_l$, $\preceq$.
The L-terms of $\mathcal{L}_n$ are the level constants of value $\leq n$; the individual terms of $\mathcal{L}_n$ coincide with $\mathcal{L}$-terms (they are generated without the clause: if $\boldsymbol{i}$ is an L-term, $LT(\boldsymbol{i})$ is a term).
The atoms of $\mathcal{L}_n$ have the following form: Nt, $t = s$, $\neg Nt$, $\neg t = s$ (e-atoms); T_it, F_it, $\neg T_it$, $\neg F_it$ (T-atoms; t, s are $\mathcal{L}_n$-terms, i is a level constant of $\mathcal{L}_n$). $\mathcal{L}_n$-formulas are inductively generated from $\mathcal{L}_n$-atoms by means of $\wedge$, $\vee$, $\forall x$, $\exists x$.

50.1.2. Since LT is omitted, we must redefine the map $A \mapsto [A]$, for $A \in \mathcal{L}_n$: we stick to the previous definition (see 36.3), except that we let $[T_{\boldsymbol{i}}t] := \langle \overline{7}, \langle \overline{i}, t\rangle\rangle$, where the boldface occurrence of i stands for a level constant of $\mathcal{L}_n$, while the overlined occurrence denotes the numeral (= closed term of $\mathcal{L}$; see Ch. I), whose value is the value of i.

CONVENTION. Unless it is unclear from the context, we keep using the same symbols i, k, j, n, m for level constants and their values.

50.1.3. *Simultaneous inductive definition of* POS_n *and* NEG_n.

(i) POS_n is the least class of $\mathcal{L}_n$-formulas, which contains every A of the form $t = s$, $\neg t = s$, Nt, $\neg Nt$, F_it, $\neg F_it$, T_it, $\neg T_it$, for $i < n$, T_nt, F_nt and is closed under $\wedge$, $\vee$, $\forall$, $\exists$; if $A \in POS_n$, A is said n-positive.

(ii) NEG_n is the least class of $\mathcal{L}_n$-formulas, which contains every A of the form $t = s$, $\neg t = s$, Nt, $\neg Nt$, F_it, $\neg F_it$, T_it, $\neg T_it$, for $i < n$, $\neg T_nt$, $\neg F_nt$ and is closed under $\wedge$, $\vee$, $\forall$, $\exists$; if $A \in NEG_n$, A is said n-negative.

(iii) A is n-separated iff $A \in POS_n$ or $A \in NEG_n$.

50.1.4. DEFINITION of $K_n(A)$ (= n-complexity of $A \in \mathcal{L}_n$).
(i) $K_n(A) = 0$ if A is n-separated; else:
(ii) $K_n(B \circ C) = max(K_n(B), K_n(C))+1$ ($\circ = \wedge, \vee$);
(iii) $K_n(QxB) = K_n(B)+1$ ($Q = \forall$, $\exists$).

For a proper statement of the axioms, it is convenient to recall the obvious finitary generalizations of $\wedge$ and $\vee$:

50.1.5. $\sqcap\{A_i : i \leq 0\} = A_0 = \sqcup\{A_i : i \leq 0\}$;

$$\sqcap\{A_i : i \leq n{+}1\} = (\sqcap\{A_i : i \leq n\}) \wedge A_{n+1};$$
$$\sqcup\{A_i : i \leq n{+}1\} = (\sqcup\{A_i : i \leq n\}) \vee A_{n+1}.$$

Then we can define the "finitary" versions of T_i- and F_i-*Clause*(t) for the language $\mathcal{L}_n$.

50.1.6.1

$$T_n\text{-}Clause^*(t) := \exists x \exists y((T = [x = y] \wedge x = y) \vee (t = [Nx] \wedge Nx) \vee$$
$$\vee ((t = [Tx] \vee t = [T_n x]) \wedge T_n x) \vee (\sqcup\{t = [T_j x] \wedge T_j x : j < n\}) \vee$$
$$\vee (t = (\neg x) \wedge F_n x) \vee (t = (x \wedge y) \wedge T_n x \wedge T_n y) \vee (t = (\forall x) \wedge \forall v. T_n(xv))).$$

50.1.6.2

$$F_n\text{-}Clause^*(t) := \exists x \exists y((t = [x = y] \wedge \neg(x = y)) \vee (t = [Nx] \wedge \neg Nx\ ) \vee$$
$$\vee ((t = [Tx] \vee t = [T_n x]) \wedge F_n x) \vee (\sqcup\{t = [T_j x] \wedge \neg T_j x : j < n\}) \vee$$
$$\vee (t = (\neg x) \wedge T_n x) \vee (t = (x \wedge y) \wedge (F_n x \vee F_n y)) \vee (t = (\forall x) \wedge \exists v. F_n(xv))).$$

50.2. *The system* IT_n^∞ is a sequent calculus in the language $\mathcal{L}_n$. In order to avoid repetitions, we shall maintain, as far as possible, the same terminology of the axioms and rules for STLR and STLR_n (§47; §49B).

50.2.1. *Axioms of* IT_n^∞. We assume the substitution closure of the following sets of sequents.

Logical axioms:

(i) $a = a$; (ii) $\neg a = b, \neg A(a), A(b)$; (iii) $\neg A, A$.

Proviso: in (ii)-(iii) A is an arbitrary $\mathcal{L}_n$-formula with $K_n(A) = 0$.

Operational axioms:

(i) $Kab = a$ and $Sabc = ac(bc)$;
(ii) $\neg C = C'$ (C, C' distinct individual constants);
(iii) $(\langle a_1, a_2 \rangle)_i = a_i$ ($i = 1, 2$);
(iv) $D\overline{m}\,\overline{m}cd = c$; (v) $D\overline{k}\,\overline{m}cd = d$;
(vi) $\neg(\overline{m+1}) = \overline{0}$; (vii) $PRED(\overline{m+1}) = \overline{m}$.

Proviso: in (iv)-(vii) $\overline{k}$, $\overline{m}$ are arbitrary distinct numerals.

Persistence: $\neg T_i a, T_j a$ ($i < j$).

Consistency: $\neg T_i a, \neg F_i a$.

Fixed point axioms:

(i) $\neg T_i$-*Clause*$^*(t)$, $T_i t$ and $\neg T_i t$, T_i-*Clause*$^*(t)$;

(ii) $\neg F_i\text{-}Clause^*(t), F_it$ and $\neg F_it, F_i\text{-}Clause^*(t)$.

NB: the formulas occurring in the IT_n^∞-axioms are all n-separated.

50.2.2. *Rules of* IT_n^∞: IT_n^∞ contains the logical inferences $(\wedge)$, $(\vee)$, $(\forall x)$, $(\exists x)$, (Cut), and, in addition, the rules $(\neg N)$, (N) below:

$$(\neg N)\quad \frac{\ldots\Gamma, \neg t = \overline{m}\ \ldots(\text{for every } m \in \omega)}{\Gamma, \neg Nt};$$

$$(N)\quad \frac{\Gamma, t = \overline{m}\ (\text{for some } m \in \omega)}{\Gamma, Nt}.$$

IT_n^∞-rules and IT_n^∞-axioms determine, as usual, IT_n^∞-derivability for finite sequents Γ of $\mathcal{L}_n$-formulas. Below we let lower case Greek letters α, $\beta, \ldots$ range over arbitrary ordinals $< \Gamma_0$ (cf. Ch. IX).

50.2.3. *Inductive definition of the derivability relation* $\mathrm{IT}_n^\infty \vdash^\alpha_\rho \Gamma$ $(n \in \omega)$.

1.1. If Γ contains (as a set) an axiom of 50.2.1, $\mathrm{IT}_n^\infty \vdash^\alpha_\rho \Gamma$, for every α and every $\rho \leq \omega$;

1.2. Assume that:

(i) Γ is the conclusion from the premises Γ_i of a logical rule, (N), $(\neg N)$ or a cut rule applied to formulas of n-complexity $< \rho \leq \omega$;

(ii) $\mathrm{IT}_n^\infty \vdash^{\alpha_i}_\rho \Gamma_i$ (for $i \leq \omega$) and $\alpha_i < \alpha$.

Then $\mathrm{IT}_n^\infty \vdash^\alpha_\rho \Gamma$.

50.2.4. NOTATIONS. (i) $\mathrm{IT}_n^\infty \vdash^\alpha_\rho \Gamma := \Gamma$ is IT_n^∞-derivable with length $\leq \alpha$ and cut complexity $< \rho$.

(ii) $\mathrm{IT}_n^\infty \vdash^{<\alpha}_{<\omega} \Gamma$ $:= \mathrm{IT}_n^\infty \vdash^\beta_k \Gamma$, for some $\beta < \alpha$ and some $k < \omega$.

{The choice of n-complexity will be clear from the partial cut elimination theorem for IT_n^∞ of Ch. XI}.

§51. Embedding $\mathbf{STLR}_n$ into $\mathbf{IT}_n^\infty$

We define a translation e_n of $\mathcal{L}_n^*$ into $\mathcal{L}_n$, which transforms provable sentences of STLR_n into provable sentences of IT_n^∞. If n is fixed, we simply write e instead of e_n.

51.1. *Inductive definition of* $e(t)$, *for each* $\mathcal{L}_n^*$-*term* t.

(i) $t^e = t$, if t is an individual parameter or an individual constant;

(ii) $(LT(\boldsymbol{j}))^e = \overline{j}$ {remind that the boldface j is an L-constant, while the overlined one is the numeral having the value of the given constant};

(iii) $(Ap(t,s)^e = Ap(t^e, s^e)$.

51.1.1. REMARK. By (ii) above and 50.1.2, if $[T_i t]$ is the term which encodes $T_i t$ in $\mathcal{L}_n^*$, $([T_i t])^e = \langle \overline{7}, \langle \overline{i}, t^e \rangle\rangle$ is the term which encodes $T_i(t^e)$ in $\mathcal{L}_n$.

51.2. *Inductive definition of* $e_n(A)$, *for each formula* A *of* $\mathcal{L}_n^*$.

We write $e(A)$ for $e_n(A)$, while $\perp$ stands for the $\mathcal{L}$-atom $\overline{0} = \overline{1}$.

(i) $(t = s)^e = (t^e = s^e)$; $(Nt)^e = N(t^e)$;
$(T_i t)^e = T_i(t^e)$ and $(F_i t)^e = F_i(t^e)$;
$(A)^e = \neg\perp$ if A is a positive L-atom which is true in the standard number-theoretic interpretation; else, $(A)^e = \perp$;
$(\neg A)^e = \neg (A)^e$ if A is a positive atom;

(ii) $(A \circ B)^e = A^e \circ B^e$ ($\circ = \vee, \wedge$); $(QxA)^e = Qx(A^e)$ ($Q = \forall, \exists$);

(iii) $(\forall j \preceq \boldsymbol{k}.B(k))^e = \sqcap\{B(\boldsymbol{j}) : j \leq k\}$ and

$(\exists j \preceq \boldsymbol{k}.B(j))^e = \sqcup\{B(\boldsymbol{j}) : j \preceq k\}$.

It is immediate to check that:

51.3. LEMMA

(i) *If* t *is a term of* $\mathcal{L}_n^*$, t^e *is a term of* $\mathcal{L}_n (= \mathcal{L})$ *with the same parameters of* t; *if* A *is a formula of* $\mathcal{L}_n^*$, *then* A^e *is a formula of* $\mathcal{L}_n$ *with the same parameters of* A.

(ii) *If* A *is a formula of* $\mathcal{L}_n$, $A^e = A$.

51.4. FACT. IT_n^∞ satisfies the analogues of substitution, inversion for ($\wedge$), ($\vee$), ($\forall x$), weakening and tautology lemmata of §48.

A proof of 51.4 will follow from §52 {and the appendix of the next chapter}.

51.5. LEMMA. (i) *If* A *is an arbitrary* $\mathcal{L}_n$-*formula and* $m \geq 2 \cdot K_n(A)$,

$$\mathrm{IT}_n^\infty \vdash_0^m A, \neg A; \quad \mathrm{IT}_n^\infty \vdash_0^m \neg t = s, \neg A[x := t], A[x := s].$$

(ii) *If* $A(x)$ *is an arbitrary* $\mathcal{L}_n$-*formula and* $k = K_n(A)+1$,

$$\mathrm{IT}_n^\infty \vdash_k^\omega \neg A(\overline{0}), \neg\forall x(A(x) \rightarrow A(x+1)), \neg Nt, A(t).$$

(iii) $\mathrm{IT}_n^\infty \vdash_{<\omega}^{<\omega} (\overline{p} = \overline{m})$ or $\mathrm{IT}_n^\infty \vdash_{<\omega}^{<\omega} \neg(\overline{p} = \overline{m})$, for every p, $m \in \omega$;

(iv) $\mathrm{IT}_n^\infty \vdash \,^{<\omega}_{<\omega}\, \Gamma$, *provided* Γ *is a sequent of the following form:*

$\{\neg Na,\ \neg Nb,\ \neg a = b,\ Dabcd = c\}$;

$\{\neg Na,\ \neg Nb,\ a = b,\ Dabcd = d\}$; $\{N\overline{0}\}$;

$\{\neg Na,\ N(a+1)\}$; $\{\neg Na,\ \neg(a+1) = \overline{0}\}$; $\{\neg Na,\ \ PRED(a+1) = a\}$.

PROOF. (i): induction on n-complexity of A.
(ii): similar to 49.9 (apply (i) and the rule $(\neg N)$ of 50.2.2).
(iii): apply the logical axiom (i) and the operational axioms (vi), (vii) of 50.2.1.
(iv): the proof is easy by use of (N), $(\neg N)$ and the operational axioms (iv)-(vii). We check $\mathrm{IT}_n^\infty \vdash \,^{<\omega}_{<\omega}\, \neg Na,\ PRED(a+1) = a$. First of all, for each $p \in \omega$, we have:

$$\mathrm{IT}_n^\infty \vdash^0_0 PRED(\overline{p+1}) = \overline{p};$$

$$\mathrm{IT}_n^\infty \vdash^0_0 \neg\overline{p} = a,\ \neg PRED(\overline{p+1}) = \overline{p},\ PRED(a+1) = a.$$

An application of (cut) yields $\mathrm{IT}_n^\infty \vdash^1_1 \neg\overline{p} = a,\ PRED(a+1) = a$, for every $p \in \omega$ and $(\neg N)$ yields the conclusion. □

51.6. THEOREM. *If* $\mathrm{STLR}_n \vdash \Gamma$, *then* $\mathrm{IT}_n^\infty \vdash^\alpha_m \Gamma^e$, *for some finite* m *and some* $\alpha < \omega 2$ *(of course* $\Gamma^e = \{A_1{}^e, \ldots, A_k{}^e\}$ *for* $\Gamma = \{A_1, \ldots, A_k\}$*).*

PROOF. By induction on k such that $\mathrm{STLR}_n \vdash^k \Gamma$.

Case 1. Assume Γ is an instance of a STLR_n-axiom.

1.1. Γ is an operational axiom (a Peano axiom, a fixed point axiom, a consistency axiom, an instance of Δ_0-N-induction, persistence, or a logical axiom): then Γ^e is IT_n^∞-derivable by 51.5 and the corresponding axioms of IT_n^∞.

1.2. Γ is a level axiom of STLR_n. Then either $\Gamma = \Delta,\ \neg LT(\boldsymbol{i}) = LT(\boldsymbol{j})$, for $i \neq j$ or $\Gamma = \Delta,\ A$, where A is a true atom of the form $(\boldsymbol{i} = \boldsymbol{j})$, $\neg(\boldsymbol{i} = \boldsymbol{j})$, $(\boldsymbol{i} \preceq \boldsymbol{j})$, $\neg(\boldsymbol{i} \preceq \boldsymbol{j})$ $(i,\ j \leq n)$. It follows that either $\Gamma^e = \Delta^e,\ \neg(\overline{i} = \overline{j})$, or else $\Gamma^e = \Delta^e,\ \neg\perp$; both sequents are derivable in IT_n^∞ by 51.5 (iii).

Case 2: Γ is not an axiom. By IH the e-translations of the premises of the logical rules and cut are IT_n^∞-derivable: since the ^{e}-map commutes with the corresponding logical operators, we conclude by use of $(\wedge)$, $(\vee)$, $(\forall)$, $(\exists)$, and (Cut). As to the bounds on the length and cut complexity, only the interpretation of the Δ_0-N-induction schema requires the transfinite ordinal ω; in the other steps, we only finitely increase length and cut complexity. □

51.7. COROLLARY

(i) *If $B(x)$ is an $\mathcal{L}$-formula which is elementary extensional in x and*

$$\mathrm{TLR} + TI(lev) \vdash \forall x(Cl(x) \to B(x)),$$

then we can effectively find a natural number n such that, for some $k < \omega$, $\alpha < \omega 2$:

$$\mathrm{IT}_n^\infty \vdash_k^\alpha \forall x(Cl_0(x) \to B_0(x));$$

(here $B_0(x)$ results from $B(x)$ by replacing T with T_0).

(ii) *In particular, if A is a formula in the operational fragment $\mathcal{L}_{op}$ and $\mathrm{TLR} + TI(lev) \vdash A$, then $\mathrm{IT}_n^\infty \vdash_k^\alpha A$, for some n, $k < \omega$, $\alpha < \omega 2$.*

PROOF. By 48.8 and 49.19, we can find n such that

$$\mathrm{STLR}_n \vdash \forall x(Cl_0(x) \to (B(x))[0,n]). \tag{1}$$

By induction on B, using consistency and persistence axioms, we have

$$\mathrm{STLR}_n \vdash \neg Cl_0(a),\ \neg B(a)[0,n],\ B_0(a). \tag{2}$$

By (1)-(2) with inversion, cut and $(\forall x)$, $\mathrm{STLR}_n \vdash \forall x(Cl_0(x) \to B_0(x))$; since $(\forall x(Cl_0(x) \to B_0(x)))^e = (\forall x(Cl_0(x) \to B_0(x))$, the conclusion follows with 51.6.
(ii): in addition to the previous argument, recall that the $[m,n]$-transform is the identity map on the operational language. □

We conclude the chapter with a model-theoretic application; we assume that the reader keeps in mind the recursion-theoretic model $\mathfrak{C}_\iota$ of Ch. VIII, § 39 and the related notions.

Let $\mathfrak{C}_\omega$ be the first order structure $\langle \mathbf{CTM}, \{\mathcal{V}_k : k \in \omega\}\rangle$, where **CTM** denotes the closed term model of OP and $\{\mathcal{V}_k : k < \omega\}$ is the function $\mathcal{V}$ of 39.14, restricted to ω. Now let $\mathcal{V}_i$, $\mathcal{V}_\omega := \bigcup\{\mathcal{V}_k : k < \omega\}$ be the intended interpretations of T_i, T (respectively), while level variables range over ω. It is clear that an obvious adaptation of 39.16 yields:

51.8. PROPOSITION. *$\mathfrak{C}_\omega$ is a model of* IT_n^∞, for each $n \in \omega$ *(i.e. if A is a sentence of $\mathcal{L}_n$ and $\mathrm{IT}_n^\infty \vdash A$, then $\mathfrak{C}_\omega \models A$).*

51.9. THEOREM

If $\mathrm{TLR} + TI(lev) \vdash \forall i \exists j \forall x(Cl_i(x) \to \exists y(Cl_j(y) \wedge A(x,y)))$, where $A(x,y)$ is an $\mathcal{L}$-formula, elementary extensional in x, y, then

$$\mathfrak{C}_\omega \models \forall i \exists j \forall x(Cl_i(x) \to \exists y(Cl_j(y) \wedge A(x,y))).$$

PROOF. Let $B := \forall i \exists j \forall x(Cl_i(x) \rightarrow \exists y(Cl_j(y) \wedge A(x,y)))$, where $A(x,y)$ is a formula of $\mathfrak{L}$, elementary extensional in x, y and $\mathrm{TLR} + TI(lev) \vdash B$. By use of 48.8, 49.8, 49.11 and 51.6, we can find, for each m, a number $n \geq m$ such that

$$\mathrm{IT}_n^\infty \vdash \forall x(Cl_m(x) \rightarrow \exists y(Cl_n(y) \wedge A_{mn}(x,y))), \tag{1}$$

where $A_{mn}(x,y)$ results from $A(x,y)$ by replacing each occurrence of $t\eta x$ ($t\eta y$) with $t\eta_m x$ ($t\eta_n y$). By 51.8 and (1) we have

$$\mathfrak{C}_\omega \models \forall i \exists j \forall x(Cl_i(x) \rightarrow \exists y(Cl_j(y) \wedge A_{ij}(x,y))). \tag{2}$$

By assumption on A, we also have, for m, $n \in \omega$:

$$\mathfrak{C}_\omega \models \forall x \forall y(Cl_m(x) \wedge Cl_n(y) \wedge A_{mn}(x,y) \rightarrow A(x,y)). \tag{3}$$

(2)-(3) yield the required conclusion. □

However, theorem 51.9 is still unsatisfactory, because it does not make any deep use of the constructive information, associated to the proof tree of the given theorem of $\mathrm{TLR} + TI(lev)$. In chapter XI, we shall give a finer proof-theoretic interpretation.

CHAPTER 11

PROOF-THEORETIC INVESTIGATION OF FINITELY ITERATED REFLECTIVE TRUTH

§52. The ramified system RS_n
§53. Cut elimination
§54. Some derivable sequents of RS_n
§55. Embedding IT_n^∞ into RS_n
§56. The upper bound theorem for IT_n^∞
§57. Upper bound theorems for TLR and its subsystems
§58. Conclusion: the conservation theorems
Appendix: primitive recursive cut elimination for RS_n

In chapter X we embedded the theory of truth with levels into infinitary systems IT_n^∞ with iterated truth predicates, where level variables and level quantifiers are explained away. We proceed further ahead in investigating the arithmetical content of the systems IT_n^∞ and TLR.

We first design an infinitary system RS_n in which T_n is split into a family $\{T_n^\alpha : \alpha < \Gamma_0\}$ of approximations. The T_n^α's are linked together by natural recursive conditions, which can be encoded by symmetric introduction rules with the cut elimination property (§§52-53).

Then we embed IT_n^∞ into RS_n by a modified version of the asymmetric interpretation technique of §49 (see §§54-55). The analysis of cut-free RS_n-derivations readily implies that *RS_n-theorems of level $< n$ are already derivable without T_n-rules*, of course at cost of greatly increasing the derivation length. This increase is given an upper bound with the main theorems of §§56-57.

As a final step, if we formalize the whole reduction procedure, we realize that the proof-theoretic analysis only involves OP-principles, except for the schema $TI(<\Gamma_0)$ of transfinite induction for operational formulas along each $\alpha < \Gamma_0$. We can conclude – with the help of the main result of Ch.VIII – that the operational consequences of TLR are already consequences of $OP + TI(<\Gamma_0)$.

Similar results hold for the systems MF, MF_p, MF_c of Ch. II. In particular, the operational consequences of MF (MF_p, MF_c) are axiomatized by $OP + TI(<\phi\epsilon_0 0)$ ($OP + TI(<\phi\omega 0)$, OP respectively).

§52. The ramified system RS_n

To a certain extent, the system RS_n is meant to be a constructive simulation of the n-th stage in the recursion-theoretic model of Ch.VIII (see §39). The truth predicates of level $< n$ are assumed as given and satisfy closure conditions corresponding to the IT_n^∞-axioms for the predicates T_i and F_i with $i < n$; on the contrary, T_n and F_n are built up in stages (here ordinals $< \Gamma_0$), and there are rules for passing from stage α to stage $\alpha+1$ and for collecting information at limit stages λ.

The essential fact is that the corresponding rules can be symmetrically arranged as introduction rules for $T^{\alpha+1}$ and $\neg T^{\alpha+1}$ (similarly for $F^{\alpha+1}$ and $\neg F^{\alpha+1}$). As a consequence, the standard predicative cut elimination procedure of Schütte applies, and this property grants elimination of level n statements from derivations, whose conclusion refers only to lower levels.

As the reader will see, each RS_n is an infinitary calculus; however, it is well-known how to represent infinite derivations by suitable *finite data structures* (Mints 1975, Schwichtenberg 1977, Buchholz 1991). It follows that the cut elimination procedure is indeed primitive recursive and "tractable", within a fragment of OP, possibly expanded by transfinite induction principles.

52.1. *The syntax of* RS_n. The ramified language $\mathcal{L}_{n,r}$ of RS_n contains the following primitive symbols:

(i) individual variables $x_0, x_1, x_2 \ldots$ (x, y, z are metavariables);

(ii) individual constants $\overline{0}$, SUC, $PRED$, $PAIR$, $LEFT$, $RIGHT$, D; the function symbol for application Ap (binary);

(iii) predicate symbols $True_n$ and $False_n$, $=$, Vr, Fl (binary) and N(unary);

(iv) level constants for each level $< n$ (i, j, k syntactical variables);

(v) ordinal constants $\overline{\alpha}$, for each ordinal $\alpha < \Gamma_0$;

(vi) the logical constants $\vee$, $\wedge$, $\neg$, $\forall$, $\exists$.

52.1.1. The L-terms are exactly the L-constants; the o-terms are exactly the ordinal constants; i, j, k range over L-terms, while we ambiguously use lower case Greek letters α, β, γ, etc..., both for o-terms and the corresponding ordinals. $\mathcal{L}_{n,r}$-terms are inductively generated from individual constants and variables by use of applications; thus they coincide with the terms of the underlying combinatory logic, and we stick to the previous conventions and definitions. Expressions of the form $t = s$, $\neg t = s$, Nt, $\neg Nt$ are called e-atoms, while $Vr(i,t)$, $\neg Vr(i,t)$, $Fl(i,t)$, $\neg Fl(i,t)$, $True_n(t,\alpha)$, $\neg True_n(t,\alpha)$, $False_n(t,\alpha)$, $\neg False_n(t,\alpha)$ are called T-atoms

(t, s range over individual terms, i is an L-term and α is an o-term).

$\mathcal{L}_{n,r}$-formulas are inductively generated from $\mathcal{L}_{n,r}$-atoms by use of the logical constants $\wedge$, $\vee$, $\forall$, $\exists$. It is understood that we adopt the variable separation property VSP of §47. We also adopt the more perspicuous abbreviations $T_i t := Vr(i,t)$ and $F_i t := Fl(i,t)$, while $T^\alpha_n(t) := True_n(t,\alpha)$ and $F^\alpha_n(t) := False_n(t,\alpha)$. We generally omit the subscript n and we simply write $T^\alpha(t)$ and $F^\alpha(t)$, whenever n is clear from the context. Similarly $t\eta^\alpha s := T^\alpha(st)$ and $t\bar{\eta}^\alpha s := F^\alpha(st)$.

NB. The map $A \mapsto [A]$ is left unchanged (see 50.1.2); so it *is well defined for* $\mathcal{L}_n$*-formulas and formulas of* $\mathcal{L}_{n,r}$ *not containing the new predicates* T^α and F^α; the expression $\{x : x\eta^\alpha y\}$ does not make sense.

52.2. If $A \in \mathcal{L}_{n,r}$, $Lc(A)$(= the logical complexity of A) is the number of logical symbols occurring in A ($\neg$ excluded).

52.3. *Level and n-stage of A (A formula* $\mathcal{L}_{n,r}$*).* Let $\circ = \wedge, \vee$, $Q = \forall, \exists$.

(i) $Lev(T_j t) = j$ and $Lev(T^\alpha_n) = n$; if A is an e-atom, $Lev(A) = 0$;
$Lev(B \circ C) = max(Lev(B), Lev(C))$ and $Lev(QxB) = Lev(B)$.

(ii) $St_n(A) = 0$, if A is an e-atom or $Lev(A) < n$;
$St_n(A) = \alpha$, if A is a T-atom of level n and α occurs in A;
$St_n(B \circ C) = max\{St_n(B), St_n(C)\}$ and $St_n(QxB) = St_n(B)$.

52.4. *n-rank of A (A formula of* $\mathcal{L}_{n,r}$*).*

$R_n(A) = 0$, if $n = 0$ and A is an e-atom or $Lev(A) < n$; else:
$R_n(T^\alpha t) = R_n(\neg T^\alpha t) = \omega\alpha$;
$R_n(B \circ C) = max\{R_n(B), R_n(C)\}+1$ and $R_n(QxB) = R_n(B)+1$.

52.5. *The* α*-transform* A_α of a formula $A \in \mathcal{L}_n$ is the $\mathcal{L}_{n,r}$-formula, which is obtained from A by replacing each occurrence of the atoms $(\neg)T_n t$, $(\neg)F_n t$ respectively with $(\neg)T^\alpha t$, $(\neg)F^\alpha t$ respectively. Similarly, the α-transform of $A \in \mathcal{L}$ ($\mathcal{L}$ = language of the basic systems of Ch. II) is obtained from A by replacing each occurrence of $(\neg)Tt$ in A with $(\neg)T^\alpha$t.

52.6. LEMMA

(i) *If A is a formula of* $\mathcal{L}_{n,r}$, $R_n(A) = \omega \cdot St_n(A)+m$, *for some* $m \in \omega$;

(ii) *If A is a* $\mathcal{L}_n$*-formula (or a formula of* $\mathcal{L}$*),* $R_n(A_\alpha) < \omega(\alpha+1)$.

Before specifying axioms and rules of the new system, we define, in analogy with 50.1.6:

52.7

(i) $T^{\alpha}\text{-}Clause(t) := \exists x \exists y((t = [x = y] \wedge x = y) \vee (t = [Nx] \wedge Nx) \vee$
$\vee ((t = [Tx] \vee t = [T_n x]) \wedge T^{\alpha}x) \vee (\sqcup \{t = [T_j x] \ \wedge \ T_j x : j < n\}) \vee$
$\vee (t = (\neg x) \wedge F^{\alpha}x) \vee (t = (x \wedge y) \wedge T^{\alpha}x \ \wedge T^{\alpha}y) \vee (t = (\forall x) \wedge \forall v. T^{\alpha}(xv))).$

(ii) $F^{\alpha}\text{-}Clause(t) := \exists x \exists y((t = [x = y] \wedge \neg x = y) \vee (t = [Nx] \wedge \neg Nx) \vee$
$\vee ((t = [Tx] \vee (t = [T_n x]) \wedge \ F^{\alpha}x) \vee (\sqcup \{t = [T_j x] \ \wedge \neg T_j x : j < n\}) \vee$
$\vee (t = (\neg x) \wedge T^{\alpha}x \ ) \vee (t = (x \wedge y) \wedge (F^{\alpha}x \vee F^{\alpha}y)) \vee (t = (\forall x) \wedge \exists v. F^{\alpha}(xv))).$

RS_n is a Tait-style sequent calculus, like the systems STLR and IT_n^{∞} of §§47-50; so, we have to specify axioms in sequent form and introduction rules for the logical and mathematical primitives.

52.8. *Axioms of* RS_n. We assume the substitution closure of the following sets of sequents.

52.8.1. *LOG* (= Logical axioms):

(i) $t = t$;
(ii) $\neg t = s, \neg A(t), A(s)$ $(R_n(A) = 0)$;
(iii) $\neg A, A$ $(R_n(A) = 0)$.

52.8.2. *OPER* (= Operational axioms):

(i) $Kab = a$ and $Sabc = ac(bc)$;
(ii) $\neg C = C'$ (C, C' distinct individual constants);
(iii) $(\langle a_1, a_2 \rangle)_i = a_i$, where $i = 1, 2$;
(iv) $D\overline{m}\,\overline{m}\,cd = c$;
(v) $D\overline{k}\,\overline{m}\,cd = d$;
(vi) $\neg(\overline{m}{+}1) = 0$;
(vii) $PRED(\overline{m}{+}1) = \overline{m}$.
Proviso: $\overline{k}$, $\overline{m}$ stand for distinct numerals.

52.8.3. $PERS_{ij}$ (= Persistence): $\neg T_i a, T_j a$, for $i < j < n$;

52.8.4. $CONS_i$ (= Consistency): $\neg T_i a, \neg F_i a$ $(i < n)$;

52.8.5. FIX_i (= Fixed point axioms for level $i < n$):

(i) $\neg T_i\text{-}Clause^*(a), T_i a$ and $\neg T_i a, T_i\text{-}Clause^*(a)$;

(ii) $\neg F_i\text{-}Clause^*(a), F_i a$ and $\neg F_i a, F_i\text{-}Clause^*(a)$.

T_i-*Clause*$^*(t)$, F_i-*Clause*$^*(t)$ are defined in the previous chapter (see § 50.1.6).

52.8.6. INI_n (= initial T_n-axioms): $\neg T^0 t$ and $\neg F^0 t$, if $n = 0$; else, if $n > 0$:

$$\neg T^0 t\ , T_{n-1} t\ ; \quad \neg F^0 t\ , F_{n-1} t;$$

$$\neg T_{n-1} t,\ T^0 t\ ; \quad \neg F_{n-1} t\ ,\ F^0 t\ .$$

NB: the level terms occurring in the axioms are all $< n$; if $n = 0$, $CONS_i$, FIX_i, $PERS_{ij}$ must be omitted, for $i < n$. Every formula occurring in the axioms has n-rank 0.

52.9. *Rules of* RS_n. They include:

(i) the standard logical inferences ($\wedge$), ($\vee$), ($\forall x$), ($\exists x$), (Cut);

(ii) the N-rules:

$$(\neg N)\ \frac{\ldots \Gamma,\ \neg t = \overline{m} \ldots \text{ (for each } m \in \omega)}{\Gamma,\ \neg N t}\ ; \quad (N)\ \frac{\Gamma,\ t = \overline{m} \text{ (for some } m \in \omega)}{\Gamma,\ N t}\ ;$$

(iii) Successor rules for T_n and F_n (remind that T^α, F^α stand for T_n^α, F_n^α):

$$(T^{\alpha+1})\ \frac{\Gamma,\ T^\alpha\text{-}Clause(t)}{\Gamma,\ T^{\alpha+1} t}; \qquad (\neg T^{\alpha+1})\ \frac{\Gamma,\ \neg T^\alpha\text{-}Clause(t)}{\Gamma,\ \neg T^{\alpha+1} t}\ ;$$

$$(F^{\alpha+1})\ \frac{\Gamma,\ F^\alpha\text{-}Clause(t)}{\Gamma,\ F^{\alpha+1} t}; \qquad (\neg F^{\alpha+1})\ \frac{\Gamma,\ \neg F^\alpha\text{-}Clause(t)}{\Gamma,\ \neg F^{\alpha+1} t}.$$

(iv) Limit rules for T_n. Let $\alpha < \Gamma_0$ be a limit:

$$(T\text{-}LIM^\alpha)\ \frac{\Gamma,\ T^\beta s}{\Gamma,\ T^\alpha s}, \text{ for some } \beta < \alpha; \quad (\neg T\text{-}LIM^\alpha)\ \frac{\ldots\ \Gamma,\ \neg T^\beta s \ldots}{\Gamma,\ \neg T^\alpha s}\ , \text{ for every } \beta < \alpha.$$

$$(F\text{-}LIM^\alpha)\ \frac{\Gamma,\ F^\beta s}{\Gamma,\ F^\alpha s}, \text{ for some } \beta < \alpha; \quad (\neg F\text{-}LIM^\alpha)\ \frac{\ldots\ \Gamma,\ \neg F^\beta s \ldots}{\Gamma,\ \neg F^\alpha s}\ , \text{ for every } \beta < \alpha.$$

The rules and axioms of RS_n induce a relation of RS_n-derivability for finite sequents Γ of $\mathcal{L}_{n,r}$-formulas. Remind that low Greek letters α, β,... range over arbitrary ordinals $< \Gamma_0$, but also over ordinal constants of RS_n.

52.10. *Inductive definition of the derivability relation* $\mathrm{RS}_n \vdash^{\alpha}_{\rho} \Gamma$ $(n \in \omega)$.

DER.1. If (the finite set) $\Gamma \supseteq \Gamma'$ and Γ' is an axiom of RS_n, $\mathrm{RS}_n \vdash^{\alpha}_{\rho} \Gamma$, for every α and every ρ;

DER.2. Assume:

(i) $\mathrm{RS}_n \vdash^{\alpha_\beta}_{\rho_\beta} \Gamma_\beta$, for every $\beta < \delta$;

(ii) Γ follows from $\{\Gamma_\beta : \beta < \delta\}$ by means of the rule $\mathfrak{I}$, where $\mathfrak{I}$ is an inference of RS_n with δ premises $(0 < \delta < \Gamma_0)$;

(iii) $\alpha_\beta < \alpha$ for every $\beta < \delta$;

(iv) $sup\{\rho_\beta : \beta < \delta\} \leq \rho$ and $\mu \leq \rho$, where $\mu := R_n(A)+1$, if $\mathfrak{I}$ is a cut with cut formula A; else $\mu := 0$.

Then $\mathrm{RS}_n \vdash^{\alpha}_{\rho} \Gamma$.

The previous inductive definition immediately implies:

52.11. LEMMA (*Monotonicity of ordinal assignments*).

If $\mathrm{RS}_n \vdash^{\alpha}_{\rho} \Gamma$ *and* $\alpha \leq \beta$, $\rho \leq \delta$, *then* $\mathrm{RS}_n \vdash^{\beta}_{\delta} \Gamma$.

52.12. NOTATION

(i) $\mathrm{RS}_n \vdash^{\alpha}_{\rho} \Gamma := \Gamma$ is RS_n-derivable with length $\leq \alpha$ and cut rank $< \rho$.

(ii) $\mathrm{RS}_n \vdash^{<\alpha}_{<\rho} \Gamma := \mathrm{RS}_n \vdash^{\beta}_{\delta} \Gamma$, for some $\beta < \alpha$ and some $\delta < \rho$.

§53. Cut elimination

Following the classical method of Schütte, we show that every sequent derivable in RS_n is already RS_n-derivable with cut rank at most 1. As a preliminary step, we collect a few simple properties of the derivability relation for RS_n.

53.1. LEMMA

(i) *Weakening: if* $\mathrm{RS}_n \vdash^{\alpha}_{\rho} \Gamma$, *then* $\mathrm{RS}_n \vdash^{\alpha}_{\rho} \Gamma, \Delta$.

(ii) *Substitution: if* $\mathrm{RS}_n \vdash^{\alpha}_{\rho} \Gamma(a)$, *then* $\mathrm{RS}_n \vdash^{\alpha}_{\rho} \Gamma[a := t]$.

PROOF: by induction on α. (i) is a consequence of clause DER.1 of the definition of derivability for RS_n; (ii) follows from the fact that RS_n-axioms are closed under substitution. □

53.2. DEFINITION

(i) A formula A is *reducible* to Δ iff one of the following conditions holds:

1. $A = B \wedge C$ and $\Delta = \{B\}$ or $\Delta = \{C\}$;
2. $A = B \vee C$ and $\Delta = \{B, C\}$;
3. $A = \forall x B$ and $\Delta = \{B[x := t]\}$, for some t free for x in B;
4. $A = (\neg)\, T^{\alpha+1}(t)$ and $\Delta = \{(\neg)T^{\alpha}\text{-}Clause(t)\}$;
5. $A = (\neg)F^{\alpha+1}(t)$ and $\Delta = \{(\neg)F^{\alpha}\text{-}Clause(t)\}$;
6. $A = \neg T^{\alpha}t$ $(\neg F^{\alpha}t)$, α limit and $\Delta = \{\neg T^{\beta}t\}$ $(\Delta = \{\neg F^{\beta}t\})$, for some $\beta < \alpha$.

(ii) A is *reducible* iff A is reducible to some Δ.

Clearly, a formula of RS_n is reducible iff it can occur as active formula in one of the following inferences: $(\wedge)$, $(\vee)$, $(\forall)$, $((\neg)T^{\alpha+1})$, $((\neg)F^{\alpha+1})$, $(\neg T\text{-}LIM^{\beta})$, $(\neg F\text{-}LIM^{\beta})$, ($\beta$ limit).

53.3. LEMMA. *If $R_n(A) > 0$, A is reducible to Δ and $B \in \Delta$, then $R_n(B) < R_n(A)$.*

The verification is obvious by definition of n-rank (52.4).

53.4. LEMMA (*Inversion*). *Let $RS_n \vdash^{\alpha}_{\rho} \Gamma, A$ with $R_n(A) > 0$ and let A be reducible to Δ. Then*

$$RS_n \vdash^{\alpha}_{\rho} \Gamma, \Delta.$$

PROOF. Induction on α.
Case 1: Γ, A is an axiom. Since no reducible formula with n-rank > 0 is active in the axioms of RS_n, Γ, Δ is still an axiom.
Case 2: $A = \forall x B$ and A is active in the inference $\mathfrak{I} = (\forall)$ which concludes to Γ, A. Then we must have, possibly by use of 52.11 and weakening lemma, $RS_n \vdash^{\beta}_{\rho} \Gamma, \forall x B, B(a)$ (where a is an eigenparameter not in Γ, A), for some $\beta < \alpha$. Moreover Δ has the form $B[x := t]$, for some t. Then by IH, $RS_n \vdash^{\beta}_{\rho} \Gamma$, $B(t)$, $B(a)$, whence $RS_n \vdash^{\alpha}_{\rho} \Gamma$, $B[x := t]$ by the substitution lemma 53.1 (ii) and monotonicity.
Case 3: A is active in the inference $\mathfrak{I}$ which concludes to Γ, A, but $\mathfrak{I} \neq (\forall)$. Then Γ, Δ follows by applying IH to the premise of Γ, A, which is obviously determined by the given reduction Δ.
Case 4: A is not active in the inference $\mathfrak{I}$ which concludes to Γ, A. Then by IH we can replace every occurrence of A in the premises of $\mathfrak{I}$ by means of Δ and finally conclude with $\mathfrak{I}$. □

We proceed to the crucial step in the proof of cut elimination; but we first need the notion of *natural ordinal sum.* We know that, by the Cantor normal form theorem 45.3, every ordinal γ is uniquely representable in the form $\gamma_1+\ldots+\gamma_n$, with $\gamma_1 \geq \ldots \geq \gamma_n$, where each γ_i has the form $\omega^{\xi(i)}$, for some $\xi(i)$. If $\alpha_1+\ldots+\alpha_k$ and $\alpha_{k+1}+\ldots+\alpha_{k+m}$ are the normal forms of α and β respectively, we define:

$$\alpha\#\beta := \alpha_{\pi(1)}+\ldots+\alpha_{\pi(k+m)},$$

π being the permutation of $\{1,\ldots,k+m\}$ such that $\alpha_{\pi(1)} \geq \ldots \geq \alpha_{\pi(k+m)}$. Clearly $\#$ is commutative and strictly increasing in each variable: i.e. $\delta < \gamma$ implies both $\alpha\#\delta < \alpha\#\gamma$ and $\delta\#\alpha < \gamma\#\alpha$.

53.5. LEMMA (*Reduction*)

If $\mathrm{RS}_n \vdash^{\alpha}_{\rho} \Gamma, A$ *and* $\mathrm{RS}_n \vdash^{\beta}_{\rho} \Delta, \neg A$ *with* $R_n(A) = \rho \geq 1$, *then*

$$\mathrm{RS}_n \vdash^{\alpha\#\beta}_{\rho} \Gamma, \Delta.$$

PROOF. We argue by induction on $\alpha\#\beta$.

Case 1: Γ, A and $\Delta, \neg A$ are axioms. Then Γ, Δ is already an axiom because neither A nor $\neg A$ can be active formulas of an axiom, having n-rank > 0.

Case 2: We may assume that, say, Γ, A is not an axiom. {This is not restrictive: since $R_n(A) = R_n(\neg A)$, the whole argument is symmetric with respect to Γ, A and $\Delta, \neg A$}.

2.1: A is not active in the application of $\mathfrak{I}$, $\mathfrak{I}$ being the rule applied to infer Γ, A. Then $\mathfrak{I}$ has the form:

$$\text{infer } \mathrm{RS}_n \vdash^{\alpha}_{\rho} \Gamma, A \ \text{ from } \mathrm{RS}_n \vdash^{\alpha_k}_{\rho_k} \Gamma', A, B_k, \ldots, \text{ for each } k \in (0,\ \delta),$$

where δ is the number of the premises of $\mathfrak{I}$. Since $\alpha_k\#\beta < \alpha\#\beta$, we get by IH, for every $k \in (0, \delta)$,

$$\mathrm{RS}_n \vdash^{\alpha_k\#\beta}_{\rho} \Gamma',\ B_k,\ \Delta.$$

The conclusion follows by application of $\mathfrak{I}$ with length $\alpha_k\#\beta\#1 \leq \alpha\#\beta$ and rank $\leq \rho$.

2.2: A, $\neg A$ are active formulas of the inferences $\mathfrak{I}_0$ and $\mathfrak{I}_1$, which conclude to Γ, A and $\Delta, \neg A$, respectively. Again by symmetry, we may assume that A is reducible and not disjunctive. Then $\mathfrak{I}_1$ can only have one premise and it has the form:

$$\text{from } \mathrm{RS}_n \vdash^{\beta_1}_{\rho} \Delta, \neg A, \neg C \text{ infer } \mathrm{RS}_n \vdash^{\beta}_{\rho} \Delta, \neg A,$$

where $\neg C$ is the minor formula of $\mathfrak{I}_1$. Since $\alpha\#\beta_1 < \alpha\#\beta$, we get by IH :

$$\mathrm{RS}_n \vdash^{\alpha\#\beta_1}_{\rho} \Gamma,\ \Delta,\ \neg C. \tag{1}$$

By inspection of the rules, it is readily seen that C must be a reduction of

A; hence by inversion, weakening, we have:

$$\mathrm{RS}_n \vdash^{\alpha}_{\rho} \Gamma,\ C,\ \Delta. \tag{2}$$

Since $R_n(C) < R_n(A)$, and $\alpha < \alpha\#\beta_1$ (as $0 < \beta$), a cut between (1) and (2) yields $\mathrm{RS}_n \vdash^{\alpha\#\beta}_{\rho} \Gamma, \Delta$. □

53.6. THEOREM (*1-step-cut elimination*)

If $\mathrm{RS}_n \vdash^{\alpha}_{\rho+1} \Gamma$ *and* $\rho > 0$, *then* $\mathrm{RS}_n \vdash^{\omega^\alpha}_{\rho} \Gamma$.

PROOF. We may assume that the last inference $\mathfrak{I}$, which concludes to Γ, is a cut of rank $\rho > 0$ on a formula A; otherwise, the claim follows by IH applied to the premises of $\mathfrak{I}$. Thus we have by monotonicity lemma, for some $\alpha_0 < \alpha$:

$$\mathrm{RS}_n \vdash^{\alpha_0}_{\rho+1} \Gamma, A \text{ and } \mathrm{RS}_n \vdash^{\alpha_0}_{\rho+1} \Gamma, \neg A \text{ with } \rho = R_n(A).$$

Hence by IH:

$$\mathrm{RS}_n \vdash^{\omega^{\alpha_0}}_{\rho} \Gamma, A \text{ and } \mathrm{RS}_n \vdash^{\omega^{\alpha_0}}_{\rho} \Gamma, \neg A.$$

By the reduction lemma, we obtain $\mathrm{RS}_n \vdash^{\gamma}_{\rho} \Gamma$, where $\gamma = \omega^{\alpha_0} \cdot 2 < \omega^\alpha$. □

53.7. THEOREM (*"Schütte's first cut elimination theorem"*)

If $\mathrm{RS}_n \vdash^{\alpha}_{\rho} \Gamma$, *then* $\mathrm{RS}_n \vdash^{\phi\rho\alpha}_{1} \Gamma$.

PROOF. By main induction on ρ and secondary induction on α. Clearly we may assume that $\rho > 1$, $\alpha > 0$.
Case 1: the last inference $\mathfrak{I}$ is not a cut or is a cut of rank zero. Then we apply secondary IH to each premise Γ_k of Γ, thus obtaining $\mathrm{RS}_n \vdash^{\phi\rho\alpha_k}_{1} \Gamma_k$ (where $k \in (0, \delta)$, δ is the number of premises of $\mathfrak{I}$ and $\alpha_k < \alpha$). Since ϕ is increasing in the second variable (§ 45), an application of $\mathfrak{I}$ yields

$$\mathrm{RS}_n \vdash^{\phi\rho\alpha}_{1} \Gamma.$$

Case 2: the last inference $\mathfrak{I}$ is a cut on a formula A with $R_n(A) > 0$, i.e.

$$\mathrm{RS}_n \vdash^{\alpha_0}_{\rho} \Gamma, A \quad \text{and} \quad \mathrm{RS}_n \vdash^{\alpha_1}_{\rho} \Gamma, \neg A.$$

Then by secondary IH and monotonicity:

$$\mathrm{RS}_n \vdash^{\phi\rho\alpha_0}_{R_n(A)} \Gamma, A \text{ and } \mathrm{RS}_n \vdash^{\phi\rho\alpha_1}_{R_n(A)} \Gamma, \neg A. \tag{$*$}$$

Hence, by reduction lemma, $\mathrm{RS}_n \vdash^{\phi\rho\alpha_0 \# \phi\rho\alpha_1}_{R_n(A)} \Gamma$. {Alternatively, we might apply a cut to $(*)$ and then 53.6.}. Since $R_n(A) < \rho$, we obtain by main IH, $\mathrm{RS}_n \vdash^{\phi(R_n(A))(\phi\rho\alpha_0 \# \phi\rho\alpha_1)}_{1} \Gamma$. But $R_n(A) < \rho$ and $\alpha_0, \alpha_1 < \alpha$, and hence with the property P4 of § 45, $\mathrm{RS}_n \vdash^{\phi\rho\alpha}_{1} \Gamma$. □

53.6-53.7 can be generalized (the proof is left to the reader).

53.8. THEOREM ("*Tait's second cut elimination theorem*")
If $\mathrm{RS}_n \vdash^{\beta}_{\delta+\omega^\nu} \Gamma$, *then* $\mathrm{RS}_n \vdash^{\phi\nu\beta}_{\delta} \Gamma$.

§ 54. Some derivable sequents of RS_n

We prove some useful sequents of RS_n, which are needed in the proof of the main interpretation theorem of § 55.

54.1. LEMMA

(i) *Tautology:* $\mathrm{RS}_n \vdash^{2\cdot R_n(A)}_{1} \neg A,\ A;$

(ii) *Substitutivity:* $\mathrm{RS}_n \vdash^{2\cdot R_n(A)}_{1} \neg t = s,\ \neg A[x := t],\ A[x := s].$

PROOF. (i)-(ii): by induction on $R_n(A)$. We apply the logical axioms *LOG* of 52.8.1 (ii)-(iii) and the rules corresponding to the main logical symbol of A. For instance, let $A = T^{\alpha+1}(t)$.
Then by IH, since $R_n(T^\alpha\text{-}Clause(t)) < \omega(\alpha+1) = R_n(A)$, we have:

$$\mathrm{RS}_n \vdash^{<\omega(\alpha+1)}_{1} \neg T^\alpha\text{-}Clause(t),\ T^\alpha\text{-}Clause(t).$$

The conclusion is immediate by use of $(T^{\alpha+1})$ and $(\neg T^{\alpha+1})$. □

54.2. LEMMA (*Independence*)

(i) *We can find* $k \in \omega$ *such that, if* C, D *are substitution instances of distinct elements of the set*

$$\mathcal{A} = \{[a = b], [Na], [Ta], [T_0 a], \dots, [T_n a], a \wedge b, \neg a, \forall a\},$$

then $\mathrm{RS}_n \vdash^{k}_{1} \neg t = C, \neg t = D$.

(ii) *We can find* $m \in \omega$ *such that, whenever* $s(a', b') = s[a := a', b := b']$ *and* $s(a, b) \in \mathcal{A}$, *then*

$$\mathrm{RS}_n \vdash^{m}_{1} \neg s(a,b) = s(a',b'),\ a = a' \ \wedge\ b = b'.$$

(iii) *We can find* $p \in \omega$ *such that, if* $i < k < n$

$$\mathrm{RS}_n \vdash^{p}_{1} \neg T_i T_k t\ (or\ \neg F_i T_k t).$$

The proof of (i)-(ii) above is a simple consequence of identity, pairing and number-theoretic axioms, like the independence lemma of 48.6; (iii) is a consequence of (i)-(ii) with the fixed point axioms of 52.8.5 (observe that we can derive $\neg T_i$-$Clause^*(T_k t)$ if $i < k$).

At this point, the reader should recollect the notions of n-positive $\mathcal{L}_n$-formula from 50.1.3 and α-transform A_α from 52.5 (for $A \in \mathcal{L}_n$).

54.3. LEMMA (*Persistence*)

(i) *Let A be an n-positive $\mathcal{L}_n$-formula. Assume*

$\mathrm{RS}_n \vdash^{\gamma}_{\rho} \neg T^\alpha t,\ T^\beta t$ *and* $\mathrm{RS}_n \vdash^{\gamma}_{\rho} \neg F^\alpha t,\ F^\beta t$ *(t arbitrary).*

Then for some finite k,

$$\mathrm{RS}_n \vdash^{\gamma+k}_{\rho} \neg A_\alpha, A_\beta.$$

(ii) $i < n \Rightarrow \mathrm{RS}_n \vdash^{<\omega}_{1} \neg T_i t,\ T^\beta t$ *(or* $\neg F_i t,\ F^\beta t$*).*

(iii) $\alpha < \beta \Rightarrow \mathrm{RS}_n \vdash^{\omega\beta}_{\omega\beta} \neg T^\alpha t,\ T^\beta t$ *(or* $\neg F^\alpha t,\ F^\beta t$*).*

PROOF. (i): immediate by induction on A, using the definition of α-transform, POS_n, and the hypothesis.

(ii) We first check by induction on β, for arbitrary t:

$$\mathrm{RS}_n \vdash^{<\omega}_{1} \neg T_{n-1}t,\ T^\beta t \text{ and } \mathrm{RS}_n \vdash^{<\omega}_{1} \neg F_{n-1}t,\ F^\beta t. \tag{1}$$

If $\beta = 0$ (β is a limit), we apply the initial axioms 52.8.6 (and the limit rules LIM^β). In the successor case observe that (1) implies, for arbitrary t,

$$\mathrm{RS}_n \vdash^{<\omega}_{1} \neg T_{n-1}\text{-}Clause^*(t),\ T^\beta\text{-}Clause(t) \tag{2}$$

$$\mathrm{RS}_n \vdash^{<\omega}_{1} \neg F_{n-1}\text{-}Clause^*(t),\ F^\beta\text{-}Clause(t).$$

From (2), the axiom FIX_{n-1} (see 52.8.5) and the rules $(T^{\beta+1})$, $(F^{\beta+1})$, we can derive:

$$\mathrm{RS}_n \vdash^{<\omega}_{1} \neg T_{n-1}t,\ T^{\beta+1}t \text{ and } \mathrm{RS}_n \vdash^{<\omega}_{1} \neg F_{n-1}t,\ F^{\beta+1}t \ (t \text{ arbitrary}).$$

To establish the general case, we apply (1) and sufficiently many cuts with the sequents $\{\neg T_i t,\ T_{n-1}t\}$ and $\{\neg F_i t,\ F_{n-1}t\}$, which are RS_n-derivable by means of the persistence axioms (52.8.3).

(iii): by induction on β. We assume that (iii) holds for every $\alpha < \gamma$, if $\gamma < \beta$. We explicitly consider the T-case, as the F-case can be simultaneously handled with similar arguments.

If $\beta = 0$, the conclusion is trivial; if β is a limit, $\mathrm{RS}_n \vdash^{\omega(\alpha+1)}_{\omega(\alpha+1)} \neg T^\alpha t,\ T^{\alpha+1}t$

and the conclusion follows by one application of (LIM^{β}). Let $\beta = \delta{+}1$ and choose $\alpha < \beta$. By tautology and IH, we get, for arbitrary t:

$$\mathrm{RS}_n \vdash^{\omega\delta}_{\omega\delta} \neg T^{\alpha}t,\ T^{\delta}t \text{ and } \mathrm{RS}_n \vdash^{\omega\delta}_{\omega\delta} \neg F^{\alpha}t,\ F^{\delta}t. \tag{3}$$

If $\alpha < \delta$, $\mathrm{RS}_n \vdash^{\omega(\alpha+1)}_{\omega(\alpha+1)} \neg T^{\alpha}t,\ T^{\alpha+1}t$ again by IH; on the other hand, (3), (i) above and the successor rules $(\neg T^{\alpha+1})$, $(T^{\delta+1})$ imply, for some finite k:

$$\mathrm{RS}_n \vdash^{\omega\delta+k}_{\omega\delta} \neg T^{\alpha+1}t,\ T^{\beta}t;$$

the required conclusion follows by a cut of rank $\omega(\alpha{+}1) < \omega\beta$.
Let $\alpha = \delta$. If $\delta = 0$, we apply (ii); if $\delta = \xi{+}1$, we apply IH and (i). If δ is a limit, we have, for every $\xi < \delta$:

$$\mathrm{RS}_n \vdash^{\omega(\xi+1)}_{\omega(\xi+1)} \neg T^{\xi}t,\ T^{\xi+1}t \tag{4}$$

$$\mathrm{RS}_n \vdash^{\omega\delta}_{\omega\delta} \neg T^{\xi}t,\ T^{\delta}t \text{ and } \mathrm{RS}_n \vdash^{\omega\delta}_{\omega\delta} \neg F^{\xi}t,\ F^{\delta}t. \tag{5}$$

(5), (i) and the successor rules $(\neg T^{\xi+1})$ $(T^{\delta+1})$ imply, for some finite k

$$\text{for every } \xi < \delta,\ \mathrm{RS}_n \vdash^{\omega\delta+k}_{\omega\delta} \neg T^{\xi+1}t,\ T^{\delta+1}t. \tag{6}$$

Then we get for each $\xi < \delta$, by use of a cut of rank $\omega(\xi{+}1) < \omega\delta$ with (4):

$$\mathrm{RS}_n \vdash^{\omega\delta+k+1}_{\omega\delta} \neg T^{\xi}t,\ T^{\delta+1}t. \tag{7}$$

Then $\mathrm{RS}_n \vdash^{\omega\beta}_{\omega\beta} \neg T^{\delta}t,\ T^{\delta+1}t$ by application of $(\neg T\text{-}LIM^{\delta})$. □

54.4. LEMMA (*Consistency*). $\mathrm{RS}_n \vdash^{2+\omega\alpha}_{\omega\alpha} \neg T^{\alpha}t,\ \neg F^{\alpha}t$.

PROOF. By induction on α. If $\alpha = 0$, the statement follows by cut with length 2 from the initial axioms 52.8.6. If α is a limit, the conclusion is an immediate consequence of IH with the limit rules for T and F, respectively. Thus it is enough to check, for t arbitrary:

$$\mathrm{RS}_n \vdash^{<\omega\alpha}_{\omega\alpha} \neg T^{\alpha}\text{-}Clause(t), \neg F^{\alpha}\text{-}Clause(t), \tag{1}$$

under the assumption

$$\mathrm{RS}_n \vdash^{\omega\alpha}_{\omega\alpha} \neg T^{\alpha}t,\ \neg F^{\alpha}t. \tag{2}$$

(1) clearly yields the conclusion by the appropriate successor rules for T and F. In order to verify (1), we define formulas $A_i(a,b,t)$ and $B_i(a',b',t)$, where $0 \le i \le n{+}6$ and a, b, a', b' are distinct parameters not occurring in t; in short, we below omit the explicit mention of t, a, a', b, b':

$$A_i := (\neg t = [T_i a] \vee \neg T_i a), \text{ if } 0 \le i < n; \tag{3.1}$$

$$A_n := (\neg t = [T_n a] \vee \neg T^{\alpha} a); \quad A_{n+1} := (\neg t = [Ta] \vee \neg T^{\alpha} a);$$

$$A_{n+2} := (\neg t = [Na] \vee \neg Na); \quad A_{n+3} := (\neg t = [a = b] \vee \neg a = b);$$

$A_{n+4} := (\neg t = (\neg a) \vee \neg F^{\alpha} a);\quad A_{n+5} := (\neg t = (a \wedge b) \vee \neg T^{\alpha} a \vee \neg T^{\alpha} b);$

$A_{n+6} := (\neg t = (\forall a) \vee \exists u. \neg T^{\alpha}(au));$

$$B_i := (\neg t = [T_i a'] \vee T_i a'), \text{ if } 0 \leq i < n; \tag{3.2}$$

$B_n := \ (\neg t = [T_n a'] \vee \neg F^{\alpha} a')\ ;\ B_{n+1} := (\neg t = [Ta'] \vee \neg F^{\alpha} a');$

$B_{n+2} := (\neg t = [Na'] \vee Na');\quad B_{n+3} := (\neg t = [a' = b'] \vee a' = b');$

$B_{n+4} := (\neg t = (\neg a') \vee \neg T^{\alpha} a');\ B_{n+5} := (\neg t = (a' \wedge b') \vee (\neg F^{\alpha} a' \wedge \neg F^{\alpha} b');$

$B_{n+6} := (\neg t = (\forall a') \vee \forall u. \neg F^{\alpha}(a'u)).$

Let $C(a,b,t) := \sqcap\{A_i : 0 \leq i \leq n\}$: if we show

$$\mathrm{RS}_n \vdash^{<\omega\alpha}_{\omega\alpha} C(a,b,t),\ B_i(a',b',t) \text{ for each } i \leq n{+}6, \tag{4}$$

(4) will imply (1) by applications of $(\wedge)$ and $(\forall)$. As to (4), it is enough to check

$$\mathrm{RS}_n \vdash^{<\omega\alpha}_{\omega\alpha} A_j(a,b,t),\ B_i(a',b',t), \text{ for each } i,\ j \leq n{+}6. \tag{5}$$

If $j \neq i$, (5) is a consequence of Lemma 54.2 (i). Let $i = j$. If $i < n$ or $i = n{+}2,\ n{+}3$, we simply apply the second part of the independence lemma above and LOG; else, we use 54.2 (ii), the identity lemma and (2). □

54.5. LEMMA. *Let* $S := T$ *(or* F*). Then we have:*

(i) $\mathrm{RS}_n \vdash_1^{<\omega(\beta+1)} \neg S^{\beta}\text{-}Clause(t), S^{\beta+1}(t);$

$\mathrm{RS}_n \vdash_1^{<\omega(\beta+1)} \neg S^{\beta+1}(t),\ S^{\beta}\text{-}Clause(t).$

(ii) $\mathrm{RS}_n \vdash^{<\omega(\beta+2)}_{<\omega(\beta+2)} \neg S^{\beta} t, S^{\beta+1}\text{-}Clause(t).$

PROOF. (i): immediate by tautology, $((\neg)T^{\beta+1})$ and $((\neg)F^{\beta+1})$.
(ii). By persistence lemma 54.3 and part (i), we have:

$$\mathrm{RS}_n \vdash^{<\omega(\beta+2)}_{\omega(\beta+1)} \neg T^{\beta}\text{-}Clause(t),\ T^{\beta+1}\text{-}Clause(t); \tag{1}$$

$$\mathrm{RS}_n \vdash_1^{<\omega(\beta+1)} \neg T^{\beta+1}(t),\ T^{\beta}\text{-}Clause(t); \tag{2}$$

$$\mathrm{RS}_n \vdash^{\omega(\beta+1)}_{\omega(\beta+1)} \neg T^{\beta} t,\ T^{\beta+1} t. \tag{3}$$

The conclusion follows with two applications of cut rule. The case of F is similar. □

§55. Embedding IT_n^∞ into RS_n

The time is ripe to call upon the family of infinitary systems IT_n^∞, which were introduced in the final section of Ch. X and where TLR can be suitably embedded. By inspecting §50 and §52, we immediately realize that IT_n^∞ is just the $\mathcal{L}_n$-fragment of RS_{n+1}; thus it should not come as a surprise that the cut elimination argument of §53 works for IT_n^∞ as well. Of course, we must have clear that in the derivability relation $\mathrm{IT}_n^\infty \vdash_\rho^\alpha \Gamma$, *$\rho$ is an upper bound for the n-complexity of the cut formulas* (*see* 50.1.4). For the reader's sake, we recall that n-complexity simply counts the usual logical complexity, except that it considers n-separated, subformulas as atoms; furthermore, a formula A of $\mathcal{L}_n$ is n-separated whenever A is n-positive or n-negative (in the sense that T_n and F_n occur only positively or negatively in A, but not both).

55.1. THEOREM (*Partial cut elimination*). *Assume that* $\mathrm{IT}_n^\infty \vdash_{k+\omega^i}^{\alpha} \Gamma$, *where* $i \in \{0,1\}$, $1 \leq k < \omega$. *Then*:

$$\textit{either } i = 0 \textit{ and } \mathrm{IT}_n^\infty \vdash_1^{\omega_k(\alpha)} \Gamma, \textit{ or } i = 1 \textit{ and } \mathrm{IT}_n^\infty \vdash_1^{\phi 1\alpha} \Gamma.$$

Here $\omega_0(\alpha) = \alpha$, $\omega_{k+1}(\alpha) = \omega^{\omega_k(\alpha)}$ and ϕ is the Veblen hierarchy of §45; for the proof, apply the elimination lemma and Tait's refinement of cut elimination in 53.8. The partial cut elimination theorem ensures that cuts can be always reduced to cuts over n-positive or n-negative conditions; and this fact is essential to establish an interpretation of IT_n^∞ into RS_n, which is based on the "separation" of *positive* and *negative* occurrences of T_n, F_n.

55.2. *Inductive definition of* $A[\beta,\gamma]$ (*where* $0 \leq \beta \leq \gamma$ *and A is a formula of* IT_n^∞).

(i) $(T_n t)[\beta,\gamma] = T^\gamma s$ and $(\neg T_n t)[\beta,\gamma] = \neg T^\beta s$;

$(F_n t)[\beta,\gamma] = F^\gamma t$ and $(\neg F_n t)[\beta,\gamma] = \neg F^\beta t$;

$A[\beta,\gamma] = A$, if either $A = T_i t,\ F_i t,\ \neg T_i t,\ \neg F_i t$ and $i < n$ or $A = (\neg)Nt,\ (\neg)t = s$;

$(B \circ C)[\beta,\gamma] = B[\beta,\gamma] \circ C[\beta,\gamma]$ and $(\mathrm{Q}xB)[\beta,\gamma] = \mathrm{Q}x(B[\beta,\gamma])$; ($\circ = \wedge, \vee$ and $\mathrm{Q} = \forall, \exists$).

(ii) Let $\Gamma = \{B_1, \ldots, B_m\}$ be a set of IT_n^∞-formulas; we set

$\Gamma\,[\beta,\gamma] = \{B_1[\beta,\gamma], \ldots, B_m\,[\beta,\gamma]\}$.

55.3. FACT. *Let* $A \in \mathcal{L}_n$ *and* $K_n(A) = 0$: *if A is n-positive,* $A[\beta,\gamma] = A_\gamma =$ *the γ-transform of A; if A is n-negative,* $A[\beta,\gamma] = A_\beta =$ *the β-transform* (*see* 52.5 *above*).

Verification: by induction on A.

55.4. LEMMA

(i) *Let $0 \le \gamma \le \delta < \xi < \alpha$. Then*

$$RS_n \vdash_{\omega\alpha}^{<\,\omega(\alpha+1)} \neg(A[\delta,\xi]),\ A[\gamma,\alpha].$$

(ii) *Assume* $RS_n \vdash_\sigma^\nu \Gamma[\delta,\xi], \Delta$ *with* $0 \le \gamma \le \delta < \xi < \alpha$ *and* ν, $\sigma < \omega(\alpha+1)$. *Then* $RS_n \vdash_{\omega\alpha}^{<\,\omega(\alpha+1)} \Gamma\,[\gamma,\alpha], \Delta$.

PROOF. (i) Induction on A. If A is an atom not of the form T_nt, F_nt, we apply the logical axioms and the N-rules; else, we use persistence lemma 54.3. If A is built up from $\wedge$, $\vee$, $\forall$, $\exists$, we apply IH and the corresponding logical rule.
(ii) By induction on the cardinality of Γ, using (i) above, cuts on formulas of n-rank $< \omega(\xi+1) \le \omega\alpha$ and hypothesis on ν, σ. □

55.5.THEOREM. *If* $IT_n^\infty \vdash_1^\alpha \Gamma$, *then* $RS_n \vdash_{\omega(\gamma+1)}^{\omega(\gamma+1)} \Gamma[\beta,\gamma]$, *for every* $\beta \ge 0$, $\gamma \ge \beta+\omega^\alpha$.

PROOF. By induction on α.
Case 1: Γ is an axiom of IT_n^∞. We neglect the side formulas (this is not restrictive by weakening lemma) and we consider a number of subcases.
1.1. Γ is an axiom which does not involve the level-n-predicates T_n, F_n. Then the conclusion is trivial, since $\Gamma = \Gamma[\beta,\gamma]$ is an axiom of RS_n as well.
1.2. Γ is a logical axiom of the form $\neg A$, A, where A can be assumed n-positive. Hence by 55.3 above, $(\neg A)[\beta,\gamma] = \neg A_\beta$ and $A[\beta,\gamma] = A_\gamma$.
Since $\beta < \gamma$, by 54.3 (iii) $\{\neg T^\beta t,\ T^\gamma t\}$ and $\{\neg F^\beta t,\ F^\gamma t\}$ are RS_n-derivable with length and cut rank $< \omega(\gamma+1)$. But A is n-positive and the conclusion is immediate by application of 54.3 (i).
1.3. $\Gamma = \{\neg T_nt, \neg F_nt\}$ or $\Gamma = \{\neg T_it,\ T_nt\}$, $\{\neg F_it,\ F_nt\}$ with $i < n$: apply 54.4 and 54.3 (ii).
1.4. Γ is a fixed point axiom involving T_n and F_n.
1.4.1. $\Gamma = \{\neg T_n\text{-}Clause^*(t), T_nt\ \}$. Then by 54.5 (i)

$$RS_n \vdash_1^{<\,\omega(\beta+1)} \neg T^\beta\text{-}Clause(t),\ T^{\beta+1}t,$$

If $\gamma = \beta+\omega^0 = \beta+1$, we are done; else, by 54.3 (iii) we also have

$$RS_n \vdash_{\omega\gamma}^{\omega\gamma} \neg T^{\beta+1}t,\ T^\gamma t.$$

By a cut of n-rank $\omega(\beta+1) < \omega\gamma$, we get $RS_n \vdash_{\omega\gamma}^{\omega\gamma+1} \neg T^\beta\text{-}Clause(t),\ T^\gamma t$.

1.4.2. $\Gamma = \{\neg T_nt, T_n\text{-}Clause^*(t)\}$. Then by 54.5 (ii)

$$RS_n \vdash_{<\,\omega(\beta+2)}^{<\,\omega(\beta+2)} \neg T^\beta t,\ T^{\beta+1}\text{-}Clause(t).$$

If $\gamma = \beta{+}1$ we are done; if $\beta{+}1 < \gamma$, we have by 54.3 (i)-(iii):

$$\mathrm{RS}_n \vdash_{\omega\gamma}^{<\omega(\gamma+1)} \neg T^{\beta+1}\text{-}Clause(t),\ T^{\gamma}\text{-}Clause(t).$$

The required conclusion follows by a cut with n-rank $< \omega(\beta{+}2) \leq \omega(\gamma{+}1)$. The interpretation of the fixed point axioms for F_n is similar.

Case 2: Γ is not an axiom and $\alpha > 0$, $\beta \geq 0$ and $\gamma \geq \beta{+}\omega^{\alpha}$. We have to distinguish seven subcases corresponding to the possible inference $\mathfrak{I} = (\neg N)$, (N), $(\wedge)$, $(\vee)$, $(\exists)$, $(\forall)$, (Cut) of IT_n^{∞}.

2.1. Let $\mathfrak{I} = (\neg N)$ and assume, for each $k \in \omega$:

$$\mathrm{IT}_n^{\infty} \vdash_1^{\alpha_k} \Gamma,\ \neg t = \bar{k}, \text{ where } \alpha_k < \alpha.$$

Now, if we put $\sigma_k = \beta{+}\omega^{\alpha_k}$, we have by IH for each k:

$$\mathrm{RS}_n \vdash_{\omega(\sigma_k+1)}^{\omega(\sigma_k+1)} \Gamma[\beta,\sigma_k], \neg t = \bar{k}.$$

Clearly $\sigma_k < \gamma$ and $\omega(\sigma_k{+}1) < \omega(\gamma{+}1)$; hence by lemma 55.4 (ii):

$$\mathrm{RS}_n \vdash_{\omega\gamma}^{<\omega(\gamma+1)} \Gamma[\beta,\gamma], \neg t = \bar{k}.$$

By application of $(\neg N)$, we conclude

$$\mathrm{RS}_n \vdash_{\omega(\gamma+1)}^{\omega(\gamma+1)} \Gamma[\beta,\gamma], \neg Nt.$$

2.2. Let $\mathfrak{I} = (Cut)$, then for some $\alpha_0 < \alpha$

$$\mathrm{IT}_n^{\infty} \vdash_1^{\alpha_0} \Gamma, A \text{ and } \mathrm{IT}_n^{\infty} \vdash_1^{\alpha_0} \Gamma, \neg A; \tag{1}$$

(by monotonicity lemma, it is not restrictive to assume that the premises have the same bound on the length). By assumption $K_n(A) = 0$; we may also suppose that T_n, F_n occur in A and that A is n-positive (hence $\neg A$ is n-negative). Choose any $\gamma \geq \beta{+}\omega^{\alpha} > \beta{+}\omega^{\alpha_0}$. By IH applied to the left premise of (1), we obtain with $\sigma_0 = \beta{+}\omega^{\alpha_0}$ and 55.3:

$$\mathrm{RS}_n \vdash_{\omega(\sigma_0+1)}^{\omega(\sigma_0+1)} \Gamma[\beta,\sigma_0],\ A_{\sigma_0}. \tag{2}$$

If we apply IH to the second premise of (1), choosing $\beta = \sigma_0$, $\sigma_1 = \sigma_0{+}\omega^{\alpha_0}$, we obtain, since $\neg A$ is n-negative and again with 55.3:

$$\mathrm{RS}_n \vdash_{\omega(\sigma_1+1)}^{\omega(\sigma_1+1)} \Gamma[\sigma_0,\sigma_1],\ \neg A_{\sigma_0}. \tag{3}$$

Since $\omega^{\alpha_0}\cdot 2 < \omega^{\alpha}$ and $\beta{+}\omega^{\alpha} \leq \gamma$, we immediately have $\omega(\sigma_0{+}1)$, $\omega(\sigma_1{+}1) < \omega(\gamma{+}1)$ and hence by application of 55.4 (ii) to (2)-(3) we have:

$$\mathrm{RS}_n \vdash_{\omega\gamma}^{<\omega(\gamma+1)} \Gamma[\beta,\gamma],\ A_{\sigma_0} \text{ and } \mathrm{RS}_n \vdash_{\omega\gamma}^{<\omega(\gamma+1)} \Gamma[\beta,\gamma],\ \neg A_{\sigma_0}.$$

An application of (Cut) with n-rank $< \omega(\gamma{+}1)$ yields the conclusion. The remaining cases do not present any additional difficulty and are left as exercises. □

§56. The upper bound theorem for IT_n^∞

We combine the strength of cut elimination for the infinitary systems IT_n^∞, RS_n with the embedding of IT_n^∞ into RS_n. The main result shows that, *if a statement A in the language $\mathcal{L}_{op}$ is IT_n^∞-derivable for some n, then not only A is true in the given ground model, but A is "cut-free" derivable with an infinitary derivation tree, whose height is bounded by Γ_0 (whence the name of the theorem).*

56.1. DEFINITION

(i) OP^∞ is the subsystem of IT_0^∞ in the language $\mathcal{L}_{op}$(= the language of the theory OP of Ch. I). More specifically:

1) OP^∞ has the same axioms of IT_0^∞ in the language $\mathcal{L}_{op}$;
2) the rules of OP^∞ include $(\wedge)$, $(\vee)$, $(\forall)$, $(\exists)$, (Cut), (N), $(\neg N)$;
3) the derivability relation $\mathrm{OP}^\infty \vdash_\rho^\alpha \Gamma$ is inductively defined as in 50.2.3, except that now ρ is a strict upper bound to the *usual standard logical complexity of cut formulas*, which assigns complexity 0 to e-atoms and counts distinct occurrences of logical symbols (cf. 47.2.6).

(ii) We put $\mathcal{T}_0 = \mathrm{OP}^\infty$ and $\mathcal{T}_{n+1} = \mathrm{IT}_n^\infty$.

56.2. LEMMA (OP^∞-*cut elimination*). *If* $\mathrm{OP}^\infty \vdash_\rho^\alpha \Gamma$, *either* $\rho = k+1$, *for some* $k \in \omega$ *and* $\mathrm{OP}^\infty \vdash_1^{\omega_k(\alpha)} \Gamma$, *or* $\rho = \omega$ *and* $\mathrm{OP}^\infty \vdash_1^{\phi 1 \alpha} \Gamma$.

The verification follows the standard pattern of §53.

If Γ is a set of formulas in RS_n, let $R_n(\Gamma) = 0$ stands for "every formula of Γ has n-rank 0". Clearly if $R_n(\Gamma) = 0$, the only formulas of Γ which are not formulas of $\mathcal{T}_n$ must have the form $T_n^0 t$, $F_n^0 t$ or $\neg T_n^0 t$, $\neg F_n^0 t$. In order to get a "level lowering lemma", we define by cases a transformation Γ^- of Γ into the language of $\mathcal{T}_n$, whenever $R_n(\Gamma) = 0$:

(i) $n = 0$: Γ^- is obtained from Γ by replacing each occurrence of $T^0 t$, $F^0 t$ by the formula $(\neg t = t)$;

(ii) $n > 0$: Γ^- is obtained from Γ by replacing each occurrence of $T^0 t$ ($F^0 t$) by the formula $T_{n-1} t$ ($F_{n-1} t$).

{According to 52.1.1 we leave the index n for the ramified predicates T and F implicit}

56.3. LEMMA. *If* $\mathrm{RS}_n \vdash_1^\alpha \Gamma$ *and* $R_n(\Gamma) = 0$, *then* $\mathcal{T}_n \vdash^{<\omega+\alpha} \Gamma^-$.

($\mathcal{T}_n \vdash^\alpha \Gamma := \mathcal{T}_n \vdash_\rho^\alpha \Gamma$, for some $\rho \le \omega$).

PROOF. Induction on α. Let Γ be an axiom of RS_n. If Γ is a logical axiom, say $\Gamma = \Delta, A, \neg A$ where $R_n(A) = 0$, then by 51.5 and monotonicity of ordinal assignment, $\mathcal{T}_n \vdash_1^{<\omega+\alpha} \Gamma^-$. As to non-logical axioms, observe that the $^-$-transform sends them into corresponding axioms of $\mathcal{T}_n$. Assume now that Γ is derivable with a cut. Then we obtain by IH and by hypothesis on the cut rank, for some A with $R_n(A) = 0$ and some $\alpha_0 < \alpha$,

$$\mathcal{T}_n \vdash^{<\omega+\alpha_0} \Gamma^-, A^- \text{ and } \mathcal{T}_n \vdash^{<\omega+\alpha_0} \Gamma^-, \neg A^-.$$

A cut on A^- yields $\mathcal{T}_n \vdash^{<\omega+\alpha} \Gamma^-$; but note that $K_n(A^-)$ may be greater than 0; this explains why the cut rank may increase up to ω. The remaining cases are immediate by IH. □

56.4. LEMMA (*Level Lowering Lemma*). *Assume that* $\mathrm{RS}_n \vdash_\rho^\alpha \Gamma$, *where* Γ *is a set of formulas in the language of* $\mathcal{T}_n$ (*so* T_n^α, F_n^α *do not occur in* Γ). *Then, if* $\rho > 0$ *or* $\alpha > 0$,

$$\mathcal{T}_n \vdash^{\phi\rho\alpha} \Gamma$$

PROOF. By 53.7 we have $\mathrm{RS}_n \vdash_1^{\phi\rho\alpha} \Gamma$. By assumption on Γ, $R_n(\Gamma) = 0$ and $\Gamma = \Gamma^-$. The conclusion is immediate by 56.3, observing that $\rho > 0$ or $\alpha > 0$ implies $k+\phi\rho\alpha = \phi\rho\alpha$, for $k \in \omega$. □

We now appeal to the notions of Ch.IX, especially to the basic order-theoretic properties of §45A, concerning the Veblen hierarchy. Recall that an ordinal α is an ϵ-number, if α has the form $\phi 1\beta$, for some β.

56.5. DEFINITION

(i) $\alpha_0 := \epsilon_0$ and $\alpha_{k+1} = \phi\alpha_k 0$;

(ii) $H_0(\beta) = \beta$ and $H_{n+1}(\beta) = \phi(H_n(\beta))0$.

56.5.1. FACT: $H_n(\alpha_{k+1}) = H_{n+1}(\alpha_k)$.

56.6. DEFINITION. A formula A of $\mathcal{L}_0$ (= the language of IT_0^∞) is T_0-positive iff A is inductively generated by means of $\forall$, $\exists$, $\wedge$, $\vee$ from atoms of the form $t = s$, $\neg t = s$, Nt, $\neg Nt$, $T_0 t$.
If A is a T_0-positive $\mathcal{L}_0$-formula, A_0^δ is the formula of RS_0 which results from A by replacing every atom of the form $T_0 t$ with $T_0^\delta t$.

56.7. THEOREM (*Upper bound*)

(i) *If* $\mathrm{IT}_n^\infty \vdash_1^\alpha A_0$, A_0 *is a* T_0*-positive formula of* $\mathcal{L}_0$ *and* $\alpha < \alpha_k$, *then* $\mathrm{RS}_0 \vdash A_0^\delta$, *for some* $\delta < H_n(\alpha_k)$.

(ii) *If* $\mathrm{IT}_n^\infty \vdash_1^\alpha A$, $\alpha < \alpha_k$ *and* $A \in \mathcal{L}_{op}$, *then* $\mathrm{OP}^\infty \vdash_1^\delta A$, *for some* $\delta < H_{n+1}(\alpha_k)$.

PROOF. Both (i) and (ii) are verified by induction on n.
Ad (i): $n = 0$. By the embedding theorem 55.5, we immediately have:

$$\mathrm{RS}_0 \vdash^{\omega(\delta+1)}_{\omega(\delta+1)} A_0^\delta,$$

where $\delta = 1+\omega^\alpha < H_0(\alpha_k) = \alpha_k$ (each α_k is closed under ω-exponentiation).
$n = m+1$. Assume

$$\mathrm{IT}_n^\infty \vdash^\alpha_1 A_0 \text{ with } \alpha < \alpha_k.$$

Again by 55.5, it follows, for $\delta = 1+\omega^\alpha$,

$$\mathrm{RS}_n \vdash^{\omega(\delta+1)}_{\omega(\delta+1)} A_0;$$

hence, if we set $\xi = \omega(\delta+1)$, we can conclude by level lowering lemma 56.4 and partial cut elimination 55.1:

$$\mathrm{IT}_m^\infty \vdash^{\phi 1(\phi\xi\xi)}_1 A_0. \tag{1}$$

But $\alpha < \alpha_k$ implies $\phi 1(\phi\xi\xi) < \phi\alpha_k 0 = \alpha_{k+1}$. Hence, by IH applied to (1),

$$\mathrm{RS}_0 \vdash A_0^\delta, \text{ for some } \delta < H_m(\alpha_{k+1}) = H_n(\alpha_k) \text{ (by 56.5)}. \tag{2}$$

Ad(ii): $n = 0$. If $\mathrm{IT}_0^\infty \vdash^\alpha_1 A$, $\alpha < \alpha_k$, $A \in \mathcal{L}_{op}$ and $\delta = 1+\omega^\alpha$, then $\mathrm{RS}_0 \vdash^{\omega(\delta+1)}_{\omega(\delta+1)} A$ by embedding. Hence by level lowering lemma and partial cut elimination 55.1, again with $\xi = \omega(\delta+1)$,

$$\mathrm{OP}^\infty \vdash^\lambda_1 A, \text{ where } \lambda = \phi 1(\phi\xi\xi) < \phi\alpha_k 0 = H_1(\alpha_k).$$

$n = m+1$: by embedding, cut elimination, level lowering, IH and 56.5.1 □

§57. Upper bound theorems for TLR and its subsystems

We lift the upper bound theorem to the finitary formal system $\mathrm{TLR} + TI(lev)$ and its subsystems MF, MF_p, MF_c. Actually, the results are corollaries of the elimination of variable levels, carried out in Ch. X.

In addition, we shall apply 56.7 for obtaining explicit semantical information, which involves the intended inductive interpretation of our basic language. To this aim, we recall that **CTM** is the closed term model of OP (§4B) and that $\mathbf{O}(\mathbf{CTM})$, $\mathbf{O}(\mathbf{CTM}, \delta)$ are the least inductive model of MF over **CTM** and the δ-th stage of the same inductive model (in the given order; see Ch. II, 7.6). We further recall from 37.1 of Ch.VIII that, if $A \in \mathcal{L}$, the i-transform of A, denoted by A_i, is the $\mathcal{L}_V$-formula, which results from A by substituting each atom of the form Tt by $T_i t$. If $A \in \mathcal{L}$, A is T-positive iff A is inductively generated by means of $\wedge, \vee, \forall, \exists$ from atoms of the form $t = s$, Nt, $\neg t = s$, $\neg Nt$, Tt.

57.1. THEOREM

(i) *If* $\mathrm{TLR} + TI(lev) \vdash A$ *and* A *is a sentence of* $\mathfrak{L}_{op}$, *then* $\mathrm{OP}^{\infty} \vdash_{1}^{\alpha} A$, *for some* $\alpha < \Gamma_0$.

(ii) *If* A *is a* T*-positive* $\mathfrak{L}$*-sentence and* $\mathrm{TLR} + TI(lev) \vdash \forall i. A_i$, *then* $\mathrm{RS}_0 \vdash A_0^{\delta}$, *for some* $\delta < \Gamma_0$ *and hence* $\mathbf{O}(\mathbf{CTM}, \Gamma_0) \models A$.

PROOF. (i): if $\mathrm{TLR} + TI(lev) \vdash A \in \mathfrak{L}_{op}$, we have

$$\Rightarrow \mathrm{IT}_n^{\infty} \vdash_{q+1}^{\alpha} A, \text{ for some } \alpha < \omega \cdot 2,\ q < \omega \text{ (by 51.7);}$$

$$\Rightarrow \mathrm{IT}_n^{\infty} \vdash_{1}^{\omega_q(\alpha)} A \text{ (by 55.1);}$$

$$\Rightarrow \mathrm{OP}^{\infty} \vdash_{1}^{\delta} A \text{ with } \delta = H_{n+1}(\epsilon_0) \text{ (by 56.7 (ii)).}$$

Ad (ii). First of all, observe that the axioms and inferences of RS_0 are trivially sound for $\mathbf{O}(\mathbf{CTM}, \Gamma_0)$, as soon as we interpret T^{α} (F^{α}) by the set $\mathbf{O}(\mathbf{CTM}, \alpha)$ (respectively by the set $\{t : t \in \mathrm{CTM} \text{ and } \mathbf{O}(\mathbf{CTM}, \alpha) \models T\neg t\}$). This remark ensures that the final part of the conclusion is true, if the first one holds. Now, if $\mathrm{TLR} + TI(lev) \vdash \forall i. A_i$ and A is a T-positive $\mathfrak{L}$-formula, we have, arguing as in (i) above, for some n and $\alpha < \epsilon_0$:

$$\mathrm{IT}_n^{\infty} \vdash_{1}^{\alpha} A_0$$

(A_0 being obtained from A by replacing every atom of the form Tt with $T_0 t$). But A_0 meets the hypothesis of 56.7 (i) and we are done. □

Theorem 57.1 (ii) immediately yields the following interesting cases:

57.2. COROLLARY

(i) *If* $\mathrm{TLR} + TI(lev) \vdash \forall i. T_i t$ (t *closed*), *then* $\mathrm{RS}_0 \vdash T^{\delta} t$ $(= T_0^{\delta}(t))$, *for some* $\delta < \Gamma_0$ *and hence* $\mathbf{O}(\mathbf{CTM}, \Gamma_0) \models Tt$.

(ii) *If* $\mathrm{TLR} + TI(lev) \vdash \forall i. Cl_i(t)$ (t closed), *then* $\mathrm{RS}_0 \vdash Cl_0^{\delta}(t)$, *for some* $\delta < \Gamma_0$ *and hence* $\mathbf{O}(\mathbf{CTM}, \Gamma_0) \models Cl(t)$.

It is not difficult to adapt the previous machinery to the proof-theoretic investigation of MF, MF_p and MF_c of Ch. II, 10.7. We begin with the theory MF with full number-theoretic induction.

57.3. THEOREM

(i) *If* $\mathrm{MF} \vdash A$ *and* A *is a sentence of* $\mathfrak{L}_{op}$, *then* $\mathrm{OP}^{\infty} \vdash_{1}^{\alpha} A$, *for some* $\alpha < \alpha_1 = \phi\epsilon_0 0$.

(ii) *If* $\mathrm{MF} \vdash Tt$ (t closed), *then* $\mathrm{RS}_0 \vdash T^{\alpha} t$, *for some* $\alpha < \epsilon_0$, *and hence* $\mathbf{O}(\mathbf{CTM}, \epsilon_0) \models Tt$.

PROOF. Ad (i)-(ii). First observe that the language $\mathcal{L}$ of MF can be translated into the language $\mathcal{L}_0$ of the system IT_0^∞, once we identify the atom Tt of $\mathcal{L}$ with $T_0 t$. It is convenient to leave the translation implicit; so we use the same symbol A to denote the $\mathcal{L}$-formula A and its cognate translation in $\mathcal{L}_0$. With this in mind, a straightforward adaptation of 51.5, 51.6 and the partial cut elimination theorem 55.1 imply:

$$\text{if MF} \vdash A, \text{ then } \mathrm{IT}_0^\infty \vdash_1^\alpha A, \text{ for some } \alpha < \epsilon_0. \tag{1}$$

But (1) yields (i)-(ii), as a consequence of the upper bound theorem 56.7. □

57.4. THEOREM. *Let A be any sentence of $\mathcal{L}_{op}$ and let t be any closed term. Then:*

(i) *if $\mathrm{MF}_p \vdash A$, then $\mathrm{OP}^\infty \vdash_1^\alpha A$, for some $\alpha < \phi\omega 0$;*

(ii) *if $\mathrm{MF}_p \vdash Tt$, then $\mathrm{RS}_0 \vdash T^\alpha t$ for some $\alpha < \omega^\omega$ and hence*

$$\mathbb{O}(\mathbf{CTM}, \omega^\omega) \models Tt;$$

(iii) *if $\mathrm{MF}_c \vdash A$, then $\mathrm{OP}^\infty \vdash_1^\alpha A$, for some $\alpha < \epsilon_0$;*

(iv) *if $\mathrm{MF}_c \vdash Tt$, then $\mathrm{RS}_0 \vdash T^k t$, for some $k < \omega$ and hence*

$$\mathbb{O}(\mathbf{CTM}, \omega) \models Tt.$$

PROOF. We stick to the convention of identifying $\mathcal{L}$ with $\mathcal{L}_0$ as in the previous proof.
Ad (i)-(ii). We exploit the fact that in these cases the *infinitary system* IT_0^∞ can be replaced by its *finitary version* $\mathrm{IT}_0(p)$.
$\mathrm{IT}_0(p)$ is obtained from IT_0^∞ by means of the following modifications:

1. we omit the N-rules (N) and $(\neg N)$;
2. the operational axioms of 50.2.1 are replaced by the corresponding axioms of A.2 and A.3 (see 47.3);
3. we add the N-induction rule in the form *P-N-IND*:

   infer Γ, $\neg Nt$, $t\eta a$ from Γ, $\bar{0}\eta a$ and Γ, $\forall x(x\eta a \rightarrow (x+1)\eta a)$.

Clearly MF_p can be embedded into $\mathrm{IT}_0(p)$.

The derivability relation "$\mathrm{IT}_0(p) \vdash_m^k \Gamma$" is inductively generated by axioms and rules of $\mathrm{IT}_0(p)$, as in 50.2.3; of course, since the rules are finitary, we have that k, $m < \omega$ (m being an upper bound on the 0-complexity of the cut formulas occurring in the given derivation of Γ). Now, due to the restricted form of *P-N-IND*, $\mathrm{IT}_0(p)$ enjoys partial cut elimination:

$$\text{if } \mathrm{IT}_0(p) \vdash_{m+1}^k A, \text{ then } \mathrm{IT}_0(p) \vdash_1^n A \text{ for some } n \in \omega. \tag{1}$$

Hence the embedding theorem 55.5 can be specialized to $\mathrm{IT}_0(p)$:

$$\text{if } \mathrm{IT}_0(p) \vdash^{k}_{1} \Gamma, \text{ then } \mathrm{RS}_0 \vdash^{\omega(\gamma+1)}_{\omega(\gamma+1)} \Gamma[\beta,\gamma], \text{ for every } \beta,\ \gamma, \tag{2}$$

such that $0 < \beta$ and $\beta + \omega^k \leq \gamma$.

Verification of (2): by induction on k. We only consider the case of positive N-induction and we assume:

$$\mathrm{IT}_0(p) \vdash^{k_0}_{1} \Gamma,\ \bar{0}\eta a \text{ and } \mathrm{IT}_0(p) \vdash^{k_0}_{1} \Gamma,\ \neg t\eta a,\ (t+1)\eta a,$$

where $k_0 < k$, while the second premise is obtained by $(\forall)$, $(\vee)$-inversion property, adapted to $\mathrm{IT}_0(p)$ from 53.4. Fix arbitrary $\beta > 0$, $\gamma \geq \beta + \omega^k$ and put $\sigma_n := \beta + \omega^{k_0}(n+1)$. We check, by secondary induction on $n \in \omega$:

$$\mathrm{RS}_0 \vdash^{<\omega(\gamma+1)}_{\omega\gamma} \Gamma[\beta,\gamma],\ \bar{n}\eta^{\sigma_n} a. \tag{2.1}$$

Indeed, by main IH we also have, for arbitrary $\beta > 0$ and $\sigma(\beta) = \beta + \omega^{k_0}$:

$$\mathrm{RS}_0 \vdash^{\omega(\sigma(\beta)+1)}_{\omega(\sigma(\beta)+1)} \Gamma[\beta,\sigma(\beta)],\ \bar{0}\eta^{\sigma(\beta)} a; \tag{2.2}$$

$$\mathrm{RS}_0 \vdash^{\omega(\sigma(\beta)+1)}_{\omega(\sigma(\beta)+1)} \Gamma[\beta,\sigma(\beta)],\ \neg t\eta^{\beta} a, (t+1)\eta^{\sigma(\beta)} a \ (t \text{ arbitrary}). \tag{2.3}$$

If $n = 0$, (2.1) is simply a consequence of (2.2) with the persistence lemma 55.4 and $\gamma > \sigma_0$. If $n = m+1$, we have by secondary IH:

$$\mathrm{RS}_0 \vdash^{<\omega(\gamma+1)}_{\omega\gamma} \Gamma[\beta,\gamma],\ \bar{m}\eta^{\sigma_m} a. \tag{2.4}$$

If we choose $\beta = \sigma_m$ in (2.3) above, we get, since $\sigma(\sigma_m) = \sigma_n$:

$$\mathrm{RS}_0 \vdash^{\omega(\sigma_n+1)}_{\omega(\sigma_n+1)} \Gamma[\sigma_m,\sigma_n],\ \neg\bar{m}\eta^{\sigma_m} a,\ \bar{n}\eta^{\sigma_n} a. \tag{2.5}$$

But $\gamma > \sigma_n$ and hence, by persistence lemma 55.4,

$$\mathrm{RS}_0 \vdash^{<\omega(\gamma+1)}_{\omega\gamma} \Gamma[\beta,\gamma],\ \neg\bar{m}\eta^{\sigma_m} a,\ \bar{n}\eta^{\sigma_n} a. \tag{2.6}$$

An application of (Cut) between (2.4) and (2.6) concludes the verification of (2.1). On the other hand, if we apply the substitution lemma, we obtain, with t arbitrary, for each n:

$$\mathrm{RS}_0 \vdash^{<\omega(\gamma+1)}_{\omega\gamma} \Gamma[\beta,\gamma],\ \neg t = \bar{n},\ t\eta^{\sigma_n} a. \tag{2.7}$$

Hence by 54.3 (iii) and a cut of rank $\omega\sigma_n < \omega\gamma$:

$$\mathrm{RS}_0 \vdash^{<\omega(\gamma+1)}_{\omega\gamma} \Gamma[\beta,\gamma],\ \neg t = \bar{n},\ t\eta^{\gamma} a. \tag{2.8}$$

A final application of $(\neg N)$ yields (2.8):

$$\mathrm{RS}_0 \vdash^{\omega(\gamma+1)}_{\omega(\gamma+1)} \Gamma[\beta,\gamma],\ \neg Nt,\ t\eta^{\gamma} a. \ \square$$

Finally, if $\mathrm{MF}_p \vdash A$, we have for some $n > 0$, by (1)-(2) with $\beta = 1$:

$$\mathrm{RS}_0 \vdash^{\omega^{n+2}}_{\omega^{n+2}} A[1,\omega^n],$$

whence by Tait's cut elimination $\mathrm{RS}_0 \vdash^{\phi(n+2)(\omega^{n+2})}_1 A[1,\omega^n]$. If we apply the upper bound theorem and the closure properties of $\phi\omega 0$, we obtain

$$\mathrm{OP}^\infty \vdash^{<\phi\omega 0}_1 A, \text{ if } A \in \mathcal{L}_{op} \text{ and } \mathbf{O}(\mathbf{CTM}, \phi\omega 0) \models A, \text{ if } A \text{ is } T\text{-positive.} \ \Box$$

Ad(iii)-(iv). We introduce a variant $\mathrm{IT}_0(c)$ of $\mathrm{IT}_0(p)$ such that, if $\mathrm{MF}_c \vdash A$, then $\mathrm{IT}_0(c) \vdash A$. $\mathrm{IT}_0(c)$ is obtained from $\mathrm{IT}_0(p)$ by replacing the rule *P-N-IND* with *CL-N-IND* below:

infer $\Gamma, \neg Nt, t\eta a$ from the premises $\Gamma, Cl(a)$; $\Gamma, \bar{0}\eta a$; $\Gamma, \forall x(x\eta a \rightarrow (x+1)\eta a)$.

As for $\mathrm{IT}_0(p)$, $\mathrm{IT}_0(c)$ enjoys partial cut elimination, i.e. if $\mathrm{IT}_0(c) \vdash^k_{m+1} A$, then $\mathrm{IT}_0(c) \vdash^n_1 A$, for some $n \in \omega$. Instead of reproving the appropriate form of embedding for $\mathrm{IT}_0(c)$, we establish the crucial case of the interpretation theorem in semantical terms; we then give directions to obtain its proof-theoretic version. First of all, we recursively define:

$$\models A[m,n] \text{ iff either } A \text{ is an e-atom and } \mathbf{CTM} \models A; \text{ or} \tag{3}$$

$A = Tt$ (Ft) and $t \in \mathbf{O}(\mathbf{CTM}, n)$ $((\neg t) \in \mathbf{O}(\mathbf{CTM}, n))$; or

$A = \neg Tt$ $(\neg Ft)$ and $t \notin \mathbf{O}(\mathbf{CTM}, m)$ $((\neg t) \notin \mathbf{O}(\mathbf{CTM}, m))$; or

$A = \forall x B$ $(\exists x B)$ and $\models B(t)[m,n]$ for every (some) $t \in \mathrm{CTM}$; or

$A = B \wedge C$ $(B \vee C)$ and $\models B[m,n]$ and (or) $\models C[m,n]$.

We also write $\models T^n t$ $(F^n t)$ for $\models (Tt)[m,n]$ $(\models (Ft)[m,n])$ and $\models \neg T^m t$ $(\neg F^m t)$ for $\models (\neg Tt)[m,n]$ $(\models (\neg Ft)[m,n])$. If $\Gamma = \{A_1, \ldots, A_k\}$, $\models \Gamma[m,n]$ is interpreted disjunctively as $\models (A_1 \vee \ldots \vee A_n)[m,n]$. As expected, we have:

$$\text{if } \models \Gamma[m',n'], \Delta \text{ and } m \le m' \le n' \le n, \text{ then } \models \Gamma[m,n], \Delta. \tag{4}$$

Now we claim:

$$\text{if } \mathrm{IT}_0(c) \vdash^k_1 \Gamma, \text{ then } \models \Gamma[m,n], \text{ for every } m > 0,\ n \ge m+2^k. \tag{5}$$

Verification of (5): by induction on k. Let us only consider the case where, for some $k_0 < k$, we have:

$$\mathrm{IT}_0(c) \vdash^{k_0}_1 \Gamma, Cl(a) \quad \mathrm{IT}_0(c) \vdash^{k_0}_1 \Gamma, \bar{0}\eta a \quad \text{and} \tag{5.1}$$

$$\mathrm{IT}_0(c) \vdash^{k_0}_1 \neg t\eta a, (t+1)\eta a \ (t \text{ arbitrary}).$$

Then by IH applied to (5.1), and (4), we have, for every $m > 0$, $n \ge m+2^{k_0}$ with $p_0 = m+2^{k_0}$ and remembering that $t\eta^{p_0} a = T^{p_0}(at)$, $t\bar{\eta}^{p_0} a = F^{p_0}(at)$:

$$\models\Gamma[m,n], t\eta^{p_0} a, t\overline{\eta}^{p_0} a \text{ (t arbitrary)}; \tag{5.2}$$

$$\models\Gamma[m,n], \overline{0}\eta^{p_0} a; \tag{5.3}$$

$$\models\Gamma[m,n], \neg t\eta^{m} a,\ (t+1)\eta^{p_0} a \text{ (t arbitrary)}. \tag{5.4}$$

Now fix $m_1 > 0$, $n_1 \geq m_1+2^k$ and set $p_1 = m_1+2^{k_0}$. We check by secondary induction on $l \in \omega$

$$\models\Gamma[m_1,n_1],\ \overline{l}\eta^{p_1} a. \tag{5.5}$$

If $l = 0$, apply (5.3) with $m = m_1$. If $l = j+1$, assume by secondary IH:

$$\models\Gamma[m_1,n_1],\ \overline{j}\eta^{p_1} a. \tag{5.6}$$

If we apply (5.4) with $m = p_1$, we get, for $q = p_1+2^{k_0}$:

$$\models\Gamma[p_1,q],\ \neg\ \overline{j}\eta^{p_1} a,\ \overline{l}\eta^{q} a. \tag{5.7}$$

By (4), since $n_1 \geq m_1+2^k \geq p_1+2^{k_0} = q \geq m_1$, we get

$$\models\Gamma[m_1,n_1],\ \neg\ \overline{j}\eta^{p_1} a,\ \overline{l}\eta^{q} a;$$

hence, with a cut:

$$\models\Gamma[m_1,n_1], \overline{l}\eta^{q} a. \tag{5.8}$$

By Ch. II, 7.7, we have $\mathbf{O}(\mathbf{CTM},n) \cap \{t\colon t \in \mathrm{CTM}, \mathbf{O}(\mathbf{CTM},n) \models T\neg t\} = \emptyset$ and $\mathbf{O}(\mathbf{CTM},n) \subseteq \mathbf{O}(\mathbf{CTM},m)$ for $n < m$, whence

$$\models\Gamma[m_1,n_1], \neg\ \overline{l}\overline{\eta}^{q} a,\ \neg\ \overline{l}\eta^{q} a. \tag{5.9}$$

$$\models\Gamma[m_1,n_1],\ \neg\ \overline{l}\overline{\eta}^{p_1} a,\ \overline{l}\overline{\eta}^{q} a. \tag{5.10}$$

The conclusion follows by application of the cut rule to (5.8), (5.9), (5.10) and (5.2) (in the last case choose $m = m_1$). □

If we inspect the verification of (5), we can observe: the levels involved are *finite*, by contrast with the corresponding step (2) for $\mathrm{IT}_0(\mathrm{p})$; the levels depend only on the given parameter m and on the derivation length. Hence *the interpretation theorem for* $\mathrm{IT}_0(c)$ *can be carried out within the fragment of* RS_0 *with finite levels only.* In particular, we can modify the notion of 0-rank for formulas of RS_0, in such a way that the 0-rank is always finite and cut elimination still works with respect to the new notion (Hint: choose $R_0(T^k) = R_0(F^k) = 30k$ and reprove 54.1-5, 55.4). Finally the embedding theorem 55.5 is refined for $\mathrm{IT}_0(c)$, to the extent that:

$$\text{if } \mathrm{IT}_0(c) \vdash^{k}_{1} A, \text{ then } \mathrm{RS}_0 \vdash^{\lambda}_{1} A[1,1+2^k] \text{ for some } \lambda < \epsilon_0, \tag{6}$$

which implies (iii)-(iv). We leave a complete check of (6) as exercise. □

§ 58. Conclusion: the conservation theorems

If we piece together the results of Ch. IX with the main theorems of § 57, we can obtain a characterization of the operational theorems of TLR + TI(lev), MF, MF_p, MF_c. In order to state the theorem, let us agree that:

(i) $\alpha < \beta$ stands for the $\mathcal{L}_{op}$-formula defining the primitive recursive well-ordering of type Γ_0 and lower case Greek letters α, β, $\gamma \ldots$ range over the field of $<$;

(ii) $Progr(<, B) = \forall\alpha(\forall\beta(\beta < \alpha \rightarrow B(\beta)) \rightarrow B(\alpha))$;

(iii) $TI(\alpha)$ is the schema $Progr(<, B) \rightarrow \forall\beta < \alpha. B(\beta)$;

$TI(<\gamma) := \cup\{TI(\alpha) : \alpha < \gamma\}$;

$TI_{op}(<\alpha)$ is $TI(<\alpha)$, restricted to $\mathcal{L}_{op}$-formulas.

58.1. THEOREM (*Conservation*). *Let A be a formula of* $\mathcal{L}_{op}$. *Then:*

(i) TLR + $TI(lev) \vdash A$ *iff* OP + $TI_{op}(<\Gamma_0) \vdash A$;

(ii) MF $\vdash A$ *iff* OP + $TI_{op}(<\phi\epsilon_0 0) \vdash A$;

(iii) $\mathrm{MF}_p \vdash A$ *iff* OP + $TI_{op}(<\phi\omega 0) \vdash A$;

(iv) $\mathrm{MF}_c \vdash A$ *iff* OP $\vdash A$.

PROOF. $\Leftarrow$: § 46 of Ch. IX contains the relevant facts for proving the implications from right to left. Indeed, observe that elementary comprehension and $PWO(\alpha)$ imply $TI_{op}(\alpha)$; then apply 46.2, 46.2.6 for (i) and (iii). As to (ii), 46.2.3 and full N-induction imply $TI(<\epsilon_0)$ for arbitrary formulas of $\mathcal{L}$, which yields the existence of the ramified hierarchy up to any $\alpha < \epsilon_0$ (by 44.3), and hence the conclusion by 46.2.5.
$\Rightarrow$: we informally sketch the argument. First, it is essential to realize that the proofs involved in chapters VIII-IX are constructive: indeed, they can be formalized in the elementary theory OP extended by the schema $TI(<\alpha)$ on an appropriate segment α of Γ_0. The major obstacle to the formalization is to find $\mathcal{L}_{op}$-formulas which adequately represent in OP the infinitary derivability relations $\mathrm{RS}_n \vdash^{\alpha}_{\rho} \Gamma$, $\mathrm{IT}^{\infty}_n \vdash^{\alpha}_{\rho} \Gamma$, $\mathrm{OP}^{\infty} \vdash^{\alpha}_{\rho} \Gamma$. Now it is possible to find adequate $\mathcal{L}_{op}$-formulas by making the derivation trees themselves explicit, and by observing that they can be effectively encoded. Then one shows, by means of the recursion theorem for primitive recursive functions, that *the operation of cut elimination is primitive recursive*, and that also the embedding operations are effective. These steps are non-trivial, but well-known; details for formalizing cut elimination can be gained from Mints (1975), Schwichtenberg (1977), Buchholz (1991) {or from the appendix}.

In order to state the formalized results, we adopt the following stipulations. First of all, we fix a Gödel numbering of basic syntax; if E (respectively "$\ldots \vdash \ldots$") is a syntactical expression (a derivability predicate), which belongs to one of the systems involved, let $\ulcorner E \urcorner$ ($\ulcorner \ldots \vdash \ldots \urcorner$ respectively) denote the corresponding arithmetized term (predicate) of $\mathcal{L}_{op}$. $Sent_{op}(x)$ defines the predicate " x is a closed formula of $\mathcal{L}_{op}$"; $Com(x,y)$ stands for "the sentence encoded by x has logical complexity $\leq y$". In addition, if SF is a finitary formal system, let $Dim_k(x,y,\mathrm{SF})$ stand for the arithmetized predicate "x encodes a proof of y in SF, having $\leq k$ lines".

If we combine formalized versions of the embedding theorems 48.8, 49.11, weak cut elimination 49.8 and the main corollary to the asymmetric interpretation of TLR + TI(lev) in STLR (see 48.8-49.19), we obtain for each given $k \in \omega$, provably in OP, if $\mathfrak{T} := \mathrm{TLR} + TI(lev)$:

$$\forall d \forall \ulcorner A \urcorner (Sent_{op}(\ulcorner A \urcorner) \wedge Dim_k(d, \ulcorner A \urcorner, \mathfrak{T}) \rightarrow (\ulcorner \mathrm{STLR}_{\phi(d)} \vdash A \urcorner)), \tag{1}$$

for a suitable term $\phi(d)$ representing an $\alpha(k)$-recursive function, and for some $\alpha(k) < \epsilon_0$ (cf. 49.20).

Now define $\alpha_0 = \epsilon_0$, $\alpha_{n+1} = \phi\alpha_n 0$. Then by formalization of 51.6, 55.1 and the upper bound theorem 56.7 (ii), we have for each given m, provably in $\mathrm{OP} + TI_{op}(\alpha_{m+1})$:

$$\forall \ulcorner A \urcorner (Sent_{op}(\ulcorner A \urcorner) \wedge \ulcorner \mathrm{STLR}_m \vdash A \urcorner \rightarrow \exists \alpha (\alpha < \alpha_{m+1} \wedge \ulcorner \mathrm{OP}^{\infty} \vdash^{\alpha}_{1} A \urcorner)). \tag{2}$$

At this stage, we can use a well-known "reflection" technique. If $\mathrm{OP}^{\infty} \vdash^{\alpha}_{1} A$, then each formula occurring in the given OP^{∞}-derivation $\mathfrak{D}$ must have logical complexity $\leq co(A)$(= the logical complexity of A), because the only possible cut formulas within $\mathfrak{D}$, by assumption, either have the form $t = s$ or Nt. Hence the correctness of a "cut-free" derivation (in the above sense) can be checked by reference to a truth predicate Tr_n for $\mathcal{L}_{op}$-sentences of logical complexity $\leq n$, for some given n; and Tr_n can be defined within OP. Hence we obtain, for each finite n and k, provably in $\mathrm{OP} + TI_{op}(\alpha_k)$:

$$\alpha < \alpha_k \wedge Com(\ulcorner A \urcorner, \overline{n}) \wedge Sent_{op}(\ulcorner A \urcorner) \wedge \mathrm{OP}^{\infty} \vdash^{\alpha}_{1} A \rightarrow Tr_n(\ulcorner A \urcorner); \tag{3}$$

$$\mathrm{OP} \vdash Tr_n(\ulcorner A \urcorner) \rightarrow A \text{ (where } A \in \mathcal{L}_{op} \text{ and has logical complexity } \leq n). \tag{4}$$

The verification of (i) is now straightforward, as soon as we note that $\mathrm{TLR} + TI(lev) \vdash A$ with $A \in \mathcal{L}_{op}$ implies $\mathrm{OP} \vdash \ulcorner \mathrm{TLR} + TI(lev) \vdash A \urcorner$ and we apply (1)-(4) above. The other results follow by the formalized versions of the corresponding upper bound theorems for MF, MF_p and MF_c. □

The previous theorem, coupled with the general conservation theorem 15.5 concerning PW_c, yields an exhaustive proof-theoretic classification of the systems introduced up till now (with the exception of the impredicative extensions of §§41-42).

58.2. *Final remarks*

(i) The techniques of this chapter adapt those applied in Cantini (1985a) to certain predicatively reducible theories ID_n^* of iterated inductive definitions. Jäger(1984, 1986) develops an elegant and uniform approach to reductive proof theory, based on admissible set theories; the exact relationship of TLR and its variants with iterated admissibility is yet to be investigated in some detail.

(ii) The results of chapters X-XI can be strengthened with the addition of local weak generalized induction principles. To be more specific, let us consider the term $I(W,a,i)$ of lemma 41.11, which defines the collection of well-founded trees recursive in the i-class a. We already know that it is consistent to assume the $\subseteq$-minimality of such collection. Formally let LGI(= local generalized induction) be the axiom:

$$Cl_i(a) \wedge \forall x(W_i(x,a,b) \rightarrow x\eta_i b) \rightarrow I(W,a,i) \subseteq b,$$

where $W_i(x,a,b)$ formalizes the operator, which inductively generates the collection of trees recursive in the set represented by the i-class a. It turns out that TLR + LGI is not stronger than TLR (the arguments parallel those of §55, but the Veblen function $\lambda\alpha\lambda\beta.\phi\alpha\beta$ is required for the asymmetric interpretation of LGI).

(iii) A further refinement concerns the direct proof-theoretical analysis of PW_c + GID; this is already in the literature for an equivalent system (Cantini 1992) and entirely analogous to the method of §49. Thus we only sketch the basic idea. The starting point is to axiomatize, in a natural ordinal theory PWO, the features of the inductive model of Ch. II, §7, that make PW_c + GID true. One can easily find a system PWO, where:

1. *the inductive generation of* T in ordinal stages is made explicit via certain local predicative closure conditions (extending those of 52.9 so that π-axioms become true);
2. *the full transfinite induction schema* TI on ordinals is accepted, together with a form TR of ordinal reflection, granting the existence of transfinite ordinals;
3. *number-theoretic induction* is needed only in a restricted, local form, i.e. for properties at a given ordinal stage.

Then PWO is reduced to OP. First of all, the schema TI is eliminated: PWO is interpreted in a system PWO_∞ where ordinal variables are forced to range over *finite ordinals* (via ω-rule). PWO_∞ is consistent and it admits an asymmetric interpretation in a ramified system with finite stages PWR. The final step shows that PWR is proof-theoretically reducible to OP using a standard cut elimination argument.

Appendix: primitive recursive cut elimination for RS_n

Preliminaries

We outline a primitive recursive cut elimination algorithm for the predicative systems RS_n. We combine the use of repetition rule, due to Mints (1975), with the encoding technique of Schwichtenberg (1977). For an elegant unified treatment of primitive recursive and continuous cut elimination, the reader is urged to consult Buchholz (1991). All the cited references only deal with Peano arithmetic with ω-rule.

Roughly speaking, RS_n-derivations are (possibly) infinite trees of inferential figures, inductively generated according to the clauses DER.1-DER.2, which define the derivability relation $\mathrm{RS}_n \vdash^{\alpha}_{\rho} \Gamma$ of 52.10. Now the problem is that we need a finitary description of the associated infinitary trees and their properties: thus we shall only consider those derivations that can be effectively described by *finite sequences of data*, the so-called *codes*. In particular, we apply an infinitary inference, only if we have a primitive recursive control over the immediate subderivations of the premises. As a preliminary step, we fix the following data:

(i) a primitive recursive Gödel numbering $\ulcorner\ \urcorner$ of the language of RS_n;

(ii) a primitive recursive indexing $\{[e] : e \in \omega\}$ for primitive recursive functions (*PR* in short); *Uni* is the recursive universal function such that $Uni(e, x) = [e](x)$; *PRI* is the set of *PR*-indices (*PRI* is primitive recursive);

(iii) a *PR*-injection $Inj : \mathbb{N}^7 \to \mathbb{N}$, with *PR*-projections; we assume that the image of Inj, $Inj[\mathbb{N}^7]$ is disjoint from the set *PRI* of *PR*-indices; if $x \in Inj[\mathbb{N}^7]$, x_i (for $0 \leq i \leq 6$) denotes the i-th coordinate of x.

A sequent $\Gamma = \{A_1, \ldots, A_n\}$ is encoded as a finite set of Gödel numbers:

if $A_1, \ldots, A_n$ are distinct, $\ulcorner\Gamma\urcorner = 2^{\ulcorner A_1 \urcorner} + \ldots + 2^{\ulcorner A_n \urcorner}$.

For notational simplicity, we systematically identify syntactical entities with their number-theoretic encodings; in general, this will cause no ambiguity. We also use the formal expression AX for referring to any possible axiom of RS_n. *REP* stands for the repetition rule: infer $\mathrm{RS}_n \vdash^{\alpha}_{\rho} \Gamma$ from $\mathrm{RS}_n \vdash^{\beta}_{\rho} \Gamma$, whenever $\beta < \alpha$. *REP* is semantically trivial, but it makes sense as a geometric operation, which adds a new node to a given derivation tree of length β and increases the height of the given tree. If a *PR*-function F operates on Γ and on an inference label, say (CUT) for the cut rule, we simply write $F(\Gamma, CUT)$ instead of $F(\ulcorner\Gamma\urcorner, \ulcorner CUT \urcorner)$. Similarly, if $S_0, \ldots, S_6$ are syntactical entities, we write $\langle S_0, \ldots, S_6 \rangle$ instead of $Inj(\ulcorner S_0 \urcorner, \ldots, \ulcorner S_6 \urcorner)$. The basic data structures we deal with for representing derivations are called *labels* and they are suitable vectors $f = \langle f_0, \dots, f_6 \rangle$ of dimension 7. Formally, we are led to introduce the following

1. DEFINITION

(i) $f \in LABEL$ iff either $f = 0$ (0 stands for the code of the empty set, too) or $f = \langle f_0, \dots, f_6 \rangle$ and:

1. $f_0 = RF(f)$ is the name of an inference $\mathfrak{I}$ of RS_n; from f_0 we can read off

   1.1. the eigenvariable of $\mathfrak{I}$, if $\mathfrak{I}$ requires it; else
   1.2. an index $\alpha \in [0, \Gamma_0)$, which gives the "position" of the premise of $\mathfrak{I}$ (see the case of the repetition rule below), or identifies the minor formula of $\mathfrak{I}$, if there is any ambiguity;

2. $f_1 = END(f)$ encodes the disjoint union of the set of side formulas $LAT(f)$ and the set of active formulas $AF(f)$;

3. $f_2 = L(f)(=$ the length of $f)$, $f_3 = R(f)(=$ the rank of $f)$ are elements of OT;

4. $f_4 = PAR(f)$ is a finite set of parameters;

5. if $f_5 = SD_1(f)$, $f_6 = SD_2(f)$, then:

   5.1. if $RF(f) = AX$, each $SD_i(f)$ is empty $(i = 1, 2)$;
   5.2. if $RF(f) = REP$ and its position index is $\delta \in OT$, or $RF(f)$ is infinitary, then $SD_2(f) = 0$ and $SD_1(f) \in PRI$ (i.e. a PR-index);
   5.3. if $RF(f) \neq REP$ and $RF(f)$ is 2-ary (1-ary), then $SD_i(f) \neq 0$, for $i = 1, 2$ ($SD_1(f) \neq 0$ and $SD_2(f) = 0$, or $SD_2(f) \neq 0$ and $SD_1(f) = 0$, respectively).

(ii) We inductively define the operation DEP as follows:

$$DEP(f) = 0, \text{ if } f = \emptyset \text{ or } f \notin LABEL; \text{ else}$$
$$DEP(f) = max(DEP(SD_1(f)), DEP(SD_2(f)))+1.$$

Clearly $LABEL$ and DEP are primitive recursive.

If $f \in LABEL$, we can build up a formal figure, which is to be regarded as *a finite stump of a (possible) derivation*; $DEP(f)$ is the depth of f. Informally, if f is a label, $RF(f)$ names the final inference of a derivation, whose conclusion is $END(f)$ and whose immediate subderivations are identified respectively by $SD_1(f)$, $SD_2(f)$. $L(f)$, $R(f)$ play the role of the length and the rank, while $PAR(f)$ lists the parameters occurring in the derivation; it can be assumed that we can effectively identify the subset $EIGEN(f)$ of eigenparameters. We set $PAR(f, g) = PAR(f) \cup PAR(g)$, if f, $g \in LABEL$. By $PAR(f)(b/b^+)$ we denote the set which is obtained from $PAR(f)$ by replacing b with b^+; the "plus"-sign is used to declare that b occurs as eigenparameter. Below we generally skip places corresponding to empty coordinates of f, unless some ambiguity arises.

We now proceed to the definition of a collection $CODE \subseteq LABEL$, whose elements represent *true* derivations.

2. *Inductive Definition of CODE*

$CODE$ is the smallest set of labels, which satisfies the initial clause below and is closed under clauses, corresponding to the rules of RS_n:

(i) If Γ, Δ is an axiom of RS_n, $\langle AX;\ \Gamma,\ \Delta;\ \alpha;\ \rho;\ PAR(\Gamma,\Delta)\rangle \in CODE$, for every α, $\rho \in OT$ (here $f_6 = f_5 = 0$);

(ii) if $f_i \in CODE$, $END(f_i) = \{\Gamma, A_i\}$, $\alpha > L(f_i)$, $\rho \geq R(f_i)$ $(i = 1,2)$, $\Gamma = \{\Gamma, A_1 \wedge A_2\}$, then

$$\langle(\wedge);\ \Gamma,\ A_1 \wedge A_2;\ \alpha;\ \rho;\ PAR(f_1, f_2);\ f_1;\ f_2\rangle \in CODE;$$

(iii) if $f_i \in CODE$, $END(f_i) = \{\Gamma,\ A_1 \vee A_2,\ A_i\}$ $(i = 1,\ 2)$, $\alpha > L(f_2)$, $\rho \geq R(f_i)$, then

$$\langle\langle(\vee), i\rangle;\ \Gamma,\ A_1 \vee A_2;\ \alpha;\ \rho;\ PAR(f_i);\ f_i\rangle \in CODE$$

{it is understood that $SD_1(f) = \emptyset$ ($SD_2(f) = \emptyset$) whenever $i = 2$ $(i = 1)$};

(iv) if $g \in CODE$, $END(g) = \{\Gamma,\ A[x := a]\}$, $a \notin PAR(\Gamma, \forall x A)$, $\alpha > L(g)$, $R(g) \leq \rho$, then

$$\langle\langle\forall, a\rangle;\ \Gamma,\ \forall x A;\ \alpha;\ \rho;\ PAR(g)(a/a^+);\ g\rangle \in\ CODE;$$

(v) if $g \in CODE$, $END(g) = \{\Gamma, A[x := t]\}$, $\alpha > L(g)$, $\rho \geq R(g)$, then

$$\langle\langle\exists,\ t\rangle;\ \Gamma,\ \exists x A;\ \alpha;\ \rho;\ PAR(g);\ g\rangle \in CODE;$$

(vi) if h, $g \in CODE$, $END(h) = \{\Gamma, A\}$, $END(g) = \{\Gamma,\ \neg A\}$, $L(h)$, $L(g) < \alpha$, $max(R_n(A){+}1,\ R(g),\ R(h)) \leq \rho$, then

$$\langle\langle CUT,\ A\rangle,\ \Gamma;\ \alpha;\ \rho;\ PAR(g,\ h);\ h;\ g\rangle \in CODE;$$

(vii) if $e \in PRI$ and $END([e](\delta)) = \Gamma$, $L([e](\delta)) < \alpha$, $R([e](\delta)) \leq \rho$, $\mathrm{V} = PAR([e](\delta))$, $[e](\delta) \in CODE$ and $[e](\delta') = \emptyset$ for $\delta' \neq \delta$, then

$$\langle REP_\delta;\ \Gamma;\ \alpha;\ \rho;\ \mathrm{V};\ e\rangle \in CODE;$$

(viii) if $g \in CODE$, $END(g) = \{\Gamma,\ Nt,\ t = \overline{k}\}$ for some $k \in \omega$, $\alpha > L(g)$, $\rho \geq R(g)$, then

$$\langle\langle(N), k\rangle;\ \Gamma,\ Nt;\ \alpha;\ \rho;\ PAR(g);\ g\rangle \in CODE;$$

(ix) let $e \in PRI$, $f_n = [e](n) \in CODE$ and $END(f_n) = \{\Gamma,\ \neg t = \overline{n}\}$, for every $n \in \omega$; assume $\alpha > L(f_n)$, $\rho \geq R(f_n)$ for every $n \in \omega$, and let $V = \cup\{PAR(f_n): n \in \omega\}$ be finite. Then

$$\langle(\neg N);\ \Gamma, \neg Nt;\ \alpha;\ \rho;\ V;\ e\rangle \in CODE;$$

(x) if $g \in CODE$ and $END(g) = \{\Gamma,\ (\neg)T^{\alpha}\text{-}Clause(s)\}$, $\alpha > L(g)$, $\rho \geq R(g)$, then

$$\langle((\neg)T^{\alpha+1});\ \Gamma,\ (\neg)T^{\alpha+1}s;\ \alpha;\ \rho;\ PAR(g);\ g\rangle \in CODE;$$

(xi) the clauses corresponding to the introduction of $F^{\alpha+1}$ are obtained from (x) by replacing T with F;

(xii) let $e \in PRI$ and $f_\alpha = [e](\alpha) \in CODE$ with $END(f_\alpha) = \{\Gamma, \neg T^{\alpha}t\}$, for every $\alpha < \delta$, δ limit; assume $\beta > L(f_\alpha)$, $\rho \geq R(f_\alpha)$ for every $\alpha < \delta$, and let $V = \cup\{PAR(f_\alpha): \alpha < \delta\}$ be finite. Then

$$\langle(\neg T\text{-}LIM^{\delta});\ \Gamma,\ \neg T^{\delta}t;\ \alpha;\ \rho;\ V;\ e\rangle \in CODE;$$

(xiii) let $g \in CODE$ and $END(g) = \{\Gamma,\ T^{\alpha}t\}$, where $\alpha < \delta$ and δ limit; if $\beta > L(g)$, $\rho \geq R(g)$, then

$$\langle\langle(T\text{-}LIM^{\delta}), \alpha\rangle;\ \Gamma,\ T^{\delta}t;\ \beta;\ \rho;\ PAR(g);\ g\rangle \in CODE;$$

(xiv) the clauses corresponding to the introduction of F^{δ}, for δ limit, are obtained from (xiii) by replacing T with F.

NB: that only derivations with a finite set of parameters are encoded, as implied by (ix), is not restrictive for our aim of embedding finitary systems into infinitary ones.

By generalized recursion theory, $CODE$ is the fixed point of a positive arithmetical operator (indeed a boolean combination of Π^0_2- and Σ^0_2-conditions) and it is generally not elementarily definable. However, we can effectively associate to each derivation label f a well-founded tree of formal figures, which is locally correct (hence a derivation in the true sense), exactly when $f \in CODE$. Since local correctness is an elementary condition, we can get an elementary representation of $CODE$.

3. DEFINITION. Let $*$ denote concatenation on OT-elements.

(i) OT^* is the smallest set X such that: $\langle\ \rangle \in X$; if $s \in X$, $s{*}\alpha \in X$, for $\alpha \in OT$.

(ii) $Length(\langle\ \rangle) = 0$; $Length(s{*}\alpha) = Length(s){+}1$.

(iii) If $s, s' \in OT^*$, $s \subseteq_1 s'$ iff $s = s'$ or $s' = s{*}\alpha$, for some $\alpha \in OT$. $\subseteq$ is the transitive closure of $\subseteq_1$; $s \subseteq s'$ iff s is a subsequence of s'.

(iv) $S \subseteq OT^*$ is an OT^*-tree if $\langle\ \rangle \in S$ and S is closed under the subsequence relation $\subseteq$ ($s' \subseteq s \in S$ implies $s' \in S$).

An OT^*-tree S is well-founded if every S-path is finite; by S-path we mean a maximal $\subseteq$-chain.

4. DEFINITION (by course of value recursion on $Length(s)$, $s \in OT^*$).

1. $Der(f, s) = \emptyset$, if $f \notin LABEL$ or $s \notin OT^*$;

2. else, assume $f \in LABEL$, $s \in OT^*$: then

2.1. $Der(f, \langle\ \rangle) = f$;

2.2. let $s = s' * \beta$, for some $\beta \in OT$; assume $\mathfrak{I} = RF(Der(f, s'))$;

2.2.1. $\mathfrak{I} = AX$: $Der(f, s) = \emptyset$;

2.2.2. $\mathfrak{I} = REP_\delta$: then $Der(f, s) = \emptyset$ if $\beta \neq \delta$;
else, $Der(f, s) = Uni(SD_1(Der(f, s')), \beta)$;

2.2.3. $\mathfrak{I}$ has δ premises $\delta \geq \omega$, δ limit: then $Der(f, s) = \emptyset$ if $\beta \geq \delta$; else,

$Der(f, s) = Uni(SD_1(Der(f, s')), \beta)$ (here Uni is a fixed recursive universal function for the unary PR-functions);

2.2.4. $\mathfrak{I} = \langle \vee, i \rangle$, $i = 1$ or 2: then $Der(f, s) = SD_i(Der(f, s'))$ if $\beta = i - 1$; else $Der(f, s) = \emptyset$;

2.2.5. $\mathfrak{I}$ is not as in 2.2.4, but is k-ary ($1 \leq k \leq 2$). Then $Der(f, s) = \emptyset$ if $\beta \geq k$; else $Der(f, s) = SD_{\beta+1}(Der(f, s'))$.

By inspection of 4, we obtain:

5. LEMMA

(i) *The operation* $\lambda x \lambda y. Der(x, y)$ *is primitive recursive in* Uni (= *the given universal function for* 1-*ary* PR-*functions*).

(ii) *We can find a* PR-*operation* DER *such that* $DER(f)$ *is an index for* $\lambda x. Der(f, x)$.

Clearly (ii) follows by the s-m-n theorem. Henceforth we write $s \in f$ for $Der(f, s) \neq \emptyset$ and we say that s is a node of f. $T(f) = \{s \in OT^* : s \in f\}$ is obviously an OT^*-tree.

6. DEFINITION. $LC(DER(f))$ holds iff it satisfies the following conditions:

1. $f \in LABEL$, $\langle\ \rangle \in f$; $Der(f, s) \in LABEL$, if $s \in f$;
2. if $s \in f$, $\mathfrak{I} = RF(Der(f, s))$, then
2.1. either $\mathfrak{I} = AX$, $END(Der(f, s))$ is an axiom and $s * \delta \notin f$ for every $\delta \in OT$; or
2.2. $\mathfrak{I} \neq AX$, $END(Der(f, s))$ follows from the set

$\{END(Der(f, s*\beta)) : s*\beta \in f\}$ by means of $\mathfrak{I}$;

3. in 2.1-2.2 the conditions on the length and the rank are met; this means in particular that if $s \in f$, $s*\beta \in f$, then
$$L(Der(f,s)) > L(Der(f,s*\beta)).$$

$\{LC(DER(f))$ is read as "*$DER(f)$ encodes a locally correct derivation*"$\}$.

7. LEMMA. *Let $f \in LABEL$.*

(i) *The predicate $LC(DER(-))$ is definable in $\mathfrak{L}_{op}$;*

(ii) *$LC(DER(f))$ implies that $T(f)$ is well-founded;*

(iii) *$f \in CODE$ iff $LC(DER(f))$ holds.*

PROOF. (i): by a straightforward formalization of 6.
(ii): by condition 6.3, any descending infinite sequence in the tree ordering would produce a descending infinite sequence in OT^*.
(iii): $\Rightarrow$: by generalized induction on the definition of $CODE$. $\Leftarrow$: by (ii), $T(f)$ is well-founded and we can apply transfinite induction to verify $f \in CODE$. □

As consequence of 7 (iii), there is an $\mathfrak{L}_{op}$-formula $A(f)$, which represents the condition $f \in CODE$; for convenience, we write $f \in DER$ instead of $A(f)$, and we simply say that f is a derivation. Lemma 7 makes possible to define elementary operations on $CODE$; more properly, when we talk about operations on DER, we mean PR-operations

$$F : LABEL \rightarrow LABEL,$$

which preserve the property $CODE$, i.e. if $f \in CODE$, $F(f) \in CODE$. As a rule, we only rely on *the surface structure of labels, without appealing to transfinite induction*; instead, we apply course-of-value recursion (on the depth of the derivation labels), PR-distinction of cases and the second recursion theorem for PR-indices (after Kleene 1958):

8. THEOREM. *There is a Kalmar elementary operation Fix such that, if e is a PR-index for a $k+1$-ary PR-function, then $Fix(e) \in PRI$ and*

$$[Fix(e)](n_1, \ldots, n_k) = [e](Fix(e), n_1, \ldots, n_k).$$

Operations on DER and cut elimination

We essentially exploit the finitary presentation of RS_n-derivations; the basic operations involved in cut elimination are shown to be primitive recursive and they naturally work on arbitrary derivation labels. Transfinite induction is needed only for checking that the basic operators work properly (correctness proofs). First of all, we see that it is possible to deform monotonically the ordinal assignment (of lengths and ranks), and to rename parameters.

If $f \in LABEL$, $f(a)$ abbreviates the condition "$a \in \mathrm{PAR}(f)$"; $f(a/t)$ denotes the result of replacing everywhere a by t. DER is not closed under the operation $t \mapsto f(a/t)$, because substitution may spoil the eigenparameter restrictions. In order to reconcile the present notations with those of § 52, we introduce the following

9. DEFINITION. $f \vdash^{\alpha}_{\rho} \Gamma := LC(DER(f)) \wedge f \in LABEL \wedge$
$$\wedge END(f) = \Gamma \wedge L(f) = \alpha \wedge R(f) = \rho.$$

Hence, as a consequence of lemma 7, we have:

9.1 $f \vdash^{\alpha}_{\rho} \Gamma$ is definable in the language $\mathcal{L}_{op}$.

Now we state a few technical lemmata, which are essential for manipulating derivations. Proofs are straightforward: however the formal definitions of the primitive recursive operations, which are claimed to exist, require lengthy PR-distinction by cases, course-of-value recursion on DEP and a final application of the second recursion theorem. The technique will be illustrated below in the case of the inversion lemma and the subsequent theorems.

10. LEMMA

(i) *We can find a PR-operation MON such that, if* $f \vdash^{\alpha}_{\rho} \Gamma$, $\alpha \le \alpha'$, $\rho \le \rho'$, *then* $MON(f, \alpha', \rho') = f' \in DER$ *and* $f' \vdash^{\alpha'}_{\rho'} \Gamma$.

(ii) *Renaming: we can find a PR-operation RENAME such that, if* $f \vdash^{\alpha}_{\rho} \Gamma$, $c \notin PAR(f)$, $a \in PAR(f)$, *then*
$$RENAME(f, a, c) = f(a/c) \in DER,\ f(a/c) \vdash^{\alpha}_{\rho} \Gamma[a := c],$$
$$PAR(f(a/c)) = PAR(f)(a/c) \text{ and } DEP(f) = DEP(f(a/c)).$$

(iii) *Renaming of eigenparameters: we can find a PR-operation EG such*
that, if $f \vdash^{\alpha}_{\rho} \Gamma$, $a \in EIGEN(f)$, $c \notin PAR(f)$, *then*
$$EG(f, c, a) = f' \in DER,\ f' \vdash^{\alpha}_{\rho} \Gamma,\ DEP(f') = DEP(f) \text{ and}$$
$$EIGEN(f') = (EIGEN(f) \cup \{c\}) - \{a\}.$$

By lemma 10 (iii), it is not restrictive to assume that, if a parameter b occurs as an eigenparameter at a certain node of a derivation f, then b does not occur below that node.

11. LEMMA (*Substitution*). *We can find a PR-derivation SOST such that, if* $a \in PAR(\Gamma)$ *and* $f \vdash^{\alpha}_{\rho} \Gamma$, *then* $SOST(f,a,t) = f(a/t) \vdash^{\alpha}_{\rho} \Gamma[a := t]$ *(t arbitrary term). Moreover* $DEP(f(a/t)) = DEP(f)$ *and*

$$PAR(f(a/t)) = PAR(f) \cup PAR(t) \cup PAR(C_0) - (V_0 \cup \{a\}),$$

where $V_0 = \{a_1, \ldots, a_n\} = EIGEN(f) \cap PAR(t)$, $C_0 = \{c_1, \ldots, c_n\}$ *is the set of the first n parameters, which do not belong to* $PAR(f)$.

The proof is a combination of eigenparameter-renaming and of simple inductive considerations; we also need the finiteness of $PAR(f)$.

12. LEMMA (*Weakening*). *We can find a PR-operation INDEB, such that, if* $PAR(\Delta) \cap EIGEN(f) = \emptyset$, *and* $f \vdash^{\alpha}_{\rho} \Gamma$, *then*

(i) $INDEB(f) \vdash^{\alpha}_{\rho} \Gamma, \Delta$;

(ii) $DEP(INDEB(f)) = DEP(f)$ *and*

$$PAR(INDEB(f)) = PAR(f) \cup PAR(\Delta).$$

The condition on $PAR(\Delta)$ is not restrictive by lemma 11 above.
We now assume *in toto* the notions of § 53; in particular, the notion of reducible formula of RS_n is primitive recursive.

13. LEMMA (*Inversion*). *We can find a PR-operation INV such that, if*

$f \vdash^{\alpha}_{\rho} \Gamma, C$, $R_n(C) > 0$, *C is reducible to* Δ, *and*

$PAR(\Delta) \cap EIGEN(f) = \emptyset$, *then*

$$INV(f, \Delta) = f' \in DER \text{ and } f' \vdash^{\alpha}_{\rho} \Gamma, \Delta;$$

$$DEP(INV(f, \Delta)) \leq DEP(f).$$

PROOF. We sketch the formal definition of INV; but the reader should keep in mind the informal argument of 53.4 (which runs by transfinite induction on $L(f)$). Recall that $\mathfrak{I}$ stands for an inference name. We set:

1. $F(a, f, \Delta) = \emptyset$ if $f \notin LABEL$, or $EIGEN(f) \cap PAR(\Delta) \neq \emptyset$ or no $C \in END(f)$ reduces to Δ;

else, if f has the form $\langle \mathfrak{I}\,;\ \Gamma, C;\ \alpha;\ \rho;\ V;\ f_1;\ f_2 \rangle$, C reduces to Δ and $V' = (V - PAR(C)) \cup PAR(\Delta)$, we define:

2. $F(a, f, \Delta) = \langle AX;\ \Gamma, \Delta;\ \alpha;\ \rho;\ V' \rangle$ provided $\mathfrak{I} = AX$ (hence $f_1 = f_2 = \emptyset$); else:

3. $F(a, f, \Delta) = \langle \mathfrak{I};\ \Gamma, \Delta;\ \alpha;\ \rho;\ V';\ F(a, f_1, \Delta);\ F(a, f_2, \Delta) \rangle$, if $\mathfrak{I}$ is finitary, C is not active; else:

4. $F(a,f,\Delta) = \langle \mathfrak{I};\ \Gamma,\Delta;\ \alpha;\ \rho;\ V';\ j(a,e,\Delta)\rangle$, provided $\mathfrak{I}$ is infinitary (hence $f_1 = e \in PRI$, $f_2 = \emptyset$), C is not active, and j is a primitive recursive function such that $j(a,e,\Delta)$ is a PR-index for $\lambda x.[a]([e](x),\Delta)$, if $a \in PRI$; else:

5. $F(a,f,\Delta) = \langle REP_1;\ \Gamma,\ \Delta;\ \alpha;\ \rho;\ V';\ comp(sost,a,f_1,b,t)\rangle$, provided $\mathfrak{I} = \langle(\forall),b\rangle$, $\Delta = A[b := t]$, $(A(b) =$ minor formula of $\mathfrak{I})$. Here $comp$ is a PR-function, which produces a primitive recursive index for the function $\lambda x.[\mathbf{1}]([sost]([a](f_1,\Delta),b,t),x)$, if $a \in PRI$; while $sost$ is a PR-index for $SOST$ (see lemma 10), and in general, if $\alpha \in OT$, $[\boldsymbol{\alpha}](h,x) = h$, if $x = \alpha$ and $[\boldsymbol{\alpha}](h,x) = \emptyset$ otherwise; else:

6. $F(a,f,\Delta) = \langle REP_i;\ \Gamma,\Delta;\ \alpha;\ \rho;\ V';\ g(i,a,f_i,\Delta)\rangle$, provided $\mathfrak{I}$ is finitary, C is active, Δ is the minor formula of the i-th premise of $\mathfrak{I}$ or the minor formula of $\mathfrak{I}$ is located at the i-th place (e.g. $\mathfrak{I} = \langle(\vee),i\rangle$), while g is a PR-function such that, if $a \in PRI$, $g(i,a,f_i,\Delta)$ is a PR-index for the function $\lambda x.[\boldsymbol{i}]([a](f_i,\Delta),x)$ (for $\lambda x.[\boldsymbol{i}](f,x)$, see item 5); else:

7. $F(a,f,\Delta) = \langle REP_\delta;\ \Gamma,\Delta;\ \alpha;\ \rho;\ V';\ h(\delta,a,f_1,\Delta)\rangle$, provided the rule $\mathfrak{I}$ is infinitary, Δ is the reduction of C determined by the δth-premise of $\mathfrak{I}$, C is active, h is a PR-function such that, if $a \in PRI$,

$h(\delta,a,f_1,\Delta)$ is a PR-index for $\lambda x.[\boldsymbol{\delta}]([a]([f_1](\delta),\Delta),x)$.

By theorem 8, we choose a PR-index inv such that:

$$F(inv,f,\Delta) = [inv](f,\Delta);$$

it is straightforward to check that $CODE$ is closed under $\lambda x.[inv](x,\Delta)$ and that $DEP([inv](f,\Delta)) \leq DEP(f)$ (by induction on $L(f)$ and $DEP(f)$ respectively). □

REMARK. (i) In the previous proof, we introduced a PR-function j, yielding a PR-index $j(a,e,\Delta)$ for $\lambda x.[a]([e](x),\Delta)$. It is tacitly understood that, if $a \notin PRI$ or $e \notin PRI$ (i.e. they are not PR-indices), then $j(a,e,\Delta)$ assumes an arbitrary fixed value (say 0). Similar assumptions are made for h, g, $comp$ and will be presupposed in the definitions of RDC and CF (to be given below).

(ii) REP_δ is essential to the primitive recursive character of INV. Consider, for instance, the code $f = \langle(\neg T^\delta);\ \Gamma,\neg T^\delta t;\ \alpha;\ \rho;\ V;\ e\rangle$, where $\beta < \delta$ and δ is a limit. Then we could have simply defined

$$INV(f,\ulcorner\neg T^\beta t\urcorner) = INV([e](\beta),\ulcorner\neg T^\beta t\urcorner). \qquad (*)$$

The point is that $(*)$ makes sense only by effective transfinite recursion on $L(f)$ and by the *application operation for PR-indices.* Hence INV would not have been primitive recursive. On the contrary, REP allows to delay the application of INV, and it is exactly this device which makes possible to consider only the "name" of the procedure enumerating the infinitely many subderivations of f.

14. LEMMA (*Reduction*). *We can define a PR-operation RDC, such that, if* $f_0 \vdash^{\alpha}_{\rho_0} \Gamma, A$ *and* $f_1 \vdash^{\beta}_{\rho_1} \Delta, \neg A$, $R(A) \geq 1$, $R(A) \geq \rho_0, \rho_1$, *then*

(i) $RDC(f_0, f_1) = f \in DER$;

(ii) $f \vdash^{\alpha \# \beta}_{\rho} \Gamma, \Delta$ and $\rho \leq R(A)$;

(iii) $DEP(RDC(f_0, f_1)) \leq DEP(f_0) + DEP(f_1)$.

(# *is the natural sum, which is well-defined on OT*).

PROOF. We only write down the specific PR-equations, whose fixed point solution is RDC. The verification proceeds by transfinite induction on $\alpha\#\beta$. By course of value recursion on $DEP(f_0)+DEP(f_1)$, we can find a PR-operation $F(a, f_0, f_1)$ such that:

$F(a, f_0, f_1) = \emptyset$, if it is false that f_0, f_1 are labels, where $END(f_0)$, $END(f_1)$ have the form Γ, A and Δ, $\neg A$ respectively, $R_n(A) \geq R(f_0)$, $R_n(A) \geq R(f_1)$, $R_n(A) > 0$;
else, suppose that $V = PAR(f_0) \cup PAR(f_1)$, $END(f_0) = \{\Gamma, A\}$, $END(f_1) = \{\Delta, \neg A\}$, $L(f_0) = \alpha$, $L(f_1) = \beta$, $R(f_0) = \rho_0$, $R(f_1) = \rho_1$; then we define:

1. $F(a, f_0, f_1) = \langle AX;\ \Gamma,\ \Delta;\ \alpha \# \beta;\ \rho;\ V \rangle$ if the final inferences of f_0, f_1 are axioms; else, let $\mathfrak{I}$ be the last inference of f_0; then:

2. $F(a, f_0, f_1) = \langle \mathfrak{I};\ \Gamma,\ \Delta;\ \alpha\#\beta;\ \rho;\ V;\ F(a, SD_1(f_0), f_1), F(a, SD_2(f_0), f_1) \rangle$, whenever $\mathfrak{I}$ is finitary, A is not active; else:

3. $F(a, f_0, f_1) = \langle \mathfrak{I};\ \Gamma, \Delta;\ \alpha\#\beta;\ \rho;\ V;\ cp(a, e, f_1) \rangle$, if $\mathfrak{I}$ is infinitary, A is not active, $cp(a, e, f_1)$ is the PR-index of $\lambda x.[a]([e](x), f_1)$ and $e = SD_1(f_0)$; else:

4. $F(a, f_0, f_1) =$

$$=\langle \langle CUT, \neg MA(f_1) \rangle;\ \Gamma, \Delta;\ \alpha\#\beta;\ \rho;\ V;\ F(a, f_0, SD_1(f_1));\ G(f_0, f_1) \rangle,$$

if $G(f_0, f_1) = INDEB(INV(f_0, \neg MA(f_1)), \Delta))$, and the formulas A, $\neg A$ are both active, A is reducible but not disjunctive, $\neg MA(f_1) =$ "negation of the minor formula of the last inference of f_1".
The symmetric cases are left to the reader. RDC is the fixed point of $F(a, f_0, f_1)$. □

15. THEOREM. *We can find a PR-operation CF such that, if* $f \vdash^{\alpha}_{\rho} \Gamma$, *then* $CF(f) \in DER$ *and* $CF(f) \vdash^{\phi\rho\alpha}_{1} \Gamma$.

PROOF. As usual, we define $C(a, f)$ by distinction of cases on PR-

conditions and by course-of-value recursion on $DEP(f)$.

1. $C(a,f) = \emptyset$ if $f \notin LABEL$; else:

2. $C(a,f) = f$, if $R(f) = 1$ or $\alpha = 0$ (in this last case replace $\rho = R(f)$ by 1, if $\rho > 1$); else, assume $\rho = R(f) > 1$, $\alpha = L(f) > 0$, $RF(f) = \mathfrak{I}$, and $f = \langle \mathfrak{I};\ \Gamma;\ \alpha;\ \rho;\ V;\ f_0;\ f_1 \rangle$. We have three subcases.

3.1. $\mathfrak{I}$ is finitary, but it is not a cut of rank ≥ 1:

$$C(a,f) = \langle \mathfrak{I};\ \Gamma;\ \phi\rho\alpha;\ 1;\ V;\ C(a,f_0);\ C(a,f_1) \rangle$$

(this is well-defined since $DEP(f_i) < DEP(f)$, $i \in \{0,1\}$).

3.2. $\mathfrak{I}$ is infinitary, so f_0 is a PR-index: we let

$$C(a,f) = \langle \mathfrak{I};\ \Gamma;\ \phi\rho\alpha;\ 1;\ V;\ comb(a,f_0) \rangle,$$

where $comb(a,f_0)$ is a PR-index for $\lambda x.[a]([f_0](x))$.

3.3. $\mathfrak{I}$ is a cut of rank $\nu = \nu(f) > 1$, $\nu < \rho$ and $L(f_i) = \alpha_i$, for $i = 0,1$. By lemma 11, there is a PR-function μ_i such that $\mu_i(a)$ satisfies $[\mu_i(a)](f) = MON([a](f_i), \phi\rho\alpha_i, \nu)$, whenever $a \in PRI$. By lemma 14, there is a PR-function h such that, for $a \in PRI$:

$$[h(a)](f) = [a](RDC([\mu_0(a)](f), [\mu_1(a)](f))).$$

Hence, recalling lemma 13 (item 5) and s-m-n theorem, there is a PR-function ψ such that, if $a \in PRI$:

$$[\psi(a,f)](x) = [\mathbf{0}]([h(a)](f), x).$$

Finally, we put:

$$C(a,f) = \langle REP_0;\ \Gamma;\ \phi\rho\alpha;\ 1;\ \psi(a,f) \rangle.$$

If we choose a PR-index CF by theorem 8, we can verify, by main induction on ρ and secondary induction on α, that $CF(f) \vdash_1^{\phi\rho\alpha} \Gamma$. □

With similar arguments, we can prove that there is a PR-operation satisfying Tait's second cut elimination theorem (and hence 1-step cut elimination). By inspecting the construction of theorem 15, we can conclude that OP+$TI_{op}(\Gamma_0)$ proves the cut elimination theorem for RS_n.

PART E

ALTERNATIVE VIEWS

"On a signalé beaucoup d'antinomies, et le désaccord a subsisté, personne n'a été convaincu; d' une contradiction, on peut toujours se tirer par un coup de pouce ! Je veux dire par un distinguo." (H. Poincaré, 1913).

CHAPTER 12

NON-REDUCTIVE SYSTEMS FOR TYPE-FREE ABSTRACTION AND TRUTH

§59. The core system VF^- and transfinite induction
§60. Supervaluation models of VF^-
§61. An abstract sequent calculus and truth
§62. Cut elimination and related properties
§63. A provability interpretation and the upper bound theorem
§64. Reconciling supervaluation models with provability interpretation

In this chapter we critically reconsider the basic truth axioms of §7 and their semantics. An essential feature of these principles is that they are *reductive*: they (roughly) presuppose a "compositional theory of meaning", in that the truth conditions of logically complex sentences are reduced to corresponding conditions for logically simpler sentences. In particular, the basic idea is *predicativistic* in spirit: a statement is justified only if its truth can be ultimately grounded upon elementary truths (see Kripke's classification in §34).

A major consequence of this general attitude is that even a tautology may not be accepted, whenever it involves ungrounded sentences. It is therefore natural to investigate the consistency of alternative views, which are well-behaved with respect to *logical consequence*. In particular, it seems reasonable to accept every classical tautology A, irrespective of its complexity and its specific content (e.g. A might have the form $r\eta r \rightarrow r\eta r$, r being the Russell property).

In §59 we present a non-reductive system VF^-, which has non-trivial mathematical content (indeed, it proves a generalized induction principle). The main theorem tells us that VF^-, even if number-theoretic induction for classes is assumed, *is proof-theoretically equivalent to* OP (and hence to PA); furthermore, the same equivalence holds between VF_p := "VF^- plus internal number-theoretic induction axiom" and the impredicative theory ID_1 of elementary inductive definition. Thus the non-reductive approach overcomes the deductive limits of the reductive notion of reflective truth.

The rest of the chapter is devoted to illustrate two types of semantics for VF^-: supervaluation models (SV-models, in short) and a provability interpretation. SV-models are introduced in §60 and they take inspiration from the *supervaluation method* (van Fraassen 1968, 1970). Indeed, we show that there is a simple monotone operator, whose fixed points provide models for VF^-. §§61-64 describe a proof-theoretic semantics for VF^-: the truth predicate T is interpreted as provability in an abstract infinitary system, which enjoys cut elimination (§62). We underline that the proof-theoretic investigation is worked out in a restricted metatheory, i.e. the theory $PW^- + GID$ of §16 with approximation operator and generalized induction. As a byproduct, we shall obtain an upper bound for VF^- and, at the same time, insights on new principles for truth (§63). In addition, §64 shows that for countable ground structures, SV-least fixed point models and provability models coincide.

§59. The core system VF^- and transfinite induction

For the sake of simplicity, we restrict our consideration to a variant of $\mathcal{L}$, which assumes $\rightarrow$, $\forall$ as primitive logical symbols; we let $\perp := (K{=}S)$ and $\neg A := (A \rightarrow \perp)$, while $\vee$, $\wedge$, $\exists$ are defined as usual in classical logic. We write TA and FA for $T[A]$ and $T[\neg A]$, $[A]$ being the term of $\mathcal{L}_{op}$, which represents A (cf.§7; we assume that $IMPLY := \lambda x \lambda y.\langle \overline{31}, \langle x, y\rangle\rangle$ encodes $\rightarrow$). In order to simplify a few arguments below, it is convenient to fix an axiomatization of classical logic with modus ponens as the only inference rule (Tarski 1965).

59.1. DEFINITION

(i) VF^- is the elementary theory (in the language $\mathcal{L}$, as modified above), which includes classical first-order logic with equality, the axioms of OP^- and the five schemata below:

T-out: $TA \rightarrow A$ (A arbitrary sentence);

T-elem: $(A \rightarrow TA)$, where A has the form $t = s$, $\neg t = s$, Nt, $\neg Nt$;

T-imp: $T(A \rightarrow B) \rightarrow (TA \rightarrow TB)$;

T-univ: $\forall x TA \rightarrow T\forall x A$;

T-log: TA, provided A is a logical axiom.

(ii) $VF_c := VF^-$ plus the class number-theoretic induction axiom *CL-NIND*, i.e. the formula

$$Cl(a) \wedge \overline{0}\eta a \wedge \forall x(x\eta a \rightarrow (x{+}1)\eta a) \rightarrow \forall x(Nx \rightarrow x\eta a);$$

$VF_p := VF^- + P\text{-}NIND := \overline{0}\eta a \wedge \forall x(x\eta a \rightarrow (x{+}1)\eta a) \rightarrow \forall x(Nx \rightarrow x\eta a)$;

P-NIND is the number-theoretic induction for properties.

59.1.1. REMARK. The T-schemata above are theorems of NMF^{-}; what really makes the difference, is the closure of T under logical deduction.

59.2. LEMMA

(i) *The LOG-rule: if A is a formula of $\mathfrak{L}$ and A is provable in pure logic, then* $\mathrm{VF}^{-} \vdash TA$.

(ii) *The internal abstraction schema:*

$\mathrm{VF}^{-} \vdash T(\forall u(u\eta\{x: A\} \leftrightarrow TA[x := u]))$;

(iii) $\mathrm{VF}^{-} \vdash TA \wedge TB \leftrightarrow T(A \wedge B)$;

(iv) $\mathrm{VF}^{-} \vdash \neg(TA \wedge FA)$;

(v) $\mathrm{VF}^{-} \vdash (TA \vee FA) \rightarrow ((A \rightarrow TB) \leftrightarrow T(A \rightarrow B))$;

(vi) $\mathrm{VF}^{-} \vdash TA \vee TB \rightarrow T(A \vee B)$;

(vii) $\mathrm{VF}^{-} \vdash T\forall xA \leftrightarrow \forall xTA$;

(viii) $\mathrm{VF}^{-} \vdash \exists xTA \rightarrow T\exists xA$;

(ix) $\mathrm{VF}^{-} \vdash TA \leftrightarrow F\neg A$;

(x) *if A is a formula of $\mathfrak{L}_{op}$ (A does not contain occurrences of T),*

$\mathrm{VF}^{-} \vdash (TA \vee FA) \wedge (TA \leftrightarrow A)$.

PROOF. (i) By induction on the derivation of A in pure logic. If A is a logical axiom, we are done by T-*log*. If A is obtained from B and $B \rightarrow A$, we get TB and $T(B \rightarrow A)$ by IH, whence $TB \rightarrow TA$ by T-*imp*, and finally TA.

(ii) By identity logic and (i) above, we have (u fresh variable):

$$T((\lambda x.[A])u = [A[x := u]] \rightarrow .(u\eta\{x: A\} \leftrightarrow TA[x := u])).$$

We then obtain, by T-*imp* and T-*elem*,

$$(\lambda x.[A])u = [A[x := u]] \rightarrow T(u\eta\{x: A\} \leftrightarrow TA[x := u]),$$

whence $T(u\eta\{x: A\} \leftrightarrow TA[x := u])$ by β-conversion. The conclusion follows by logic and T-*univ*.

(iii) $A \wedge B \rightarrow A$, $A \wedge B \rightarrow B$ hold by logic, whence by LOG-rule $T(A \wedge B) \rightarrow TA \wedge TB$; in the other direction we apply the tautology $A \rightarrow (B \rightarrow (A \wedge B))$ and T-*imp*.

(iv): trivial consequence of T-*out*.

(v) Assume $TA \vee FA$, $A \rightarrow TB$. If TA holds, then A holds by T-*out*, whence TB; but $T(B \rightarrow (A \rightarrow B))$ by LOG-rule; hence $TB \rightarrow T(A \rightarrow B)$ with T-*imp*, i.e. $T(A \rightarrow B)$. If FA is assumed, we have $T(\neg A \rightarrow (A \rightarrow B))$ by LOG-rule; the conclusion again follows by T-*imp*. The reverse direction

follows by observing that $TA \vee FA$ implies $A \rightarrow TA$ (use *T-imp*, *T-out*).
(vi)-(ix): by LOG-rule and *T-imp*.
(x): by induction on A, applying *T-elem*, LOG-rule and the previous facts. □

59.2.1. REMARK. (i) $T(TA \rightarrow A)$ is inconsistent with *T-out*, *T-imp* and *T-log*. If we choose $A := \neg TL$, where $L = [\neg TL]$, we have the following chain of implications:

$$T(TL \rightarrow L) \Rightarrow T(\neg L \rightarrow \neg TL) \Rightarrow T(\neg L \rightarrow L) \Rightarrow TL \Rightarrow T\neg TL \Rightarrow \neg TL.$$

(ii) Assume the schema $T(\exists xA) \rightarrow \exists xTA$. Then we have in pure logic, with $\neg K = S$, $\exists x(x = K \leftrightarrow A)$, hence $T(\exists x(x = K \leftrightarrow A))$, i.e. for some c, $T(c = K \leftrightarrow A)$, which implies by 59.2 either $(c = K \wedge TA)$ or $(\neg c = K \wedge T\neg A)$, i.e. $TA \vee T\neg A$: absurd.
To sum up, $\mathrm{VF}^- + \{T\exists xA \leftrightarrow \exists xTA\}$ is inconsistent. But a special case of the schema is consistent with VF^- (see 63.9).

Fix an enumeration $\{A_i\}_{i \in \omega}$ of $\mathfrak{L}_{op}$-formulas, which have exactly two distinct free variables. Let $x \prec y$ stand for any formula $A_i(x, y)$ in the given list.

59.3. DEFINITION

(i) $Progr(\prec, B) := \forall x(\forall y \prec x.B(y) \rightarrow B(x))$ (= the property defined by B is $\prec$ - progressive);

(ii) $W(\prec, x) := \forall z(Progr(\prec, z) \rightarrow x\eta z)$; ($Progr(\prec, z)$ stands for $Progr(\prec, B)$ with $B(x) := x\eta z$ and we simply say that z is $\prec$-progressive);

(iii) $WF(\prec) := \{u: W(\prec, u)\}$.

$WF(\prec)$ is clearly suggested by the set-theoretic definition of the largest well-founded part of a relation; remarkably, VF^- justifies the corresponding transfinite induction schema.

59.4. THEOREM (*Transfinite induction*). VF^- *proves:*

(i) $Progr\ (\prec, WF(\prec))$;

(ii) $Progr(\prec, B) \rightarrow \forall x(x\eta WF(\prec) \rightarrow B(x))$,
where B is an arbitrary $\mathfrak{L}$-formula.

PROOF. (i) $Progr(\prec, W(\prec, -))$ is clearly derivable in pure logic; hence by LOG-rule we can infer

$$T(\forall x(\forall y \prec x.W(\prec, y) \rightarrow W(\prec, x))). \tag{1}$$

Since $\prec$ is defined by a formula of $\mathfrak{L}_{op}$, we have $T(x \prec y) \vee F(x \prec y)$ by

59.2 (x). If we apply 59.2 (vii), 59.3 and 59.2 (v), (1) implies

$$\forall x(\forall y \prec x. TW(\prec, y) \to TW(\prec, x)), \tag{2}$$

which yields $Progr(\prec, WF(\prec))$ by means of *T-out* and 59.2 (ii).
(ii) It is enough to check

$$\forall x(x\eta WF(\prec) \to B'(x)),$$

where $B'(x) := Progr(\prec, B) \to B(x)$. In pure logic, we have

$$Progr(\prec, B').$$

Then we can repeat the argument for (2), thus obtaining

$$\forall x(\forall y(y \prec x \to TB'(y)) \to TB'(x)),$$

whence by abstraction,

$$Progr(\prec, \{u: B'(u)\}). \tag{3}$$

If we assume $x\eta WF(\prec)$ and we choose $z := \{u: B'(u)\}$, we get

$$Progr(\prec, \{u: B'(u)\}) \to x\eta\{u: B'(u)\}. \tag{4}$$

From (3)-(4) and abstraction, it follows $TB'(x)$, hence $B'(x)$ with *T-out*. □

59.4.1. REMARK. The previous argument only requires that $\{\langle x, y\rangle : x \prec y\}$ is a class, and not the stronger $\mathcal{L}_{op}$-definability.

To appreciate the strength of VF^-, the reader with a "logicistic" inclination may be willing to verify the following theorem. Let $\mathcal{L}^-$ be $\mathcal{L}$ without the predicate N and with the combinators K and S as the only individual symbols; let VF_0 be the subsystem of VF^-, formalized in the fragment $\mathcal{L}^-$. Then we have:

59.5. THEOREM. *Peano arithmetic is interpretable in* VF_0.

PROOF (Hint). Simply define $(x+1) := \{x\}$, $\overline{0} := \emptyset$ and

$$Nx := x\eta\{u: \forall y(Clos_N(y) \to u\eta y)\},$$

where $Clos_N(y) := \overline{0}\eta y \wedge \forall u(u\eta y \to (u+1)\eta y)$. Plus and times are introduced à la Dedekind as the least relations satisfying the obvious recursive clauses. By adapting 59.4, we can verify the appropriate induction schemata (for details, see Cantini 1991). □

It will follow from the main result of § 63 that VF_0 is not stronger than OP. In contrast to MF, VF_p (= VF^- + property N-induction axiom) goes beyond the limits of predicative mathematics. This is most easily seen by appealing to the theory $\mathrm{ID}_1(\mathrm{acc})$ of *accessibility elementary inductive definitions*, which proves the 1-consistency of Predicative Analysis (see

Buchholz et al., 1981).

For the reader's sake, we outline $\mathrm{ID}_1(\mathrm{acc})$. If $\mathcal{L}(\mathrm{PA})$ is the language of Peano arithmetic, fix an effective enumeration $\{A_i : i \in \omega\}$ of all $\mathcal{L}(\mathrm{PA})$-formulas, containing two distinct free variables. The language of $\mathrm{ID}_1(\mathrm{acc})$ is $\mathcal{L}(\mathrm{PA})$, expanded by a countable sequence of unary predicate symbols, say $\{I_i : i \in \omega\}$. As above, let $\prec$ stand for any $A_i(x, y)$; we use $WF(\prec)(x)$ as a more suggestive notation for $I_i(x)$ (whenever $\prec$ is any A_i). Formulas are inductively generated as usual; atoms obviously have the form $t = s$, $WF(\prec)(t)$. The axioms of $\mathrm{ID}_1(\mathrm{acc})$ contain: (i) Peano axioms; (ii) number-theoretic induction for the full language; (iii) for each $\prec$ and arbitrary $B(x)$, the axioms:

$WF(\prec).1$ $\quad Progr(\prec, WF(\prec))$;

$WF(\prec).2$ $\quad Progr(\prec, B) \rightarrow \forall x(WF(\prec)(x) \rightarrow B(x))$.

59.6. THEOREM. $\mathrm{ID}_1(\mathrm{acc})$ *is interpretable in* VF_p.

PROOF. By the theorem 59.4, it only remains to check that VF_p proves the number-theoretic induction schema for arbitrary formulas of $\mathcal{L}$. Set $Clos_N(A) := A(\bar{0}) \wedge \forall x(A(x) \rightarrow A(x+1))$ and $A'(x) := Clos_N(A) \rightarrow A(x)$, where A is a given arbitrary formula of $\mathcal{L}$; then $Clos_N(A')$ is trivially derivable in pure logic.
Thus $\mathrm{VF}^- \vdash TClos_N(A')$, which implies $\mathrm{VF}^- \vdash Clos_N(\{x : A'(x)\})$, whence by property N-induction $\forall u(Nu \rightarrow u\eta\{x : A'(x)\})$; *T-out* and exchange of premises imply $\mathrm{VF}_p \vdash Clos_N(A) \rightarrow \forall u(Nu \rightarrow A(u))$. □

59.6.1. REMARK. Conversely, VF_p has a model in a set theory, which is proof-theoretically equivalent to ID_1; the basic steps are similar to those of the main theorem of § 63 and the result is essentially contained in Cantini (1990). Thus we concentrate on VF_c, in accord with our choice of stressing systems not stronger than PA.

To conclude, the reader may naturally wonder whether the strength increase sensibly affects the structure of classes in VF^-. For instance, does any of the properties $WF(\prec)$ define a class? The answer is negative and it can be readily obtained with a recursion-theoretic investigation of the inductive models of § 60. By the way, the fundamental closure properties of $CL := \{x : Cl(x)\}$ in VF^- are best summarized by the non-surprising

59.7. THEOREM. VF^- *proves that* CL *is closed under the join principle and the elementary comprehension schema* (*see* Ch. II, 9.7-9.9).

The proofs are straightforward and make use of the elementary facts of 59.2.

§60. Supervaluation models of VF$^-$

We keep using the conventions and notations of §7 and §30; we fix a standard operational structure $\mathcal{M}\models \mathrm{OP}^-$; $\mathcal{L}_{op}(\mathcal{M})$ and $\mathcal{L}(\mathcal{M})$ are the usual languages expanded with distinct constants for elements of M, M being the universe of $\mathcal{M}$. If t is a closed term of $\mathcal{L}_{op}(\mathcal{M})$, $\mathcal{M}(t)$ is the value of t in $\mathcal{M}$.

60.1. DEFINITION

(i) Once $\mathcal{M}$ is given, $X \subseteq M$ and A is an arbitrary sentence of $\mathcal{L}(\mathcal{M})$, $X \models A$ stands for "A holds true in the structure $\langle \mathcal{M}, X\rangle$", whenever $\mathcal{L}_{op}$ receives its usual interpretation in $\mathcal{M}$ and T is assigned the subset X, i.e. $\langle \mathcal{M}, X\rangle \models Tt$ iff $\mathcal{M}(t) \in X$.

(ii) $X \Vdash A$ iff for every $Y \subseteq M$, if $X \subseteq Y$, then $Y \models A$ (where A is an arbitrary $\mathcal{L}(\mathcal{M})$-sentence).

(iii) If $X \subseteq M$, we let $\Phi_0(X) := \{\mathcal{M}([A]) : X \Vdash A,\ A\ \mathcal{L}(\mathcal{M})\text{-sentence}\}$.

(iv) X is Φ_0-*dense* iff $X \subseteq \Phi_0(X)$; X is Φ_0-*closed* iff $\Phi_0(X) \subseteq X$; X is *consistent* (*complete*) iff for no $b \in M$, $b \in X$ and $(\neg b) \in X$ (for every $b \in M$, $b \in X$ or $(\neg b) \in X$). $CONS(\mathcal{M}) := \{X \subseteq M : X \text{ consistent}\}$.

(v) $SENT(\mathcal{M}) := \{\mathcal{M}([A]) : A\ \mathcal{L}(\mathcal{M})\text{-sentence}\}$.

Since $\mathcal{M}$ is fixed, we generally omit the explicit indication of $\mathcal{M}$ and we simply speak of "sentence" and "consistent" tout court.

60.1.1. REMARK. Variants of the relation $X \Vdash A$ are obtained by imposing additional constraints on the possible extensions of X (cf.§67). For instance, if we define the relation $X \Vdash A$ by quantifying over all *consistent* and *complete* extensions of X, we obtain van Fraassen's notion of *supervaluation for* $\mathcal{L}(\mathcal{M})$. For this reason and because of theorem 60.3 below, the fixed points of the operator Φ_0 are also called *supervaluation models* (in short *SV-models*).

60.2. LEMMA

(i) $X \Vdash A$ *implies* $X \models A$ (A $\mathcal{L}(\mathcal{M})$-sentence);

(ii) $X \Vdash A \rightarrow B$ *and* $X \Vdash A$ *imply* $X \Vdash B$;

(iii) *If* $X \Vdash A(a)$ *for every* $a \in M$, *then* $X \Vdash \forall x A$;

(iv) Φ_0 *is monotone*: $X \subseteq Y \Rightarrow \Phi_0(X) \subseteq \Phi_0(Y)$ $(X, Y \subseteq M)$.

Hence $FIX_0(\mathcal{M}) = \{X \subseteq M : X = \Phi_0(X)\}$ *is non-empty*.

(v) *If* X *is* Φ_0-*dense, then* $X \subseteq SENT(\mathcal{M})$ *and* $X \models TA \rightarrow A$; *hence* $FIX_0(\mathcal{M}) \subseteq CONS(\mathcal{M})$.

PROOF: (i)-(iv) are trivial by definition of $||\!-$ and the Knaster-Tarski theorem. As to (v), if $X \subseteq \Phi_0(X)$, *T-out* holds by (i) and trivially $X \subseteq SENT(\mathcal{M})$. If $\mathcal{M}([A]) \in X$ and $\mathcal{M}([\neg A]) \in X$, then $X||\!- A$ and $X||\!- \neg A$, which yield a contradiction by (i). □

60.3. THEOREM

(i) *If* $X \in FIX_0(\mathcal{M})$ *and* $\mathcal{M} \models \mathrm{OP}^-$, *then* $X \models \mathrm{VF}^-$. *In addition we have, for arbitrary* $a \in M$:

T-rep $\qquad X \models Ta \rightarrow TTa;$

(ii) *If* $\mathcal{M}$ *is an* ω-*model of* OP *and* $X \in FIX_0(\mathcal{M})$, *then* $X \models \mathrm{VF}_p$ (*cf.* 59.1). *In particular, if* A *is an arbitrary instance of* N-*induction in the language* $\mathcal{L}$, $X \models TA$.

PROOF. (i). *T-out*, *T-imp*, *T-univ*: apply 60.2 (v), (ii), (iii).
T-elem: if A has the form Nt, $t = s$ or the negation thereof, $X \models A$ implies $\mathcal{M} \models A$, whence $Y \models A$, for every $Y \supseteq X$, i.e. $X||\!- A$, i.e. $X \models TA$ as X is Φ_0-closed. The converse is similar. *T-log*: if A is a logical axiom, $X \models A$, for arbitrary $X \subseteq M$, i.e. $X||\!- A$. Hence, if $X \in FIX_0(\mathcal{M})$, $\mathcal{M}([A]) \in X$ by Φ_0-closure, i.e. $X \models TA$. *T-rep*: if $X \models Ta$, also $a \in Y$ and $Y \models Ta$, for every $Y \supseteq X$, whence $X||\!- Ta$, i.e. $X \models TTa$ (Φ_0-closure).
(ii) If A is an $\mathcal{L}(\mathcal{M})$-instance of N-induction, $Y \models A$, for every $Y \subseteq M$. The conclusion follows by Φ_0-closure of X. □

60.4. COROLLARY. $\mathrm{VF}^- + T\text{-}rep + \{TA : A$ *is a logical axiom or an axiom of* OP^-, *or an arbitrary instance of* N-*induction in the language* $\mathcal{L}\}$ *is consistent.*

At this stage we shall not undertake the investigation of the lattice-theoretic structure of the fixed point models of Φ_0: suffice it to say that also in the present situation the encoding techniques of Ch. VII can be profitably applied. For instance, the reader can verify:

60.5. THEOREM. $Card(FIX_0(\mathcal{M})) = 2^{card(M)}$.

However, we warn against mechanical repetitions of the old arguments.

§61. An abstract sequent calculus and truth

We consider the problem of giving a more constructive semantics for VF^-; in particular, we show how to avoid universal quantification over arbitrary subsets in the definition 60.1. To this aim, we shall define a generalized sequent calculus $\mathcal{K}$, in such a way that provability in $\mathcal{K}$ yields an interpretation of the predicate T of VF^-. This step is rewarding in two

respects. First of all, provability semantics validates new schemata and hence we shall obtain a stronger consistency result. Furthermore, the definition of $\mathcal{K}$ and the derivation of its main properties can be easily carried out in the system $PW_c + GID$ with N-induction for classes, which is conservative over OP (and hence PA; see Ch. III, 15.5).
Since the $\mathcal{K}$-provability interpretation is the identity over $\mathcal{L}_{op}$-formulas, it will follow that VF^-, VF_c are conservative extensions of OP; in addition, VF_p turns out to be interpretable in $PW + GID$, the system with full induction schema, which can be shown proof-theoretically equivalent to the theory $ID_1(acc)$ of § 59.

The crux of the construction lies in devising a sequent calculus $\mathcal{K}$, which enjoys cut elimination and hence is consistent, provably in $PW_c + GID$. Clearly $\mathcal{K}$ has to be infinitary (by axiom *T-univ*). However, the problem with the usual cut elimination proofs is that they require induction on cut formulas (of maximal complexity), i.e. *forms of number-theoretic induction that may not be available in weak systems* like $PW_c + GID$. In essence, the solution we present here is simply *to replace the usual finitary sentences with natural abstract counterparts, which are introduced by generalized induction.* It follows that N-induction can be avoided by means of ordinal transfinite induction, which is available in unrestricted form using GID (see § 23).

Henceforth, we use the abbreviations: $(\forall a) := \langle\overline{6}, a\rangle$, $(a \to b) := \langle\overline{31}, \langle a, b\rangle\rangle$, and $(\neg a) := (a \to [K = S])$; $\forall x . t$ stands for $\forall(\lambda x . t)$ The other logical operations are introduced by mimicking the definitions of the corresponding logical operators:

$$(a \wedge b) := (\neg(a \to \neg b)); \quad (a \vee b) := (\neg a \to b); \quad (\exists a) := \neg(\forall(\lambda x(\neg(ax)))).$$

Also the map $A \mapsto [A]$ on atomic formulas is defined as in § 7. We put:

$$At_\infty(a) := \exists x \exists y(a = [x = y] \vee [Nx] \vee [Tx]);$$

$$Eat_\infty(a) := At_\infty(a) \wedge \neg\exists x(a = [Tx]).$$

We usually omit outer square brackets to distinguish an element of $\{x : At_\infty(x)\}$ from the corresponding formula, whenever the distinction is clear from the context. Thus $Ta = a$, $(Ta \to a) = a$ abbreviate $[Ta] = a$ and $([Ta] \to a) = a$ (respectively).

61.1. LEMMA. *We can find a formula $AST(x, v)$ operative in v, such that, if $Sent_\infty := Ixv. AST(x, v)$ (= the fixed point of $AST(x, v)$, cf.* 10.1), *then we have, provably in* $PW^- + GID$:

1.1 $\quad At_\infty(a) \to a \eta Sent_\infty$;

1.2. $\quad (a \to b) \eta Sent_\infty \leftrightarrow a \eta Sent_\infty \wedge b \eta Sent_\infty$;

1.3. $(\forall a)\eta Sent_\infty \leftrightarrow \forall x((ax)\eta Sent_\infty)$;

1.4. $\forall x(AST(x,B) \rightarrow B(x)) \rightarrow \forall x(x\eta Sent_\infty \rightarrow B(x))$, *B arbitrary.*

PROOF: clearly $AST(x,v)$ formalizes the clauses corresponding to 1.1-1.3, which are positive in $Sent_\infty$. Then we apply the fixed point theorem for predicates and GID to get 1.4. □

The elements of $Sent_\infty$ are naturally viewed as *abstract sentences* and they suffice for the main result below (cut elimination).

61.2. LEMMA. *If* $A(x_1,\ldots,x_n)$ *is a formula of* $\mathfrak{L}$ *with the free variable shown, then*

$$\mathrm{PW}^- + \mathrm{GID} \vdash \forall x_1 \ldots \forall x_n([A(x_1,\ldots,x_n)]\eta Sent_\infty).$$

PROOF: by metamathematical induction on A, using 61.1. □

An adequate notion of ordinal in the sense of von Neumann is available in $\mathrm{PW}^- + \mathrm{GID}$ (see §§21-23); in particular, we can define a predicate $Ord(x)$, two binary relations $<, \equiv$, such that if the lower case Greek letters α, β, ρ, ... range over $ON = \{x: Ord(x)\}$, then $\mathrm{PW}^- + \mathrm{GID}$ proves (see 23.4 for notations):

61.3. LEMMA

(i) $x\eta ON \leftrightarrow Ord(x)$; $\forall x(Ord(x) \rightarrow x\eta AD)$ (*every ordinal is a set*);
(ii) $<$ *is linear, i.e.* $(\alpha < \beta) \vee (\alpha \equiv \beta) \vee (\beta < \alpha)$;
(iii) $TI(ON,B) := \forall\alpha(\forall\beta(\beta < \alpha \rightarrow B(\beta)) \rightarrow B(\alpha)) \rightarrow \forall\alpha B(\alpha)$.

Since the operator $fa := \{x: AST(x,a)\}$ is $\subseteq$-monotone, we exploit the ordinal analysis of inductively defined predicates and rephrase 23.6-23.10 in the special case of $Sent_\infty$.

61.4. LEMMA. *We can find a term* $\lambda x.Sent_\infty(x)$ *such that* $\mathrm{PW}^- + \mathrm{GID}$ *proves:*

(i) $\alpha \leq \beta \rightarrow Sent_\infty(\alpha) \subseteq Sent_\infty(\beta)$ (*where* $\alpha \leq \beta := (\alpha \equiv \beta \vee \alpha < \beta)$);

(ii) $\forall x(x\eta Sent_\infty(\alpha) \leftrightarrow AST(x, \{y: \exists\beta(\beta < \alpha \wedge y\eta Sent_\infty(\beta))\}))$;

(iii) $\forall x(x\eta Sent_\infty \leftrightarrow \exists\alpha(x\eta Sent_\infty(\alpha)))$;

(iv) $\forall x(At_\infty(x) \leftrightarrow x\eta Sent_\infty(0))$ (*where* $0 = \emptyset$);

(v) $\forall x\forall y((x \rightarrow y)\eta Sent_\infty(\alpha) \leftrightarrow \exists\beta(\beta < \alpha \wedge x\eta Sent_\infty(\beta) \wedge y\eta Sent_\infty(\beta)))$;

(vi) $\forall z((\forall z)\eta Sent_\infty(\alpha) \leftrightarrow \forall x\exists\beta(\beta < \alpha \wedge (zx)\eta Sent_\infty(\beta)))$.

61.4.1. REMARK. If $a\eta Sent_\infty(\alpha)$ is read as "a has complexity α", then the associated notion of complexity has the expected properties: the atomic elements of $Sent_\infty$ have complexity 0 (by 61.4 (iv)); the immediate constituents of a non-atomic element b of $Sent_\infty$ have strictly lower complexity than b itself (61.4 (v)-(vi)).

We proceed to describe the announced generalized sequent calculus $\mathfrak{K}$. In order to enhance readability, we content ourselves with an informal presentation; however, we underline with appropriate remarks the steps that require non-trivial considerations for the formalization in $\mathrm{PW}^- + \mathrm{GID}$.

Notation. $\Gamma, \Delta := \Gamma \cup \Delta$; $\Gamma, a := \Gamma, \{a\}$; Γ and Δ will stand for *classes of abstract sentences.*

61.5. DEFINITION. $\vdash^\alpha_\rho \Gamma \Rightarrow \Delta$ is the least relation such that

(i) α, ρ are ordinals and Γ, Δ are classes $\subseteq Sent_\infty$;
(ii) it is closed under the axioms and rules below.

61.5.1. *Axioms*

AX.1: if $b\eta\Gamma \cap \Delta$ and $At_\infty(b)$, i.e. b has the form $[x = y]$, $[Tx]$, $[Nx]$, then $\vdash^\alpha_\rho \Gamma \Rightarrow \Delta$;

AX.2.1: if $A = (a = b)$, (Na), A is true and $[A]\eta\Delta$, then $\vdash^\alpha_\rho \Gamma \Rightarrow \Delta$;

AX.2.2: if $A = (a = b)$, (Na), A is false and $[A]\eta\Gamma$, then $\vdash^\alpha_\rho \Gamma \Rightarrow \Delta$;

AX.3: if $\{[Ta], [T\neg a]\} \subseteq \Gamma$, then $\vdash^\alpha_\rho \Gamma \Rightarrow \Delta$.

61.5.2. *Rules*

(R$\rightarrow$): if $(\vdash^\beta_\rho \Gamma, a \Rightarrow \Delta, b)$, $(a \rightarrow b)\eta\Delta$ and $\beta < \alpha$, then $\vdash^\alpha_\rho \Gamma \Rightarrow \Delta$;

(L$\rightarrow$): if $(\vdash^\beta_\rho \Gamma \Rightarrow \Delta, a)$, $\vdash^\gamma_\rho \Gamma, b \Rightarrow \Delta$ and $(a \rightarrow b)\eta\Gamma$ with $\beta < \alpha$, $\gamma < \alpha$, then $\vdash^\alpha_\rho \Gamma \Rightarrow \Delta$;

(R$\forall$): assume that for every x and some $\beta < \alpha$ $\vdash^\beta_\rho \Gamma \Rightarrow \Delta, ax$ and let $(\forall a)\eta\Delta$; then $\vdash^\alpha_\rho \Gamma \Rightarrow \Delta$;

(L$\forall$): if $\vdash^\beta_\rho \Gamma, ax \Rightarrow \Delta$ for some x, $\beta < \alpha$, $(\forall a)\eta\Gamma$, then $\vdash^\alpha_\rho \Gamma \Rightarrow \Delta$;

(RT): if $(\vdash^\beta_\delta \Rightarrow b)$ and $\beta < \alpha$, $\delta < \rho$, $[Tb]\eta\Delta$, then $\vdash^\alpha_\rho \Gamma \Rightarrow \Delta$;

(LT): if $(\vdash^\beta_\delta b \Rightarrow)$ and $\beta < \alpha$, $\delta < \rho$, $[Tb]\eta\Gamma$, then $\vdash^\alpha_\rho \Gamma \Rightarrow \Delta$.

61.6. LEMMA. *We can find a formula $D(z, w)$, such that*

(i) *D is operative in w;*

(ii) *D formalizes the inductive clauses corresponding to the axioms and the rules of* 61.5;

(iii) $\mathrm{PW}^- \vdash TD(z,w) \leftrightarrow D(z,w)$.

PROOF. By straightforward formalization of 61.5 and application of lemmas 61.3-61.4; note that, in order to get (iii), it is essential to have condition 61.5 (i). □

GID and the fixed point theorem for predicates imply:

61.7. LEMMA. *We can find a term* Der_∞ *such that* $\mathrm{PW}^- + \mathrm{GID}$ *proves*

(i) $\forall z(z\eta Der_\infty \leftrightarrow D(z, Der_\infty))$;

(ii) $z\eta Der_\infty \rightarrow \exists x \exists y \exists u \exists v(z = \langle x,y,u,v\rangle \wedge$

$\wedge\, x\eta ON \wedge y\eta ON \wedge Cl(u) \wedge Cl(v) \wedge u \subseteq Sent_\infty \wedge v \subseteq Sent_\infty)$.

(iii) *If B is arbitrary,*

$\forall z(D(z,B) \rightarrow B(z)) \rightarrow \forall z(z\eta Der_\infty \rightarrow B(z))$.

(iv) *The formula* $\langle\alpha,\rho,u,v\rangle\eta Der_\infty$ *is extensional in* α, ρ, *i.e.*

$$\alpha \equiv \alpha' \wedge \rho \equiv \rho' \rightarrow (\langle\alpha,\rho,u,v\rangle\eta Der_\infty \rightarrow \langle\alpha',\rho',u,v\rangle\eta Der_\infty).$$

61.8. CONVENTION

(i) We henceforth write $\vdash^\alpha_\rho \Gamma \Rightarrow \Delta$, in place of $\langle\alpha,\rho,\Gamma,\Delta\rangle\eta Der_\infty$. Accordingly, $\vdash^\alpha_\rho \Gamma, a \Rightarrow \Delta, b$ is an abbreviation for the formula

$$\langle\alpha,\rho,\Gamma\cup\{a\},\Delta\cup\{b\}\rangle\eta Der_\infty. \tag{$*$}$$

By lemma 61.7 (ii), $(*)$ implies $Cl(\Gamma)$, $Cl(\Delta)$, $\Gamma \subseteq Sent_\infty$, $\Delta \subseteq Sent_\infty$, $a\eta Sent_\infty$, $b\eta Sent_\infty$, $\alpha\eta ON$, $\rho\eta ON$. So we keep using capital Greek letters Γ, Δ, etc., as variables for classes $\subseteq Sent_\infty$ and

$$\forall\Gamma(\ldots) := \forall x(Cl(x) \wedge \forall u(u\eta x \rightarrow u\eta Sent_\infty)\ldots);$$

$$\forall\alpha(\ldots) := \forall x(x\eta ON \rightarrow \ldots).$$

(ii) We finally let:

$$\vdash_\rho \Gamma \Rightarrow \Delta := \exists\alpha(\vdash^\alpha_\rho \Gamma \Rightarrow \Delta);$$

$$\vdash \Gamma \Rightarrow \Delta := \exists\rho(\vdash_\rho \Gamma \Rightarrow \Delta).$$

We repeatedly use the notational conventions of 61.8 in stating the propositions below.

61.9. LEMMA ($\mathrm{PW}^- + \mathrm{GID}$).

(i) *Monotonicity*: $(\vdash^\alpha_\rho \Gamma \Rightarrow \Delta) \wedge \alpha \le \beta \wedge \rho \le \sigma. \rightarrow \vdash^\beta_\sigma \Gamma \Rightarrow \Delta$.

(ii) *Weakening*: $(\vdash^{\alpha}_{\rho} \Gamma \Rightarrow \Delta) \wedge \Gamma \subseteq \Gamma' \wedge \Delta \subseteq \Delta'. \rightarrow \vdash^{\alpha}_{\rho} \Gamma' \Rightarrow \Delta'.$

(iii) *Consistency*: $(\vdash \Gamma \Rightarrow \Delta) \rightarrow \neg(\Gamma \cup \Delta =_{e} \emptyset).$

PROOF: by transfinite induction on α (61.3 (iii)) using the definitions and the elementary properties of ordinals. □

61.10. LEMMA (*Inversion*; PW^{-} + GID)

(i) $(\vdash^{\alpha}_{\rho} \Gamma, a \rightarrow b \Rightarrow \Delta) \rightarrow \vdash^{\alpha}_{\rho} \Gamma, b \Rightarrow \Delta;$

$(\vdash^{\alpha}_{\rho} \Gamma, a \rightarrow b \Rightarrow \Delta) \rightarrow \vdash^{\alpha}_{\rho} \Gamma \Rightarrow \Delta, a;$

(ii) $(\vdash^{\alpha}_{\rho} \Gamma \Rightarrow \Delta, a \rightarrow b) \rightarrow \vdash^{\alpha}_{\rho} \Gamma, a \Rightarrow \Delta, b;$

(iii) $(\vdash^{\alpha}_{\rho} \Gamma \Rightarrow \Delta, \forall a) \rightarrow \forall x (\vdash^{\alpha}_{\rho} \Gamma \Rightarrow \Delta, ax);$

(iv) $(\vdash^{\alpha}_{\rho} \Rightarrow Ta) \rightarrow \exists \sigma (\sigma < \rho \wedge \vdash^{\alpha}_{\sigma} \Rightarrow a);$

$a \eta Sent_{\infty} \wedge (\vdash^{\alpha}_{\rho} Ta \Rightarrow). \rightarrow \exists \sigma (\sigma < \rho \wedge \vdash^{\alpha}_{\sigma} a \Rightarrow);$

(v) $Eat_{\infty}(a) \wedge Ta \wedge (\vdash^{\alpha}_{\rho} \Gamma, a \Rightarrow \Delta). \rightarrow \vdash^{\alpha}_{\rho} \Gamma \Rightarrow \Delta;$

$Eat_{\infty}(a) \wedge Fa \wedge (\vdash^{\alpha}_{\rho} \Gamma \Rightarrow \Delta, a). \rightarrow \vdash^{\alpha}_{\rho} \Gamma \Rightarrow \Delta;$

(vi) $(\vdash^{\alpha}_{\rho} \Gamma, T\neg a \Rightarrow Ta, \Delta) \rightarrow \vdash_{\rho} \Gamma, T\neg a \Rightarrow \Delta.$

PROOF. Standard induction on the length α. We only check (vi).
Case 1: $\Gamma, T\neg a \Rightarrow Ta, \Delta$ is an axiom. Then we may assume that either Ta is active or $T\neg a$ (else the conclusion is trivial). If Ta is in Γ, then $\Gamma, T\neg a \Rightarrow \Delta$ is already an instance of the axiom AX.3; if $T\neg a$ is in Δ, $\Gamma, T\neg a \Rightarrow \Delta$ is an instance of AX.1.
Case 2: $\Gamma, T\neg a \Rightarrow Ta, \Delta$ is not an axiom. If $T\neg a, Ta$ are not active, we simply apply IH to the premises of $\Gamma, T\neg a \Rightarrow Ta, \Delta$. If $T\neg a$ is active, we simply erase Ta from the conclusion of (LT). If Ta is active, $\vdash^{\beta}_{\sigma} \Rightarrow a$, with $\sigma < \rho$, $\beta < \alpha$; hence by (L→) and AX.2.2 we have $\vdash_{\sigma} \neg a \Rightarrow$. An application of (LT) yields $\vdash_{\rho} \Gamma, T\neg a \Rightarrow \Delta$. □

REMARK. 61.10.1. If $a \eta Sent_{\infty}$, the equation $a = \neg a$ has no solution (see 62.4 below); this is essential for the soundness of the second part of (iv).

61.11. DEFINITION. $Sent^{-}_{\infty} \subseteq Sent_{\infty}$ is the least property, which contains $\{x : Eat_{\infty}(x)\}$ and is closed under → and ∀. Clearly, the elements of $Sent^{-}_{\infty}$ are "abstract" counterparts of T-free sentences.

61.12. LEMMA (PW^{-} + GID)

(i) *Tautology*: $\vdash_{0} \Gamma, a \Rightarrow \Delta, a.$

(ii) $\mathcal{L}_{op}$*-completeness*: $a \eta Sent^{-}_{\infty} \rightarrow (\vdash_{0} \Rightarrow a) \vee (\vdash_{0} a \Rightarrow).$

PROOF. (i)-(ii): standard induction on α with the assumption $a\eta Sent_{\infty}(\alpha)$, AX.1, AX.2 and logical rules. □

§62. Cut elimination and related properties

$\mathcal{H}$ is closed under cut rule:

62.1. THEOREM (*Cut elimination*). PW$^-$ + GID *proves*:

$$a\eta Sent_{\infty}(\delta) \wedge (\vdash^{\alpha}_{\rho} \Gamma \Rightarrow \Delta, a) \wedge (\vdash^{\beta}_{\rho} \Gamma, a \Rightarrow \Delta). \rightarrow \vdash_{\rho} \Gamma \Rightarrow \Delta. \qquad (*)$$

PROOF. Let $C(\rho, \delta, \alpha, \beta, a, \Gamma, \Delta)$ stand for $(*)$; let $C'(\rho, \delta, \alpha, \beta) := \forall a \forall \Gamma \forall \Delta\, C(\rho, \delta, \alpha, \beta, a, \Gamma, \Delta)$. We apply transfinite ordinal induction to check

$$\forall \rho \forall \delta \forall \alpha \forall \beta C'(\rho, \delta, \alpha, \beta). \qquad (**)$$

As usual, we can rephrase the proof of $(**)$ as a multiple induction on the well-ordering defined by: $(\rho', \delta', \alpha', \beta') < (\rho, \delta, \alpha, \beta)$ iff $(\rho' < \rho)$ or $(\rho' = \rho$ and $\delta' < \delta)$ or $(\rho' = \rho,\ \delta = \delta',\ \alpha' < \alpha)$ or $(\rho' = \rho,\ \delta' = \delta,\ \alpha' = \alpha$ and $\beta' < \beta)$.
The induction hypothesis is:

$$\text{if } \vdash^{\alpha'}_{\rho'} \Gamma' \Rightarrow \Delta', b \text{ and } \vdash^{\beta'}_{\rho'} b, \Gamma' \Rightarrow \Delta', \text{ then } \vdash_{\rho'} \Gamma' \Rightarrow \Delta', \qquad \text{(IH)}$$

whenever $\Gamma', \Delta' \subseteq Sent_{\infty}$, $b\eta Sent_{\infty}(\delta')$, $(\rho', \delta', \alpha', \beta') < (\rho, \delta, \alpha, \beta)$ and Γ', Δ' are classes. Then we assume that

$$a\eta Sent_{\infty}(\delta) \wedge (\vdash^{\alpha}_{\rho} \Gamma \Rightarrow \Delta, a) \wedge (\vdash^{\beta}_{\rho} \Gamma, a \Rightarrow \Delta). \qquad \text{(Ass)}$$

Case 1: $Eat_{\infty}(a)$ holds. i.e. $a = [c = d]$ or $a = [Nc]$. Then $\vdash_{\rho} \Gamma \Rightarrow \Delta$ follows by 61.10 (v).

Case 2: $a = Tc$. We distinguish the following subcases.
2.1.1: Tc is not active and either $\Gamma \Rightarrow \Delta, Tc$ is an axiom, or $\Gamma, Tc \Rightarrow \Delta$ is an axiom. Then $\Gamma \Rightarrow \Delta$ is trivially an axiom.
2.1.2: Tc is active and $\Gamma \Rightarrow \Delta, Tc$ is an axiom. Then $\Gamma =_{e} \Gamma', Tc$, for some Γ' and hence $\vdash_{\rho} \Gamma \Rightarrow \Delta$ follows from the other assumption.
2.1.3: Tc is active and $\Gamma, Tc \Rightarrow \Delta$ is an axiom. Then either $\Delta =_{e} \Delta', Tc$ or $\Gamma =_{e} \Gamma', T\neg c$. In the first case, we get $\vdash_{\rho} \Gamma \Rightarrow \Delta$ from the first assumption, while 61.10 (vi) applies in the second case.
2.1.4: $a = Tc$ and Tc is active in either premise. Then we have applied (RT) and (LT). Hence, for some γ, $c\eta Sent_{\infty}(\gamma)$ and

$$(\vdash^{\alpha'}_{\rho'} \Rightarrow c) \text{ and } (\vdash^{\beta'}_{\rho'} c \Rightarrow), \text{ for some } \rho' < \rho.$$

As $(\rho', \gamma, \alpha', \beta') < (\rho, 0, \alpha, \beta)$, the empty sequent $\Rightarrow$ would be derivable in $\mathcal{H}$ by IH, against 61.9 (iii). Thus, either Tc is not active in $\vdash^{\alpha}_{\rho} \Gamma \Rightarrow \Delta, Tc$, or

Tc is not active in $\vdash^{\beta}_{\rho} \Gamma, Tc \Rightarrow \Delta$. Since the argument is symmetric, let us consider only the first case. If $\Gamma \Rightarrow \Delta, Tc$ follows with (RT) or (LT) applied to some $d \neq Tc$, we simply erase Tc in the conclusion and we get $\Gamma \Rightarrow \Delta$. Else, the premises of $\Gamma \Rightarrow \Delta, Tc$ follow with a rule $\mathfrak{R} \neq$ (RT), (LT). For instance, let $\mathfrak{R} =$ (R$\forall$) and assume that $\Delta =_e \Delta' \cup \{\forall d\}$ and

$$\forall x \exists \gamma (\gamma < \alpha \wedge \vdash^{\gamma}_{\rho} \Gamma \Rightarrow \Delta, Tc, dx).$$

By weakening (61.9 (ii)), we also get $\vdash^{\beta}_{\rho} \Gamma, Tc \Rightarrow \Delta, dx$, for each x. By IH, if $\gamma < \alpha$ and x is arbitrary, we get $\vdash^{\xi}_{\rho} \Gamma \Rightarrow \Delta, dx$, where $\xi = max(\alpha, \beta) + 1$ (as $(\rho, 0, \gamma, \beta) < (\rho, 0, \alpha, \beta)$ holds). Hence $\vdash_{\rho} \Gamma \Rightarrow \Delta$ by (R$\forall$). The extant cases are similar.

Case 3: $\neg At_{\infty}(a)$. We again consider a few subcases.

3.1. either $\Gamma \Rightarrow \Delta, a$ is an axiom or $\Gamma, a \Rightarrow \Delta$ is an axiom. Then $\Gamma \Rightarrow \Delta$ is already an axiom, as a cannot be active by assumption on a and 61.5.1.

3.2. $\Gamma \Rightarrow \Delta, a$ and $\Gamma, a \Rightarrow \Delta$ are not axioms and a is active in either premise.

3.2.1. Let $a = \forall b$. Then we must have $\vdash^{\beta'}_{\rho} \Gamma, \forall b, bc \Rightarrow \Delta$, for some c and $\beta' < \beta$; by weakening (61.9 (iii)), also $\vdash^{\alpha}_{\rho} \Gamma, bc \Rightarrow \Delta, \forall b$. Then we obtain $\vdash^{\xi}_{\rho} \Gamma, bc \Rightarrow \Delta$, for some ξ (we use IH because $(\rho, \delta, \alpha, \beta') < (\rho, \delta, \alpha, \beta)$), and by inversion $\vdash^{\alpha}_{\rho} \Gamma \Rightarrow \Delta, bc$. But $(\forall b) \in Sent_{\infty}(\delta)$ yields $(bc) \eta Sent_{\infty}(\delta')$, for some $\delta' < \delta$ by lemma 61.4 (vi); hence $(\rho, \delta', \alpha, \xi) < (\rho, \delta, \alpha, \xi)$ and by IH $\vdash_{\rho} \Gamma \Rightarrow \Delta$.

3.2.2. Let $a = (b \rightarrow c)$: then we must have $\vdash^{\alpha'}_{\rho} \Gamma, b \Rightarrow (b \rightarrow c), c, \Delta$ for some $\alpha' < \alpha$; hence, by weakening, $\vdash^{\beta}_{\rho} \Gamma, b \rightarrow c, b \Rightarrow c, \Delta$ and by IH, since $(\rho, \delta, \alpha', \beta) < (\rho, \delta, \alpha, \beta)$, $\vdash^{\xi}_{\rho} \Gamma, b \Rightarrow \Delta, c$, for some ξ. Now 61.10 (i) and weakening yield $\vdash^{\beta}_{\rho} \Gamma, c, b \Rightarrow \Delta$ and $\vdash^{\beta}_{\rho} \Gamma \Rightarrow b, \Delta$. But b, $c \eta Sent_{\infty}(\delta')$, for some $\delta' < \delta$ by Lemma 61.4 (v): then $(\rho, \delta', \xi, \beta) < (\rho, \delta, \alpha, \beta)$, whence IH implies $\vdash^{\gamma}_{\rho} \Gamma, b \Rightarrow \Delta$, for some γ. But $(\rho, \delta', \beta, \gamma) < (\rho, \delta, \alpha, \beta)$ and finally $\vdash_{\rho} \Gamma \Rightarrow \Delta$.

3.3. $\Gamma \Rightarrow \Delta, a$ and $\Gamma, a \Rightarrow \Delta$ are not axioms, a is not active in one premise. We apply IH to the premises of the final inference $\mathfrak{R}$ and then we use $\mathfrak{R}$. □

Not surprisingly, theorem 62.1 leads to a form of Herbrand's theorem for $\mathfrak{K}$, which is required by subsequent consistency results. However, we have to establish that certain equations have no solutions in $Sent_{\infty}$; but this project encounters difficulties, derived from the clause that introduces $(\forall f)$ in $Sent_{\infty}$. In contrast to the case where $f = \lambda x[A]$ and A is a usual sentence, we are not entitled to infer that fx and fy have the same logical form (e.g. if fx is $(bx \rightarrow cx)$, then also $fy = by \rightarrow cy$). This suggests a suitable notion $HSent_{\infty}(f)$ of *hereditary sentential function.*

$HSent_\infty$ is inductively generated by the following clauses:

(i) $\lambda x[T(ax)]$, $\lambda x[ax = bx]$, $\lambda x[N(ax)]$ are in $HSent_\infty$;

(ii) if a, b are in $HSent_\infty$, then $\lambda x(ax \to bx)$ is in $HSent_\infty$;

(iii) if $(\lambda y. axy)$ is in $HSent_\infty$, for every x, then $\lambda x. \forall(\lambda y. ayx)$ is in $HSent_\infty$.

Obviously, $HSent_\infty$ is formally representable; if we further assume extensionality for operations, we notice that we can associate elements to every formula of $HSent_\infty$:

62.2. LEMMA. *We can find a formula $ASF(x,u)$, operative in u, such that, if $HSent_\infty := Ixu.ASF(x,u)$, then we have provably, in* $\mathrm{PW}^- + \mathrm{GID}$:

(i) $f\eta HSent_\infty \leftrightarrow \exists h \exists g \forall x((fx = [T(hx)]) \vee (fx = [N(hx)]) \vee$
$\vee (fx = [hx = gx]) \vee (fx = (hx \to gx) \wedge h\eta HSent_\infty \wedge g\eta HSent_\infty) \vee$
$\vee (fx = \forall(\lambda y. hyx) \wedge \forall z(\lambda x. (hzx)\eta HSent_\infty)))$.

(ii) $\forall x(ASF(x, B) \to B(x)) \to \forall x(x\eta HSent_\infty \to B(x))$, ($B(x)$ arbitrary).

Moreover:

(iii) $\mathrm{PW}^- + \mathrm{GID} \vdash f\eta HSent_\infty \to \forall x((fx)\eta Sent_\infty)$.

(iv) *If A is a formula of $\mathfrak{L}$ with free variables in the list $x, w_1, \ldots, w_n$ then*

$$\mathrm{PW}^- + \mathrm{GID} + Ext_{op} \vdash \forall w_1 \ldots \forall w_n(\lambda x\,[A(x, w_1, \ldots, w_n)])\eta HSent_\infty).$$

(Ext_{op} = extensionality for operations, see §2; *$Ixu.ASF(x,u)$ is the fixed point given by* 10.1).

PROOF. (i)-(ii): by 61.1 and GID.
(iii): straightforward induction on $HSent_\infty$, using the closure properties of $Sent_\infty$.
(iv): metamathematical induction on A with repeated applications of ξ-conversion (2.2) for operations. For simplicity, let w also denote a finite list $w_1, \ldots, w_n$ of variables. If A is atomic and $A(x,w) = (t(x,w) = s(x,w))$, choose $h := \lambda x. t(x,w)$, $g := \lambda x. s(x,w)$, $f := \lambda x. [t(x,w) = s(x,w)]$. Then we get $fu = [hu = gu]$ by (β)-conversion, hence $f = \lambda x. [hx = gx]\eta HSent_\infty$. The other atomic cases are similar. Assume $A(x,w) = (B(x,w) \to C(x,w))$. Then by IH, $h = \lambda x. [B(x,w)]$ and $g = \lambda x. [C(x,w)]$ are in $HSent_\infty$; if we choose $f = (\lambda x. [B(x,w) \to C(x,w)])$, then $f\eta HSent_\infty$ (as $fu = (hu \to gu)$). Assume $A = \forall y B(y,x,w)$; by IH, $\lambda x[B(y,x,w)]\eta HSent_\infty$, for every y. Then by (i), $\lambda x. [\forall y B(y,x,w)]\eta HSent_\infty$. □

62.3. LEMMA ("*No solution lemma*")

(i) $\mathrm{PW}^{-}+\mathrm{GID}\vdash f\eta HSent_{\infty}\to\forall z\forall x\forall y(\neg(fx=(fy\to z)))$.

(ii) $\mathrm{PW}^{-}+\mathrm{GID}$ proves

$$\forall x\forall y(x\eta Sent_{\infty}\wedge y\eta Sent_{\infty}\to\neg((x\to y)=x)\wedge\neg((y\to x)=x)).$$

(iii) $\mathrm{PW}^{-}+\mathrm{GID}\vdash\ \forall x(x\eta Sent_{\infty}\to\ \neg((Tx\to x)=x))$.

(iv) *If A is an $\mathfrak{L}$-formula,*

$$\mathrm{PW}^{-}+\mathrm{GID}\vdash\neg([A(u,v_1,\ldots,v_n)]=[\forall xA(x,w_1,\ldots,w_n)]).$$

PROOF. (i) If $f\eta HSent_{\infty}$, $(fx)\eta Sent_{\infty}$, for every x by 62.3 (iii). Then by 61.3-61.4, we show by transfinite induction on α:

$$\forall x((fx)\eta Sent_{\infty}(\alpha)\to\forall y\forall z(fx\neq(fy\to z))). \qquad (*)$$

If $At_{\infty}(fx)$ holds or $fx=(\forall g)$, $(*)$ is clear, by the independence of $\to$, N, $=$, T, $\forall$. If $fx=(b\to c)$, there exist h, g in $HSent_{\infty}$ such that $fx=(hx\to gx)$. Were $fx=(fy\to z)$, we ought to have, by pairing and by definition of the operation $\lambda x\lambda y.(x\to y)$:

$$hx=(hy\to gy), \qquad (**)$$

where $(hx)\eta Sent_{\infty}(\beta)$, for some $\beta<\alpha$; but $(**)$ contradicts IH.
(ii): similar to (i), arguing by induction on $Sent_{\infty}$ with the independence properties of $\to$, $\forall$ and the pairing axioms.
(iii): immediate by (ii).
(iv) We verify, by induction on A, that the assumption

$$[A(u,v_1,\ldots,v_n)]=[\forall xA(x,w_1,\ldots,w_n)]$$

leads to a contradiction; in the case $A(u,v_1,\ldots,v_n)=\forall yB(y,u,w_1,\ldots,w_n)$ we apply IH. □.

62.4. THEOREM. $\mathrm{PW}^{-}+\mathrm{GID}$ *proves:*

(i) $Eat_{\infty}(a)\to(\vdash Ta\Rightarrow a)\wedge(\vdash a\Rightarrow Ta)$;

(ii) $a\eta Sent_{\infty}\wedge b\eta Sent_{\infty}\wedge(a\neq\neg b)\wedge(b\neq\neg a)\to$
$\to((\vdash Ta,\ Tb\Rightarrow)\to(\vdash a\Rightarrow)\vee(\vdash b\Rightarrow))$;

(iii) $a\eta Sent_{\infty}\wedge b\eta Sent_{\infty}\wedge a\neq b\to((\vdash Ta\Rightarrow Tb)\to.(\vdash a\Rightarrow)\vee(\vdash\Rightarrow b))$;

(iv) *Herbrand:* $f\eta HSent_{\infty}\to((\vdash\forall xT(fx)\Rightarrow)\to\exists x(\vdash fx\Rightarrow))$.

(v) *If A is an arbitrary formula of $\mathfrak{L}$,* $\mathrm{PW}^{-}+\mathrm{GID}+Ext_{op}$ *proves*

$$((\vdash\forall xTA(x)\Rightarrow T\forall xA(x))\to.(\vdash\Rightarrow\forall xA(x))\vee\exists x(\vdash A(x)\Rightarrow).$$

PROOF. (i) If a encodes a true atom of the form $t = s$, Nt, $\vdash Ta \Rightarrow a$ is an axiom (by AX.2.1) and $\vdash a \Rightarrow Ta$ is derivable from $\vdash \Rightarrow a$ by 61.5.2 (RT). A dual argument works if the atom in question is false.
(ii) By hypothesis, the sequent cannot be an axiom; hence, either Ta is active and $\vdash a \Rightarrow$, or Tb is active and hence $\vdash b \Rightarrow$.
(iii): similar to (ii).
(iv) First of all, let us say that Γ is a class of Tf-instances (in symbols $Tf\text{-}Instan(\Gamma)$) iff Γ satisfies the following condition:

$$Cl(\Gamma) \wedge \forall u(u\eta\Gamma \rightarrow \exists x(u = [T(fx)])). \tag{1}$$

Then observe that, if f is in $HSent_\infty$, and Γ is a class of Tf-instances,

$$\text{for no } x, \text{ both } T(fx) \text{ and } T\neg(fx) \text{ are in } \Gamma. \tag{2}$$

Indeed, if there is some y such that $T(fy)$ and $T\neg(fy)$ are in Γ, then for some z, $T(fz) = T\neg(fy) = T(fy \rightarrow \bot)$, whence $fz = fy \rightarrow \bot$, against 62.3 (i). We now claim:

$$f\eta HSent_\infty \wedge Tf\text{-}Instan(\Gamma) \wedge (\vdash \Gamma \cup \{\forall(\lambda x T(fx))\} \Rightarrow) \rightarrow \exists x(\vdash fx \Rightarrow). \tag{3}$$

Clearly (iv) is an immediate consequence of (3); (3) is verified by induction on α such that $(\vdash^\alpha \Gamma \cup \{\forall(\lambda x.T(fx))\} \Rightarrow)$. If the given sequent is inferred by (L$\forall$), we have $(\vdash^\beta \Gamma \cup \{\forall(\lambda x.T(fx))\} \cup \{T(fc)\} \Rightarrow)$, for some c and $\beta < \alpha$: the conclusion follows by IH. If $\Gamma \cup \{\forall(\lambda x.T(fx))\} \Rightarrow$ is inferred by (LT), then for some c, $\vdash \Rightarrow fc$, and we are done.
(v) If Γ is a class of Tf-instances and $f = \lambda x[A]$, then the sequent $\{\forall xTA(x)\} \cup \Gamma \Rightarrow T\forall xA(x))$ cannot be an instance of an axiom (use 62.2 (iv), (2) above and 62.3 (iv)). Hence we obtain, by induction on α:

$$Tf\text{-}Instan(\Gamma) \wedge (\vdash \Gamma \cup \{\forall x.TA(x)\} \Rightarrow T\forall xA) \rightarrow$$
$$\rightarrow \exists x(\vdash A(x) \Rightarrow) \vee (\vdash \Rightarrow \forall yA). \qquad \square$$

The cut elimination theorem implies that those predicates, which are decidable in the sense of $\mathcal{K}$, are equiextensional with classes; more precisely :

62.5. LEMMA. *We can find a term $\lambda u.t(u)$ such that:*

$$\mathrm{PW}^- + \mathrm{GID} \vdash \forall x((\vdash ax \Rightarrow) \vee (\vdash \Rightarrow ax)) \rightarrow (Cl(t(a)) \wedge t(a) =_e \{x : \vdash \Rightarrow ax\}).$$

PROOF. $a(1) = \{x : \vdash \Rightarrow ax\}$ and $a(0) = \{x : \vdash ax \Rightarrow\}$ are disjoint by cut elimination; by dual representation (theorem 16.11), we can choose $t(a) = ER(a(1), a(0)) =_e a(1)$ such that $-t(a) =_e a(0)$. By assumption on a, $t(a)$ is a class. $\square$

62.6. THEOREM (*Internal N-induction*)

(i) *$\mathcal{K}$ derives N-induction for arbitrary conditions: if $A(x)$ is an* *arbitrary* $\mathfrak{L}$*-formula (with the free variable shown only),*

$$\mathrm{PW}_p + \mathrm{GID} \vdash (\vdash A(\overline{0}), \forall x(A(x) \to A(x+1)) \Rightarrow \forall x(Nx \to A(x))).$$

(ii) $\mathcal{K}$ *derives the* N*-induction principle for* $\mathcal{K}$*-decidable predicates, provably in* $\mathrm{PW}_c + \mathrm{GID}$:

$$\vdash \forall x((\vdash \Rightarrow ax) \vee (\vdash ax \Rightarrow)) \to (\vdash a\overline{0}, \forall x(ax \to a(x+1)) \Rightarrow \forall x(Nx \to ax)).$$

PROOF. (i) Set $Clos(A) := \{A(\overline{0}), \forall x(A(x) \to A(x+1))\}$ and $Ind(A) := \{x : \vdash \Rightarrow Clos(A) \to A(x)\}$. By AX.2 and P-$NIND$ (see 10.7) it suffices to check that $N \subseteq Ind(A)$. But the tautology lemma yields $\vdash Clos(A) \Rightarrow A(\overline{0})$; $\vdash Clos(A) \Rightarrow A(a)$ implies $\vdash Clos(A) \Rightarrow A(a+1)$ by tautology, inversion and cut.
(ii) For every x, assume:

$$\vdash \Rightarrow ax \text{ or } \vdash ax \Rightarrow. \tag{1}$$

Then we claim:

$$\text{either } \vdash \Rightarrow a\overline{0} \text{ and } \vdash \Rightarrow \forall x(ax \to a(x+1)), \text{ or} \tag{2.1}$$

$$\vdash a\overline{0}, \forall x(ax \to a(x+1)) \Rightarrow. \tag{2.2}$$

If (2.1) does not hold, either not $\vdash \Rightarrow a\overline{0}$ or not $\vdash \Rightarrow \forall x(ax \to a(x+1))$. In the first case, (1) implies $\vdash a\overline{0} \Rightarrow$, and (2.2) follows by weakening. As to the second case, we have by inversion 61.10 (ii)-(iii):

$$\text{for some } b, \text{ not } \vdash ab \Rightarrow a(b+1).$$

By (1), $\vdash \Rightarrow ab$ and $\vdash a(b+1) \Rightarrow$, for some b, hence (2.2) by (L$\to$), (L$\forall$) and weakening. We can prove for any x, if $Clos(a) := a\overline{0} \wedge \forall x(ax \to a(x+1))$,

$$(\vdash \Rightarrow Clos(a) \to ax) \text{ or } (\vdash Clos(a) \to ax \Rightarrow). \tag{3}$$

(3) is trivial if $(\vdash \Rightarrow ax)$ or $(\vdash Clos(a) \Rightarrow)$. Else, we must have $(\vdash ax \Rightarrow)$ and $(\vdash \Rightarrow Clos(a))$ by (1)-(2); hence $(\vdash Clos(a) \to ax \Rightarrow)$. Now consider $Ind(a) := \{x : (\vdash \Rightarrow Clos(a) \to ax)\}$, $Ind(a)^{-} := \{x : (\vdash Clos(a) \to ax \Rightarrow)\}$: by lemma 62.5 and (3), there exists a class c such that $c =_e Ind(a)$ and $-c =_e Ind(a)^{-}$. So we can apply class induction to check $N \subseteq c =_e Ind(a)$ and we finally get with AX.2, $(\vdash \Rightarrow Clos(A) \to \forall x(Nx \to ax))$. □

§63. A provability interpretation and the upper bound theorem

Provability in the sequent calculus $\mathcal{K}$ provides an interpretation, which validates VF^{-}-axioms plus certain additional T-schemata, directly inspired by closure properties of $\mathcal{K}$. The pay-off of the main interpretation theorem is that it gives precise information on the proof-theoretic strength of the object theories.

First of all, we define the announced provability interpretation.

63.1. DEFINITION (Induction on the definition of formula).

$$A_\infty = A, \text{ if } A = (Nt),\ (t = s); \quad (Tt)_\infty = (\vdash \Rightarrow t);$$
$$(A \to B)_\infty = A_\infty \to B_\infty; \quad (\forall x A)_\infty = \forall x (A_\infty).$$

If k is a natural number, it is understood that in the context $Sent_\infty(k)$, k stands for the (closed term representing the) corresponding finite ordinal.

63.2. LEMMA (*Local truth*). *For each finite k, we can define a predicate $TR_k(x)$ such that* $\mathrm{PW}^- + \mathrm{GID}$ *proves:*

(i) $TR_0([Ta]) \leftrightarrow (\vdash \Rightarrow a);\quad TR_0([x = y]) \leftrightarrow x = y;\quad TR_0([Nx]) \leftrightarrow Nx;$

(ii) $TR_{k+1}(a \to b) \leftrightarrow a\eta Sent_\infty(k) \wedge b\eta Sent_\infty(k) \wedge (TR_k(a) \to TR_k(b));$

$TR_{k+1}(\forall a) \leftrightarrow \forall x((ax)\eta Sent_\infty(k) \wedge TR_k(ax));$

(iii) $At_\infty(a) \to (TR_{k+1}(a) \leftrightarrow TR_k(a));$

if $k \leq n$, $a\eta Sent_\infty(k) \to (TR_k(a) \leftrightarrow TR_n(a));$

(iv) $\forall x(x\eta Sent_\infty(k) \to \neg(TR_k(x) \wedge TR_k(\neg x));$

(v) $TR_k([A(x_1, \ldots, x_n]) \leftrightarrow A_\infty(x_1, \ldots, x_n),$

for every formula of logical complexity $\leq k$.

The proof is well-known; (iv) requires outer induction on k; as to (v), we proceed by induction on A, using 61.10 (v), whenever $A := Tt$.

63.3. NOTATION. Let Γ, Δ be classes of abstract sentences; we set:

(i) $TR_k(\bigwedge \Gamma) := \forall x(x\eta\Gamma \to TR_k(x));$

(ii) $TR_k(\bigvee \Delta) := \exists x(x\eta\Delta \wedge TR_k(x)).$

63.4. THEOREM (*Formalized $\mathcal{K}$-soundness*). *For each finite k,* $\mathrm{PW}^- + \mathrm{GID}$ *proves:*

(i) $\forall\Gamma\forall\Delta(\Gamma \cup \Delta \subseteq Sent_\infty(k) \wedge (\vdash \Gamma \Rightarrow \Delta) \to (TR_k(\bigwedge \Gamma) \to TR_k(\bigvee \Delta))).$

(ii) *If A is an arbitrary formula of complexity $\leq k$,*

$\mathrm{PW}^- + \mathrm{GID} \vdash (\vdash \Rightarrow A(x_1, \ldots, x_n)) \to A_\infty(x_1, \ldots, x_n).$

PROOF. (ii) is immediate from (i) and 63.2 (v).
(i) is checked by induction on the definition of $\vdash$ (see 61.7 (iii)). We rely without explicit mention upon the truth conditions of 63.2. and the fact that TR_k is monotone in k (i.e. $TR_k(a)$ implies $TR_m(a)$, whenever $k \leq m$).

1. If $\Gamma \Rightarrow \Delta$ is an instance of AX.1, the conclusion is trivial. If $\Gamma \Rightarrow \Delta$ is an instance of AX.2, either $TR_k(\wedge \Gamma)$ is trivially false or $TR_k(\vee \Delta)$ is trivially true. If $\Gamma \Rightarrow \Delta$ is an instance of AX.3, $\Gamma = {}_e\Gamma' \cup \{Ta, T\neg a\}$, for some a. It suffices to show that $TR_k(\wedge \Gamma)$ leads to contradiction. If $TR_k(\wedge \Gamma)$ is assumed, then $(\vdash \Rightarrow a)$ and $(\vdash a \Rightarrow)$ (apply 61.10 and 63.2 (i)). Hence by cut elimination, we get $(\vdash \Rightarrow)$, against the consistency lemma 61.9 (iii) !

2.1. $\Gamma \Rightarrow \Delta$ is inferred by (RT): then for some a, such that $[Ta]\eta\Delta$, $(\vdash \Rightarrow a)$, which implies $TR_0([Ta])$ (63.2(i)) and hence $TR_k(\vee \Delta)$. If $\Gamma \Rightarrow \Delta$ is inferred by (LT), then $(\vdash a \Rightarrow)$, which implies, by 62.1, not $\vdash \Rightarrow a$, i.e. $\neg TR_k([Ta])$, whence $\neg TR_k(\wedge \Gamma)$ since $[Ta]\eta\Gamma$.

2.2. $\Gamma \Rightarrow \Delta$ is inferred from $\Gamma', a \rightarrow b, b \Rightarrow \Delta$ and $\Gamma', a \rightarrow b \Rightarrow \Delta, a$ where $\Gamma = {}_e\Gamma' \cup \{a \rightarrow b\}$. We suppose

$$TR_k(\wedge \Gamma), \text{ i.e. } TR_k(\wedge \Gamma') \text{ and } TR_k(a \rightarrow b); \qquad (*)$$

in addition we can assume by IH:

$$TR_k(\wedge (\Gamma \cup \{b\})) \rightarrow TR_k(\vee \Delta); \qquad (**)$$

$$TR_k(\wedge \Gamma) \rightarrow TR_k(\vee (\Delta \cup \{a\}). \qquad (***)$$

By $(*)$ and $(***)$, either $TR_k(\vee \Delta)$ and we are done, or $TR_k(a)$, which implies $TR_k(b)$, by the second part of $(*)$; hence, again by $(*)$, $TR_k(\wedge (\Gamma \cup \{b\}))$, and finally $TR_k(\vee \Delta)$ with $(**)$. The cases where $\Gamma \Rightarrow \Delta$ is the conclusion of an application of $(R\rightarrow), (R\forall), (L\forall)$ are easily handled and left to the reader. □

63.5. LEMMA ($\mathcal{K}$-*Reflection*) $PW^- + GID \vdash \forall x(\vdash \Rightarrow fx) \rightarrow (\vdash \Rightarrow \forall f)$.

PROOF. The argument is not new (see 23.8); we apply CL-compactness 16.3. First consider $R(f,x) = \{\alpha : \vdash^{\alpha} \Rightarrow fx\}$ and assume $(\vdash \Rightarrow fx)$. Then for every x, $\xi R(f,x)$ is a non-empty class of ordinals $\subseteq R(f,x)$; hence $d = \{y : \exists x(y\eta\xi R(f,x))\}$ is a non-empty set of ordinals, such that if $\gamma = \cup d$, then

$$\forall x \exists \rho \ (\vdash^{\gamma}_{\rho} \Rightarrow (fx)).$$

Finally, we find a bound σ such that $\forall x(\vdash^{\gamma}_{\sigma} \Rightarrow (fx))$ and we conclude by $(R\forall)$. □

We are now ready to state the promised strengthened interpretation result.

63.6. DEFINITION

(i) VFH^- is the elementary theory, based on classical predicate calculus, which includes the schema *T-out* of 59.1, and, in addition:

1. the *axioms* corresponding to *T-imp* and *T-univ* of 59.1:

$T\text{-}imp_{ax}$: $T(x \to y) \to (Tx \to Ty)$;

$T\text{-}univ_{ax}$: $\forall x T(fx) \to T\forall f$;

2. the new principles:

$T^+\text{-}elem$: $T(A \leftrightarrow TA)$, where $A = (x = y), \neg(x = y), Nx, \neg Nx$;

$T^+\text{-}log$: TA, if A is any axiom of OP^-;

$T\text{-}rep$: $Tx \to TTx$;

$T^+\text{-}cons$: $T(T\neg x \to \neg Tx)$;

$T(\neg T \wedge)$: $([A] \neq [\neg B]) \wedge ([B] \neq [\neg A]) \to . T(\neg(TA \wedge TB)) \to (T\neg A \vee T\neg B)$.

$T(T\to)$: $([A] \neq [B]) \to . T(TA \to TB) \to (T\neg A \vee TB)$.

$T\text{-}Herb$: $T(\forall x TA(x) \to T\forall x A(x)) \to . \exists y T(\neg A(y)) \vee T(\forall x A(x))$.

($T\text{-}Herb$:= the Herbrand schema; the acronym is justified by 63.9 (viii) below).

(ii) $\mathrm{VFH}_c := \mathrm{VFH}^-$ plus the axiom of internal class-N-induction

I-CL-NIND $\quad Cl(a) \to T[a\bar{0} \wedge \forall x(ax \to a(x+1)) \to \forall x(Nx \to ax)]$;

(iii) $\mathrm{VFH}_p := \mathrm{VFH}^-$ plus the axiom of internal induction I-NIND

$$T[a\bar{0} \wedge \forall x(ax \to a(x+1)) \to \forall x(Nx \to ax)].$$

Of course $\mathrm{VF}^- \subseteq \mathrm{VFH}^-$.

63.7. MAIN THEOREM (*Provability Interpretation + Upper Bound*).

Let $\mathcal{T} = \mathrm{VFH}^-(\mathrm{VFH}_c, \mathrm{VFH}_p)$, and $\mathcal{S} = \mathrm{PW}^- + \mathrm{GID} + \mathrm{Ext}_{op}$ (*respectively* $\mathrm{PW}_c + \mathrm{GID} + \mathrm{Ext}_{op}$, $\mathrm{PW}_p + \mathrm{GID} + \mathrm{Ext}_{op}$).

(i) *If* $\mathcal{T} \vdash A$, *then* $\mathcal{S} \vdash A_\infty$.

(ii) *If* A *is a formula of* $\mathcal{L}_{op}$, $\mathcal{T} = \mathrm{VFH}^-$ (VFH_c), $\mathcal{S} = \mathrm{OP}^- + \mathrm{Ext}_{op}$ $(\mathrm{OP} + \mathrm{Ext}_{op})$, *then* $\mathcal{T} \vdash A$ *implies* $\mathcal{S} \vdash A$.

PROOF. (i) First assume $\mathcal{T} = \mathrm{VFH}^-$. It suffices to derive the translation A_∞ of each VFH^--axiom in the metatheory $\mathrm{PW}^- + \mathrm{GID}$.

$(T\text{-}out)_\infty$: apply the $\mathcal{K}$-soundness theorem and 63.4 (ii).

$(T^+\text{-}elem)_\infty$: by theorem 62.4 (i), 61.2.

$(T\text{-}imp_{ax})_\infty$: by 61.10 (ii) and the cut elimination theorem 62.1.

$(T\text{-}univ_{ax})_\infty$: by the $\mathcal{K}$-reflection lemma 63.5.

$(T^+\text{-}log)_\infty$: apply 61.12.

$(T\text{-}rep)_\infty$: by closure of $\vdash$ under (RT).

$(T^+\text{-}cons)_\infty$: by axiom AX.3 for $\vdash$.

$(T(\neg T\wedge))_\infty$, $(T(T\rightarrow))_\infty$, $(T\text{-}Herb)_\infty$: by 62.4 (ii), (iii), (v) (in the given order).

If $\mathfrak{T} = \mathrm{VFH}_c, \mathrm{VFH}_p$, we apply the N-induction theorem of 62.6.

(ii): notice that $A_\infty = A$, if $A \in \mathcal{L}_{op}$, and use the conservation theorem of 15.5. □

63.7.1. REMARK. (i) A provability interpretation was already exploited by Cantini (1990) in the context of a proof-theoretic investigation of truth theories over Peano arithmetic, proposed by Friedman and Sheard (1987). The idea was suggested by W. Buchholz, in order to simplify a formalized model-theoretic construction of the present author.

(ii) The statement of 63.7 (ii) leaves the case of VFH_p open; but it can easily be settled by building a model of $\mathrm{PW}_p + \mathrm{GID} + \mathrm{Ext}_{op}$ in the admissible set theory of Jäger (1982), which is proof-theoretically equivalent to $\mathrm{ID}_1(\mathrm{acc})$.

We conclude with a few consequences of the extended systems.

63.8. DEFINITION. WVF^- is the (weak) subsystem of VF^-, whose only T-axioms are $T\neg a \rightarrow \neg Ta$ and $TA \leftrightarrow A$, where A is an e-atom; $\mathrm{WVF}_c := \mathrm{WVF}^- + \text{I-Cl-NIND}$, $\mathrm{WVF}_p := \mathrm{WVF}^- + \text{I-NIND}$ (see 63.6 above for notations).

63.9. PROPOSITION. *The following schemata are provable in* VFH^-:

(i) $T(TA \rightarrow \neg TB) \rightarrow T(A \rightarrow \neg B)$;

(ii) $T(TA \rightarrow TB) \rightarrow T(A \rightarrow B)$;

(iii) $T\neg TA \leftrightarrow T\neg A$;

(iv) $TA \leftrightarrow TTA \leftrightarrow FFA$;

(v) $FA \leftrightarrow FTA \leftrightarrow TFA$;

(vi) *the Meta-Löb principle:*

$T(T(TA \rightarrow A) \rightarrow TA) \leftrightarrow TA$;

(vii) $T(T(A \rightarrow TA) \rightarrow TA) \leftrightarrow TA$;

(viii) $T(\neg \forall x TA) \rightarrow \exists x T \neg A$.

(ix) *If* A *is a sentence of* $\mathcal{L}$ *and* $\mathrm{WVF}^- \vdash A$, *then* $\mathrm{VFH}^- \vdash TA$. *The same holds if we replace* WVF^- *with* WVF_c (WVF_p) *and* VF^- *with* VF_c (VF_p).

PROOF. (i) If $[A] \neq [\neg B]$ and $[B] \neq [\neg A]$, $T(\neg T \wedge)$ implies $T\neg A$ or $T\neg B$: in either case $T(A \rightarrow \neg B)$ is derivable by using the LOG-rule (see 59.2) and *T-imp*. Otherwise, $T(A \rightarrow \neg B)$ has the form $T(C \rightarrow C)$ or $T(C \rightarrow \neg\neg C)$ and we are done.

(ii) Assume $T(TA \rightarrow TB)$; since $T(TB \rightarrow \neg T \neg B)$ (T^+-*cons*), we get $T(TA \rightarrow \neg T \neg B)$, whence by the previous step $T(A \rightarrow \neg\neg B)$ and finally $T(A \rightarrow B)$.

(iii) Assume $T\neg TA$; by LOG-rule $T(\neg TA \rightarrow (TA \rightarrow \neg TA))$, whence $T(TA \rightarrow \neg TA)$, i.e. by (i) above and LOG-rule, $T\neg A$. Conversely, apply *T-rep*, *T-imp*, T^+-*cons*.

(iv)-(v): use (iii), *T-rep*, *T-out*.

(vi) Assume $T(T(TA \rightarrow A) \rightarrow TA)$. Then $[TA \rightarrow A] \neq [A]$ by lemma 62.3 (ii). Hence by $T(T\rightarrow)$ we get $T(\neg(TA \rightarrow A))$ or TA. If the first case holds, we get $T(TA \wedge \neg A)$, which finally yields a contradiction. The reverse direction easily follows in VF^-.

(vii) Let $T(T(A \rightarrow TA) \rightarrow TA)$. Then $T((A \rightarrow TA) \rightarrow A)$ by (ii) above. But Peirce 's law and LOG-rule yield $T(((A \rightarrow TA) \rightarrow A) \rightarrow A)$, whence TA by *T-imp*. For the converse, apply $T(A \rightarrow (B \rightarrow A))$.

(viii) If $T\neg \forall x TA$, then $T(\forall x TA \rightarrow T \forall x A)$ (by *T-imp* and *T-log*), whence $T\forall x A$ or $\exists y T \neg A(y)$ (*T-Herb*). But $T\forall x A$ and $T\neg \forall x TA$ together imply a contradiction (again by *T-imp*, *T-log* and *T-out*). Hence $\exists y FA(y)$.
(ix): apply *T-univ*, *T-imp*, T^+-*log*, T^+-*elem*, T^+-*cons*. □

63.9.1. REMARK. (i) By 63.9 (vi) VFH^- system is incompatible with Löb schema

$$T(T(TA \rightarrow A) \rightarrow TA).$$

(ii) VFH^- refutes the schema T^+-*univ* $:= T(\forall x TA \rightarrow T\forall x A)$: apply *T-Herb* with $A(x) := (x = x \wedge \neg TL)$, where L is the Liar (i.e. $L = [\neg TL]$).

§ 64. Reconciling supervaluation models with provability interpretation

The reader will certainly notice that the definition of the operator $X \mapsto \Phi_0(X)$ of § 60 is highly impredicative: *we quantify over all possible interpretations of* T, and we do not have a set of constructive rules to produce $\Phi_0(X)$ from X. This situation is in sharp contrast with the case of the Γ-operator of Ch. II. Thus it is natural to wonder whether there is any constructive access to the least fixed point of Φ_0. In particular, is it possible to reconcile the provability interpretation with the supervaluation models of § 61?

We show that the answer is positive, at least in the case where we deal with ground *countable structures.* Most work has been carried out in §§ 59-60; the essential step is to relativize the derivability relation $\vdash \Gamma \Rightarrow \Delta$ of § 61 to finite sets of usual $\mathfrak{L}(\mathcal{M})$-sentences and to observe that the new derivability relation characterizes the least fixed point model. Actually, we shall consider the relation $||\!\!-$, restricted to consistent subsets of $\mathcal{M}$.

64.1. DEFINITION

(i) $X ||\!\!-_1 A$ iff "$Y \models A$, for every $Y \in CONS(\mathcal{M})$ such that $Y \supseteq X$" (recall that $X \in CONS(\mathcal{M}) := X \subseteq M$ and for every $a \in M$, either $a \notin X$ or $(\neg a) \notin X$).

(ii) Φ_1 is the operator defined by $\Phi_1(X) := \{\mathcal{M}([A]) : X ||\!\!-_1 A\}$. Let

$$FIX_1(\mathcal{M}) := \{X \subseteq M : \Phi_1(X) = X\}.$$

Clearly Φ_1 is monotone and we set $V_\infty(\mathcal{M})$ = the least fixed point of Φ_1.

(iii) $\vdash^*_M \Gamma \Rightarrow \Delta$ is inductively defined as $\vdash \Gamma \Rightarrow \Delta$ in 61.5, except that we require:

1. Γ, Δ are finite subsets of $\mathfrak{L}(\mathcal{M})$-sentences;

2. the axioms AX.2.1-AX.2.2 of 61.5 are replaced by

$\mathcal{M}$-AX.2.1: $\vdash^*_M \Gamma \Rightarrow \Delta$, provided $A \in \Gamma$ and $\mathcal{M} \models \neg A$, $A = Nt$, $t = s$;

$\mathcal{M}$-AX.2.2: $\vdash^*_M \Gamma \Rightarrow \Delta$, provided $A \in \Delta$ and $\mathcal{M} \models A$, $A = Nt$, $t = s$.

(iv) If $X \subseteq M$, the relation $X \vdash^*_M \Gamma \Rightarrow \Delta$ (to be read as "$\Gamma \Rightarrow \Delta$ is M-derivable with X-axioms") is inductively defined by modifying the definition of $\vdash^*_M \Gamma \Rightarrow \Delta$ as follows:

1. we omit the T-rules (RT) and (LT);

2. we add the X-axioms:

$X \vdash^*_M \Gamma \Rightarrow \Delta$, provided for some t, $(Tt) \in \Delta$ and $\mathcal{M}(t) \in X$;

$X \vdash^*_M \Gamma \Rightarrow \Delta$, provided for some t, $(Tt) \in \Gamma$ and $\mathcal{M}(\neg t) \in X$;

3. we add the cut rule:

infer $X \vdash^*_M \Gamma \Rightarrow \Delta$ from $X \vdash^*_M \Gamma \Rightarrow \Delta, A$ and $X \vdash^*_M A, \Gamma \Rightarrow \Delta$.

Clearly, $X \vdash^*_M \Gamma \Rightarrow \Delta$ defines a certain (non-effective) theory in M-logic (= first order logic extended with the infinitary M-rule: if $A(a)$ is derivable for each $a \in M$, so is $\forall x A$). If we specialize 62.1 and 61.9-61.12 to $\vdash^*_M$, we immediately have:

64.2. PROPOSITION

(i) $\vdash^*_M \Gamma \Rightarrow \Delta, A$ *and* $\vdash^*_M A, \Gamma \Rightarrow \Delta$ *imply* $\vdash^*_M \Gamma \Rightarrow \Delta$.

(ii) *If A is a sentence of $\mathfrak{L}_{op}(\mathcal{M})$, either* $\vdash^*_M \Rightarrow A$ *or* $\vdash^*_M A \Rightarrow$.

(iii) *The set* $DER(\mathcal{M}) := \{[A] : A$ *is an* $\mathfrak{L}(\mathcal{M})$*-sentence with* $\vdash^*_M \Rightarrow A\}$ *is consistent (i.e.* $DER(\mathcal{M}) \in CONS(\mathcal{M})$*).*

64.3. LEMMA. *If* $\mathcal{M}$ *is countable,* $X \Vdash_1 A$ *iff* $X \vdash^*_M \Rightarrow A$ *(A arbitrary* $\mathfrak{L}(\mathcal{M})$*-sentence).*

PROOF. If not $X \vdash^*_M \Rightarrow A$, we have, by the Henkin-Orey ω-completeness theorem (Shoenfield 1967), a subset Y of M such that $Y \models \neg A$, Y satisfies the X-axioms and the consistency axiom AX.3; hence $Y \in CONS(\mathcal{M})$ and $Y \supseteq X$, which implies not $X \Vdash_1 A$. Conversely, it is straightforward to check $X \Vdash_1 (\bigwedge \Gamma) \to (\bigvee \Delta)$ by induction on the definition of $X \vdash^*_M \Gamma \Rightarrow \Delta$ (here $\bigwedge \Gamma :=$ conjunction over Γ; $\bigvee \Delta :=$ disjunction over Δ). □

64.4. THEOREM (*Characterization*)
If $\mathcal{M}$ *is countable,* $V_\infty(\mathcal{M}) = DER(\mathcal{M})$.

PROOF. (i) Closure of $DER(\mathcal{M})$. By lemma 64.3, it suffices to check:

$$DER(\mathcal{M}) \vdash^*_M \Gamma \Rightarrow \Delta \text{ implies } \vdash^*_M \Gamma \Rightarrow \Delta. \tag{1}$$

Inductive verification of (1). If $\Gamma \Rightarrow \Delta$ is an instance of AX.1, AX.3, $\mathcal{M}$-AX.2, we are done by the corresponding axioms of $\vdash^*_M$. If $\Delta := \Delta', Tt$ with $\mathcal{M}(t) \in DER(\mathcal{M})$, then $\mathcal{M} \models t = [B]$, for some sentence $B \in \mathfrak{L}(\mathcal{M})$ such that $\vdash^*_M \Rightarrow B$. Hence $\vdash^*_M \Gamma \Rightarrow \Delta$ by (RT). The case of the extant $DER(\mathcal{M})$-axiom is disposed of by means of (LT). In the induction step, we apply 64.2 (cut elimination theorem) and IH.
(ii) If $\Phi_1(X) \subseteq X$, then $DER(\mathcal{M}) \subseteq X$. It suffices to check

$$\vdash^*_M \Gamma \Rightarrow \Delta \text{ implies } X \vdash^*_M \Gamma \Rightarrow \Delta, \tag{2}$$

which implies $X \Vdash_1 \bigwedge \Gamma \to \bigvee \Delta$ (by 64.3); we then apply Φ_1-closure of X.

(2) is inductively checked. We only consider the case where $\vdash^*_M \Gamma \Rightarrow \Delta$ is obtained via (RT) and (LT). Assume $\vdash^*_M \Rightarrow A$ and let $\Delta = \Delta', Tt$ with $\mathcal{M}(t) = \mathcal{M}([A])$. Then, by IH, $X \vdash^*_M \Rightarrow A$, i.e. by 64.3 $X \| \!-_1 A$, whence $\mathcal{M}(t) \in X$ (Φ_1-closure of X) and $X \vdash^*_M \Gamma \Rightarrow Tt, \Delta'$ by (X)-axiom. Assume $\vdash^*_M A \Rightarrow$ and let $\Gamma = \Gamma', Tt$ (t as above). Then by IH, 64.3 and Φ_1-closure, $\mathcal{M}([\neg A]) \in X$. Hence $X \vdash^*_M \Gamma', TA \Rightarrow \Delta$ is an X-axiom. □

CHAPTER 13

THE VARIETY OF NON-REDUCTIVE APPROACHES

§ 65. An inconsistency
§ 66. On a truth theory of Friedman and Sheard
§ 67. Fitch's models
§ 68. Introducing semi-inductive definitions
§ 69. Semi-inductive models for reflective truth

We address the problem of strengthening the internal logic of T with truth principles suggested by modal logic, according to the natural idea of reading T as a necessity operator. Of course, according to Montague (1963) and subsequent work up to recent classification results by Friedman and Sheard (1987), there are severe limitations to this move.

Be as it may, we know from Ch. XII that there is a consistent notion T of self-referential truth that "believes" in its own consistency and is closed under logical consequence. Can we coherently maintain that, in addition to its consistency, T believes in its own closure under infinitary logical consequence? Formally: can we add to $T^+\text{-}cons := T(T\neg a \rightarrow \neg Ta)$ the schemata:

$$T^+\text{-}imp := T(T(A \rightarrow B) \rightarrow (TA \rightarrow TB));$$

$$T^+\text{-}univ := T(\forall x TA(x) \rightarrow T(\forall x A))?$$

Of course, we already know by 63.9.1 that $T^+\text{-}imp$ and $T^+\text{-}univ$ force us to give up the provability interpretation of § 63. But, more surprisingly, as we shall see in § 65, there is no chance to find a model for theories extended with $T^+\text{-}imp$ and $T^+\text{-}univ$, insofar as we stick to the core system VF_p of Ch. XII with internal number-theoretic induction. The point is that the number-theoretic induction of VF_p is strong enough to exclude certain simple ω-inconsistencies, which are, on the other hand, called into existence by the above mentioned axioms.

Nevertheless, as § 66 shows, the internal logic of truth can be greatly expanded, up to include $T^+\text{-}univ$, $T^+\text{-}imp$, $T^+\text{-}cons$ and even internal completeness $T^+\text{-}comp := T(TA \vee T\neg A)$, once we give up the soundness schema $T\text{-}out := TA \rightarrow A$ together with $T\text{-}rep := Tx \rightarrow TTx$. The resulting theory, which appears in Friedman and Sheard (1987), turns out to be ω-inconsistent, and yet consistent by means of a semi-inductive procedure, to be developed in § 68. In § 67 we refine the supervaluation models, according to an idea of Fitch (1963), in order to obtain a system which extends VF_p and is compatible with T^+-*imp*, T^+-*cons*, but not with T^+-*univ*. The final sections outline a radically new solution of the problem. We introduce models for VF^-, which use *non-monotone semantic valuation schemata* and are based on the notion of *truth revision*.

In particular, we first define in § 68 Herzberger's semi-inductive definitions with their basic properties (stabilization and periodicity theorems). The new machinery produces strengthened internal logics (§ 69), which were first investigated (as far as we know) by Aczel and Turner. These systems differ from those of Ch. XII and § 67, and they are subsumed under the Friedman-Sheard theory. However, we are quite far from having a satisfactory, complete axiomatic treatment of the two approaches.

§ 65. An inconsistency

We observe that a theorem of Mc Gee (1985), concerning the ω-inconsistency of certain theories of truth over Peano arithmetic, extends to VF_p. As immediate consequence, we obtain the announced inconsistency result. Below we stick to the notations of the previous chapter. Henceforth, we presuppose an axiomatization of predicate calculus with *modus ponens as only inference rule* (*cf. p.*352).

65.1. DEFINITION

(i) IL (=internal logic) is the least set $\mathfrak{S}$ of $\mathcal{L}$-formulas such that: $\mathfrak{S}$ is closed under modus ponens, and the rule

T-intro: infer TA from A;

$\mathfrak{S}$ contains the axioms of classical predicate calculus, the axioms of OP^- and the following schemata:

T-elem: $A \leftrightarrow TA$, where A has the form $t = s$, $\neg t = s$, Nt, $\neg Nt$;

T-imp: $T(A \rightarrow B) \rightarrow (TA \rightarrow TB)$;

T-univ: $\forall x TA \rightarrow T\forall x A$;

T-cons: $T\neg x \rightarrow \neg Tx$;

NIND: $A(\bar{0}) \wedge \forall x(A(x) \rightarrow A(x+1)) \rightarrow \forall x(Nx \rightarrow A(x))$.

The acronyms *T-elem*, *T-imp*, *T-cons* correspond to the schemata of § 59 and § 63; *T-elem* is the adequacy schema for elementary atoms. We write $\mathrm{IL} \vdash A$, instead of $A \in \mathrm{IL}$.

(ii) VF^+ is the elementary theory (in the language $\mathcal{L}$ of § 59), which includes: classical first-order logic with equality; the axioms of the theory of operations OP^-; the T-axioms:

T^+-*elem*:	$T(A \leftrightarrow TA)$, where $A = (t = s)$, $\neg t = s$, Nt, $\neg Nt$;
T^+-*imp*:	$T(T(A \rightarrow B) \rightarrow (TA \rightarrow TB))$;
T-*rep*:	$Tx \rightarrow TTx$;
T^+-*cons*:	$T(T\neg x \rightarrow \neg Tx)$;
T^+-*univ*:	$T(\forall x TA \rightarrow T\forall x A)$;
I-NIND:	$T[A(\overline{0}) \wedge \forall x(A(x) \rightarrow A(x{+}1)) \rightarrow \forall x(Nx \rightarrow A(x))]$;

and the T-rules;

T^+-*log*:	TA , provided A is a logical axiom, an axiom of OP^-;
T-*elim*:	from TA infer A (A arbitrary sentence).

VF^+ is closed under a strengthened *LOG*-rule (see 59.2 (i)), namely:

65.2. LEMMA. *If* IL $\vdash A$, *then* $VF^+ \vdash TA$ (*and hence* $VF^+ \vdash A$).

PROOF (Induction on IL-definition). If A is an axiom of IL, by inspection $VF^+ \vdash TA$. If $A = TB$ has been obtained by T-*intro* from IL $\vdash B$, $VF^+ \vdash TB$ by induction hypothesis and hence $VF^+ \vdash TA$ by means of T-*rep*. The remaining cases only require the obvious provability of T-*imp*, T-*univ* (via T-*elim*) and the application of induction hypothesis. □

The following remark is essential to the inconsistency argument:

65.3. LEMMA

$$\text{IL} \vdash \forall x(Nx \rightarrow TA(x)) \leftrightarrow \forall x T(Nx \rightarrow A(x)) \leftrightarrow T(\forall x(Nx \rightarrow A(x))).$$

PROOF. ⇒: apply T-*intro*, T-*imp*, T-*elem* in the form $\neg Nx \rightarrow T[\neg Nx]$, and T-*univ*.
⇐: apply T-*imp*, T-*intro* and $Nx \rightarrow T[Nx]$. □

65.4. THEOREM. IL *is* ω-*inconsistent, i.e. we can find a formula* $H(x)$ *such that*:

(i) IL $\vdash \neg\forall x(Nx \rightarrow H(x))$;

(ii) IL $\vdash H(\overline{n})$, *for each* $n \in \omega$.

PROOF. We lift Mc Gee's trick to the present framework. By recursion on N (Ch. I, 3.2), we can find a term G such that,

$$\mathrm{IL} \vdash G\bar{0}y = y \text{ and } G(k{+}1)y = [T(Gky)]. \tag{1}$$

By fixed point theorem, there is a solution M to the equation

$$\mathrm{IL} \vdash M = [\neg\forall x(Nx \rightarrow T(GxM))]. \tag{2}$$

As a consequence of *T-cons*,

$$\mathrm{IL} \vdash T(M) \rightarrow \neg T(\forall x(Nx \rightarrow T(GxM))). \tag{3}$$

By lemma 65.3 and (1),

$$\mathrm{IL} \vdash T(M) \rightarrow \neg\forall x(Nx \rightarrow T(G(x{+}1)M)), \tag{4}$$

whence, with the trivial $\forall x(Nx \rightarrow T(GxM)) \rightarrow \forall x(Nx \rightarrow T(G(x{+}1)M))$ (which requires the axiom $\forall x(Nx \rightarrow N(x{+}1)))$,

$$\mathrm{IL} \vdash T(M) \rightarrow \neg\forall x(Nx \rightarrow T(GxM)). \tag{5}$$

But (1) again yields, using the axiom $N(\bar{0})$,

$$\mathrm{IL} \vdash \forall x(Nx \rightarrow T(GxM)) \rightarrow T(M), \text{ whence } \mathrm{IL} \vdash \neg\forall x(Nx \rightarrow T(GxM)),$$

i.e. $\mathrm{IL} \vdash T(M)$ (by T-rule), or equivalently, $\mathrm{IL} \vdash T(G\bar{0}M)$. By iterating *T-intro*, we inductively get $\mathrm{IL} \vdash T(G\bar{k}M)$, for each $k \in \omega$. Hence the choice $H(x) = T(GxM)$ proves the theorem. □

We are now in the position to refute VF^+:

65.5. COROLLARY. VF^+ *is inconsistent.*

PROOF. If we choose M and G as in Theorem 65.4, from Lemma 65.2 we have $\mathrm{VF}^+ \vdash T(M)$ and hence $\mathrm{VF}^+ \vdash \neg\forall x(Nx \rightarrow T(GxM))$ by *T-elim*. By definition of G,

$$\mathrm{VF}^+ \vdash T(G\bar{0}M).$$

But $\mathrm{VF}^+ \vdash T(GxM) \rightarrow TT(GxM)$ (by *T-rep*) and hence, by choice of G:

$$\mathrm{VF}^+ \vdash \forall x(T(GxM) \rightarrow T(G(x{+}1)M)).$$

We conclude by internal N-induction that

$$\mathrm{VF}^+ \vdash \forall x(Nx \rightarrow T(GxM))): \text{contradiction ! } \square$$

Of course, it may be asked whether IL is nevertheless consistent: a positive answer is offered in the next section. We do not know whether the argument above works with number-theoretic induction for classes.

§ 66. On a truth theory of Friedman and Sheard

If we omit *T-rep* from VF^+, we can produce a theory of truth $FSL \supseteq IL$, which is consistent but ω-inconsistent. The theory was (essentially) introduced as an extension of Peano arithmetic by Friedman and Sheard (1987). The construction below shows that Mc Gee's construction is sharp: in the theorem 65.4 "ω-inconsistent" cannot be simply replaced by "inconsistent". Moreover, as we shall see, there are negative applications to the case of the logic of truth revision.

66.1. DEFINITION

(i) Fix any structure $\mathcal{M}\models OP^-$; M is the domain of $\mathcal{M}$. If $X \subseteq M$, A is a sentence of $\mathcal{L}(\mathcal{M})$, $X\models A$ means "A holds true in the structure $\langle \mathcal{M}, X\rangle$", i.e. A is true whenever $\mathcal{L}_{op}(\mathcal{M})$ receives its usual interpretation in $\mathcal{M}$ and $\langle \mathcal{M}, X\rangle \models Tt$ iff $\mathcal{M}(t) \in X$ ($\mathcal{M}(t)$ being the value of t in $\mathcal{M}$). As usual, we systematically rely on abuse of notation: a, b, c, d, e... stand both for elements of $\mathcal{M}$ and the corresponding constants. Once $\mathcal{M}$ is fixed and $b, c \in M$, we shall write bc, $\neg b$, $\forall b$, $b \rightarrow c$, instead of the proper $\mathcal{M}(Ap(b,c))$, $\mathcal{M}(NEGb)$, $\mathcal{M}(ALLb)$, $\mathcal{M}(IMPLYbc)$ (in the given order; for $IMPLY$, see § 59).

(ii) If $X \subseteq M$, $SENT(\mathcal{M}) := \{\mathcal{M}([A]) : A\ \mathcal{L}(\mathcal{M})\text{-sentence}\}$;

(iii) $J(X) := \{\mathcal{M}([A]) : \mathcal{M}([A]) \in SENT(\mathcal{M}) \text{ and } X\models A\}$.

(iv) We set:

$$X(0) = \emptyset \text{ and } X(k+1) = J(X(k));$$

$$Th_\infty(\mathcal{M}) := \{\mathcal{M}([A]) : \mathcal{M}([A]) \in SENT(\mathcal{M}) \text{ and, for some } k,$$
$$X(m)\models A, \text{ for every } m > k\}.$$

If $\mathcal{M}([A]) \in Th_\infty(\mathcal{M})$, $ko(A) :=$ least number m such that $X(n)\models A$, for every $n > m$.

(v) $Th(\mathcal{M}) = \{\mathcal{M}([A]) : A \text{ is an } \mathcal{L}_{op}(\mathcal{M})\text{-sentence such that } \mathcal{M}\models A\}$.

(vi) We say that $Th_\infty(\mathcal{M})$ is closed under an inference rule $\mathcal{R}$, if $Th_\infty(\mathcal{M})$ contains the $\mathcal{M}$-term encoding the conclusion of $\mathcal{R}$, whenever $Th_\infty(\mathcal{M})$ contains the $\mathcal{M}$-terms encoding the premises of $\mathcal{R}$ (e.g. "$Th_\infty(\mathcal{M})$ is closed under modus ponens" means $b \in Th_\infty(\mathcal{M})$, whenever $a \in Th_\infty(\mathcal{M})$ and $(a\rightarrow b) \in Th_\infty(\mathcal{M})$).

Now observe that we cannot have both $\mathcal{M}([A]) \in Th_\infty(\mathcal{M})$ and $\mathcal{M}([\neg A]) \in Th_\infty(\mathcal{M})$ (otherwise $X(m)\models A$ and $X(m)\models\neg A$, for every $m > max(ko(A), ko(\neg A))$: contradiction). A similar argument shows that $Th_\infty(\mathcal{M})$ is closed under modus ponens. If A is any valid formula of $\mathcal{L}(\mathcal{M})$, or $\mathcal{M} \models A$ and A does not contain T, $X(m) \models A$ for arbitrary m. Hence we have:

66.2. LEMMA

(i) $Th_\infty(\mathcal{M})$ *is consistent and closed under (first-order) logical consequence.*

(ii) $Th(\mathcal{M}) \subseteq Th_\infty(\mathcal{M})$.

66.3. DEFINITION

(i) *T-Comp*: $TA \vee T\neg A$; *T-exist*: $T\exists xA \to \exists xTA$ (A $\mathcal{L}$-formula);

(ii) Let us consider the following finitary inferences:

(*T-intro*): from A infer TA; (*T-elim*): from TA infer A;

($\neg$*T-intro*): from $\neg A$ infer $\neg TA$; ($\neg$*T-elim*): from $\neg TA$ infer $\neg A$.

(iii) FSL(="the Friedman-Sheard logic") is the least set of $\mathcal{L}$-formulas such that:
FSL contains the axioms of classical predicate logic with $=$, the non-logical axioms of OP^-, *T-elem*, *T-cons*, *T-comp*, *T-imp*, *T-univ*, *T-exist* and *I*-NIND (see 65.1 (ii));
FSL is closed under modus ponens and the rules *T-intro*, *T-elim*.

We write FSL $\vdash A$ for $A \in$ FSL.

66.3.1. REMARK. FSL is closed under $\neg$*T-intro* and $\neg$*T-elim*.
(Verification: assume FSL $\vdash \neg A$; then FSL $\vdash T\neg A$ (by *T-intro*). But FSL $\vdash T\neg A \to \neg TA$ (by the axiom *T-cons* of IL), whence FSL $\vdash \neg TA$.
Closure under $\neg$*T-elim*: assume FSL $\vdash \neg TA$: then FSL $\vdash T\neg A$ (by *T-comp*) and FSL $\vdash \neg A$ by *T-elim*.)

66.4. THEOREM (*Soundness*). *Assume that* $\mathcal{M}$ *is an* ω*-standard model of* OP. *Then*

$$\mathrm{FSL} \vdash A \textit{ implies } \mathcal{M}([A]) \in Th_\infty(\mathcal{M}).$$

PROOF. We argue by induction on the definition of FSL. Thus it suffices to check: (i) every axiom of FSL belongs to $Th_\infty(\mathcal{M})$; (ii) $Th_\infty(\mathcal{M})$ is closed under logical consequence, *T-intro* and *T-elim*.

(i). By lemma 66.2, all valid sentences are in $Th_\infty(\mathcal{M})$. Observe that if $A \in \mathcal{L}_{op}(\mathcal{M})$, then for every m, $X(m{+}1) \models A$ iff $X(m) \models A$ iff $X(m{+}1) \models TA$, and hence $\mathcal{M}([A \leftrightarrow TA]) \in Th_\infty(\mathcal{M})$, which takes care of *T-elem* .

Ad *T-imp*. If $X(m{+}2) \models T(A \to B)$ and $X(m{+}2) \models TA$, then $X(m{+}1) \models A$

and $X(m+1) \models A \to B$; hence $X(m+1) \models B$, i.e. $X(m+2) \models TB$. Therefore we obtain, for each $n \geq 2$, $X(n) \models T(A \to B) \to (TA \to TB)$.

Ad *T-univ*. We assume $X(m+2) \models \forall x TA(x)$, i.e. for each $c \in M$, $X(m+1) \models A(c)$, which implies $\mathcal{M}([\forall x A]) \in X(m+2)$, i.e. $X(m+2) \models T\forall x A$. Thus for each $n \geq 2$, $X(n) \models \forall x TA \to T\forall x A$.

Ad *T-cons*. If $X(m+1) \models T\neg A$ holds, then not $X(m) \models A$, which implies $\mathcal{M}([A]) \notin X(m+1)$, i.e. $X(m+1) \models \neg TA$. Therefore for each $n > 0$, we obtain $X(n) \models T\neg A \to \neg TA$.

I-NIND: note that, since $\mathcal{M}$ is ω-standard, we have, for every m,

$$X(m) \models A(\bar{0}) \wedge \forall x(A(x) \to A(x+1)) \to \forall x(Nx \to A(x)).$$

Ad *T-comp*. We have, for each m, either $X(m) \models A$ or $X(m) \models \neg A$, i.e. $\mathcal{M}([A]) \in X(m+1)$ or $\mathcal{M}([\neg A]) \in X(m+1)$. Then it follows, for each $n > 0$, $X(n) \models TA \vee T\neg A$, i.e. $\mathcal{M}([TA \vee T\neg A]) \in Th_\infty(\mathcal{M})$. The case of *T-exist* is trivial as well.

(ii). By lemma 66.2, it suffices to check closure under *T-intro* and *T-elim*. Assume $\mathcal{M}([A]) \in Th_\infty(\mathcal{M})$. Pick any $m > ko(A)+1$: then $X(m-1) \models A$ and hence $X(m) \models TA$; so $\mathcal{M}([TA]) \in Th_\infty(\mathcal{M})$. Thus $Th_\infty(\mathcal{M})$ is closed under *T-intro*. Assume $\mathcal{M}([TA]) \in Th_\infty(\mathcal{M})$ and let $m > ko(TA)$. Were $X(m) \models \neg A$, then $\mathcal{M}([A]) \notin X(m+1)$ by definition of J, i.e. $X(m+1) \models \neg TA$. But $m+1 > ko(TA)$: hence $X(m+1) \models TA$, absurd.
It follows that $\mathcal{M}([A]) \in X(m)$, for every $m > ko(TA)$, whence

$$\mathcal{M}([A]) \in Th_\infty(\mathcal{M}).$$

Thus $Th_\infty(\mathcal{M})$ is closed under *T-elim*. □

66.5. COROLLARY. FSL *is consistent, but ω-inconsistent.*

PROOF. FSL is consistent by 66.4 and 66.2 (i); but FSL contains IL and hence it is ω-inconsistent by 65.4. □

66.5.1. REMARK. (i) It is immediate to see that FSL proves the consistency of arithmetical analysis (see 40.3.1). By a result of Halbach (1994), FSL is proof-theoretically equivalent to ramified analysis up to any level $< \omega$.

(ii) The addition of $TTx \leftrightarrow Tx$ would clearly imply in FSL the unrestricted Tarski schema $TA \leftrightarrow A$ (A arbitrary). However, FSL becomes inconsistent, even if we add either $TTx \to Tx$ or $Tx \to TTx$; in addition the Löb axiom $T(Tx \to x) \to Tx$ is inconsistent with FSL, since it implies $Tx \to TTx$.

§ 67. Fitch's models

Despite the negative result of § 65, we can adapt the supervaluation semantics of § 60 to validate the assumption that T recognizes its consistency and its closure under logical consequences, plus the adequacy schema with respect to T-free atomic sentences.

The result is implicit in Fitch (1963), and so we speak of Fitch's models. We prove that VF^- can be consistently enlarged by accepting, as new schemata, $T(T\neg A \to \neg TA)$, $T(T(A \to B) \to (TA \to TB))$, and closure under a stronger T-introduction inference. In particular, we can infer TA, whenever the formula obtained from A by replacing T with the necessity operator $\Box$, is derivable in a quantified extension of deontic logic (see Bull-Segerberg 1983).

Technically, we first refine the basic relation $\Vdash_0$ of § 60.

67.1. DEFINITION. Fix $\mathcal{M}\models \mathrm{OP}^-$; recall that an e-atom has the form $t = s$, Nt or the negation thereof.

(i) $Diag(\mathcal{M}) := \{\mathcal{M}([A]) : \mathcal{M}\models A,\ A \text{ e-atom of } \mathcal{L}(\mathcal{M})\}$;

(ii) Recall that $SENT(\mathcal{M}) = \{\mathcal{M}([A]) : A\ \mathcal{L}(\mathcal{M})\text{-sentence}\}$.

$X \subseteq SENT(\mathcal{M})$ is $\mathcal{M}$-*normal* iff the following closure conditions are met:

1) X contains $Diag(\mathcal{M})$;
2) $X \in CONS(\mathcal{M})$;
3) X is closed under logical consequence:
3.1) if A is a logical truth in the language $\mathcal{L}(\mathcal{M})$, then $\mathcal{M}([A]) \in X$;
3.2) X is closed under modus ponens: $a \in X$ and $(a \to b) \in X$ imply $b \in X$.

(iii) $\mathcal{M}\text{-}NOR := \{X \subseteq M : X \text{ is } \mathcal{M}\text{-normal}\}$;
$\mathcal{M}\text{-}NOR(Y) := \{X \subseteq M : Y \subseteq X \text{ and } X \in \mathcal{M}\text{-}NOR\}$.

If $Y \in \mathcal{M}\text{-}NOR(X)$, we say that Y is a normal extension of X.

(iv) Once $\mathcal{M}$ is fixed, recall that

$$J(X) := \{\mathcal{M}([A]) : A \text{ is an } \mathcal{L}(\mathcal{M})\text{-sentence with } \langle \mathcal{M}, X\rangle \models A\}.$$

We observe a number of useful facts on $\mathcal{M}$-normal sets. The verification is an easy exercise.

67.2. LEMMA

(i) *Existence of $\mathcal{M}$-normal sets: if $X \subseteq M$, then $J(X)$ is $\mathcal{M}$-normal.*

(ii) *If $\mathcal{S} \subseteq \mathcal{M}\text{-}NOR$ and $\mathcal{S}$ is non-empty, then $\cap\mathcal{S}$ is $\mathcal{M}$-normal.*

(iii) *If $\mathcal{S} \subseteq \mathcal{M}$-$NOR$ and $\mathcal{S}$ is $\subseteq$-directed (i.e. for every X, $Y \in \mathcal{S}$, there is some $Z \in \mathcal{S}$ with $X \subseteq Z$ and $Y \subseteq Z$), then $\cup\mathcal{S}$ is $\mathcal{M}$-normal.*

67.3. DEFINITION. Let $X \subseteq M$, A $\mathcal{L}(\mathcal{M})$-sentence:

(i) $X \Vdash_2 A$ iff for every $Y \in \mathcal{M}$-$NOR(X)$, then $Y \models A$;

(ii) $\Phi_2(X) := \{\mathcal{M}([A]) : X \Vdash_2 A\}$; $FIX_2(\mathcal{M}) := \{X \subseteq M : X = \Phi_2(X)\}$.

Clearly Φ_2 is monotone and hence $FIX_2(\mathcal{M})$ is non-empty.

(iii) We set $N_\infty(\mathcal{M}) :=$ the $\subseteq$-least element of $FIX_2(\mathcal{M})$. As usual $X \subseteq M$ is Φ_2-dense iff $X \subseteq \Phi_2(X)$.

67.4. LEMMA

(i) *If $\mathcal{S}$ is a $\subseteq$-directed family of $\mathcal{M}$-normal Φ_2-dense subsets of M, then $\cup\mathcal{S}$ is $\mathcal{M}$-normal and Φ_2-dense;*

(ii) *if X is $\mathcal{M}$-normal, $\Phi_2(X)$ is $\mathcal{M}$-normal;*

(iii) *$\Phi_2(\emptyset)$ is $\mathcal{M}$-normal.*

PROOF. (i): immediate from 67.2 (iii) and Φ_2-monotonicity.
(ii): consider the family $\mathcal{S}(X) = \{J(Y) : Y \in \mathcal{M}\text{-}NOR(X)\}$. Then $\mathcal{S}(X)$ is a non-empty family of $\mathcal{M}$-normal subsets by 67.2 (i). But $\Phi_2(X) = \cap\mathcal{S}(X)$ and the conclusion follows by 67.2 (ii).
(iii): $\Phi_2(\emptyset) = \cap\mathcal{S}(\emptyset)$ and $\mathcal{S}(\emptyset)$ is a non-empty family of $\mathcal{M}$-normal subsets (apply 67.2 (i)-(ii)). □

67.5. LEMMA. *Let $X \in \mathcal{M}$-NOR; then*

(i) *$X \Vdash_2 A$ implies $X \models A$ (A arbitrary sentence of $\mathcal{L}(\mathcal{M})$);*

(ii) *$X \models T(A \to B) \to (TA \to TB)$;*

(iii) *if A is a sentence of $\mathcal{L}_{op}(\mathcal{M})$ of the form $t = s$, $\neg t = s$, Nt, $\neg Nt$, $X \models A \leftrightarrow TA$;*

(iv) *if X is Φ_2-dense, $X \models TA \to A$;*

(v) *if $X \in FIX_2(\mathcal{M})$, $X \models \forall x TA \to T\forall x A$.*

PROOF. Assume that X is $\mathcal{M}$-normal. (i)-(ii) are trivial by definition and assumption, while (iv) is a consequence of (i) with Φ_2-density.
(iii). Let $A = \neg(a = b)$. If $X \models \neg(a = b)$, then $\mathcal{M} \models \neg(a = b)$ and hence

$$\mathcal{M}([\neg(a = b)]) \in Diag(\mathcal{M}) \subseteq X$$

by normality of X, i.e. $X \models T\neg(a = b)$. Conversely, if $X \models T\neg(a = b)$ and

$X \models (a = b)$, $\mathcal{M}([a = b]) \in Diag(\mathcal{M}) \subseteq X$, against the consistency of X. The other atomic cases are similar.
(v). Assume $X \models \forall x TA$; by Φ_2-density, $X \Vdash_2 A(a)$, for each $a \in M$, whence $X \Vdash_2 \forall x A$, which finally implies, by Φ_2-closure, $X \models T\forall x A$. □

67.6. LEMMA. *$N_\infty(\mathcal{M})$ is $\mathcal{M}$-normal. Hence $FIX_2(\mathcal{M}) \cap \mathcal{M}$-$NOR$ is non-empty.*

PROOF. Observe that $N_\infty(\mathcal{M}) = \cup\{N(\alpha): \alpha \in ON \text{ and } \alpha > 0\}$, where $N(\emptyset) = \emptyset$, $N(\beta+1) = \Phi_2(N(\beta))$ and $N(\lambda) = \cup\{N(\delta): \delta < \lambda\}$, λ limit. $N_\infty(\mathcal{M})$ is $\mathcal{M}$-normal by 67.4. □

67.7. DEFINITION

(i) FT (= "Fitch's theory") is the theory which contains classical predicate logic with =, the non-logical axioms of OP, *T-out*, *T^+-elem*, *T^+-cons*, *T^+-imp*, *T-rep*, *T-univ*, *T^+-log* (cf. 65.1 and 59.1) and the schema *T-negT*: $T(\neg TA) \rightarrow T\neg A$

(ii) FT^- is FT minus number-theoretic induction.

67.7.1. REMARK. Clearly $\mathrm{VF}^- \subseteq \mathrm{FT}^-$ and $\mathrm{VF}_p \subseteq \mathrm{FT}$; thus the results of § 59 can be lifted to the new systems.

67.8. THEOREM. *If $\mathcal{M} \models \mathrm{OP}^-$, $X \in FIX_2(\mathcal{M})$ and X is $\mathcal{M}$-normal, then X is a model of* FT^-. *If $\mathcal{M}$ is ω-standard, $X \models \mathrm{FT}$ and X falsifies the schema $T(\forall x TA \rightarrow T\forall x A)$.*

PROOF. We only check *T-negT*; the remaining principles are consequences of 67.5 or use arguments already applied in § 60. Assume $X \models \neg T\neg B$, then $\mathcal{M}([\neg B] \notin X$. Hence $X' = X \cup \{\mathcal{M}([B])\}$ is consistent, by closure of X under logical consequence.
Let X'' be the closure of X' under logical consequence: clearly X'' is a normal extension of X such that $X'' \models TB$. Hence we cannot have $X \Vdash_2 \neg TB$; therefore $X \models \neg T\neg TB$. Finally, if $\mathcal{M}$ is ω-standard, observe:

$$X \Vdash_2 A(\overline{0}) \wedge \forall x(A(x) \rightarrow A(x+1)) \rightarrow \forall x(Nx \rightarrow A), \text{ for arbitrary } A.$$

The final claim is immediate from 65.5. □

67.9. DEFINITION. IFT(= Fitch's internal theory) is the least set of $\mathcal{L}$-formulas satisfying the conditions:

(i) IFT is closed under modus ponens, and the rules *T-intro*, *T-elim*, *¬T-intro*, *¬T-elim*;

(ii) IFT contains logical axioms for classical predicate calculus, the axioms of OP^-, the schema of N-induction and T-*elem*, T-*imp*, T-*cons*.

As usual, we write $\mathrm{IFT} \vdash A$ instead of $A \in \mathrm{IFT}$. $\mathrm{IFT}^- \vdash A$ means that the induction schema has not been applied in deriving A.

67.10. PROPOSITION. FT^- *is closed under the following rule*: if $\mathrm{IFT}^- \vdash A$, *then* $\mathrm{FT}^- \vdash TA$ (*the same holds for* FT *and* IFT).

PROOF (cf. lemma 65.2). We use T-*negT*, if A follows by $\neg T$-*elim*, while T-*out*, T^+-*cons*, T^+-*imp*, T-*rep* apply in the case of $\neg T$-*intro*. □

67.11. PROPOSITION. FT^- *proves the schemata*:

(i) $T(TA \to TB) \to T(A \to B)$;

(ii) $T\neg\forall xTA \to T\neg\forall xA$;

(iii) $(\neg TA \to T\neg TA) \to TA \vee T\neg A$ (T-S5*comp*).

PROOF. (i) T-*cons* implies:

$$\mathrm{IFT}^- \vdash (TA \to TB) \to \neg T(A \wedge \neg B).$$

Hence $\mathrm{FT}^- \vdash T(TA \to TB) \to T\neg T(A \wedge \neg B))$ by 67.10 and T-*imp*. Then T-*negT* implies:

$$\mathrm{FT}^- \vdash T(TA \to TB) \to T(\neg(A \wedge \neg B)),$$

whence the required conclusion with $T(\neg(A \wedge \neg B)) \to T(A \to B)$.

(ii) $\mathrm{IFT}^- \vdash T\forall xA \to \forall xTA$, whence $\mathrm{FT}^- \vdash T(\neg\forall xTA \to \neg T\forall xA)$ by 67.10. Thus we conclude that $\mathrm{FT}^- \vdash T\neg\forall xTA \to T\neg T\forall xA$, which immediately yields with T-*negT* the provability of (ii).
(iii): trivial with T-*negT*. □

67.12. REMARK

(i) Let $T^+\text{-}negT := T(T\neg TA \to T\neg A)$. Then $\mathrm{FT}^- + T^+\text{-}negT$ is inconsistent (note that 67.11 (i) and T-*out* immediately yield $\neg TA \to \neg A$, whence $TA \leftrightarrow A$, for arbitrary A). A similar argument shows that $\mathrm{FT}^- + T(TA \to TTA)$ is inconsistent.

(ii) The following schemata are not provable in FT^-:

$T\text{-}Rcomp := T(TA \to A) \to TA \vee T\neg A$;

$T\text{-S4}comp := T(TA \to TTA) \to TA \vee T\neg A$.

Verification: consider a term S such that $S = [TS]$, provably in OP^-. Then trivially $N_\infty(\mathcal{M}) \models T(TS \to S)$, but $N_\infty(\mathcal{M}) \models \neg TS \wedge \neg T\neg S$ (inductive argument, using the fact that each $N_\infty(\alpha)$ is $\mathcal{M}$-normal, for $\alpha > 0$). The proof for T-S4*comp* is similar.

§68. Introducing semi-inductive definitions

There is a natural extension of the theory of inductive definitions, which involves arbitrary *non-monotone operators*. Below we state and prove the basic results of semi-inductive definitions, which concern *the existence of stabilization ordinals and the periodicity phenomenon*. These results can be found in Herzberger (1982), Burgess (1986), Visser (1989).
We keep assuming that M is an infinite set (for more general results, see Visser, cit.).

68.1. DEFINITION. Let $\Delta: \mathcal{P}(M) \to \mathcal{P}(M)$ be an arbitrary operator on the subsets of M. We call the pair $\langle \Delta, X\rangle$ *Δ-process based on X*.

(i) If $\{G(\beta): \beta < \lambda\}$ is a family of subsets of M indexed by an ordinal λ,

$$liminf\{G(\beta): \beta < \lambda\} = \{a \in M: \exists\gamma < \lambda. \forall\beta(\gamma \leq \beta < \lambda \to a \in G(\beta))\}.$$

(ii) *Δ-iteration based on $X \subseteq M$* (by transfinite recursion):

$$It(\Delta, X, 0) = X;$$

$$It(\Delta, X, \alpha{+}1) = \Delta(It(\Delta, X, \alpha));$$

$$It(\Delta, X, \lambda) = liminf\{It(\Delta, X, \beta): \beta < \lambda\}, \text{ if } \lambda \text{ is a limit.}$$

(iii) $In(\Delta, X) := \{a \in M: \exists\gamma\forall\beta(\gamma \leq \beta \to a \in It(\Delta, X, \beta))\}$;

$In(\Delta, X)$ is the set of those $a \in M$ that are *stably inside* the family of α-iterations (modulo Δ and X);

$$Out(\Delta, X) := \{a \in M: \exists\gamma\forall\beta(\gamma \leq \beta \to a \notin It(\Delta, X, \beta))\};$$

$Out(\Delta, X)$ is the set of those $a \in M$ that are *stably outside* the α-iterations (mod Δ and X).

(iv) $Stab(\Delta, X) := In(\Delta, X) \cup Out(\Delta, X)$ (= set of *stable elements mod* Δ, X);

$Unstab(\Delta, X) := M - Stab(\Delta, X) =$ set of *unstable elements* (*mod* Δ, X).

For instance, if we consider the complement operator $\Delta(P) = M - P$, we have:

$$It(\Delta, X, 2n) = X,\ It(\Delta, X, 2n{+}1) = M - X \text{ and } It(\Delta, X, \omega) = \emptyset;$$

hence $Stab(\Delta, X) = \emptyset$.

If Δ is monotone, $In(\Delta, X)$ is the least fixed point of Δ, and *liminf* and $\cup$ coincide.

68.2. CONVENTION: once Δ is fixed, we set

$$X(\alpha) := It(\Delta, X, \alpha); \quad In(X, \alpha) := liminf\{X(\beta) : \beta < \alpha\};$$

$$Out(X, \alpha) := \{a \in M : \forall \beta(\alpha \leq \beta \rightarrow a \notin X(\beta))\}.$$

We keep using ON for the class of ordinals (in set-theoretic sense). Clearly, if α is a limit ordinal, $X(\alpha) = In(X, \alpha) = It(\Delta, X, \alpha)$. We shall verify that $In(X) = In(X, \alpha)$ and $Out(X) = Out(X, \alpha)$, for some $\alpha < \aleph(M) =$ the least cardinal $> card(M)$.

68.3. DEFINITION

(i) A limit ordinal δ *covers* $\langle\Delta, X\rangle$ iff $In(X) \subseteq X(\gamma)$ and $Out(X) \cap X(\gamma) = \emptyset$, for every $\gamma \geq \delta$.

(ii) A limit ordinal δ *stabilizes* $\langle\Delta, X\rangle$ iff δ covers $\langle\Delta, X\rangle$, $In(X) = X(\delta)$ and $Out(X) = Out(X, \delta)$.

Clearly, if δ covers $\langle\Delta, X\rangle$ and $\delta \leq \gamma$, $X(\gamma) \subseteq In(X) \cup Unstab(X)$; moreover, $Out(X) \subseteq Out(X, \delta)$.

68.4. LEMMA (*Covering*). *Let $\langle\Delta, X\rangle$ be a Δ-process on $X \subseteq M$: then for every $\alpha < \aleph(M)$, there exists a limit ordinal $\delta > \alpha$ with $\delta < \aleph(M)$, which covers $\langle\Delta, X\rangle$.*

PROOF. If $a \in In(X)$ $(a \in Out(X))$, we set:

$$Height(a) := \text{the least } \alpha \text{ such that } a \in X(\beta)(a \notin X(\beta)), \text{ for all } \beta \geq \alpha.$$

Then $card(\{Height(a) : a \in Stab(X)\}) < \aleph(M)$ and hence we can choose the minimum limit $\delta < \aleph(M)$ such that

$$\delta > \alpha \text{ and } \delta > sup\{Height(a) : a \in Stab(X)\}.$$

By choice, δ covers $\langle\Delta, X\rangle$. □

68.5. THEOREM (*Stabilization*). *Let $\langle\Delta, X\rangle$ be a process on M. Then for every $\alpha < \aleph(M)$, there exists an ordinal δ with $\alpha < \delta < \aleph(M)$, which stabilizes $\langle\Delta, X\rangle$.*

PROOF. We show that it is possible to filter out all the unstable elements which possibly enter $X(\delta)$ (δ being an ordinal given by covering). To this aim, we choose a limit $\lambda < \aleph(M)$ and an enumeration $\{a(\beta) : \beta < \lambda\}$ of $Unstab(X)$, where each element occurs infinitely often (for every $a = a(\xi)$ with $\xi < \lambda$, there always exists $\nu < \lambda$ with $a(\xi) = a(\nu)$ and $\xi < \nu$).
We recursively define a strictly increasing ordinal sequence $\{f(\xi) : \xi < \lambda\}$ of length λ, whose terms are $< \aleph(M)$:

$$f(0) = min\{\gamma : \alpha < \gamma \text{ and } \gamma \text{ covers } \langle\Delta, X\rangle\};$$

$f(\mu+1) =$ the least $\gamma > f(\mu)$ such that $a(\mu) \in X(\gamma)$ $(a(\mu) \notin X(\gamma))$ if $a(\mu) \notin X(f(\mu))$ (if $a(\mu) \in X(f(\mu))$, respectively);

$f(\xi) = least\ \gamma > sup\{f(\beta) : \beta < \xi\}$, if ξ is a limit.

The sequence is well-defined by covering lemma and the choice of the enumeration of unstable elements; moreover, by definition, if $\beta < \xi < \lambda$, $f(\beta) < f(\xi)$. Hence $\delta = sup\{f(\xi) : \xi < \lambda\}$ is a limit $< \aleph(M)$ and it trivially covers $\langle \Delta, X\rangle$. It is enough to check that for every $a \in X(\delta)$, $a \notin Unstab(X)$. By contradiction, assume $a \in Unstab(X)$ and $a \in \bigcap\{X(\beta) : \sigma \leq \beta < \delta\}$, for some $\sigma < \delta$. Since f is increasing, there is some $\xi < \lambda$ such that $\sigma \leq f(\xi)$ and hence:

$$\forall\beta(f(\xi) \leq \beta < \delta \rightarrow a \in X(\beta)). \tag{1}$$

Using the enumeration of unstable elements with infinitely many occurrences of each term, we must have $a = a(\eta)$ for some η with $\xi < \eta < \lambda$, whence $\sigma \leq f(\xi) < f(\eta) < \delta$. But (1) implies $a \in X(f(\eta))$ and by construction of f, $a \notin X(f(\eta+1))$, against $a \in \bigcap\{X(\beta) : \sigma \leq \beta < \delta\}$. □

68.6. DEFINITION. If $\langle \Delta, X\rangle$ is a process on M,

$\sigma(\Delta, X) :=$ *the least stabilizing ordinal or the closure ordinal of* $\langle \Delta, X\rangle$.

68.7. LEMMA

(i) $(X(\alpha))(\beta) = X(\alpha+\beta)$;

(ii) If $X(\alpha) = X(\beta)$, then $X(\alpha+\gamma) = X(\beta+\gamma)$.

PROOF. (i): induction on β. (ii): immediate by (i). □

68.8. THEOREM (*Periodicity*). *Let σ be the closure ordinal of $\langle \Delta, X\rangle$ on M. Then there exists exactly one ordinal $\psi = \psi(\Delta, X) < \aleph(M)$, the so-called "period", such that:*

(i) *$X(\sigma) = X(\sigma+\psi\delta)$, for every ordinal δ;*

(ii) *if $\sigma < \alpha$, there is an ordinal $\nu < \psi$ with $X(\alpha) = X(\sigma+\nu)$.*

PROOF. Let ξ be the least ordinal $> \sigma$, which stabilizes $\langle \Delta, X\rangle$ and let $\psi = \xi - \sigma$. We check (i) by induction on δ. Let $\delta = \beta+1$: then, since $X(\sigma) = X(\xi) = In(X)$, we have by 68.7 (i) and IH:

$$\begin{aligned} X(\sigma+\psi\delta) &= X(\sigma+\psi\beta+\psi) = (X(\sigma+\psi\beta))(\psi) = \\ &= (X(\sigma))(\psi) = X(\sigma+\psi) = X(\sigma). \end{aligned}$$

Let $\delta = \lambda$ be a limit and assume that for all $\nu < \lambda$, $X(\sigma) = X(\sigma+\psi\nu)$. $X(\sigma) \subseteq X(\sigma+\psi\lambda)$ holds, since σ stabilizes $\langle \Delta, X\rangle$.

Conversely, if $a \in X(\sigma+\psi\lambda)$, there is some $\xi < \sigma+\psi\lambda$, such that $a \in X(\beta)$ for every β satisfying $\xi \leq \beta < \sigma+\psi\lambda$. Since $\xi \leq \sigma+\psi\eta$ for some $\eta < \lambda$, $a \in X(\sigma+\psi\eta)$ and by IH $a \in X(\sigma)$. As to (ii), if $\sigma < \alpha$, we apply ordinal division and we can find $\nu < \psi$ and ξ such that $\alpha = \sigma+\psi\xi+\nu$. Hence, with (i) above and 68.7 (ii), we get $X(\alpha) = X(\sigma+\psi\xi+\nu) = X(\sigma+\nu)$. □

68.9. DEFINITION. Set $Im(\Delta, X) := \{X(\alpha) : \alpha \in ON\} =$ *image* of the process $\langle\Delta, X\rangle$; we define:

$$Conf(X) := \{Y \in Im(\Delta, X) : \forall\beta\exists\gamma(\beta \leq \gamma \wedge Y = X(\gamma))\};$$

$$Cycle(X) := \{X(\alpha) : \sigma \leq \alpha < \sigma+\psi\} \ (\psi \text{ period});$$

$$Init(X) := \{Y \in Im(\Delta, X) : \exists\gamma\forall\beta(\gamma \leq \beta \rightarrow Y \neq X(\beta))\}.$$

68.10. LEMMA. *Let $\langle\Delta, X\rangle$ be any process on M.*

(i) *If $X(\alpha) \in Init(X)$ and $\beta < \alpha$, then $X(\beta) \in Init(X)$;*
(ii) *If $X(\alpha) \in Conf(X)$ and $\alpha < \beta$, then $X(\beta) \in Conf(X)$.*

PROOF. (i) Let η be such that for every δ, if $\eta \leq \delta$, then $X(\delta) \neq X(\alpha)$. Were $\beta < \alpha$ and $X(\beta) \notin Init(X)$, there would exist $\xi \geq \eta$ with $X(\beta) = X(\xi)$. But $\beta < \alpha$ implies $\alpha = \beta+\nu$, for some ν; hence by 68.7 (ii), $X(\alpha) = X(\xi+\nu)$, which is absurd by assumption on α and since $\xi+\nu > \eta$.
(ii): similar argument. □

68.11. THEOREM (*Decomposition*). *Let $\langle\Delta, X\rangle$ be a process with closure ordinal σ and period ψ. Then:*

(i) $Conf(X) = \{X(\alpha) : \sigma \leq \alpha < \sigma+\psi\} = Cycle(X)$.
(ii) $Init(X) \subseteq \{X(\alpha) : \alpha < \sigma\}$.
(iii) $Im(\Delta, X) = \{X(\alpha) : \alpha < \sigma+\psi\}$.

PROOF (i): by stabilization theorem, we have $X(\sigma) \in Conf(X)$, and hence $Cycle(X) \subseteq Conf(X)$ by 68.10 (ii). Conversely, if $Y \in Conf(X)$, there is $\alpha > \sigma+\psi$ such that $Y = X(\alpha)$. By periodicity, $Y = X(\alpha) = X(\sigma+\eta)$, for some $\eta < \psi$. Hence $Y \in Cycle(X)$.
(ii): if $X(\alpha) \in Init(X)$, $\alpha \neq \sigma$; were $\sigma < \alpha$, for some $\nu < \psi$, $X(\alpha) = X(\sigma+\nu)$ (by periodicity) and $X(\alpha) \in Cycle(X)$; (i) implies a contradiction.
(iii): from (i) and (ii). □

To sum up, the behaviour of $Im(\Delta, X)$ is already determined below the ordinal $\sigma+\psi$; $Im(\Delta, X)$ splits into: 1) an initial piece below σ; 2) a cycle consisting exactly of the cofinal sets of $Conf(X)$, $In(X)$ being among them. Clearly, if Δ is monotone and X is Δ-dense (i.e. $X \subseteq \Delta(X)$), the period is zero and the cycle is empty. As to the elements of M, which are unstable with respect to $\langle\Delta, X\rangle$, they can be characterized as follows:

68.12. THEOREM (*Characterization*)

(i) $\cap Cycle(X) = In(X) = X(\sigma)$;

(ii) $\cup Cycle(X) = In(X) \cup Unstab(X)$;

(iii) $Out(X) = M - \cup Cycle(X)$ *and* $Unstab(X) = \cup Cycle(X) - X(\sigma)$.

PROOF. (i) $In(X) = X(\sigma) \subseteq \cap Cycle(X)$ is an immediate consequence of stabilization theorem and 68.11 (i). The periodicity theorem guarantees that if $a \notin X(\sigma)$, there is β such that $\sigma \leq \beta < \sigma + \psi$ with $a \notin X(\beta)$; hence $a \notin \cap Cycle(X)$ (for arbitrary $a \in M$) and $\cap Cycle(X) \subseteq X(\sigma)$.
(ii). From left to right, the inclusion is immediate by the properties of σ.
If $a \notin Unstab(X)$ and $a \notin In(X)$, then $a \in Out(X)$, whence there is some $\gamma > \sigma$ such that $a \notin X(\beta)$ for all $\beta \geq \gamma$. By periodicity, $X(\gamma) = X(\sigma+\eta)$, for some $\eta < \psi$ and hence $a \notin X(\sigma+\eta)$, i.e. $a \notin \cup Cycle(X)$. Conversely, let $a \notin \cup Cycle(X)$, then $a \notin In(X)$; were $a \in Unstab(X)$, then $a \in X(\sigma+\eta)$, for some $\eta < \psi$ (ψ period, apply 68.8). Hence $a \in \cup Cycle(X)$: contradiction!
(ii) immediately yields (iii). □

§69. Semi-inductive models for reflective truth.

The application of semi-inductive definitions to the semantics of self-referential systems is due independently to Herzberger (1982) and Gupta (1982). Applications to the modelling of axiomatic systems for truth and property theory can be found in Turner (1987) and Friedman-Sheard (1987). Turner (1990) observes that the internal logic of truth, which is sound with respect to semi-inductive interpretations, is rather rich. However, in view of the inconsistency theorem of §65, the logic of stable truth cannot in general contain the schema T^+-*univ*, as claimed by Turner (1990) (1990a). As far as we know, there is at present no completeness result, which fully characterizes the logic of truth revision (possibly involving some form of infinitary logic).

The aim of this section is quite modest: we apply the new tools of §68 to make clear that there are a few principles, which separate the logics of truth (sound for the supervaluation models and the provability interpretation of Ch. XII), from the logics of truth based on semi-inductive models.

69.1. DEFINITION

(i) As in §60, if $\mathcal{M} \models OP^-$, M is the domain of $\mathcal{M}$, $X \subseteq M$, A is a sentence of $\mathcal{L}(\mathcal{M})$, $X \models A$ means "A holds true in the structure $\langle \mathcal{M}, X \rangle$", i.e. A is true, whenever $\mathcal{L}_{op}(\mathcal{M})$ receives its usual interpretation in $\mathcal{M}$ and

$\langle \mathcal{M}, X\rangle \models Tt$ iff $\mathcal{M}(t) \in X$ ($\mathcal{M}(t)$ being the value of t in $\mathcal{M}$).

(ii) If $\mathcal{M}$ is fixed and $X \subseteq M$,

$$J(X) := \{\mathcal{M}([A]) : A \text{ sentence of } \mathcal{L}(\mathcal{M}) \text{ such that } X \models A\}.$$

69.1.1. CONVENTION. We restrict our investigation to processes of the form $\langle J, X\rangle$, where $J: \mathcal{P}(M) \to \mathcal{P}(M)$ is defined as above, $\mathcal{M} \models \mathrm{OP}^-$, $X \subseteq M$. The notions of *stable*, *unstable*, *stabilization ordinal*, etc. are referred to the process $\langle J, X\rangle$.

Typically, $In(X)$ represents the set of those sentences of $\mathcal{L}(\mathcal{M})$, which are stably true, insofar as we choose X as initial value for the truth predicate T. For simplicity, we identify the elements of $In(X)$ with the corresponding sentences; the elements of $In(X)$ are simply called X-stably true sentences.

We shall proceed to check that $In(X)$ is a model of VF^- (see § 59), plus suitable additional T-schemata.

69.2. LEMMA

(i) *Soundness of stable truth: for every $X \subseteq M$ and every $\mathcal{L}(\mathcal{M})$-sentence, $In(X) \models TA \to A$.*

(ii) *Consistency: for every $X \subseteq M$, $\alpha > 0$, $X(\alpha)$ is consistent.*

(iii) *$In(X) \models T\neg A$ implies $A \in Out(X)$.*

(iv) *$In(X) \models T\neg TA \to T\neg$A.*

(v) *$Out(X) = \{\mathcal{M}([\neg A]) : In(X) \models T\neg A\}$.*

(vi) *$Stab(X) = \{\mathcal{M}([A]) : A\ \mathcal{L}(\mathcal{M})$-sentence with $In(X) \models TA \vee T\neg A\}$.*

PROOF. (i) If σ is the closure ordinal of $\langle J, X\rangle$ on $M, In(X) = X(\sigma)$; so $X(\sigma) \models TA$ implies that A is X-stably true, whence $\mathcal{M}([A]) \in X(\sigma+1)$, i.e. $X(\sigma) \models A$.

(ii): immediate by induction on α.

(iii) If $X(\sigma) = In(X) \models T\neg A$, then $\mathcal{M}([\neg A]) \in X(\delta)$, for every $\delta \geq \sigma$. Hence by (ii) $\mathcal{M}([A]) \notin X(\delta)$, for every $\delta \geq \sigma$, i.e. by definition $\mathcal{M}([A]) \in Out(X)$.

(iv) By assumption, there exists some δ such that for every $\beta \geq \delta$, $\mathcal{M}([\neg TA]) \in X(\beta)$. We inductively verify that $\mathcal{M}([\neg A]) \in X(\beta)$ for all $\beta \geq \delta+1$. Let $\beta = \gamma+1 \geq \delta+1$. By assumption $\mathcal{M}([\neg TA]) \in X(\beta+1)$, whence by definition of the process, $X(\beta) \models \neg TA$ and $X(\gamma) \models \neg A$, which implies $\mathcal{M}([\neg A]) \in X(\beta)$. If β is a limit $> \delta+1$, we have by IH $\mathcal{M}([\neg A]) \in X(\gamma)$ for every γ with $\delta+1 \leq \gamma < \beta$, i.e. $\mathcal{M}([\neg A]) \in X(\beta)$.

(v) The second set is contained in the first one by (iii) above. Let us inductively check that $\mathcal{M}([A]) \in Out(X)$ implies $\mathcal{M}([\neg TA]) \in X(\alpha)$ for every $\alpha \geq \sigma+1$. Indeed, if $\mathcal{M}([A]) \in Out(X)$, then, by definition of stabilization ordinal $\mathcal{M}([A]) \notin X(\alpha)$ for every $\alpha \geq \sigma$ (here $X(\sigma) = In(X)$) and hence $X(\alpha) \models \neg TA$, i.e. $\mathcal{M}([\neg TA]) \in X(\alpha+1)$. If $\alpha = \lambda$ is a limit $\geq \sigma+1$, by IH $\mathcal{M}([\neg TA]) \in X(\beta)$ for every β, $\sigma+1 \leq \beta < \lambda$; by definition of $X(\lambda)$, this is sufficient for $\mathcal{M}([\neg TA]) \in X(\lambda)$. In conclusion, $\mathcal{M}([\neg TA]) \in In(X)$ and $In(X) \models T\neg TA$, whence $In(X) \models T\neg A$ by (iv).

(vi): by (v) and definition of X-stable element (with respect to J). □

We now come to the problem of characterizing the set

$$\mathrm{ILST} = \{A : \mathcal{M}([A]) \in In(X), \text{ for every } \mathcal{M} \models \mathrm{OP}^- \text{ and every } X \subseteq M\}$$

(recall § 30 for comparison). ILST can be regarded as *the internal logic of stable truth* for the language $\mathcal{L}$. It is not known to the present author whether ILST is recursively axiomatizable. The best we can offer is to show that ILST has non-trivial closure properties, which are embodied in the system below.

69.3. DEFINITION

(i) LIS^- is the least set of $\mathcal{L}$-formulas with the following properties:

1) LIS^- is closed under modus ponens, and the rules *T-intro*, *T-elim*, *¬T-intro*, *¬T-elim*;

2) LIS^- contains the axioms for classical predicate logic for $\mathcal{L}$, OP^-, *T-elem*, *T-cons*, *T-imp*;

3) LIS^- contains Turner's schemata (see also 67.11):

T-Rcomp	$T(TA \to A) \to TA \vee T\neg A$;
T-S4comp	$T(TA \to TTA) \to TA \vee T\neg A$;
T-S5comp	$(\neg TA \to T\neg TA) \to TA \vee T\neg A$;

(ii) LIS is LIS^- plus the number-theoretic induction schema NIND.

$\mathrm{LIS} \vdash A := A \in \mathrm{LIS}$; $\quad \mathrm{LIS}^- \vdash A := A \in \mathrm{LIS}^-$.

69.3.1. REMARK. (i) $\mathrm{LIS} \subseteq \mathrm{FSL}$ (where FSL is the logic of 66.3): indeed, *T-Rcomp*, *T-S4comp*, *T-S5comp* are trivial consequences of *T-comp*. Turner's schemata imply that the characteristic principles of the familiar modal logics T, S4, S5 cannot be stably true.

(ii) $\mathrm{LIS}^- \vdash (TA \vee T\neg A) \leftrightarrow T(TA \to A)$. (Verification: apply *T-Rcomp* from right to left; the opposite direction is already derivable in Fitch's internal logic of 67.9).

69.4. THEOREM. *If* $\mathrm{LIS}^- \vdash A$, *then* $A \in \mathrm{ILST}$ (*i.e.* $In(X) \models TA$, *for every* $\mathcal{M} \models \mathrm{OP}^-$, $X \subseteq M$).

PROOF. Let $\mathcal{M} \models \mathrm{OP}^-$ and assume that $X \subseteq M$. As above, $X(\sigma) = In(X)$. We first check the following claim:

$$In(X) \text{ contains the axioms of LIS.} \tag{1}$$

If A is a logical truth of $\mathcal{L}$ or an axiom of OP^-, then $X(\alpha) \models A$ for all α, and hence $X(\sigma) \models TA$; also note that, if A is an atom or the negation of an e-atom, then for every α, $X(\alpha+1) \models A \leftrightarrow TA$. These remarks show that logical axioms and OP^--axioms, together with *T-elem*, are stably true.

T-cons. By induction on $\alpha > 2$, we check $(T\neg A \rightarrow \neg TA) \in X(\alpha)$. Indeed, $(T\neg A \rightarrow \neg TA) \in X(\delta+1)$, as $X(\delta) \models T\neg A \rightarrow \neg TA$ by 69.2 (ii). If λ is a limit, $(T\neg A \rightarrow \neg TA) \in X(\beta)$ by IH for $2 < \beta < \lambda$; so $(T\neg A \rightarrow \neg TA) \in X(\lambda)$ by definition.

T-imp. By induction on $\alpha > 1$, $(T(A \rightarrow B) \rightarrow (TA \rightarrow TB)) \in X(\alpha)$.
As to the case $\alpha = 2$, if $X(1) \models T(A \rightarrow B)$ and $X(1) \models TA$, it follows $X(0) \models (A \rightarrow B) \wedge A$, whence $X(0) \models B$ and $X(1) \models TB$. So

$$X(1) \models T(A \rightarrow B) \rightarrow (TA \rightarrow TB), \text{ i.e. } (T(A \rightarrow B) \rightarrow (TA \rightarrow TB)) \in X(2).$$

The limit case is trivial by IH. Let $\alpha = \delta+1$, with $\delta \geq 2$: it suffices to see that $A \rightarrow B \in X(\delta)$ and $A \in X(\delta)$ imply $B \in X(\delta)$. We can assume $\delta > 2$; if δ is a successor, the conclusion follows from the definition of $X(\delta)$. If δ is a limit, there exists $\gamma < \delta$ with $A \rightarrow B$, $A \in X(\beta)$, for all β with $\gamma \leq \beta < \delta$; by IH and definition of J, $B \in X(\beta)$, for all β with $\gamma \leq \beta < \delta$, whence $B \in X(\delta)$.

T-Rcomp. We check, by induction on $\beta > 1$

$$T((TA \rightarrow A) \rightarrow (TA \vee T\neg A)) \in X(\beta).$$

As usual, the limit case is trivial by IH. As to the successor, we consider two cases.
Case 1: $\beta = \gamma+2$. Then $X(\gamma) \models A$ or $X(\gamma) \models \neg A$ yield

$$X(\gamma+1) \models T(TA \rightarrow A) \rightarrow (TA \vee T\neg A),$$

whence the required conclusion for β.
Case 2: $\beta = \lambda+1$ with λ limit: assume $X(\lambda) \models T(TA \rightarrow A)$, i.e. there is some $\gamma < \lambda$ such that $TA \rightarrow A \in X(\beta')$ for all β' such that $\gamma \leq \beta' < \lambda$.
Subcase 2.1: $X(\delta) \models A$ for some δ such that $\gamma \leq \delta < \lambda$. Then we can check, by secondary induction on β', that $A \in X(\beta')$, for all β' with $\delta+1 \leq \beta' < \lambda$, and hence $X(\lambda) \models TA$: indeed, $A \in X(\delta+1)$ by assumption; if $\beta' = \xi+1$ and

$A \in X(\xi)$ by IH, $X(\xi) \models TA \rightarrow A$ and $X(\xi) \models TA$, whence $A \in X(\xi+1)$; the limit case is trivial by IH. Hence $X(\lambda) \models TA \vee T\neg A$.
Subcase 2.2: assume that, for all β' such that $\gamma+1 \leq \beta' < \lambda$, $(\neg A) \in X(\beta')$. By definition $X(\lambda) \models T\neg A$ and hence $X(\lambda) \models TA \vee T\neg A$.

T-S4*comp*: first observe that

$$\text{if } \lambda \text{ is a limit, } X(\lambda) \models T(TA \rightarrow TTA) \text{ implies } X(\lambda) \models T(A \rightarrow TA). \quad (*)$$

Verification of $(*)$. Let δ such that $(TA \rightarrow TTA) \in X(\beta)$, for every β satisfying $\delta \leq \beta < \lambda$. We check by induction:

$$(A \rightarrow TA) \in X(\beta), \text{ for every } \beta \text{ such that } \delta+1 \leq \beta < \lambda.$$

If β is a limit in the prescribed interval, the claim follows by definition of *liminf* and IH.
Let $\beta = \gamma+1$ and $\delta+1 \leq \beta < \lambda$. Then $(TA \rightarrow TTA) \in X(\gamma+2)$, whence $X(\gamma+1) \models TA \rightarrow TTA$, which in turn yields $X(\gamma) \models A \rightarrow TA$, i.e.

$$(A \rightarrow TA) \in X(\gamma+1).$$

Let us check that, if λ is a limit,

$$X(\lambda) \models T(A \rightarrow TA) \rightarrow . TA \vee T\neg A. \quad (**)$$

Let δ be such that, for every $\delta \leq \beta < \lambda$, $(A \rightarrow TA) \in X(\beta)$. If $X(\beta) \models A$ for every β with $\delta \leq \beta < \lambda$, we immediately have $X(\lambda) \models TA$ and hence the required conclusion. Assume $\delta \leq \gamma < \lambda$ and $X(\gamma) \models \neg A$. Then we can inductively see that if $\gamma \leq \beta < \lambda$, $X(\beta) \models \neg A$. (The only non-trivial case is the successor case $\beta = \xi+1$: by IH $X(\xi) \models \neg A$; were $X(\xi+1) \models A$, we should get, with $X(\xi+1) \models A \rightarrow TA$, that $X(\xi+1) \models TA$, i.e. $X(\xi) \models A$: absurd!). Hence, $\forall\beta(\gamma+1 \leq \beta < \lambda \rightarrow (\neg A) \in X(\beta))$, i.e. $X(\lambda) \models T\neg A$ and finally $X(\lambda) \models TA \vee T\neg A$. $(*)$-$(**)$, together with the fact that T-S4*comp* holds in every $X(\alpha+1)$, imply that T-S4*comp* is stably true.

T-S5*comp*: as in the previous cases, the schema is true in $X(\alpha+1)$.
We verify that it holds in $X(\lambda)$, for λ limit. Assume $X(\lambda) \models \neg TA \rightarrow T\neg TA$ and $X(\lambda) \models \neg TA$. Then there is some $\delta < \lambda$ such that $(\neg TA) \in X(\beta)$ for every β with $\delta \leq \beta < \lambda$, which also implies $\forall\beta(\delta \leq \beta < \lambda \rightarrow X(\beta) \models \neg TA)$.
If there were some β such that $\delta \leq \beta < \lambda$ and $X(\beta) \models A$, then also $A \in X(\beta+1)$, i.e. $X(\beta+1) \models TA$: absurd! Hence

$$\forall\beta(\delta \leq \beta < \lambda \rightarrow X(\beta) \models \neg A),$$

which implies $\forall\beta(\delta+1 \leq \beta < \lambda \rightarrow (\neg A) \in X(\beta))$, i.e. $X(\lambda) \models T\neg A$.
This completes the verification of the opening claim (1).

We now proceed to verify:

$In(X)$ is closed under logical consequence, *T-intro*, *T-elim*, $\neg$*T-intro*, $\neg$*T-elim* and the M-rule: infer $\forall x A$ from $A(a)$, for each $a \in M$. (2)

Modus ponens: apply 69.2 (i) to the fact that

$$In(X) \models T(T(A \rightarrow B) \rightarrow .TA \rightarrow TB).$$

M-rule. It suffices to check inductively that $X(\sigma) \models \forall x TA$ implies $X(\beta) \models T\forall x A$, for all $\beta \geq \sigma+1$, whence it will follow by choice of σ, $X(\sigma) \models T\forall x A$. The case of β limit is trivial by IH. Let $\beta = \gamma+1 \geq \sigma+1$. By assumption, for all $a \in M$, $A(a) \in X(\sigma)$; as σ is the stabilization ordinal, $A(a) \in X(\gamma+1)$ for every $a \in M$, whence $X(\gamma) \models \forall x A$, i.e. $\forall x A \in X(\gamma+1)$.

T-intro: it suffices to check $X(\sigma) \models TA \rightarrow TTA$. Assume $X(\sigma) \models TA$; we inductively prove that $TA \in X(\beta)$, for every $\beta \geq \sigma+1$. The limit case is trivial by IH. If $\beta = \delta+1$, observe that the assumption $A \in X(\sigma)$ implies $A \in X(\delta)$, if $\sigma < \delta$, i.e. $X(\delta) \models TA$, whence $TA \in X(\delta+1)$.

T-elim: by 69.2 (i) $X(\sigma) \models TTa \rightarrow Ta$.

$\neg$*T-intro*: apply *T-intro*, closure of $X(\sigma)$ under modus ponens and the fact that *T-cons* $\in X(\sigma)$.

$\neg$*T-elim*: if $X(\sigma) \models T\neg TB$, we inductively verify $\forall \beta \geq \alpha+1.(\neg B) \in X(\beta)$, where α is an ordinal such that $\forall \beta \geq \alpha.(\neg TB) \in X(\beta)$. The limit case is trivial by IH. If $\beta = \delta+1 \geq \alpha+1$, $(\neg TB) \in X(\delta+2)$, hence $X(\delta+1) \models \neg TB$, i.e. $X(\delta) \models \neg B$, and finally $(\neg B) \in X(\delta+1)$.
As a consequence of claims (1)-(2), we have that $\text{LIS}^- \subseteq \text{ILST}$. □

69.5. COROLLARY. *Let* $\mathcal{M}$ *be an* ω*-standard model of* OP^- *and let* $X \subseteq M$. *Then* $In(X) \models \text{LIS}$, *but* $In(X) \models \neg T(\forall x TA \rightarrow T\forall x A)$, *for some sentence* A.

PROOF. The first part is obvious by 69.4 and assumption. As to the negative claim, observe that if $In(X) \models T(\forall x TA \rightarrow T\forall x A)$, then $In(X)$ would be ω-inconsistent by 65.4. But $In(X)$ is ω-consistent, by closure under M-rule and the fact that the extension of the predicate N is a class !
□

We now turn to the question of characterizing the external logic of stable truth, namely the set

$$\text{ELST} := \{A : A\ \mathcal{L}\text{-sentence such that } In(X) \models A, \text{ for every } \mathcal{M} \models \text{OP}^- \text{ and } X \subseteq M\}.$$

A natural (finitary) approximation to ELST is suggested by the previous results.

69.6. DEFINITION

(i) Let T^+-R*comp* (T^+-S4*comp*, T^+-S5*comp*) be the schema which is obtained by prefixing T to the schema T-R*comp* (T-S4*comp*, T-S5*comp* respectively) of 69.3. For instance, T^+-R*comp* has the form:

$$T[T(TA \rightarrow A) \rightarrow (TA \vee T\neg A)].$$

(ii) LES^- is the theory which contains classical predicate logic with $=$, the non-logical axioms of OP^-, T-*out*, T^+-*elem*, T^+-*cons*, T^+-*imp*, T-*rep*, T-*univ*, T-*negT*, T^+-*log*, T^+-R*comp*, T^+-S4*comp*, T^+-S5*comp* (see 65.1, 59.1, 69.3).

(iii) $\mathrm{LES} := \mathrm{LES}^- +$ the internal N-induction schema

$$T[(A(\overline{0}) \wedge \forall x(A(x) \rightarrow A(x+1))) \rightarrow \forall x(Nx \rightarrow A(x))].$$

Notice that LES extends the system FT of 67.7. By a straightforward argument, we obtain:

69.7. LEMMA. LES^- (LES) *is closed under the rule: if* $\mathrm{LIS}^- \vdash A$, *then* $\mathrm{LES}^- \vdash TA$ ($\mathrm{LES} \vdash TA$).

69.8. THEOREM. *If* $\mathcal{M} \models \mathrm{OP}^-$, $X \subseteq M$, *then* $In(X) \models \mathrm{LES}^-$. *In addition,* $In(X) \models \mathrm{LES}$, *if* $\mathcal{M}$ *is* ω*-standard.*

PROOF. All the relevant work has been done above: we simply apply 69.2 and 69.4. □

69.9. REMARK. FT^- is strictly contained in LES^- by 67.12. In particular, if S is a term such that $S = [TS]$, then $\mathrm{LES}^- \vdash TS \vee T\neg S$, while we know that $TS \vee T\neg S$ is unprovable in FT^-.

69.10. *Problems.* Is T^+-S5*comp* independent from FT^-? What is the proof-theoretic strength of FT^-, LES^- ? We conjecture that they are equivalent to VF^-.

CHAPTER 14

EPILOGUE: APPLICATIONS AND PERSPECTIVES

§ 70A. A logical theory of constructions: informal motivations
§ 70B. A logical theory of constructions: basic syntax
§ 71. Axioms for the computation relations
§ 72. Extending the logical theory of constructions with higher reflection
§ 73. Proof-theoretic reduction
§ 74. Perspectives: related work in Artificial Intelligence and Theoretical Linguistics
§ 75. Sense and denotation as algorithm and value: subsuming theories of reflective truth under abstract recursion theory

Confronting a theoretical piece of work with applications is always useful for a critical assessment. For this reason, we address the question of relating the systems of reflective truth that we have been investigating so far, with applications in Theoretical Computer Science (TCS), Artificial Intelligence (AI), Linguistics.

We are concerned only with *potential connections, and not with direct, well-established applications*, already available in the literature. We shall consider three examples: a) a logical theory of constructions, arising from TCS, and its modeling in the systems of chapters X-XI; b) some logics, motivated by knowledge representation and the semantics of natural languages; c) Moschovakis's intensional approach to the foundation of the theory of algorithms. We underline that our choice is largely a matter of taste and strongly bound to the limited competence of the writer. Thereby, the aim of the present chapter is rather that of putting the content of the book in a wider perspective and suggesting new problems; there will be no attempt of systematization, nor we try to supply complete details. The only relative exception is the first example, dealing with the logical theory of constructions, LTC_ω; but this is due to the fact that LTC_ω fits nicely with the material of chapters VIII-XI. As to the examples of part b), we hope that some applicative-minded reader will find the results of chapters XII-XIII, as well as those of chapter VI, of some interest. The final example, which discusses Moschovakis's lower predicate calculus with reflection, is to us highly suggestive: it should lead to reflections, embracing both the foundations of recursion theory and formal semantics.

§70A. A logical theory of constructions: informal motivations

In TCS new logical formalisms are currently investigated:

(i) as tools for representing, stating and establishing properties of programs (e.g. equivalence, termination and correctness);

(ii) as tools for program extraction;

(iii) as tools for reasoning about the specification of programs and their typing (foundations of type theories);

(iv) as abstract theories of computation over abstract data types.

This list is not exhaustive and the single aims (i)-(iv) are usually integrated, being a distinctive feature of the logic methodology its unifying power. In this respect, we may mention Martin-Löf's type theories (Martin-Löf 1984, B.Nordström et al. 1990), the ELF-approach (Harper-Honsell-Plotkin 1987), the theories of constructions (Coquand 1985), NUPRL (Constable et al. 1986), the logical theories of constructions of Aczel et al. (1991), Feferman's theories and its outcomes (Feferman 1979, Hayashi-Nakano 1988, Feferman 1990, 1991a, 1992, Talcott 1992), the proofs-as-programs approach, as developed by Schwichtenberg (1991).

We concentrate upon a single example, which appears close to the spirit of this work: the logical theory of constructions LTC, as it is outlined in Aczel-Carlisle-Mendler (1991). On the conceptual side, LTC-theories are motivated by "the idea that the notions of proposition and truth are, after all, the fundamental ones for logic and that the logical notions are the fundamental ones for a deductive system for mathematics. According to this idea, although the notion of type is also essential for mathematics and computer science, it is less fundamental conceptually" (see p. 5, cit.).

Technically, we can summarize the basic features of LTC in the following points:

1) LTC includes *the values of a functional programming language, as well as the propositions of a reflective logic*; in particular, in LTC there is a truth predicate, which expresses the fact that a proposition, as an object of our universe, is true; since the underlying logic is constructive, LTC-higher systems are endowed with predicates expressing the fact that certain objects are propositions (of a given level);

2) the functional language is untyped; but the semantics is operational, explicitly controlled by a "lazy evaluation" relation;

3) the basic equality relation should be decidable; however, in the strongest theory of Aczel et al., there is a mixed approach: recursive aspects

are handled as equalities, while discrimination and selection aspects are maintained at an operational level; we do not know whether the resulting conversion relation is decidable.

An interesting feature of LTC is that it points to possible refinements of the underlying combinatory logic of OP, which make sense of important distinctions for applications, apparently inaccessible within MF^- and its extensions. In fact, a limit of our systems concerns equality: we only deal with a single basic equality $=$, which is interpreted as equality in combinatory algebras; hence $=$ is generally undecidable.
Moreover, one would like to have a notion of "value" and hence a predicate of definedness, in order to explicitly control the main properties of programs. As we shall see, the system LTC_0 of the next section offers a viable alternative, by introducing a different semantics underlying the theory of programs. A final point of interest is that LTC-theories establish a sort of natural bridge between Martin-Löf's type theories and the predicative systems of reflective truth with variable levels of part D.

§70B. A logical theory of constructions: basic syntax

In this section we are going to introduce expressions with arities and their basic definitional equality $\equiv$, together with the notion of canonical realization (term models) for the resulting formalism.

70.1. (i) *Arities*: they are inductively generated by the following clauses:
OB (individuals), $BOOL$ (formulas) are basic arities; if α, β are arities, so is $(\alpha \rightarrow \beta)$.

By currying we also write $(\alpha_1 \ldots \alpha_n) \rightarrow \alpha$ for $(\alpha_1 \rightarrow (\alpha_2 \ldots (\alpha_n \rightarrow \alpha) \ldots)$. Every symbol is assigned an arity. If we understand $BOOL$ as the arity of formulas, $(OB \rightarrow BOOL)$, $(OB, OB) \rightarrow BOOL$ will obviously represent the arities of unary and binary predicates (in the given order); on the other hand $(OB, OB) \rightarrow OB$ is the arity assigned to binary function symbols. It is clear that the stock of basic arities can be conveniently expanded, insofar as we need new basic sorts of entities.

(ii) The formal language $\mathcal{L}_{TC}$ is given by specifying a list of primitive symbols, together with their corresponding arities.

Individual symbols (arity OB): a denumerable list of individual variables (x, y, z, u syntactical variables); the constants $\overline{0}$ and $\dot{\bot}$ (the object representing the absurd proposition $\bot$);

Propositional symbols (arity $BOOL$): $\bot$ (absurd proposition);

Logical symbols: $\rightarrow$ (implication of arity $(BOOL, BOOL) \rightarrow BOOL$); $\forall_0$ (universal individual quantification of arity $(OB \rightarrow BOOL) \rightarrow BOOL$); $\forall_1$ (universal function quantifier of arity $((OB \rightarrow OB) \rightarrow BOOL) \rightarrow BOOL$);

Predicate symbols:
arity $(OB, OB) \rightarrow BOOL$: $=$ (equality), LEV (lazy evaluation), NEV (full evaluation to numbers);
arity $(OB \rightarrow BOOL)$: T (truth); P_i (proposition of level i), for any $i > 0$.

Function symbols:
arity $(OB \rightarrow OB)$: a denumerable list of unary function variables f (f, g, h syntactical variables); S (successor); Inl, Inr (projections); $\dot{\forall}_0$; $\dot{\forall}_1$ (internal quantifications); $\dot{P}_i$, for each $i > 0$; $\dot{T}$;
arity $(OB, OB) \rightarrow OB$: $Pair$; $\dot{\rightarrow}$; $LE\dot{V}$, $NE\dot{V}$, $\dot{=}$;
arity $(OB \rightarrow OB) \rightarrow OB$: λ (abstraction);
arity $(OB, (OB, OB) \rightarrow OB) \rightarrow OB$: $Spread$;
arity $(OB, OB \rightarrow OB, OB \rightarrow OB) \rightarrow OB$: $Decide$;
arity $(OB, OB, OB \rightarrow OB) \rightarrow OB$: $Decidenat$;
arity $(OB, OB, OB \rightarrow OB) \rightarrow OB$: Ind (primitive recursion operator);
arity $(OB, (OB \rightarrow OB) \rightarrow OB) \rightarrow OB$: Pa (permuted application).

70.2. *Expressions of* $\mathcal{L}_{TC}$: they are inductively generated from the set of basic symbols by means of the inductive clauses for abstraction and application:

(i) if E is an expression of arity α and x is a variable of arity β (hence $\beta = OB$ or $\beta = OB \rightarrow OB$), then $(x)E$ is an expression of arity $(\beta \rightarrow \alpha)$;

(ii) if E is an expression of arity $\alpha \rightarrow \beta$ and E' is an expression of arity α, then $E(E')$ is an expression of arity β.

70.3. *Notations.* Expressions of arity $BOOL$ are identified with formulas and A, B, C play the role of metavariables for them; $\forall x A := \forall_0((x)A)$ and $\forall_1 f.A := \forall((f)A)$. Expressions of arity OB are the usual individual terms; we let $\lambda x.t$ stand for $\lambda((x)t)$. Multiple abstraction is reduced to iterated abstraction (currying) in the usual fashion; for instance $(xy)f(x,y)$ is an abbreviation for $(x)((y)f(x,y))$.

70.4. *Dotted symbols*: they are function symbols that allow to associate to each expression of arity $BOOL$ an expression of arity OB. In particular, we have:

70.4.1. FACT. To each formula A of $\mathcal{L}_{TC}$ we can effectively associate a term $\dot{A}$ such that $\dot{A}$ has exactly the same free variables of A.

The intuitive meaning of the basic function symbols can be clarified by anticipating that the following defining equations are valid in the standard denotational semantics:

70.5

(Pa):	$Pa(\lambda(f), h) = h(f)$;
$(Decidenat)$:	$Decidenat(\overline{0}, x, f) = x$ and $Decidenat(S(y), x, f) = f(y)$;
$(Spread)$:	$Spread(Pair(x_1, x_2), h) = h(x_1, x_2)$;
$(Decide)$:	$Decide(inl(x), f, g) = f(x)$; $Decide(inr(x), f, g) = g(x)$;
(Ind):	$Ind(x, y, h) = Decidenat(x, y, (x)h(x, Ind(x, y, h)))$.

It is understood that the terms involved have the appropriate arity. If we define

$$Ap(x, y) := Pa(x, (f)f(y)),$$

then (Pa) implies (β)-conversion. $Ap(\lambda(f), y) = f(y)$.

In the present proposal denotational equality is split into finer relations, which also take care of the operational level. More precisely, while the Ind-equation is integrated in a suitable definitional equality on expressions, the remaining equations are transformed into inductive clauses, which define an appropriate evaluation relation. The first step takes inspiration from Martin-Löf's theory of expressions.

70.6. DEFINITION

(i) If σ is an arity and E, E' are expression of arity σ, we inductively define the (ternary) relation $E \equiv E' : \sigma$, to be read as *E and E' are equal expressions* of arity σ. We write $E : \sigma$ as an abbreviation of $E \equiv E : \sigma$; $E : \sigma$ means that E is of arity σ.
$E \equiv E' : \sigma$ is the smallest relation, which meets the following conditions:

1. Reflexivity: $E \equiv E : \sigma$;
2. Symmetry: $E \equiv E' : \sigma$ implies $E' \equiv E : \sigma$;
3. Transitivity: $E \equiv E' : \sigma$ and $E' \equiv E'' : \sigma$ imply $E \equiv E'' : \sigma$;
4. α-conversion: if x and y are variables of arity σ and y does not occur in E, then $(x)E \equiv (y)E[x := y] : \sigma$;
5. β-conversion: if $E' : \sigma$, $x : \sigma$, $E : \delta$, then $((x)E)(E') \equiv E[x := E'] : \delta$;
6. η-conversion: if $E : \sigma \rightarrow \delta$, $x : \sigma$, then $(x)E(x) \equiv E : \sigma \rightarrow \delta$;
7. ξ-conversion: if $x : \sigma$ and $E \equiv E' : \delta$, then $(x)E \equiv (x)E' : \sigma \rightarrow \delta$;
8. application: if $E \equiv E' : \sigma \rightarrow \delta$ and $H \equiv H' : \sigma$, then $E(H) \equiv E'(H') : \delta$;
9. *definiendum* $\equiv$ *definiens*: $E \equiv E' : OB$, provided E, E' are the expressions involved in the (Ind)-equation of 70.5 above.

(ii) Let $\equiv_\sigma := \{\langle E, E'\rangle : E \equiv E' : \sigma\}$; $[E]_\sigma := \{E' : E \equiv_\sigma E'\}$;

$M_\sigma := \{[E]_\sigma : E \text{ closed expression}\}$.

If $\sigma := \delta \to \gamma$, $[E]_\delta \in M_\delta$, and $[F]_\sigma \in M_\sigma$, we put $[F]_\sigma([E]_\delta) = [F(E)]_\gamma$.

According to the last definition, each element of $M_{\delta \to \gamma}$ represents a unique function from M_δ into M_γ. Henceforth we use the families M_σ, $\equiv_\sigma$, where σ is an arity, in order to define a Tarskian semantics for formulas of LTC_0.

70.7. DEFINITION. (i) A *canonical realization* for $\mathcal{L}_{TC}$ is a triple $\mathcal{M} := \langle M_\delta, M_\sigma, \Phi\rangle$, such that:

1. $\delta := OB$, $\sigma := OB \to OB$;
2. Φ is an interpretation function satisfying the conditions:
   - 2.1. for each expression E of arity σ, $\Phi(E) = [E]_\sigma$;
   - 2.2. $\Phi(=) =$ the relation $\equiv_\delta$;
   - 2.3 for each $i \in \omega$, $\Phi(P_i)$ and $\Phi(T_i)$ are subsets of M_δ;
   - 2.4. $\Phi(NEV)$, $\Phi(LEV)$ are subsets of $M_\delta \times M_\delta$.

(ii) The relation $\mathcal{M} \models A$ (A sentence of LTC_0) is inductively defined according to the standard Tarskian clauses, once we stipulate that:

1. variables of arity OB range over M_δ, while variables of arity $OB \to OB$ range over M_σ;
2. $\mathcal{M} \models (t = s)$ iff $t \equiv s : \delta$; $\mathcal{M} \models Tt$ iff $[t]_\delta \in \Phi(T)$;

   $\mathcal{M} \models P_i t$ iff $[t]_\delta \in \Phi(P_i)$ ($i \in \omega$, $\delta = OB$, t, s closed terms, i.e. closed expressions of arity OB).

We conclude with a remark: if we omit the defining equations for *Ind*, the expressions with the family $\{\equiv_\sigma : \sigma \text{ arity}\}$ yield a (version of) typed λ-calculus. As usual, the definitional equality relations can be generated by the corresponding natural reduction relations, which are introduced by regarding the basic $\equiv$-clauses as contractions. Moreover, every expression reduces to a unique normal form (and actually every reduction sequence is finite; strong normalization). As a corollary, one has:

70.8. PROPOSITION. *The relations* $\equiv_\sigma$ *without the Ind-defining clause are decidable.*

It is not known to us whether the unrestricted relation with *Ind*-clause $E \equiv E' : OB$ is decidable. The problem is left open by Aczel, Carlisle, Mendler (1991); according to them, let $\Rightarrow^+$ be the reduction, generated by

(α)-, (β)-, (η)-contractions, extended with the (Ind)-contraction:

$$Ind(t,s,h) \Rightarrow^{+} Decidenat(t,s,(x)Ind(x,s,h)).$$

Then $\Rightarrow^{+}$ is still consistent, in the sense that a corresponding Church-Rosser property holds (if $E \Rightarrow^{+} E'$ and $E \Rightarrow^{+} E''$, then $E' \Rightarrow^{+} H$ and $E'' \Rightarrow^{+} H$, for some expression H). As a consequence:

70.9. The following *basic special equality axioms* become true in the corresponding term model:

$$F(f_1,\ldots,f_n) \equiv F(g_1,\ldots,g_n): OB \Rightarrow f_i(u_1,\ldots,u_m) \equiv g_i(u_1,\ldots,u_m): OB;$$

$$\neg F(c_1\ldots c_n) \equiv G(h_1\ldots h_n): OB;$$

here F, G are primitive function symbols $\neq Ind$, and f_i, g_i, u_j have appropriate arities.

§ 71. Axioms for the computation relations

The basic syntax contains a functional programming notation; so we have to explain how to compute programs and how to identify them. Of course, one could stick to a denotational semantics: the resulting interpretations would be essentially the (many sorted versions of) models of λ-calculus, which were introduced in Aczel (1980) under the name of lambda structures.

By contrast, following Aczel-Carlisle-Mendler (1991), we outline a semantics which lies between denotational and operational semantics. First of all, we specify *the space of values*. This is done by exploiting the idea of *lazy evaluation* and Martin-Löf's distinction between *canonical* and *non-canonical expressions*. Roughly speaking, a canonical expression is an expression, which directly manifests the data type it belongs to and can be immediately understood in terms of the givens we are dealing with (numbers, functions, lists, etc...); as such, it is a *static* object. On the contrary, non-canonical expressions involve control features, that have to be eliminated, in order to understand the direct meaning in terms of the givens. Of course, this is vague, but it will be made precise by specifying canonical forms and by inductively defining the appropriate evaluation relations.

71.1. DEFINITION. (i) Canonical symbols: $\overline{0}$, S, λ, $Pair$, Inl, Inr, $\dot{\rightarrow}$, $\dot{\bot}$, $\dot{\forall}_0$, $\dot{\forall}_1$, $\dot{=}$, $LE\dot{V}$, $NE\dot{V}$, $\dot{P}_i$ $(i > 0)$, $\dot{T}$;

(ii) a term t of arity OB is *canonical* (or is in canonical form) iff either t is $\overline{0}$, or else its outermost function symbol is canonical.

The computation process is formalized by the predicate LEV of lazy evaluation: $LEV(t, s)$ holds if t evaluates to s and s is in canonical form. LEV is inductively defined by appropriate inductive clauses that involve non-canonical symbols. But LEV does not suffice in general: for instance, $S(t)$ is canonical, but we cannot directly read off from $S(t)$ that $S(t)$ truly represents a natural number, unless we already know that t represents a natural number, i.e. either t is $\overline{0}$ or has the form $S(r)$. The second case may require further evaluation and so on. To sum up, we also need a primitive predicate $NEV(t, s)$ which holds exactly when t *fully evaluates to a numeral* and hence we properly have that t represents a natural number. Again, this leads to an inductive specification of NEV, which involves primitive recursion and allows to introduce a natural number predicate. Since the reduction relations associated to LEV and NEV are Church-Rosser, we shall be in the position to define the appropriate equivalence relations on programs and numbers.

71.2. DEFINITION (*Computation triples*). We say that (a, b, k) is *a computation triple* (in short $(a, b, k) \in$ COMPT) iff one of the following (mutually exclusive) cases holds:

(i) $a = \lambda(f)$, $b = h(f)$ and $k = (x)Pa(x, h)$;

(ii) $a = Pair(x_1, x_2)$, $b = h(x_1, x_2)$ and $k = (x)Spread(x, h)$;

(iii) $a = Inl(u)$, $b = f(u)$ and $k = (x)Decide(x, f, g)$;

(iv) $a = Inr(u)$, $b = g(u)$ and $k = (x)Decide(x, f, g)$;

(v) $a = \overline{0}$, $b = u$ and $k = (x)Decidenat(x, u, g)$;

(vi) $a = S(z)$, $b = g(z)$ and $k = (x)Decidenat(x, u, g)$.

{For a motivation of 71.2 simply recall the equations of 70.5}.

71.3. *The ground system* LTC_0 (*without reflection*)

LTC_0 is the theory in the given language $\mathcal{L}_{TC}$ above, which contains the following axioms:

71.3.1. two sorted intuitionistic logic with standard equality axioms (for objects); recall that the basic sorts correspond to variables of arity OB (x, y, $z\ldots$) and to arity of $OB{\rightarrow}OB$ (f, g, $h,\ldots$), respectively.

In particular, we have the axiom schema:

$\forall f A(f) \rightarrow A[f := g]$ (g term of arity $OB{\rightarrow}OB$, free for f in A).

We also postulate two special principles for $=$, which are related to the fact that the intended model is a term model (see 70.9):

SE1: $F(f_1,\ldots,f_n) = F(g_1,\ldots,g_n) \to f_i(u_1,\ldots,u_m) = g_i(u_1,\ldots,u_m)$;

SE2: $\neg F(c_1,\ldots,c_n) = G(h_1,\ldots,h_n)$;

here F, G are primitive canonical function symbols; f_i, g_i, u_j have appropriate arities (remember that $=$ only applies to expressions of arity OB).

71.3.2. Closure under explicit definitions for arities $OB \to OB$ and $(OB \to OB) \to OB$:

(α) $\forall w(((v)t)(w) = ((y)t[v := y])(w))$ (provided y does not occur in t);

$\forall f(((g)t)(f) = ((h)t[g := h])(f))$ (provided h does not occur in t);

(β) $\forall u(((x)t)(u) = t[x := u])$ (provided u is free for x in t);

$\forall f(((g)t)(f) = t[g := f])$ (provided f is free for g in t).

71.3.3. LEV-axioms:

LE1: $LEV(c, c)$, provided c is a canonical term;

$LEV(x, y) \to LEV(y, y)$; (if x is evaluated to y, y is canonical);

LE2: $\forall x \forall y(LEV(x, y) \wedge LEV(x, z) \to y = z)$;

LE3: $LEV(x, a) \to (LEV(k(x), z) \leftrightarrow LEV(b, z))$, for $(a, b, k) \in$ COMPT.

71.3.4. NEV-axioms:

NE1: $LEV(x, \overline{0}) \to NEV(x, \overline{0})$;

NE2: $LEV(x, S(y)) \wedge NEV(y, z) \to NEV(x, S(z))$.

NEV-Induction: if $A(x, y)$ is a formula,

$$\forall x(LEV(x, \overline{0}) \to A(x, \overline{0})) \wedge \forall x \forall y \forall z(LEV(x, S(y)) \wedge A(y, z) \to A(x, S(z))). \to$$
$$\to \forall u \forall v(NEV(u, v) \to A(u, v)).$$

(NEV-induction states that NEV is the least predicate closed under the inductive clauses formalized by NE1, NE2).

71.3.5. Peano axioms:

$$\forall x \forall y(\neg S(x) = \overline{0} \wedge (S(x) = S(y) \to x = y));$$

71.3.6. Primitive recursion:

$$Ind(x, y, f) = Decidenat(x, y, (x)h(x, Ind(x, y, h))).$$

71.4. DEFINITION

$$N(x) := NEV(x, x) \text{ (} x \text{ is a numerical value) ;}$$

$$Nat(x) := \exists y NEV(x,y) \text{ (x denotes a natural number)};$$

$$EqNat(x,y) := \exists z(NEV(x,z) \wedge NEV(y,z));$$

$$x\downarrow := \exists y\, LEV(x,y) \text{ ($= x$ has a value or x is defined)}.$$

71.4.1. REMARK. LTC_0 can easily interpret a logic of partial existence à la Scott and distinguish between quantifying over all the entities of the universe (just the usual $\forall$ and $\exists$) and quantification restricted to values :

$$\forall x^+(\ldots) := \forall x(x\downarrow \rightarrow \ldots) \text{ and } \exists x^+(\ldots) := \exists x(x\downarrow \wedge \ldots).$$

The intended models of LTC_0 are obtained as special canonical realizations.

71.5. DEFINITION. Let $\delta = OB$:

(i) $\mathbb{LEV}$ is the least relation $\subseteq M_\delta \times M_\delta$ which is closed under the following clauses:

1. if t is canonical, then $\langle [t]_\delta, [t]_\delta \rangle \in \mathbb{LEV}$;
2. if (t,s,r) is a computation triple (71.2), $\langle [p]_\delta, [t]_\delta \rangle \in \mathbb{LEV}$ and $\langle [s]_\delta, [q]_\delta \rangle \in \mathbb{LEV}$, then $\langle [r(p)]_\delta, [q]_\delta \rangle \in \mathbb{LEV}$.

(ii) $\mathbb{NEV}$ is the least relation $\subseteq M_\delta \times M_\delta$ which is closed under the following clauses:

$\langle [t]_\delta, [\overline{0}]_\delta \rangle \in \mathbb{LEV}$ implies $\langle [t]_\delta, [\overline{0}]_\delta \rangle \in \mathbb{NEV}$;

$\langle [t]_\delta, [S(r)]_\delta \rangle \in \mathbb{LEV}$ and $\langle [r]_\delta, [p]_\delta \rangle \in \mathbb{NEV}$ imply $\langle [t]_\delta, S[p]_\delta \rangle \in \mathbb{NEV}$.

(iii) The interpretation function Φ_0 satisfies:

$\Phi_0(T) = \Phi_0(P_i) = \emptyset$, for each $i \in \omega$;

$\Phi_0(LEV) = \mathbb{LEV}$ and $\Phi_0(NEV) = \mathbb{NEV}$.

71.6. THEOREM. *If* $\mathcal{M}_0 := \langle M_\delta, M_{\delta\rightarrow\delta}, \Phi_0 \rangle$, *then* $\mathcal{M}_0 \models \mathrm{LTC}_0$ (*i.e.* $\mathcal{M}_0$ *is a canonical model of* LTC_0-axioms).

As to the proof, the crux is to extend the Church-Rosser theorem (§4) to the definitional equality relation, in order to verify the special equality axioms SE1-SE2 (see Aczel-Carlisle-Mendler 1991). The equality axioms, (α), (β), the Peano axioms and primitive recursion are immediately verified, as Φ_0 is a canonical interpretation. By choice of $\mathbb{LEV}$ and $\mathbb{NEV}$, LEV- and NEV-axioms are valid in $\mathcal{M}_0$.

What can be said about the strength of LTC_0? An answer is given by a simple observation:

71.7. PROPOSITION

(i) PA, *the first-order system of Peano arithmetic, is interpretable into* LTC_0.

(ii) LTC_0 *is interpretable in* PA.

PROOF (hint). (i): the domain of the interpretation is the defined predicate Nat, which is closed under successor and contains zero; $EqNat$ interprets equality. The Ind-axioms imply that there are functions $+$ and $\cdot$ under which Nat is closed; NEV-induction implies Nat-induction.
(ii) One has to formalize the term model construction of §70 with the corresponding Church-Rosser theorem (see appendix to chapter I). This is possible since the inductive definitions of definitional equality $\equiv$, and the relations $\mathbb{LEV}$ and $\mathbb{NEV}$, being given by existential positive clauses, can be explicitly defined and arithmetized in PA. □

71.7.1. REMARK. Assume that $\mathrm{LTC}_0 \vdash \forall x(Nat(x) \to Nat(t(x)))$: then 71.7 implies that the number-theoretic function defined by the term $(x)t(x)$ is provably recursive in PA.

§72. Extending the logical theory of constructions with higher reflection

Up to now, the predicate symbols P_i and T have been left undetermined. We wish to add axioms interpreting $P_i t$ as "t is a proposition of level i" and Tt as "t is a true proposition".

72.1. DEFINITION. LTC_ω is the extension of LTC_0 with the following propositional and truth axioms (for each i with $1 \le i < \omega$):

$\mathrm{PT1}_i$ If $c = [A]$ and $A = (r = s)$, $LEV(r,s)$, $NEV(r,s)$, $\bot$,

$$LEV(t,c) \to P_i(t);$$

$$LEV(t,c) \to (A \leftrightarrow Tt);$$

$\mathrm{PT2}_i$ $(LEV(r, t \dot{\to} s) \wedge P_i(t) \wedge (T(t) \to P_i(s))) \to P_i(r);$

$$((LEV(r, t \dot{\to} s) \wedge P_i(t) \wedge (T(t) \to P_i(s))) \to (T(r) \leftrightarrow (T(t) \to T(s)));$$

$\mathrm{PT.3}_i$ $(LEV(r, \dot{\forall}(f)) \wedge \forall x P_i(f(x))) \to P_i(r);$

$$(LEV(r, \dot{\forall}(f)) \wedge \forall x P_i(f(x))) \to (\forall x T(f(x)) \leftrightarrow T(\dot{\forall}(f))).$$

$\mathrm{PT.4}_{i,j}$ if $1 \le j < i$,

$$LEV(r, \dot{P}_j(t)) \to (P_i(r) \wedge (T(r) \leftrightarrow P_j(t)));$$

$$P_j(t) \to P_i(t).$$

72.2. DEFINITION. Let $1 \leq i \leq k$.

(i) *Internal truth of level* i:

$$T_i(t) := T(t) \wedge P_i(t).$$

(ii) LTC_k *is the subsystem of* LTC_ω, *which only contains the predicate symbols* $P_1, \ldots, P_k$.

At first sight, one might conjecture that each fragment LTC_k is directly interpretable in the system STLR_k with reflective truth predicates up to k (see §49B): one would simply be tempted to identify the truth predicate T with T_k, the truth predicate of level k, and to define the notion of proposition of level i (for $1 \leq i \leq k$) with the classical $P_i t := T_i t \vee F_i t$. The idea, though roughly sound, is not viable, since P_i and T must be *well-behaved* with respect to LEV and NEV; but these predicates cannot be simply reduced to usual conversion equality, due to the special equality axioms. Of course, we can define new systems corresponding to TLR, IT_n^∞, RS_n of Chapters VIII-XI, which are based on LTC_0 instead of OP. Then the proof-theoretic reduction of Chapters X-XI can be adapted to the new systems without any difficulty, in order to show:

72.3. THEOREM. *For each* k, LTC_k *is proof-theoretically reducible to* TLR (*i.e. the formal consistency of* LTC_k *is provable in* $\mathrm{OP} + \mathrm{TI}(\alpha)$, *for some* $\alpha < \Gamma_0$ *and hence in* TLR; *see* Ch. XI)

The reader not interested in further dreary details, can directly skip to §74. In the rest of §72 and in §73 we illustrate a different route to the theorem, which consists of building up a model of LTC_k, directly in the available systems. Since the verification of 72.3 is rather lengthy, we split it in a few steps; in this section we restrict our attention to the interpretation of LTC_0.

First Step. We simulate the term model of LTC_0 within an untyped model of OP by direct use of the combinatory structure available. Abstractions $(x)t$, $(f)t$ are represented by λ-abstraction; the basic identity of LTC_k is sent into combinatory equality: thus the interpretation is certainly *unfaithful.* Nevertheless, the interpretations of LEV, NEV and the basic function symbols are chosen in such a way that the special equality axioms, together with LEV, NEV-axioms, become true.

72.4. DEFINITION

(i) We associate to each individual constant and each primitive function symbol G of LTC_k a corresponding combinator G^* (of OP).

$$\lambda^* := \lambda f.\langle \overline{8}, f\rangle;\ Pa^* := \lambda x \lambda h.\langle \overline{9}, \langle x, h\rangle\rangle;\ (\perp)^* := \langle \overline{20}, \overline{1}\rangle;$$

$$Inl^* := \lambda x.\langle \overline{10}, x\rangle;\ Inr^* := \lambda x.\langle \overline{11}, x\rangle;\ Pair^* := \lambda x \lambda y.\langle \overline{12}, \langle x, y\rangle\rangle;$$

$Spread^* := \lambda x \lambda h.\langle \overline{13}, \langle x, h\rangle\rangle$; $Decide^* := \lambda x \lambda f \lambda g.\langle \overline{14}, \langle x, f, g\rangle\rangle$;

$Decidenat^* := \lambda x \lambda y \lambda f.\langle \overline{15}, \langle x, y, f\rangle\rangle$; $(L\dot{EV})^* := \lambda x \lambda y.\langle \overline{16}, \langle x, y\rangle\rangle$;

$(N\dot{EV})^* := \lambda x \lambda y.\langle \overline{17}, \langle x, y\rangle\rangle$; $(\dot{\forall}_1)^* := \lambda x.\langle \overline{18}, x\rangle$; $S^* := \lambda x.\langle \overline{19}, x\rangle$;

$\overline{0}^* := \langle \overline{20}, \overline{0}\rangle$; $\dot{=}^* := ID$; $(\dot{\forall}_0)^* := ALL$; $\rightarrow^* := \lambda x \lambda y.IMPxy$;

$(\dot{P}_i)^* := \lambda x.[P_i^*(x)]$; $(\dot{T})^* := \lambda x.[T_k^*(x)]$,
where $1 \le i \le k$ and P_i^*, T_i^* are defined in 73.1 below;

$Ind^* := FP(\lambda g \lambda x \lambda y \lambda h.Decidenat^*(x, y, \lambda u.g(u, y, h)))$.

N.B. It is understood that $IMP := \lambda x \lambda y.NEG(ANDx(NEGy))$; NEG, AND, ID, ALL are the combinators of Ch. II, 7.1. FP is the fixed point combinator of Ch. I,§ 2. Notice that * actually depends on k.

(ii) We inductively extend * to arbitrary (also non-primitive) function symbols and terms of LTC_k of arity OB:

1. if c is a constant $(c)^* = c^*$, as defined in 72.4 (i) ; $(x_i)^* = x_{2i+1}$ if x is a variable of arity OB; $(f_i)^* = x_{2i}$ if f is a variable of arity $OB \rightarrow OB$;

2. $F(t_1, \ldots, t_n)^* := F^* t_1^* \ldots t_n^*$ (where F^* is defined in 72.4 (i), if F is a primitive function symbol of LTC_k; F^* is defined according to the preceding clause, if F is a function variable f).

3. $(x)t^* := \lambda x.(t^*)$ and $(f)t^* = \lambda f.(t^*)$.

N.B. The use of variables of odd and even index, in order to interpret variables of arity OB and $(OB \rightarrow OB)$ respectively, is only required to avoid undue identification of variables (e.g. we want that the translation $(f(x))^*$ has two distinct variables corresponding to f and x). Henceforth we omit explicit mention of indices and we still keep using f, g, h as metavariables for variables in function position.

By inspection of the definitions above, we obtain the expected independence properties, which also imply the counterparts of the special axioms SE1-SE2 and the standard Peano Axioms:

72.5. LEMMA. *If C and E are distinct function symbols of* LTC_k, *then* $\mathrm{OP} \vdash \neg C^* = E^*$. *Moreover* OP *proves:*

SE1*: $F(f_1, \ldots, f_n)^* = F(g_1, \ldots, g_n)^* \rightarrow f_i(u_1, \ldots, u_m)^* = g_i(u_1, \ldots, u_m)^*$;

SE2*: $\neg F(c_1 \ldots c_n)^* = G(h_1 \ldots h_n)^*$;

here F, G are distinct primitive function symbols $\neq Ind$, and f_i, g_i, u_j have the appropriate arities (remember that $=$ only applies to expressions of arity OB);

SUC*: $\forall x(\neg S^*x = \overline{0}^*) \wedge \forall x \forall y(S^*x = S^*y \rightarrow x = y)$.

Second step. We introduce predicates LEV_k^* and NEV_k^*, which work as "interpreters" of LTC_k-terms. The notion of canonical object is explicitly definable by means of the following formula of STLR_k:

72.6. DEFINITION (using 72.4; $k \geq 1$)

$$\begin{aligned}
Can_k(x) := {} & \exists y \exists z((x = \lambda^* y) \vee (x = S^* y) \vee (x = \overline{0}^*) \vee (x = Inl^* y) \vee \\
& \vee (x = Inr^* y) \vee (x = (\dot{\perp})^*) \vee (x = NEGy) \vee (x = (\dot{\forall}_0)^* y) \vee \\
& \vee (x = (\dot{\forall}_1)^* y) \vee (x = (\dot{=})^* yz) \vee (x = (LE\dot{V})^* yz) \vee \\
& \vee (x = (NE\dot{V})^* yz) \vee (x = ANDyz) \vee (x = Pair^* yz) \vee \\
& \vee (x = [T_1(y)]) \vee \ldots \vee (x = [T_k(y)]) \vee \\
& \vee \ (x = [F_1(y)]) \vee \ldots \vee (x = [F_k(y)])).
\end{aligned}$$

72.6.1. FACT: we can find a formula $CPT^*(y, w, z)$, which translates the condition "(y, w, z) is a computation triple" (in the sense of 71.2 above) into the language of STLR_k:

$$\begin{aligned}
& CPT^*(y, w, z) := \exists f \exists h(y = \lambda^* f \wedge hf = w \wedge z = \lambda x.Pa^* xh) \vee \\
& \vee \exists u \exists v \exists h(y = Pair^* uv \wedge w = huv \wedge z = \lambda x.Spread^* xh) \vee \\
& \vee \exists u \exists v \exists n \exists g(z{=}\lambda x Decidenat^* xvg \wedge ((y{=}\overline{0}^* \wedge w{=}v) \vee (y{=}S^* n \wedge w{=}gn))) \vee \\
& \vee \exists u \exists f \exists g(z{=}\lambda x.Decide^* xfg \wedge ((y{=}Inl^* u \wedge w{=}fu) \vee (y{=}Inr^* u \wedge w{=}gu))).
\end{aligned}$$

72.7. LEMMA. $\text{STLR}_k \vdash CPT^*(y, w, z) \wedge CPT^*(y, w', z) \rightarrow w = w'$.

PROOF. We essentially apply lemma 72.5. Assume the antecedent of 72.7 and also $z = \lambda x.Decide^* xfg$, $y = Inl^* u$, $w = fu$, for some u, f, g. Then $CPT^*(y, w', z)$ implies by SE1*: $z = \lambda x.Decide^* xf'g'$, $y = Inl(u')$, $w' = f'u'$, for some f', g', u'. Therefore $Decide^* xfg = Decide^* xf'g'$ and $Inl^* u = Inl^* u'$, which yield by SE2* $f = f'$, $g = g'$, $u = u'$, whence $fu = f'u'$, i.e. $w = w'$. The other cases are similar. □

Let $B_k(x, y, v)$ be the formula:

$$\exists z \exists a \exists c \exists w(CPT^*(c, w, z) \wedge x = z(a) \wedge \langle a, c \rangle \eta_1 v \wedge \langle w, y \rangle \eta_1 v)$$

and define $LV_k(x, y, v) := (Can_k(x) \wedge x = y) \vee B_k(x, y, v)$.

By inspection, we see that $\lambda v.\{\langle x, y \rangle : LV_k(x, y, v)\}$ is an existential operator (see Ch. II, 10.9) and actually that $LV_k(x, y, v)$ is elementary in v. By 10.9.1 we find a term $\lambda u.LEV_k^*(u)$ such that:

$$LEV^*_k(0) = \emptyset \text{ and } LEV^*_k(m+1) = \{\langle x,y\rangle : LV_k(x,y,LEV^*_k(m))\};$$

finally we put $LEV^*_k := \{\langle x,y\rangle : \exists m.\, \langle x,y\rangle \eta_1 LEV^*_k(m)\}$.

Now we claim:

72.8. LEMMA. *Let* $k \geq 1$.

(i) $\mathrm{STLR}_k \vdash Cl_1(LEV^*_k)$;

(ii) $\mathrm{STLR}_k \vdash \langle x,y\rangle \eta_1 LEV^*_k \leftrightarrow LV_k(x,y,LEV^*_k)$;

(iii) $\mathrm{STLR}_k \vdash (Can_k(x) \to \langle x,x\rangle \eta_1 LEV^*_k) \wedge (\langle x,y\rangle \eta_1 LEV^*_k \to Can_k(y))$;

(iv) $\mathrm{STLR}_k \vdash \forall x \forall y (\langle x,y\rangle \eta_1 LEV^*_k \wedge \langle x,z\rangle \eta_1 LEV^*_k \to y = z)$;

(v) $\mathrm{STLR}_k \vdash CPT^*(y,w,g) \wedge \langle x,y\rangle \eta_1 LEV^*_k \to$

$$\to (\langle gx,z\rangle \eta_1 LEV^*_k \leftrightarrow \langle w,z\rangle \eta_1 LEV^*_k).$$

PROOF. As to (i)-(ii), STLR_k has Δ_0-N-induction, i.e. number-theoretic induction for formulas built up from atoms of the form $T_i t$, $F_i t$ $(1 \leq i \leq k)$, $t = s$, Nt, by means of standard logical operations. Now $Cl_1(LEV^*_k(y))$ is Δ_0 and we can verify with elementary comprehension for classes of level 1 and Δ_0-N-induction that for every m, $LEV^*_k(m)$ is a class of level 1. Then we apply 10.8-10.9.1.
(iii): the first conjunct is a corollary of (ii), while the second is obtained by checking $\forall x \forall y (\langle x,y\rangle \eta_1 LEV^*_k(m) \to Can_k(y))$ by Δ_0-N-induction on m.
(iv): Δ_0-N-induction and lemma 72.7.
(v): apply (ii), (iii). □

72.9. DEFINITION. $Nv_k(x,y,v)$ is the formula:

$$(y{=}\bar{0}^* \wedge \langle x,\bar{0}^*\rangle \eta_1 LEV^*_k) \vee \exists z \exists w (y{=}S^*w \wedge \langle x,S^*z\rangle \eta_1 LEV^*_k \wedge \langle z,w\rangle \eta_1 v).$$

Clearly $\lambda v\{\langle x,y\rangle : Nv_k(x,y,v)\}$ is an existential operator; hence elementary comprehension and 72.8 (i) imply, for $k \geq 1$:

$$\mathrm{STLR}_k \vdash Cl_1(v) \to Cl_1(\{\langle x,y\rangle : Nv_k(x,y,v)\}.$$

If we recursively define

$$NEV^*_k(0) = \emptyset \text{ and } NEV^*_k(m+1) = \{\langle x,y\rangle : Nv_k(x,y,NEV^*_k(m))\},$$

we can again apply Δ_0-induction on N, in order to check that $Cl_1(NEV^*_k(m))$, for every m. If we argue as in 72.8 above, we conclude:

72.10. LEMMA. *Let* $k \geq 1$.

(i) $\mathrm{STLR}_k \vdash Cl_1(NEV_k^*)$,

where $NEV_k^* := \{\langle x,y\rangle : \exists m(\langle x,y\rangle \eta_1 NEV_k^*(m))\}$;

(ii) $\mathrm{STLR}_k \vdash \langle x,y\rangle \eta_1 NEV_k^* \leftrightarrow Nv_k(x,y,NEV_k^*)$;

(iii) $\mathrm{STLR}_k \vdash \langle x,\overline{0}^*\rangle \eta_1 LEV_k^* \rightarrow \langle x,\overline{0}^*\rangle \eta_1 NEV_k^*$;

(iv) $\mathrm{STLR}_k \vdash \langle x,S^*y\rangle \eta_1 LEV_k^* \wedge \langle y,z\rangle \eta_1 NEV_k^* \rightarrow \langle x,S^*z\rangle \eta_1 NEV_k^*$.

72.10.1. CONVENTION: henceforth we adopt the abbreviations

$$LEV_k^*(x,y) := \langle x,y\rangle \eta_1 LEV_k^*; \quad NEV_k^*(x,y) := \langle x,y\rangle \eta_1 NEV_k^*.$$

72.11. LEMMA (NEV_k^**-induction*). *If* $k \geq 1$ *and* $A(x,y)$ *is a* Δ_0*-formula of* STLR_k *(i.e. built up by means of* $\wedge$, $\neg$, $\forall$ *from atoms of the form* Nt, $t = s$, $T_i t$, $F_i t$, *where* $1 \leq i \leq k$*),* STRL_k *proves:*

$$\forall x\forall y\forall z((LEV_k^*(x,\overline{0}^*) \rightarrow A(x,\overline{0}^*)) \wedge (LEV_k^*(x,S^*y) \wedge A(y,z) \rightarrow A(x,S^*z))) \rightarrow$$
$$\rightarrow \forall u\forall v(NEV_k^*(u,v) \rightarrow A(u,v)).$$

PROOF. By hypothesis on A, we can apply induction on m:

$$\forall u\forall v(\langle u,v\rangle \eta_1 NEV_k^*(m) \rightarrow A(u,v)) \qquad (*)$$

but $(*)$ is immediate under the assumption of the antecedent of 72.11. □

§ 73. Proof-theoretic reduction

We now proceed to the

Third Step

As a preliminary move, we introduce a term $V(i)$ for each $1 \leq i \leq k$, in STLR_k, such that, if we set:

73.1.1 $\quad T_i^*(x) := \langle x,\overline{0}\rangle \eta_i V(i)$ and $F_i^*(x) := \langle x,\overline{1}\rangle \eta_i V(i)$,

73.1.2 $\quad P_i^* t := (T_i^* t \vee F_i^* t)$,

then T_k^* and P_i^* are the required interpretations of the truth and proposition predicates of LTC_k within a slight extension of system STLR_k. This extension turns out to be proof-theoretically reducible to the ramified system RS_k of Ch. XI.

We first apply the fixed point theorem for predicates relativized to level i (use 10.1 and 37.2):

73.2. LEMMA. *We can find a term* $V(i)$, *and a formula* $A(a,u,v)$ *such that the following conditions on* $V(i)$, T_i^*, F_i^* *(as defined in 73.1.1-73.1.2 above) hold, provably in* STLR_k (for $k \geq 1$):

1. $\langle a,u\rangle\eta_i V(i) \leftrightarrow A(a,u,V(i))$;
2. *if* t *is a term of the form* $(=^*xy), ((L\dot{EV})^*xy), (N\dot{EV})^*xy)$, $[T_j^*x], [F_j^*x]$, *for* $1 \leq j < i$) *and* AT *is the formula* $x = y$ $(LEV_k^*(x,y),\ NEV_k^*(x,y),\ T_j^*x,\ F_j^*x$ *respectively, for* $1 \leq j < i$), *then* $LEV_k^*(a,t) \rightarrow (T_i^*a \leftrightarrow AT) \wedge (F_i^*a \leftrightarrow \neg AT)$;
3. *if* $m = 0, 1$,
   $LEV_k^*(a, \forall_m^* x) \rightarrow (T_i^*a \leftrightarrow \forall u T_i^*(xu)) \wedge (F_i^*a \leftrightarrow \exists u F_i^*(xu))$;
4. $LEV_k^*(a, ANDxy) \rightarrow (T_i^*a \leftrightarrow T_i^*x \wedge T_i^*y) \wedge (F_i^*a \leftrightarrow F_i^*x \vee F_i^*y)$;
5. $LEV_k^*(a, NEGx) \rightarrow (T_i^*a \leftrightarrow F_i^*x) \wedge (F_i^*a \leftrightarrow T_i^*x)$;
6. *if* $1 \leq j < i$, *then* $(T_j^*a \rightarrow T_i^*a\ ) \wedge (F_j^*a \rightarrow F_i^*a)$.
7. *Let* STLR_k^* *be* STLR_k *plus* $CONS^*(1), \ldots, CONS^*(k)$, *where*
   $CONS^*(i) := \forall a \forall u \forall v(\langle a,u\rangle\eta_i V(i) \wedge \langle a,v\rangle\eta_i V(i) \rightarrow u = v)$
   $(i = 1, \ldots, k)$. *Then* $\mathrm{STLR}_k^* \vdash \forall x\ \neg(T_i^*x \wedge F_i^*x)$.

As to the proof, we simply define truth and falsehood of level i, in such a way that they are invariant under the computation predicates LEV_k^* and NEV_k^* and meet the conditions (2)-(6); we then repeatedly apply 72.8-72.10 and $CONS^*(i)$.

73.3. THEOREM. *For each* $k \geq 1$, *we can define a translation* $A \mapsto (A)_k$ *of* LTC_k*-formulas into* STLR_k*-formulas such that:*

$$\mathrm{LTC}_k \vdash A \textit{ implies } \mathrm{STLR}_k^* \vdash (A)_k.$$

PROOF. If t is a term of LTC_ω of arity OB, let t^* be the term of STLR_k, obtained, according to definition 72.4 (ii). Then we inductively set:

1. $(t = s)_k := (t^* = s^*)$; $(LEV(t,s))_k := LEV_k^*(t^*, s^*)$;
   $(NEV(t,s))_k := NEV_k^*(t^*, s^*)$;
2. $(P_i t)_k := P_i^*(t^*)$ ($1 \leq i \leq k$) and $(Tt)_k := T_k^*(t^*)$; $(\ \bot\ )_k := (\overline{0} = \overline{1})$;
3. $(\forall_0 x A)_k := \forall x (A)_k$; $(\forall_1 f.A)_k := \forall y (A)_k$; $(B \rightarrow C)_k = (B)_k \rightarrow (C)_k$.

If A is a successor axiom, a special equality axiom SE.1-SE.2 of LTC_0 or the defining equation for Ind, $(A)_k$ is provable by lemma 72.5 and the fixed point theorem for operations. If A is any LEV- or NEV-axiom, $\mathrm{STLR}_k \vdash (A)_k$ is a straightforward consequence of 72.8, 72.10-72.11. If A is an axiom among PT1–PT4, we apply lemma 73.2 above. As a sample, let us check $(\mathrm{PT.2})_k$, which has the form:

$$(LEV_k^*(a, x \dot{\rightarrow} y) \wedge P_i^*(x) \wedge (T_k^*(x) \rightarrow P_i^*(y))) \rightarrow P_i^*(a); \tag{1}$$

$$(LEV_k^*(a, x \dot{\rightarrow} y) \wedge P_i^*(x) \wedge (T_k^*(x) \rightarrow P_i^*(y))) \rightarrow$$
$$\rightarrow (T_k^* a \leftrightarrow (T_k^* x \rightarrow T_k^* y)). \tag{2}$$

First of all, observe that lemma 73.2 implies:

$$LEV_k^*(a, x \dot{\rightarrow} y) \rightarrow (T_i^* a \leftrightarrow (F_i^* x \vee T_i^* y)); \tag{3}$$

$$LEV_k^*(a, x \dot{\rightarrow} y) \rightarrow (F_i^* a \leftrightarrow (T_i^* x \wedge F_i^* y)). \tag{4}$$

(1): assume the antecedent of (1) and let $\neg T_i^* a$: then $\neg F_i^* x \wedge \neg T_i^* y$ by (3), whence $T_i^* x \wedge F_i^* y$, i.e. $F_i^* a$ by (4).
(2): again assume the antecedent of (2) with $T_k^* a$, $T_k^* x$ and by contradiction $\neg T_k^* y$. By persistence $\neg T_i^* y$, hence $F_i^* y$ (with the assumption $P_i^* y$). Were $F_i^* x$, also $F_k^* x$: against consistency (lemma 73.2.7). Hence $T_i^* x$, i.e. $F_i^* a$ by (4), and finally $F_k^* a$ (persistence), which contradicts $T_k^* a$. Conversely, assume the antecedent of (2) and $T_k^* x \rightarrow T_k^* y$ together with $\neg F_k^* x$. Then by persistence $\neg F_i^* x$, whence $T_i^* x$ with the assumption $P_i^* x$ and finally $T_k^* x$ by persistence, i.e. by assumption $T_k^* y$. The conclusion follows by (3). □

Final step: concluding the proof of theorem 72.3. We rely upon the proof theory of chapters X-XI. First of all, let $\mathrm{IT}_k^\infty(*)$ be the system IT_k^∞ of § 50 extended by initial sequents, which correspond to the axioms $CONS^*(i)$, written in the language of IT_k^∞ $(1 \leq i \leq k)$. They have the form

S-$CONS^*(i)$: $\qquad \neg\langle t, s\rangle \eta_i V(i),\ \neg\langle t, r\rangle \eta_i V(i),\ s = r.$

By inspection we see:

73.4.1. the embedding theorem 51.6 trivially holds for the pair STLR_k^* and $\mathrm{IT}_k^\infty(*)$ as well: hence the first system is interpretable in the second with finite cut rank and length $< \omega 2$.

73.4.2. The partial cut elimination theorem 55.1 carries over unchanged to $\mathrm{IT}_k^\infty(*)$, due to the form of S-$CONS^*(i)$.

73.4.3. S-$CONS(k)$ is sound for the asymmetric interpretation of $\mathrm{IT}_k^\infty(*)$ into RS_k, the ramified system of § 52; more precisely, we obtain:

$$\mathrm{RS}_k \vdash^{\omega(\alpha+1)}_{\omega(\alpha+1)} \neg\langle t, s\rangle \eta_k^\alpha\, V(i),\ \neg\langle t, r\rangle \eta_k^\alpha\, V(i),\ s = r.$$

(The verification runs by induction on α; if $k = 1$, as a preliminary step, one checks the RS_k-derivability of sequents of the form:

$$\neg\langle t,s\rangle\eta_k^\alpha\ LEV_k^*,\ \neg\langle t,r\rangle\eta_k^\alpha\ LEV_k^*,\ s=r.$$

In addition, if $k=i+1$, one needs the derivability of the sequents:

$$\neg\langle t,s\rangle\eta_i V(i), \langle t,s\rangle\eta_k^\alpha\ V(i).$$

73.4.4. $\mathrm{OP}+\mathrm{TI}_{op}(<\Gamma_0)$ proves the consistency of LTC_k.
In fact, any derivation of $\perp$ in LTC_k is sent into an RS_0-derivation of $\bar{0}^*=\bar{1}^*$. But this is impossible by §§57-58, provably in $\mathrm{OP}+\mathrm{TI}(\alpha)$, for α sufficiently high $<\Gamma_0$. □

The same method might be used to classify the number-theoretic functions, which are provably total in LTC_ω, and to obtain additional recursion-theoretic information. Theorem 72.3 can be inverted as well: as sketched by Aczel-Carlisle-Mendler(1991), LTC_ω interprets Martin-Löf's theory TT_ω with arbitrary finitely many universes and without well-ordering types. It is known that TT_ω carries out the well-ordering proof for Γ_0 (see Jervell 1978).

§74. Perspectives: related work in Artificial Intelligence and Theoretical Linguistics

We conclude with a short survey of contributions, which point out applications of type-free systems to areas related with Theoretical Computer Science, Artificial Intelligence and Theoretical Linguistics. Once more, we do not offer well-established results, but only a few hints that should stimulate towards further investigation.

A. *Illative Theories of Relations.* Plotkin (1990) describes a system, in which one can formalize ideas from Situation Theory (cf. Barwise-Etchemendy 1987), namely the notions of *relation*, *assignment*, *predication*, *state-of-affairs* (soa, in short), *fact*. Plotkin's system – which we name PS for simplicity – is actually a type-free theory, formalized in a language of expressions with arities (see §70.B). Expressions are meant to (possibly) describe soas and true soas (= facts). PS is given in the style of an illative calculus (recall the system with levels of implications of §25), which is devised to produce two sorts of judgments: 1) *E is a fact* ; 2) *E is a soa.* Of course, the rules for PS depend on the basic constructors, which include, among other logical operations, *restriction.* The specific feature of PS consists of extending the basic theory of soas and facts with the notion of predication, relation, assignment, relation abstraction. It turns out that the theory derives a fixed point theorem and the undefinability of the notions of fact and soa, very much in analogy with the well-known results from the theory of propositions and truth à la Aczel.
From the metamathematical point of view, Plotkin shows that PS can be

embedded into a conservative first-order standard theory of facts and soas. This is achieved by expanding the illative language with a new predicate $Fact(x)$ and then by axiomatizing the properties of $Fact(x)$, as established by the PS-rules. On the other hand, he also sketches a three-valued logic approach PS_3, which is much in the spirit of the present work; not surprisingly, the notion of "facticity" becomes internally definable in PS_3, but the predicate of being a fact has to be partial. The interested reader might try to construct a model for PS or PS_3 within MF^-. Of course, one might wonder whether similar frameworks are useful as foundations of situation semantics, in alternative to non-wellfounded sets. As far as we can see, Plotkin's work is a move in this direction.

B. *Artificial Intelligence, semantics, linguistics.* It is clear from the heading that the problems one has to cope with, are intertwined with highly controversial themes: the logic of propositional attitudes, such as "to know", "to believe", and, more generally, the foundations of property theory and intensional logic. Consequently, this research trend is extremely sensitive to the conceptual analysis of the basic notions involved, and to philosophical discussions. It should be mentioned that there is a clear-cut distinction between the modal approach, which sticks to the idea that attitudes are to be rendered by modal operators, and the idea that propositional attitudes can be expressed by means of predicates of objects with some kind of logical structure (sentences, propositions in the sense of Barwise and Etchemendy 1987, i.e. suitable non-wellfounded sets, representations in the sense of Discourse Representation Theory). Be as it may, we only refer to contributions that are immediately related to the technical aspects discussed in this book.

Formally, the underlying problem is to design languages which are capable of representing, to a certain extent, truth, knowledge, belief, as well as reasoning about them by "reasoning agents". But the inherent "reflective" aspect apparently calls for systems, that can consistently live with circularity and self-reference. This need has naturally led some AI-researchers to consider, as a reasonable tool for the logics of knowledge representation, formalisms which are primarily languages with self-referential truth, supported either by a a partial semantics à la Kripke or by semi-inductive semantics à la Herzberger. There is a hope that a suitable combination of inductive and semi-inductive methods, and even modal logic, can lead to a natural integration of self-referential truth, belief and knowledge predicates, possibly constrained with indexical coordinates, depending upon the different knowing agents. Typical attempts in the axiomatic vein can be found in Perlis (1985), (1988), Turner (1990), Davies (1991).

Perlis experiments with several systems, which at least include first-order logic with partial truth axioms à la Kripke-Feferman and a substitution predicate ensuring self-reference. These systems can be readily considered as subsystems, say, of MF^{-}(Ch. II). In the context of a theory of reasoning agents, Perlis proceeds to enrich the ground system with predicates for knowledge and belief ($Know(x)$, $Bel(x)$). He then singles out some options, trying to discover reasonable sets of axioms for $Know(x)$, $Bel(x)$, also guided by analogies with extant work on provability predicates. According to Perlis, an advantage of the first-order approach is that no problem seems to arise with undesired substitution in non-extensional contexts (e.g. $t = s$ and $Know(t)$ do not imply in general $Know(s)$).

Turner (1990) develops a systematic proposal for integrating problems, stemming from knowledge representation and semantic theory, with applications to linguistics. In contrast to Perlis, Turner somewhat inclines towards a notion of truth, supported by semi-inductive definitions (see Ch. XII) and refined with modal operators. Like the present work, self-reference is basically granted by the underlying lambda calculus. The influence of Aczel's views is important here: propositions are essentially objects, to be characterized axiomatically (either directly or via truth); in any case, they should not simply be identified with sentences of any given formal language; properties are explained away in favour of propositional functions, defined in the underlined lambda structures. Further developments are sketched in Davies (1991).

On the side of linguistic applications and the foundations of intensional logic, we mention an attempt by Chierchia and Turner (1988). The authors present theories of properties, relations and propositions, that should be so powerful and expressive to support the semantics of natural languages and to explain specific linguistic phenomena (e.g. nominalization). Technically speaking, the theories in question can be considered specializations of the formalisms in § 70: they are formulated in a language of expressions with arities. In particular, there is a universal arity, together with special basic arities for *information units* (= propositional objects), *urelements*, *nominalized functions* (i.e. denotations of $\lambda((x)t)$), *possible worlds* and *time instants*. Complex arities are generated by arrow application, while expressions are inductively defined by means of applications of the standard logical constructors (including quantification on variables of different sorts), modal and tense operators, truth. The underlying logic is classical logic, enriched by Turner's axioms and rules for truth, plus possibly special axioms and rules for modal and tense operators. The intended interpretations are suitable expansions of lambda calculus models in the sense of Aczel, that satisfy the closure conditions required by the axioms. The authors sketch a formalization of a fragment of English in such a

framework, thus extending Montague's type-theoretic approach. They also argue that certain technical features of their model directly correspond to specific semantic problems. In particular, the distinction between a propositional function f and its individual correlate $\lambda(f)$ is motivated:
1) by the need of a reasonable semantic for infinitivals and gerunds;
2) by the search for an explanation of syntactical phenomena (e.g. why in English verbs sometimes are flected and sometimes not).

§75. Sense and denotation as algorithm and value: subsuming theories of reflective truth under abstract recursion theory

It is in harmony with the role of recursion-theoretic intuitions in this book that we conclude by mentioning a recent paper, which leads us again into the realm of generalized recursion. In a series of papers Moschovakis (see Moschovakis 1984, 1989a, 1989b) has developed a foundation for the theory of computation, which is based on the notion of recursive algorithm and which is sufficiently strong to subsume recursion on finite structures, as well as generalized recursions on abstract infinite structures.

The starting point is that recursive definitions, as specified by systems of partial monotone functionals (with respect to the appropriate partial orderings generated by the relation "to be more defined than"), are taken as primitive; a crucial and original aspect of Moschovakis's approach is that it yields, together with the usual denotational semantics, a mathematical definition of *intension for recursive algorithms* by means of certain set-theoretic objects, the so-called *recursors.* Recursors are directly related to the denotations of the (uniquely determined) terms, resulting from an abstract "compilation process" of the given recursive algorithm. Of course, this idea requires a formal language of recursion (FLR), in which algorithms are codified and "compiled" by means of a precise *reduction calculus* and for which a unique termination property is established. Syntactically, recursors are immediately read off from the normal forms of the FLR-formalizations of the given algorithms (see Moschovakis 1989a).

A stimulating feature of the theory of recursive algorithms is that the recursion-theoretic definition of intension can be readily applied to logical languages and to the modeling of *the Fregean notion of sense.* The idea is that the sense of a sentence is essentially the method we follow, in order to establish the truth value of the sentence itself, and that such a method reduces to compute a generalized recursive algorithm; but this algorithm has an associated *referential intension*, which can be precisely defined in terms of Moschovakis theory. Thus, if we identify the Fregean notion of sense with the referential intension, we have a mathematically precise notion of intensional identity at hand.

The connection with the present work is made apparent by Moschovakis' extension of lower predicate calculus with reflection, i.e. the language LPCR. In fact, LPCR easily subsumes Kripke's languages for self-referential truth and it can be shown that LPCR has enough expressive power to characterize inductive and hyperelementary sets on a given structure $\mathcal{M}$ (see Ch. III, § 13-15). Let us briefly and informally sketch LPCR. First of all, assume we are given a first-order language LPC of a given relational signature, which includes $=$ and the standard logical first-order operators (for definiteness, think of the OP-language, formalized without function symbols). The language of LPCR is obtained form LPC by first adding a new denumerable list of predicate variables (in each finite arity, including 0; P, Q, are used as metavariables). P, Q, R are to be interpreted as partially defined relations on the given interpretation domain; if P is 0-ary, P plays the role of a propositional variable, whose truth value is possibly undefined. Then LPC is further extended with a new formula constructor WHERE for introducing direct self-reference; formally the inductive definition of LPCR-formulas also includes the following clause:

if $A_0, A_1, \ldots, A_n$ are LPCR-formulas, $P_1, \ldots, P_n$ are predicate variables, each P_i has arity k_i, $u(i)$ is a string of k_i individual variables $(1 \leq i \leq n)$, then the string

$$A := A_0 \text{ WHERE } \{P_1(u(\mathrm{i})) \simeq A_1, \ldots, P_n(u(n)) \simeq A_n\}$$

is a formula of LPCR; the bound occurrences of A are the bound occurrences in the head A_0 and in the parts $A_1, \ldots, A_n$, all the occurrences of $P_1, \ldots, P_n$ and the occurrences of the individual variables in each list $u(i)$ (for $1 \leq i \leq k$). Note that the list $u(i)$ may be empty, if P_i is 0-ary, i.e. a propositional variable; this is important to get self-reference at a propositional level.

On the semantic side, since partial predicates are around, one adopts Kleene's strong three-valued logic for $\neg$, $\wedge$, $\forall$, in order to evaluate complex sentences; therefore the corresponding truth functions (for sentential connectives) and functionals (for individual quantification) are monotone. The interpretation of the WHERE-constructor is readily explained by fixed point semantics. To each formula A_i, we can canonically associate a partial functional $F[A_i]$ which is built up from the monotone partial functions and functionals corresponding: 1) to logical operators; 2) to the partial predicates $P_1, \ldots, P_n$ and eventually to the basic predicates of LPC. $F[A_i]$ will possibly depend on the free individual variables of A_i and the partial characteristic functions $p_1, \ldots, p_n$ of $P_1, \ldots, P_n$. Since each $F[A_i]$ is monotone, we can apply the Knaster-Tarski theorem and show that there are simultaneous minimal solutions $I(A_1), \ldots, I(A_n)$ to the system of equations:

$$F_1[A_1](p_1, \ldots, p_n, a(1)) \simeq p_1(a(1));$$

....

....

$$F_n[A_n](p_1, \ldots, p_n, a(n)) \simeq p_n(a(n)).$$

The truth value of $A := A_0\text{WHERE}\{P_1(u(1)) \simeq A_1, \ldots, P_n(u(n)) \simeq A_n\}$ is then essentially given by computing $F[A](I(A_1), \ldots, I(A_n), a(0))$.

In the frame of LPCR, one verifies that the usual self-referential constructions of the Liar, the SameSayer, etc., can be *directly* carried out; moreover, Kripke's partial self-referential truth can be adequately defined by simply writing down the appropriate WHERE-formulas by means of the operator of Ch. II, § 7: LPCR is able to model reflective truth. In addition, let us mention a central result of Moschovakis (1990): the notion of intensional identity, as defined for LPCR-formulas, is *decidable* (though at least hard as the isomorphism problem for finite graphs). Thus, given two arbitrary expressions of LPCR, we can effectively compute their referential intensions and decide if they coincide. On the other hand, the problem of axiomatizing LPCR and the corresponding logic of intensions is left open.

As to the relation with the present framework, the WHERE-construct can be simulated, say in MF^-, by choosing a suitable simultaneous fixed point operator, which is obtained by well-known techniques (e.g. see Barendregt 1984, p.142). However, we warn that in MF^- we only get an *unfaithful* modeling of WHERE. In fact, like the more general formalism FLR, LPCR is based on a *primitive* idea of recursive computation, *not involving a sequential order* in the computation of the fixed points (Moschovakis 's theory is motivated by applications to the analysis of concurrency).

We wish to conclude with a note of self-criticism. If we look backwards, in the light of the previous considerations, we have to admit that the systems based on combinatory logic are *at the same time too simple and strong* for capturing important distinctions one would like to have at hand for computational applications. This can be partly remedied by adding more structure at the ground level, as suggested by LTC-systems, and there seems to be room for experimentation and research here. On the other hand, at a higher level, the consideration of LPCR suggests that there are important aspects, which are left untouched by the present work. One might try to integrate the systems of reflective truth with subtler tools coming from Moschovakis (1990), that could be relevant toward a deeper foundation of a theory of abstraction and properties, and the mathematical treatment of intensional constructions. We hope that these notes will encourage someone to new investigations.

BIBLIOGRAPHY

Abbreviations for journals and book series, to be used below:

JSL Journal of Symbolic Logic;
JPL Journal of Philosophical Logic;
ZML Zeitschrift für Mathematische Logik und die Grundlagen der Mathematik;
MLQ Mathematical Logic Quarterly (formerly ZML);
AML Archive for Mathematical Logic (formerly Archiv für Mathematische Logik und die Grundlagenforschung);
APAL Annals of Pure and Applied Logic (formerly Annals of Mathematical Logic);
SLNM Lecture Notes in Mathematics, (Springer, Berlin Heidelberg);
SLNCS Lecture Notes in Computer Science (Springer, Berlin Heidelberg);
TCS Theoretical Computer Science;
NDJFL Notre Dame Journal of Formal Logic.

ACKERMANN, W.

(1941) System der typenfreier Logik I, *Forschungen zur Logik und zur exakten Wissenschaften,* Heft 7.

(1950) Widerspruchsfreier Aufbau der Logik. Typenfreies System ohne Tertium non datur, JSL, 15, 33-37.

(1952-53) Widerspruchsfreier Aufbau einer typenfreier Logik I, II, *Mathematische Zeitschrift*, 55, 364-384, 57, 155-166.

ACZEL, P.

(1977) The strength of Martin-Löf's intuitionistic type theory with one universe, in: S. Mietissen and J. Väänänen (eds.), *Proceedings of the Symposiums in Mathematical Logic in Oulu* 1974 *and Helsinki* 1975, (Report no.2, Dept.of Philosophy, University of Helsinki), 1-32.

(1977a) An introduction to inductive definitions, in: J. Barwise (ed.), *Handbook of Mathematical Logik* (North Holland, Amsterdam), 867-895.

(1980) Frege structures and the notions of proposition, truth and set, in: J. Barwise, H.J. Keisler, K. Kunen(eds.), *The Kleene Symposium* (North Holland, Amsterdam), 31-59.

ACZEL, P., P. CARLISLE and N. MENDLER

(1991) Two frameworks of theories and their implementation in Isabelle, in:

G.M.Plotkin et al. (eds.), *Logical Frameworks* (Cambridge University Press, Cambridge), 5-39.

ACZEL, P. and S. FEFERMAN

(1980) Consistency of the unrestricted abstraction principle using an intensional equivalence operator, in: J.P. Seldin and R. Hindley, (eds.), *To H.B.Curry: Essays on Combinatory Logic, Lambda Calculus and Formalism* (Academic Press, New York), 67-98.

ACZEL, P. and W. RICHTER

(1974) Inductive definitions and reflecting properties of admissible ordinals, in: J.E.Fenstad and P.G.Hinman(eds.), *Generalized Recursion Theory* (North Holland, Amsterdam), 301-381.

APT, K. and W. MAREK

(1974) Second order arithmetic and related systems, APAL, 6, 177-229.

ASPERTI, A. and G. LONGO

(1990) *Categories, Types and Structures* (Foundations of Computing Series, M.I.T. Press, Cambridge Mass.).

BARENDREGT, H.

(1973) Combinatory logic and the axiom of choice, *Indagationes Mathematicae,* 35, 203-221.

(1984) *The Lambda Calculus: its Syntax and Semantics* (North Holland, Amsterdam).

BARWISE, J.

(1975) *Admissible Sets and Structures* (Springer, Heidelberg) .

BARWISE, J., (ed.)

(1977) *Handbook of Mathematical Logic* (North Holland, Amsterdam).

BARWISE, J. and J. ETCHEMENDY

(1987) *The Liar. An Essay on Truth and Circularity* (Oxford University Press, Oxford).

BARWISE, J., R. GANDY and Y. N. MOSCHOVAKIS

(1971) The next admissible set, JSL, 36, 108-120.

BARWISE, J. and J. SCHLIPF

(1975) On recursively saturated models of arithmetic, SLNM 498, 42-55.

BEALER, G. and U. MÖNNICH

(1989) Property theories, in : D.Gabbay and F.Guenthner (eds.), *Handbook of Philosophical Logic* IV (Reidel, Dordrecht), 133-251.

BEESON, M.

(1985) *Foundations of Constructive Mathematics* (Springer, Berlin).

BEHMANN, H.

(1931) Zu den Widersprüchen der Logik und der Mengenlehre, *Jahresberichte der Deutschen Mathematischen Vereinigung*, 40 , 37-48.

(1937) The paradoxes of logic, *Mind*, 46,218-221

(1959) Der Prädikatenkalkül mit limitierten Variablen : Grundlegung einer natürlichen exakten Logik, JSL, 24, 112-140.

BELNAP, N.

(1977) A useful four-valued logic, in: M. Dunn and R. Epstein (eds.), *Modern Uses of Multiple-Valued Logic* (Reidel, Dordrecht), 9-37

BIRKHOFF, G.

(1967) *Lattice Theory* (American Mathematical Society, Colloquium Publications, vol.XXV, 3rd edition Providence, R.I).

BISHOP, E.

(1967) *Foundations of Constructive Analysis* (Mc Graw Hill, New York).

BOYD, R., G. HENSEL and H. PUTNAM

(1969) A recursion-theoretic characterization of the ramified analytical hierarchy, *Trans. Amer. Math. Soc.*, 141, 37-62.

BUCHHOLZ, W.

(1991) Notation systems for infinitary derivations, AML, 30, 277-296.

(1992) A simplified version of local predicativity, in: P. Aczel, H. Simmons, S. Wainer, (eds.), *Proof Theory* (Cambridge University Press, Cambridge), 115-147.

BUCHHOLZ, W., A. CICHON and A. WEIERMANN

(1994) A uniform approach to fundamental sequences anf hierarchies, MLQ, 40, 273-286.

BUCHHOLZ, W., S. FEFERMAN, W. POHLERS and W. SIEG

(1981) *Iterated Inductive Definitions and Subsystems of Analysis: Recent Proof Theoretical Studies*, SLNM 897.

BUCHHOLZ, W. and S. WAINER

(1987) Provably computable functions and the fast growing hierarchy, in: S. G. Simpson (ed.), *Mathematical Logic and Combinatorics* (AMS, Contemporary Mathematics, vol.65), 179-198.

BULL, R. and K. SEGERBERG

(1983) Basic Modal Logic, in: D. Gabbay and F. Guenthner (eds.), *Handbook of Philosophical Logic* I (Reidel, Dordrecht), 133-188.

BURGE, T.

(1979) Semantical paradox, *Journal of Philosophy,* 76, 169-198.

BURGESS, J.

(1986) The truth is never simple, JSL, 51, 663-681.

CANTINI, A.

(1980) A note on three-valued logic and Tarski theorem on truth definitions, *Studia Logica,* 39, 405-415.

(1982) Non-extensional theories of predicative classes over PA, *Rendiconti del Seminario Matematico dell' Università e del Politecnico di Torino,* 40, 47 -79.

(1983) A note on the theory of admissible sets with $\in$-induction restricted to formulas with one quantifier, *Bollett. Un. Mat. Italiana,* 6 (2-b), 721-737.

(1983a) *Proprieta' e Operazioni* (Bibliopolis, Naples).

(1985) On weak theories of sets and classes which are based on strict Π^1_1-reflection, ZMLG, 31,321-332.

(1985a) A note on a predicatively reducible theory of elementary iterated induction, *Bollett. Un. Mat. Italiana,* 4-b, 413-430.

(1987) Su una teoria generale delle proprietà, basata su schemi di comprensione iterati e privi di tipi, in: R. FERRO, A. ZANARDO (eds.), *Atti degli Incontri di Logica Matematica, vol.3* (Cluep, Padova), 53-57.

(1988) Two impredicative theories of sets and properties, ZMLG, 34, 403-420.

(1989) Notes on formal theories of truth, ZMLG, 35, 1989, 97-130.

(1990) A theory of formal truth arithmetically equivalent to ID_1 , JSL, 55 , 244 -259.

(1991) A logic of abstraction related to finite constructive number classes, AML, 31,69-83

(1992) Levels of implications and type-free theories of partial classifications with approximation operator, ZMLG, 38, 107-141.

(1993) Extending first order theory of combinators with self-referential truth, JSL, 58, 477-513.

(1995) Levels of Truth, NDJFL, to appear

CHANG, C.C. and H. J. KEISLER
(1990) *Model Theory* (North Holland, Amsterdam; third edition)

CHIERCHIA, G. and R. TURNER
(1988) Semantic and property theory, *Linguistics and Philosophy* 11,261-302.

CHURCH, A.
(1941) *The Calculi of λ-Conversion* (Ann. Math. Studies, no.6, Princeton University Press, Princeton, NJ).

CHURCH, A. and J. B. ROSSER
(1936) Some properties of conversion, *Trans. Amer. Math. Soc.*, 39, 472-482.

CONSTABLE, R., S. F. ALLEN, H. M. BROMLEY, W. R. CLEVELAND, J. F. CREMER, R.W. HARPER, D. J. HOWE, T. KNOBLOCK, N.P. MENDLER, P. PANANGADEN, T. SASAKI and SMITH, F.S.
(1986) *Implementing Mathematics with the Nuprl Proof Development System* (Prentice-Hall, Englewood Cliffs NJ)

COQUAND, T.
(1985) *Une Théorie des Constructions* (Thèse de troisième Cycle, Université Paris VII).

CROSSLEY, J.N.
(1969) *Constructive Order Types* (North Holland, Amsterdam).

CURRY, H.B.
(1942) The inconsistency of certain formal logics, JSL, 7, 115-117.

CURRY, H. and R. FEYS
(1958) *Combinatory Logic I* (North Holland, Amsterdam).

CURRY, H., R. HINDLEY and J. SELDIN
(1972) *Combinatory Logic II* (North Holland, Amsterdam).

DAVIES, N.
(1991) A first-order theory of truth, knowledge and belief, in: J.van Eijck, (ed.), *Logic in AI*, (Springer, Berlin), 170-179.

DERSHOWITZ, N. and J. P. JOUANNAUD
(1990) Rewrite systems, in : J. van Leeuwen (ed.), *Handbook of Theoretical Computer Science,* chapter VI (North Holland, Amsterdam).

DERSHOWITZ, N. and M. OKADA
(1988) Proof theoretic tecniques for term rewriting theory, *3rd Annual Symposium on Logic in Computer Science* (Edinburgh, Scotland, July 1988, IEEE Computer Science Press, Silver Spring, MD), 104-111.

DEVLIN, K.
(1977) Constructibility, in: J. Barwise (ed.), *Handbook of Mathematical Logic* (North Holland, Amsterdam), 453-502.

ENGELER, E.
(1981) Algebras and combinators, *Algebra Universalis*, 13, 389-392.
(1988) Cumulative logic programs and modelling, in: F. R. Drake, J. K. Truss (eds.), *Logic Colloquium '86* (North Holland, Amsterdam), 83-93.

ERSHOV, Y. L.
(1977) Model C of partial continuous functionals, in: R. O. Gandy, M. Hyland (eds), *Logic Colloquium '76* (North Holland, Amsterdam), 455-467.
(1985) Σ-predicates of finite type over an admissible set, *Algebra i Logika*, 24, 499 -536 (English translation: *Algebra and Logic*, 1986, 326-351)

FEFERMAN, S.
(1964) Systems of predicative analysis I, JSL, 29, 1-30.
(1968) Systems of predicative analysis II, JSL, 33, 193-200.
(1974) Non-extensional type free theories of partial operations and classifications I, in: J. Diller and G. H. Müller (eds.), *Proof Theory Symposium*, SLNM 500, 73-118.
(1975) A language and axioms for explicit mathematics, in : SLNM, 450, 87-139.
(1977) Theories of finite type related to mathematical practice, in: J. Barwise (ed.), *Handbook of Mathematical Logic* (North Holland, Amsterdam), 913 -971.
(1977a) Categorical foundations and foundations fo category theory, in : R. Butts, J. Hintikka (eds), *Logic, Foundations of Mathematics and Computability* (Reidel, Dordrecht), 149-169.
(1979) Constructive theories of functions and classes, in: D. Van Dalen, M. Boffa (eds.), *Logic Colloquium '78* (North Holland, Amsterdam), 159-224.
(1982) Iterated inductive fixed point theories: application to Hancock's conjecture, in: G.Metakides (ed.), *Patras Logic Symposion* (North Holland, Amsterdam), 171-195.
(1984) Towards useful type-free theories I, JSL, 49, 75-111.
(1985) Intensionality in mathematics, JPL, 14, 41-55.
(1985a) A theory of variable types, *Revista Colombiana de Matematicas*, vol.XIX, 95-105.
(1990) Polymorphic typed lambda-calculi in a type-free axiomatic framework, in:

Logic and Computation (Contemporary Mathematics 104, AMS, Providence RI), 101-136.

(1991) Reflecting on incompleteness, JSL, 56, 1-49.

(1991a) Logics for termination and correctness of functional programs II. Logics of strength PRA, in: P. Aczel, H. Simmons, S. Wainer eds. *Proof Theory* (Cambridge University Press, Cambridge), 195-225.

(1992) Logics for termination and correctness of functional programs, in: Y. N. Moschovakis (ed.), *Logic from Computer Science* (Springer, Berlin -Heidelberg).

FEFERMAN, S. and W. SIEG

(1981) Iterated inductive definitions and subsystems of analysis , in: W. Buchholz, S. Feferman, W. Pohlers and W. Sieg, SLNM 897, 16-77.

FITCH, F. B.

(1936) A system of formal logic without an analogue to Curry W-operator, JSL, 1, 92-100.

(1942) A basic logic, JSL, 7, 105,-114.

(1946) Self-reference in philosophy, *Mind*, 55, 64-73.

(1948) An extension of basic logic, JSL, 13, 95-106.

(1950) A demonstrably consistent mathematics I, JSL, 15, 17-24.

(1951) A demonstrably consistent mathematics II, JSL, 16, 121-124.

(1963) The system CΔ of combinatory logic , JSL, 28, 87-97.

(1964) Universal metalanguages in philosophy, *Review of Metaphysics,* 17, 396-402

(1967) A complete and consistent modal set theory, JSL, 32, 93-103.

(1980) A consistent combinatory logic with an inverse to equality, JSL, 45, 529--543.

FITTING, M.

(1981) *Fundamentals of Generalized Recursion Theory* (North Holland, Amsterdam).

(1986) Notes on the mathematical aspects of Kripke's theory of truth, NDJFL, 27, 75-88.

FLAGG, R.

(1987) κ-continuous lattices and comprehension principles for Frege structures, APAL, 36, 1-16.

FLAGG, R. and MYHILL, J.

(1987) Implications and analysis in classical Frege structures, APAL, 34, 33-85.

(1987a) An extension of Frege structures, *Lecture Notes in Pure and Applied Mathematics*, 106, (M.Dekker, New York), 197-217

(1989) A type-free system extending (ZFC), APAL, 43, 79-97.

FRAASSEN, B. van

(1968) Presupposition, implication and self-reference, *Journal of Philosophy*, 65, 135-152

(1970) Inference and self-reference, *Synthèse*, 21, 425-438.

FREGE, G.

(1903) *Grundgesetze der Arithmetik* I, II (Olms , Hidesheim, 1962).

(1975) *Funktion, Begriff, Bedeutung* (G.Patzig, ed., Vandenhoek & Ruprecht, Göttingen)

(1984^2) *Nachgelassene Schriften* (H. Hermes, F. Kambartel, F. Kaulbach, eds., F.Meiner, Hamburg).

FRIEDMAN, H.

(1969) Bar Induction and Π^1_1-CA, JSL, 34, 353-362.

(1975) Some systems of second order arithmetic and their use, *Proc. 1974 International Congress of Mathematicians*, Vancouver, vol.1, 235-242.

(1976) Uniformly defined descending sequences of degrees, JSL, 41, 363-367.

(1976a) Subsystems of second order arithmetic with restricted induction I, II (abstracts), JSL, 41, 557-559.

(1980) A strong conservative extension of Peano arithmetic, in : J. Barwise, H. J. Keisler, K. Kunen (eds.), *The Kleene Symposium* (North Holland, Amsterdam), 113-122.

(1981) On the necessary use of abstract set theory, *Advances in Mathematics*, 41, 209-280.

FRIEDMAN, H., K. MC ALOON and S. G. SIMPSON

(1982) A finite combinatorial principle equivalent to the 1-consistency of predicative analysis, in: G. Metakides (ed.), *Patras Logic Symposion* (North Holland, Amsterdam), 197-230.

FRIEDMAN, H. and M. SHEARD

(1987) An axiomatic approach to self-referential truth, APAL, 33, 1-21.

FRIEDMAN, H., S. G. SIMPSON and R. SMITH

(1983) Countable algebra and set existence axioms, APAL, 25, 141-181.

GALLIER, J.

(1991) What's so special about Kruskal's theorem and the ordinal Γ_0 ? A survey of some results in proof theory, APAL, 53, 199-260.

GANDY, R.

(1974) Inductive definitons, in: J.E.Fenstad, P.G.Hinman (eds.), *Generalized Recursion Theory* (North Holland, Amsterdam), 265-299.

GILMORE, P. C.

(1974) The consistency of partial set theory without extensionality, in: T. Jech(ed.), *Axiomatic Set Theory* II (Proceedings of Symposia in Pure Mathematics, 13, A.M.S, Providence R.I), 147-153.

(1980) Combining unrestricted abstraction with universal quantification, in: J.Seldin, R.Hindley(eds), *To H.B.Curry: Essays on Combinatory Logic, Lambda Calculus and Formalism* (Academic Press, New York), 99-123.

(1986) Natural deduction based set theories : a new resolution of the old paradoxes, JSL, 51, 393-411.

(1990) How many real numbers are there ?, (Amsterdam, preprint)

GIRARD, J. Y.

(1982) A survey of Π^1_2-logic, in: L. Cohen, J.Los, H. Pfeiffer, K. P. Podewski (eds.), *Logic, Methodology and Philosophy of Science VI* (North Holland, Amsterdam), 89-107.

(1987) *Proof Theory and Logical Complexity* (Bibliopolis, Napoli, 1987, vol.1).

GIRARD, J. Y. and J. VAUZEILLES

(1981) Les premières récursivement inaccessible et Mahlo et la théorie des dilatateurs, AML, 24, 167-191.

GÖDEL, K.

(1944) Russell's mathematical logic, in: P.A.SCHILPP (ed.), *The Philosophy of Bertrand Russell* (Northwestern University Press, Evanston, Ill.), 125-153.

GORDEEV, L.

(1989) Proof theoretic analysis: weak systems of functions and sets, APAL, 38, 1-121.

GUPTA, A.

(1982) Truth and paradox, JPL, 11, 1-60

HALBACH, V.

(1994) A system of complete and consistent truth, NDJFL (to appear)

HAREL, D. and D. KOZEN

(1984) A programming language for inductive sets and applications, *Inform. Control*, 63, 118-139.

HARPER,R., F. HONSELL and G. PLOTKIN

(1987) A framework for defining logics, *Proceedings of the 2nd annual conference on Logic in Computer Science* (IEEE Computer Press, Washington DC)

HARRISON, J.

(1968) Recursive pseudowellorderings, *Trans. Amer. Math. Soc.*, 131, 526-543.

HAYASHI, S. and H. NAKANO

(1988) *PX: A Computational Logic* (Foundations of Computing Series, MIT Press, Cambridge Mass.).

HERZBERGER, H.

(1982) Notes on naive semantics, JPL, 11, 1-60.

HILBERT, D. and P. BERNAYS

(1938) *Die Grundlagen der Mathematik* I, II (Springer, Berlin-Heidelberg).

HINDLEY, R. and J. SELDIN

(1986) *Introduction to Combinators and λ-Calculus* (London Mathematical Society Student Texts 1, Cambridge University Press, Cambridge).

HINMAN, P. G

(1978) *Recursion-Theoretic Hierarchies* (Springer, Berlin Heidelberg).

JÄGER, G.

(1980) Beweistheorie von KPN, AML, 20, 53-64.

(1982) Zur Beweistheorie der Kripke Platek Mengenlehre über den naturlichen Zahlen, AML, 22, 121-141.

(1984) A version of Kripke Platek set theory which is conservative over Peano arithmetic, ZMLG, 30, 2-9.

(1984a) The strength of admissibility without foundation, JSL, 49, 867-879.

(1986) *Theories for Admissible Sets* (Bibliopolis, Naples).

(1987) Induction in the elementary theory of types and names, SLNCS, 329.

JÄGER, G. and W. POHLERS

(1982) Eine beweistheoretische Untersuchung von Δ^1_2-CA+BI und verwandter Systeme, *Sitzungsberichten der bayerischen Akademie der Wissenschaften*, 1-28.

JÄGER, G. and T. A. STRAHM

(1994) Totality in applicative theories, APAL, to appear

JERVELL, H.R.

(1978) Constructive universes I, in: G. H. Müller and D. Scott, (eds.), *Higher Set Theory*, SLNM 69, 73-98.

JOCKUSCH, C.jr. and S. G. SIMPSON

(1975) A degree theoretic definition of the ramified analytical hierarchy, APAL,10, 1-32.

KELLEY, J. L.

(1955) *General Topology* (Van Nostrand, New York).

KLEENE, S. C.

(1958) Extension of an effectively generated class of functions by enumeration, *Colloq. Math.*, 7, 67-78.

(1959) Quantification of number theoretic functions, *Compositio Mathematica*, 7, 67-78.

(1952) *Introduction to Metamathematics* (Wolters-Nordhoff and North Holland, Amsterdam).

KOYMANS, K.

(1984) *Models of the Lambda Calculus* (Math. Center, Tracts, no.9, Amsterdam).

KREISEL, G.

(1967) Informal rigour and completenes proofs, in: I.Lakatos (ed.), *Problems in the Philosophy of Mathematics* (North Holland, Amsterdam), 138-171.

(1968) A survey of proof theory I, JSL, 33, 321-88.

(1971) Some reasons for generalizing recursion theory, in: R. O. Gandy, C. M. Yates (eds.), *Logic Colloquium '69* (North Holland, Amsterdam), 139-198.

KREISEL, G. and G. SACKS

(1965) Metarecursive sets, JSL, 30, 318-338.

KRIPKE, S.

(1975) Outline of a theory of truth, *Journal of Philosophy*, 72, 690-716.

LAMBEK, J. and P. SCOTT

(1986) *Introduction to Higher Order Categorical Logic* (Cambridge University Press, Cambridge).

LONGO, G.

(1983) Set theoretical models of λ-calculus: theories, expansions, isomorphisms, APAL, 24, 153-188

LORENZEN, P. and J.MYHILL

(1959), Constructive definitions of certain analytic sets of numbers, JSL, 24, 37-49

MARTIN, R. M. (editor)
(1984) *Recent Essays on Truth and the Liar Paradox* (Oxford University Press, Oxford).

MARTIN-LÖF, P.
(1984) *Intuitionistic Type Theory* (Bibliopolis, Naples).

MARZETTA, M.
(1993) Universes in the theories of types and names, in: E.Börger et alii (eds.), *Computer Science Logic '92*, SLNCS 702, 340-351.

MC-GEE, V.
(1985) How truthlike can a predicate be? A negative result, JPL, 14, 399-410.
(1991) *Truth, Vagueness, and Paradox* (Hackett, Indianapolis-Cambridge).

MEYER, A.
(1982) What is a model of the lambda calculus, *Inform.Control*, 52, 87-122.

MINTS, G.
(1975) Finite investigations of transfinite derivations, *J.Soviet Math.*, 10 (1978), 548-596 (English translation from : Zap. Nauchn. Semin. LOMI, 49).

MINARI, P.
(1987) Some aspects of Kripke's theory of truth, (unpublished notes, Florence)

MONTAGUE, R.
(1963) Syntactic treatment of modality, with corollaries on reflection principles and finite axiomatizability, *Acta Philosophica Fennica,* 16, 153-167.

MOSCHOVAKIS, Y. N.
(1974) *Elementary Induction on Abstract Structures* (North Holland, Amsterdam).
(1984) Abstract recursion as a foundation of the theory of the algorithms, *Computation and Proof Theory,* SLNM 1104, 289-364.
(1989a) A formal language of recursion, JSL 54, 1216-1252.
(1989b) A mathematical modeling of pure, recursive algorithms, in : A. R. Meyer and M.A.Taitslin, *Logic at Botik '89*, SLNCS 363, 208-229.
(1990) Sense and reference as algorithm and value, *Logic Colloquium '90*, to appear.

MUSIL, R.
(1913) Der mathematische Mensch, in: R.Musil, *Gesammelte Werke*, Bd.8 (Rowohlt, Reinbek bei Hamburg, 1978), 1004-1008.

MYHILL, J.

(1950) A system which can define its own truth, *Fund.Math.*, 37, 190-92.

(1972) Levels of implications, in: A. Ross, R. Barcan Marcus, R. M. Martin (eds.), *The Logical Enterprise* (Yale University Press, New Haven).

(1984) Paradoxes, *Synthèse*, 60 , 129-142.

NEPEIVODA, N.

(1973) A new concept of predicative truth and definability, English translation from: *Matematischskie Zametki*, 13, 735-745.

NÖRDSTROM, B., K. PETERSSON and J. SMITH

(1990) *Programming in Martin-Löf's Type Theory* (Oxford University Press, Oxford).

ODIFREDDI, P. G.

(1989) *Classical Recursion Theory* (North Holland Amsterdam).

PARSONS, C.

(1972) On n-quantifier induction, JSL, 36, 466-482.

PERLIS, D.

(1985) Languages with self-reference I. foundations, *Artificial Intelligence* 25, 301-322.

(1988) Languages with selfreference II: Knowledge, belief and modality, *Artificial Intelligence*, 34, 179-212.

PLOTKIN, G.

(1990) An illative theory of relations, in : R. Cooper, K. Mukai, J.Perry (eds.), *Situation Theory and Its Applications* (CSLI, Stanford 1990), 134-146.

POHLERS, W.

(1989) *Proof Theory. An Introduction*, SLNM, 1407.

POINCARÉ, H.

(1913) *Dernières Pensées* (Flammarion, Paris).

RATHJEN, M.

(1991) Proof-theoretic analysis of KPM, AML, 30, 377-403.

(1992) Fragments of Kripke-Platek set theory with infinity, in: P. Aczel, H. Simmons, S. Wainer (eds.), *Proof Theory* (Cambridge University Press, Cambridge), 251-273

RESSAYRE, J.P.

(1982) Bounding generalized recursive functions of ordinals by effective functors: a complement to the Girard theorem, in: J.Stern, *Logic Colloquium '82* (North Holland, Amsterdam), 251-280.

RICHTER, W.

(1971) Recursively Mahlo ordinals and inductive definitions, in: R. O. Gandy, C. M. Yates (eds.), *Logic Colloquium '69* (North Holland, Amsterdam), 273-288.

ROGERS, H. jr.

(1967) *Theory of Recursive Functions and Effective Computability* (Mc Graw-Hill, New York).

SCHELLINX, H.

(1991) Isomorphisms and nonisomorphisms of graph models, JSL, 56, 227-249.

SCHÖNFINKEL, M.

(1924) Über die Bausteine der mathematischen Logik, *Mathematische Annalen,* 92, 305-316.

SCHÜTTE, K.

(1960) *Beweistheorie* (Springer, Heidelberg).

(1977) *Proof Theory* (Springer, Heidelberg).

SCHWICHTENBERG, H.

(1971) Eine Klassifikation der ϵ_0-rekursiven Funktionen, ZMLG, 17, 61-74.

(1977) Some applications of cut elimination, in: J. Barwise (ed.), *Handbook of Mathematical Logic* (North Holland, Amsterdam), 867-895.

(1992) Proofs as programs, in: P. Aczel, H. Simmons, S. Wainer (eds). *Proof Theory* (Cambridge University Press), 79-113.

SCOTT, D.

(1975) Combinators and classes, in: C. Böhm (ed.), *λ-Calculus and Computer Science,* SLNCS 37, 1-26.

(1975a) Some philosophical issues concerning theories of combinators, ibid. 346-366.

(1976) Data types as lattices, *SIAM Journal of Computing,* 51, 522-587.

(1980) Lambda calculus: some models, some philosophy, in: J. Barwise, H. J. Keisler, K. Kunen, (eds.), *The Kleene Symposium* (North Holland, Amsterdam), 223-265.

(1980a) Relating theories of λ-calculus, in : J. P. Seldin and J. R. Hindley, (eds.), *To H.B.Curry: Essays on Combinatory Logic, Lambda Calculus and Formalism* (Academic Press, New York), 403-450.

(1982) Domains for denotational semantics, SLNCS 140, 577-613.

SELDIN, J. R.

(1976) Recent advances in Curry's program, *Rendiconti del Seminario Matematico dell' Università e del Politecnico di Torino,* 35,77-88.

(1980) Curry's program, in : J. P. Seldin and J. R. Hindley (eds.), *To H.B.Curry:Essays on Combinatory Logic, Lambda Calculus and Formalism* (Academic Press, New York), 3-33.

SHOENFIELD, J.

(1967) *Mathematical Logic* (Addison Wesley, Menlo Park).

SIMPSON, S. G.

(1982) Set theoretic aspects of ATR_0, in : D. van Dalen, D. Lascar, J. Smiley (eds.), *Logic Colloquium '80* (North Holland, Amsterdam), 255-271.

(1984) Which set existence axioms are needed to prove the Cauchy- Peano theorem for ordinary differential equations?, JSL, 49, 783-802.

(1987) Subsystems of Z_2 and reverse mathematics, in: G.Takeuti, *Proof Theory* (North Holland, Amsterdam, 2nd edition), 432-446.

(198?) *Subsystems of Second Order Arithmetic*, typewritten notes, chapters 1-5; 6-9.

SMITH, J.

(1984) An interpretation of Martin-Löf's type theory in a type-free theory of propositions, JSL, 49, 730-753.

SMULLYAN, R.

(1961) *Theory of Formal Systems* (Annals of Mathematical Studies, Princeton University Press, Princeton NJ).

TAIT, W.

(1968) Normal derivability in classical logic, in: J. Barwise, (ed.), *The Syntax and The Semantic of Infinitary Languages*, SLNM 72 , 204-236.

TAKEUTI, G.

(1975) *Proof Theory* (North Holland, Amsterdam; 2nd edition 1985).

(1978) *Two Applications of Logic to Mathematics* (Princeton University Press, Princeton, NJ and Publicat. Math.Soc.Japan) .

TALCOTT, C.

(1992) A theory of program and data type specification, TCS 104, 129-159.

TARSKI, A.

(1956) *Logic, Semantics, Metamathematics: Papers from 1923 to 1938* (J.Woodger, ed., Clarendon Press, Oxford).

(1965) A simplified version of predicate logic with identity, AML 7, 61-79.

TROELSTRA, A. S.

(1973) *Metamathematical Investigation of Intuitionistic Arithmetic and Analysis,* SLNM 344.

TROELSTRA, A. S. and D. van DALEN

(1988) *Constructivism in Mathematics I, II* (North Holland, Amsterdam).

TURNER, R.

(1987) A theory of properties, JSL, 52, 455-471.

(1990) *Truth and Modality for Knowledge Representation* (Pitman, London).

(1990a) Logics of truth, NDJFL, 31, 308-329.

VEBLEN, O.

(1908) Continuous, increasing functions of finite and transfinite ordinals, *Trans. Amer. Math. Soc.*, 9, 280-293.

VISSER, A.

(1984) Four-valued semantics and the liar, JPL, 13, 181-212.

(1989) Semantics and the liar paradox, in: D. Gabbay and F. Guenthner (eds.), *Handbook of Philosophical Logic* IV (Reidel, Dordrecht), 617-706.

WEYL, H.

(1918) *Das Kontinuum* (Von Veit, Leipzig).

WHITMAN, P.M.

(1943) Splitting of lattices, *American Journal of Mathematics*, 65, 179-196.

WOODRUFF, P.

(1984) Paradox, truth and logic. Part I: paradox and truth, JPL, 13, 213-233.

INDEX

The bibliography is not covered by the following index.

abstraction
- explicit 60, 77
- extended 222
- internal 353
- lambda (λ-) 15, 30, 34, 36, 121, 219
- local, 222
- operation 2, 44
- principle 1, 56
- unrestricted 1

Ackermann 217

Aczel 5, 6, 8, 43, 53, 66, 163, 217, 230, 254, 380, 402

admissible hull 125, 131

admissible ordinal 230

analysis
- arithmetical 240
- hyperarithmetical 240
- predicative 7, 241
- recursive 241

application (operation) 14, 29, 30, 33, 121

approximation
- αth 181
- axioms 99, 104
- operation (π) 98
- theorem 99, 109

Apt 264

arithmetic
- Peano 14, 27, 28, 39, 66, 70, 90, 91, 101, 131, 151, 241, 243, 338, 355, 356, 373, 380, 383, 411
- primitive recursive 39, 40
- second order 28, 33, 65, 239, 248

arithmetical
- comprehension 240, 243
- formula 239
- operator 66, 341
- set 240

arity 403

Asperti 34

axiom(s)
- bounded quantifier 250
- choice 128
- class N-induction 66, 352
- combinatory 139
- completeness 54
- connection 220
- consistency 49, 290 304, 314
- enumeration 27, 32, 91, 151, 152, 163
- extensional choice 129
  - dependent choice, 129
- extensionality (for operations), 16
- extensionality (for sets) 199
- independence 164
- N-induction 15
- level 220
- limit 220
- local consistency 219
- local N-induction 219
- local truth 219
- localization 220
- Loeb 389
- natural number (also

number-theoretic, Peano)
15, 39, 219 , 290, 304
negative soundness 220
ontological 139
operational 15, 219, 289, 314
pair 139
pairing 15
persistence 220
π-axioms 104
positive soundness 220
potential completeness 220
projectibility 219
property N-induction 66, 357
restriction 49, 50, 54, 209
schema, *see* schema
T-axioms 49, 50
union 139

Baire 243
bar induction, 241
Barendregt 17, 18, 20, 21, 22, 25, 28, 32, 34, 38, 39, 424
Barwise 131, 207, 230, 232, 233, 235, 256
basis
change of, 81
for a topology, 114, 117, 121
Beeson 27, 59, 75
Behmann 151, 162, 163
Belnap 55
Bendixson 243
Bernays 238
Birkhoff 11, 190, 192
Bishop 75
bounded formula, *see* formula
bounded level quantifier, 237, 249, 250, 287
bounded number quantifier 39, 64, 239
bounded set quantifier, 138, 198, 230, 238
boundedness theorem 147
Boyd 262
Buchholz 278, 279, 312, 335, 338, 373
Burge 218
Burgess 205, 206

calculus
lambda (λ-) 13, 16, 34, 35
sequent 286, 288, 289, 290, 291, 294, 303-305, 314 351, 358, 361, 369
canonical
enumeration 28
expression 407
model, realization 406, 410
term 407
term (for logical operators) 45
Cantini 101, 144, 178, 199, 337, 355, 356, 373.
Cantor 42, 243
Cauchy 243
characterization theorem 376
Church 22, 25, 26
Church's thesis 20, 27, 28, 32
Church-Rosser theorem 13, 22, 23, 26, 38, 41, 151, 152, 162, 165, 169
class 55, 56, 58-64, 71, 73, 74, 75
explicit 79
of level i 218, 219, 222, 224,225
class compactness 104, 105, 111, 122
class finiteness 114
closed
unbounded set 269
Γ-closed 47, 48, 50, 179-181
Φ_0-closed 357, 358
club, *see* closed unbounded set
CODE 340
coherent
element 204
poset 182
coinduction 177, 178, 207
combinator 14, 16, 20, 21, 27
K, S 14, 15

paradoxical 17
combinatorial operation 60, 77
combinatory algebra 2, 3, 5, 6, 13, 16
combinatory logic 13, 14, 15, 16, 18, 21, 25, 34
complement 59
complete
coherent poset (ccpo) 182
lattice 183, 185, 186
set 47, 208, 357
completeness
$\mathcal{L}_{op}$-completeness 363
ω-completeness 376
complexity
logical 10, 257, 264-68, 281 285, 293, 295, 313, 327, 336, 370
n-complexity 303, 305, 307, 324
comprehension
analytical 203
elementary 58, 60, 61, 62, 67, 223, 229, 242, 243, 253
second order 63, 178, 200, 239
type-free 44
computation relations 407
computation triples 408
concatenation 244, 272, 341
connection axioms 220
conservation, *see* theorem
consistency axiom, *see* axiom
consistency lemma 322, 363,
consistency theorem 156
consistent set 47, 179
Constable 402
constructive completeness 115
continuity 27, 29, 30, 31, 36, 103, 109, 112, 118-120, 144
continuous operator, *see* operator
conversion 16
covering 147
creative set (property) 71-72
Crossley 259
Curry 2, 17, 21, 25, 43, 47, 53, 151
currying 403
cut elimination theorem
for OP^{∞} 327
for the calculus $\mathcal{H}$ 364
partial 324
primitive recursive 347
Schütte's 319
Tait's 320
weak 295, 301
cut rule 155, 291, 315

Dalen, D. van 27
decomposition theorem 393
Dedekind 152, 243
deducibility
abstract 153
of level n, 157
definability theory 85
definition by cases 14, 15
denotation 422
dense, *see* set
dependent choice, *see* schema
derivability relation for
IT_n^{∞} 305
OP^{∞} 327
RS_n316
Σ_{∞} 158
STLR 291
$STLR^{\infty}$ 294
$STLR_n$ 297
derivative 270
Dershowitz 283
Devlin 232
diagonalization 76, 77
domain operation 59
dual extensional membership 132, 133

effective
inseparabilty 108
operator 118
elementary
atom 46
comprehension 58
extensional 130
formula 57
predicate 60
elimination lemma 296
embedding
of admissible sets 135
theorem 190
encoding 68
of logical operators 43, 80
Engeler 13, 28, 33
enumeration
axiom 27, 126, 144, 151, 152, 163
theorem 32, 42
envelope 88
equality 14, 15
definitional 403-407
extensional 28, 60, 70, 108, 110,
intensional 429, 431
level 220
pointwise 113, 121,
set-theoretic 132, 133
Ershov 104, 114, 121
exact representation 108, 122
expansion (operation) 60, 78
explicitly CL-continuous 118
explicitly open 116
exponential (ω-) 274
exponentiation 61
extensional
choice, *see* choice
equality, *see* equality
membership 134
model 34
extensional_1 (for properties) 70, 110
extensional_2 (operation) 110, 128
2-extensional 129
extensionality axiom, *see* axiom
extensionality
for properties, classes 73-74
for sets 135, 199

family
$\mathcal{M}$-normal 387
RS- 116
Feferman 2, 5, 6, 7, 44, 50, 55, 57, 58, 59, 66, 75, 80, 151, 152, 162, 163, 165, 166, 199, 203, 217, 226, 238, 241, 243, 253, 257, 273, 278, 402, 421
field (of a relation) 126
first recursion theorem
(analogue of) 104, 120
Fitch 2, 5, 43, 217, 379, 380, 386, 388, 396
Fitch's internal logic 388
Fitch's models 386
Fitch's theory 388
Fitting 70, 71, 110
fixed point
axiom 50, 66
complete 47, 179
consistent 47, 179
dual 181
Γ-, 48
intrinsic 186,
largest 50, 181
least 48, 88, 181
maximal 186
Φ_0-, 357
Φ_1-, 375,
Φ_2-, 387.
model 179, 186, 352, 358, 379
fixed point theorem
for operations 16
for monotone operations 112
for predicates or
properties 63

fixed point theory 50
Flagg 33, 39, 103, 151, 152,
156, 162,
formula
analytical 199
arithmetical 239
bounded 39, 40, 138, 230, 239
elementary 57
elementary extensional 130
η-free 199
operative 64
positive elementary 88
quasi-elementary 57
stratified 199
T-negative 53,
T-positive 53, 220
Fraassen, van B. 352, 357
Frege 1, 2, 422
Frege structures 2, 43, 53
Friedman 7, 217, 227, 238,
241, 243, 258, 261, 282-83
379, 380, 383, 384, 394
function
continuous on ordinals 269
$\subseteq$-increasing 49, 181, 208
increasing on ordinals 269, 391
normal 269
uniformly Δ_1-, Σ_1-, 231
fundamental sequence 275, 276

Gale 243
Gallier 282, 283
Gandy 91, 261, 262
Gilmore 73, 75, 198
Girard 7, 9, 253, 271
global consistency 222
Gödel 1, 60
Gödel numbering 42, 68, 82,
336, 338
Gordeev 73, 74
graph model 13, 28, 34
graph of a continuous operator
28, 121
grounded element 204
Gupta 394

Halbach 385
Harper 402
Harrison 258
Hayashi 402
Hensel 262
Herbrand 365, 367, 372
Herzberger 380, 390, 394
hierarchy
ramified 261, 263, 267
Veblen 270
Hilbert 15, 238
Hindley 18, 21, 22, 34
Hinman 91, 230, 233, 237, 256
Honsell 402
hyperarithmetical analysis 240
hyperelementary set 87
hyperjump 237, 258

ideal completion 115
implication
levels of, 151, 152
R- 153
incoherent element 204
independence 45, 80, 81, 164,
292, 320, 407, 413
induction, *see* axiom, schema
inductive
definition 63-67, 85, 87,
232, 257, 351, 355
model 43, 58, 85, 86,
88, 91, 97
set 88
intensional equivalence 151,
162, 174
interpretation
asymmetric 7, 299, 311,
324, 325, 336
provability 369, 370, 372
intersection 59

generalized 61
intrinsic element 204
intrinsic fixed point 186
inversion lemma 234, 295, 317, 345, 363
involution 182
iteration Δ-, 390
Iterationsprinzip 217, 229

Jäger 27, 74, 144, 217, 226, 337
Jockusch 262
join 58, 59, 79, 80, 223, 225, 226, 228, 229, 253

Kalmar 343
Kelley 114
Kleene 3, 7, 20, 28, 80, 91, 98, 104, 244, 245, 246, 248, 258, 262, 343, 423
Knaster 113, 358, 424
König 243
Koymans 34
Kreisel 4, 114, 238
Kripke 5, 6, 44, 139, 177, 178, 186, 203, 231, 351, 420, 421, 423, 424
Kripke's classification 203
Kripke-Platek set theory 139
Kruskal 282

label 339
lambda calculus 16, 17, 18
lattice 179
complete 183
involutive 182
non-modular 192
lazy evaluation 407, 408
level
axioms 220, 290
induction, 293
lowering lemma 328
of implication 151
of truth 215
Levy 206
liar (sentence) L 52, 204, 374, 424
local consistency 219
local truth axioms 219
local truth lemma 370
LOG- rule 353
logic
combinatory 13-16
deontic 386
external 180, 399
Fitch's internal 388
Friedman-Sheard 384
internal 180, 380, 394, 396, 400
modal 396
M-logic 376
type free 6, 151, 174
logical consequence 351, 355, 379, 384, 386, 388, 398
logical theory of constructions 401-403, 411
Longo 34, 35
Lorenzen 217
Lusin 243

Marek 262
Martin-Löf 217, 226, 402, 403, 405, 407, 419
Marzetta 74
Mc Gee 7, 380, 381, 383
Meyer 17, 28
Minari 74, 112, 122, 194, 196
minimalization 20
Mints 312, 335, 338
model
β-model 258, 262
closed term 26
D_∞ 35
Engeler's D-, 33
Fitch's, 379, 386
fixed point, *see* inductive
inductive, *see* inductive
open term 26
ω-model (ω-standard) 47, 49,

180, 194, 230, 237, 258, 358, 388
Pω 28-32
recursion-theoretic 215, 230
recursive graph RE, 28, 32
recursively saturated 90
semi-inductive 394
supervaluation 357
monotonicity
of deducibility 153
of ordinal assignement 316, 362
monotone, *see* operation, operator
Montague 383
Moschovakis 6, 8, 64, 80, 85, 88, 98, 131, 206, 207, 262, 401, 422 425
Musil 149
Myhill 1, 4, 6, 39, 44, 68, 71, 103, 109, 112, 147, 151, 152, 153, 156, 160, 162, 217
Myhill's theorem 71
Myhill-Shepherdson's theorem (analogue of) 112

Nakano 402
natural ordinal sum 318
nesting property 277
Nordström 402
normal form
Cantor 271
for combinatory terms 22
no solution lemma 367

Odifreddi 82
operation
basic (for predicates) 59
choice 128
combinatorial 60
$extensional_2$ 110
Kalmar elementary 343
lattice-theoretic 181
monotone 62, 109-110, 112, 113, 117
number-theoretic 20
of ordinal arithmetic 274-275,
selection 127
ξ- 107
operator (formula) 64
closure 29, 36
continuous 28
CL-continuous 117
ECL-continuous 118
effectively continuous 28
RS-continuous 119
elementary 64
existential 65
monotone 29, 88, 120, 144, 207, 234, 371, 387
non-monotone 394
ordering
connected (linear) 126, 227
directed unbounded partial 220
on ω 20
ordinal
admissible 230
closure 253, 392
constructive 64
notation 272
number 145
of predicative analysis Γ_0 271
projectible 231
recursively inaccessible 231
recursively Mahlo 255
stabilization 392

pair 61, 77
axiom 15, 139
ordered 14
pairing
combinator 21
function 42
surjective 31
paradox
Curry's 53, 151
Gordeev's 74

Russell's 56
paradoxical
combinator 17
element 204
parametrization 56
Park 37
Parsons 40, 41, 42
Peano 243, *see* arithmetic
period 392
periodicity 392
Perlis 420, 421
persistence
axioms 220, 290, 296, 304, 314
lemma 298, 321
Plato 11
Plotkin 13, 28, 402, 419
Pohlers 9, 217
Poincarè 75, 349
power set 10, 62, 161
prewellordering 104
primitive recursion 19, 20, 27, 39-40, 404, 409
principle
abstraction 56
choice 126
construction 80
join 58
meta-loeb 373
reducibility 244
reflection 224
CL-reflection 106
$\mathfrak{H}$-reflection 371
process 390
product
cartesian 61
generalized 58
progressive property 126, 227, 259, 354
proposition 51
propositional function 51
provability interpretation 370
pseudo-well-ordering (pwo) 126, 259
acceptable 126, 259
locally decidable 259
unbounded 259
Putnam 262

quasi-well-ordering (qwo) 259
quasi-elementary formula 57

ramified hierarchy 261-263
bounded 267
n-rank 313
Rathjen 256
recursion
arithmetical transfinite 240
Δ_1-, Σ_1-, 232 , 235-37
formal language of, 422
on natural numbers 19
on ordinals 48, 146
special transfinite 260
ω_k-, 302
recursion theorem
first 120
second 343
recursive functions
partial 19, 244-45
primitive 19, 20, 27, 39, 40, 93, 101, 246, 303, 338, 348
provably 40, 91, 101
representability of, 20
recursor 19, 146, 260, 261
reducible formula 317
m-reducible property 71
reducibility, *see* principle
reduction lemma 295, 318, 346
reduction relation 22, 23
infinitary 165, 166-169
reduction theorem 109
reflection 224
$\mathfrak{H}$-reflection 371
repetition rule 338
representable
function 20

set 88
representation theorem for
extensional operations 117
Ressayre 253
reverse mathematics 217, 238, 241
Rice 44, 68, 70, 71, 72, 110, 116,
122
Rice's theorem 71
Rice-Shapiro
family 116
theorem 110, 122
Richter 230, 232, 253, 254, 255
Rogers 70, 71, 82, 258
Russell 56, 70, 90, 151, 162, 215,
222, 223, 351

satisfaction 68, 69
Schellinx 34
schema
ATR 240
bar induction 241
bounded collection 139
bounded complete induction
139
bounded separation 139
choice 33
Δ-comprehension 108
Δ_1^1-comprehension 240
Π_1^1-comprehension 241
Σ_1^1-dependent choice 240
elementary choice 130
elementary comprehension 58
elementary dependent choice
130
explicit abstraction 60
extended abstraction 222
generalized coinduction 207
generalized induction 87
Herbrand 372
N-induction 15, 50
$\exists(+)$-N-induction 27
internal abstraction 353
level transfinite induction 293
local abstraction 222
CL-reflection 106
REFL^+ 237
second-order comprehension
200, 239
soundness 54
Tarski's 53, 385
transfinite induction 259,
278, 354
$\in$-transfinite induction 139
transfinite recursion 227-229
Turner's 396
type-free abstraction 56

Schütte 7, 9, 217, 241, 243, 257,
269, 270, 271, 273, 278, 280,
312, 316, 319
Schwichtenberg 302, 304, 312, 335,
338, 402
Scott 5, 13, 17, 28, 32, 33, 34, 37,
104, 114, 217, 410
Scott's extension theorem, 32
Scott topology 28, 104, 114
section 90
Seldin 18, 21, 22, 34
selection 127
semi-inductive
definition 390
model 394
sense 422
separation 98, 108,
CL-, 107
sequent calculus, 286, 303-304,
314, 361
set
admissible 125, 137
arithmetical 240
bounded (of ordinals) 269
closed (of ordinals) 269
Γ-closed 47
Φ_0-closed 357

coinductive 89
complete 47
consistent 47
$\mathcal{J}$-definable 262
Σ_1-, Δ_1-definable 231
Δ-dense 393
Γ-dense 179
Φ_0-dense 357
Φ_2-dense 387
hyperelementary 89
inductive 89
iterative 63
representable 88
Shapiro 110, 116, 117, 122
Shepherdson 103, 109, 112
Shoenfield 14, 246, 376
Sierpinski 243
Simpson 217, 238, 241, 243, 262
Smullyan 109
soundness 153
formalized $\mathcal{H}$-, 370
positive, negative 220
Δ^+-, 221
Σ_∞-, 160
splitting pair 195
stabilization theorem 391
stably inside 390
stably outside 390
n-stage 313
Stewart 243
Strahm 27
subsequence relation 244, 341
substitution
closure 289
instance 289
lemma 292, 316, 344
substitutivity 292, 295, 320
subsystems of second order arithmetic 238-239
sum
direct 61
generalized 58
supervaluation model 357
Suslin 80, 98

Tait 286, 289, 314, 320, 324, 333, 348
Tait's 2nd cut elimination 320
Takeuti 9, 238, 243
Talcott 402
Tarski 37, 44, 53, 65, 70, 113, 125, 201, 215, 352, 358, 385, 412, 424
tautology lemma 292, 295, 320, 363
term model, *see* model
theorem
approximation 99
boundedness 147
cardinality 191
characterization 376
conservation 101, 335
decomposition 393
embedding 190
fixed point for operations 16
fixed point for predicates 63
generalized induction 87
internal N-induction 368
Kleene basis 246
Knaster-Tarski 113
Levy absoluteness 206
Myhill-Shepherdson 112
perfect set 206
periodicity 392
reduction 107
representation 119
Rice 71
Rice-Shapiro 110
separation 107, 108
stabilization 391
Suslin-Kleene 80, 98
transfinite induction 354
tree 246
uniform ordinal comparison 94
upper bound 328

theory
admissible set 139
minimal frame MF 50
of operations OP 15
prewellordering PW 106
truth with levels TL 219
VF 356
topology
class 114
positive information 28
RS-topology 117
translation 140, 174, 201, 242, 247, 305, 372, 413, 417
lemma 247
transpose 60
tree, recursive wellfounded 245
Troelstra 27
T-rules 361
truth
reflective, self-referential 2, 5, 6, 7, 43, 44, 50, 51, 85, 103, 104, 120, 125, 151, 177, 178, 180, 196, 198, 206, 215, 216, 217, 218, 220, 223, 230, 249, 257, 258, 285, 286, 303, 311, 351, 379, 394, 401, 403, 412, 420, 422, 423, 424, 425
stable 394, 395, 396, 399
Turner 380, 394, 396
type 402, 403, 406
finite 64, 65, 75

Ulm 243
uniform ordinal comparison 94
ungrounded element 204
union 71
generalized 61
universe 61, 226
unparadoxical element 204

variable 9
individual 14
for levels 218
for set 198, 239
Vaught 201
Vauzeilles 253
Veblen 269, 270
Visser 194, 390

weakening 292, 316, 345, 363
well-founded, *see* tree
well-ordering 258-259
predicative 269, 277
Weyl 83, 213, 215, 217, 225, 229, 257, 260
Weyl's principle 229
Zorn's lemma 187

LIST OF SYMBOLS

Part I lists the abbreviations designating formal systems, arranged in order of appearance. Part II contains abbreviations for axioms, axiom schemata and rules, while Part III contains basic abbreviations and symbols.
In parts II-III, the list is arranged per chapters and, within each chapter, in order of appearance. We give the page number of the first occurrence of the each symbol we consider.

I. Formal Systems

PC, I, 15	classical predicate logic
OP, I, 15	theory of operation
OP^-, I, 16	" " without N-induction
$OP\lambda^-$, I, 17	OP^- based on λ-calculus
CL, I, 21	pure combinatory logic
PA, I, 27, 40	Peano arithmetic
PRA, I, 27, 39	primitive recursive arithmetic
PA_1, I, 40	Peano arithmetic based on Σ_1-induction
MF^-, II, 43, 50	minimal framework without N-induction
MF, II, 50	MF with full induction
NMF, II, 54	neutral minimal framework
$\widehat{ID}_1$, II, 66	fixed point theory of elementary inductive definitions
MF_c, II, 67	MF with class N-induction
MF_p, II, 67	MF with property N-induction
PT, II, 77	pure property theory
PW_c (PW^-, PW_p), IV, 104, 105	MF_c (MF^-, MF_p) +approximation axioms
KPU(op), V, 139	admissible set theory above combinatory logic
Σ_∞, VI, 158	Myhill's system with levels of implication
$\Gamma \vdash^n t$, VI, 158	formal deducibility with levels of implication
BL_c, VI, 163	Behmann's logic with class-N-induction
MFS^-, VII, 199	minimal framework with sets
TL (TL^-), VIII, 219, 220	theory of truth with levels (without N-induction)
TLR, VIII, 224	theory of truth with levels and reflection
TLR^-, VIII, 224	TLR without N-induction reflection
TLR^*, 250	TLR plus axioms ONT+BLQ
ATR_0, VIII, 241	arithmetic transfinite recursion

Δ^1_1-CA_0, Σ^1_1-AC_0 Π^1_1-CA_0, Δ^1_2-CA_0, VIII, 241	basic subsystems of 2nd order arithmetic
MFR(p), IX, 278	MF_c plus $RAM(\alpha, p)$ for each $\alpha < \Gamma_0$
STLR, X, 289	sequent calculus for truth with levels
$STLR^{\infty}$, X, 294	infinitary STLR
$STLR_n$, X, 297	sequent calculus for truth up to level $\leq n$ with bounded level quantifiers
$STLR_{\omega}$, X, 297	union over $STLR_n$, $n \in \omega$
IT^{∞}_n, X, 304	infinitary sequent calculus for truth up to level n
RS_n, XI, 314	ramified system for truth of level n
OP^{∞}, XI, 327	OP based on ω-logic
VF^-, XII, 352	basic non-reductive theory for self-referential truth without N-induction
VF_c (VF_p), XII, 352	VF^-+ class (resp. property) N-induction
VF_0, XII, 355	VF^- in the language of pure combinatory logic
ID_1(acc), XII, 356	theory of accessibility inductive definitions
VFH^-, VFH_c, VFH_p, XII, 372	extensions of VF^-
IL, XIII, 380	internal non-reductive T-logic
FSL, XIII, 384	Friedman-Sheard system
IFT, XIII, 388	internal Fitch's logic
FT (FT^-), XIII, 388	Fitch's theory (without N-induction)
LIS, XIII, 396	internal axioms for stable truth
LES, XIII, 400	external axioms for stable truth
LTC_0, XIV, 408	logical theory of constructions (without proposition and truth predicates)
LTC_{ω}, XIV, 411	LTC with propositions and truth of arbitrary finite level
LPCR, XIV, 423	lowest predicate calculus with reflection

II. Axioms, rules and other symbols

Chapter 1

COMB, 15	combinatory logic
PAIR, 15	pairing
NAT, 15	natural numbers
NIND, 15	number-theoretic induction schema
Ext_{op}, 16	extensionality for operations
MS.1-MS.4, 17	Meyer-Scott axioms
CT, 20	Church's thesis
EA, 27	enumeration axiom

$\exists(+)$-NIND, 27	NIND for positive existential formulas
AC_N, 33	axiom of choice restricted to N
$AC_N!$, 33	comprehension for operations on N
Σ_1-IND, 39	NIND for Σ_1-formulas

Chapter 2

T.1-T.5, 49	axioms for reflective truth
RES, 49	restriction axiom
CONS, 49	consistency axiom
COMP, 54	completeness axiom
AP, 56	abstraction principle
EC, 58	elementary comprehension
J, 58	join principle
P-NIND, 66	property N-induction
CL-NIND, 66	class N-induction
CP, 80	construction principle

Chapter 3

GID, 87	generalized induction schema
π, 98	approximation operation
ΠAX, 100	π-axioms (or approximation axioms)

Chapter 5

AC_V(op), 128	choice axiom for operations on V
Ext-AC, 129	extensional choice axiom
Ext-DC, 129	extensional dependent choice axiom
EAC, 130	elementary choice schema
EDC, 130	elementary dependent choice schema

Chapter 6

Hyp/Tnd, Lift, D, N, Eq, K, S } 159	axioms for Myhill's system
$I\wedge$, $E\wedge$, $I\neg\wedge$, $E\neg\wedge$, $I\vee$, $E\vee$, $I\neg\vee$, $E\neg\vee$, Red, $\forall\vee$; $I\forall$, $E\forall$, $I\neg\forall$, $E\neg\forall$ } 159-160	logical rules for Myhill's system
$I_n\supset$, $E_n\supset$, 160	rules for level n implication
$I_n\neg\supset$, $E_n\neg\supset$, 160	rules for level n negated implication
E.1-E.7, 164	axioms for Behmann's logic

Chapter 7

Set.1-Set.3, 199	set axioms
ℵ, 199	"anti-cantorian axiom"
$\mathrm{GID}^{\wedge}$, 207	generalized coinduction principle

Chapter 8

LIND, 219	local N-induction
PROJ, 219	projectibility axiom
REF, 224	reflection principle
LIM, 226	limit axiom for universes
WP, 229	Weyl's principle
$\mathfrak{F}$-CA, 239	$\mathfrak{F}$-comprehension schema for analysis
$\mathfrak{F}$-DC, 240	$\mathfrak{F}$-dependent choice schema
BI, 241	bar induction
RPC, 244	reducibility principle for classes
ONT, 250	ontological axiom
BLQ, 250	bounded level quantifier axiom

Chapter 9

$RAM(p,\alpha)$, 278	existence axiom of bounded ramified hierarchy
$I^U(\beta)$, 278	transfinite induction for classes of U up to β

Chapter 10

$TI(lev)$, 290, 293	Level induction
$(\wedge)$, $(\vee)$, 290	logical rules
$(\forall x)$, $(\exists x)$, $(\forall i)$, $(\exists j)$, 291	quantifier rules
(Cut), 291	cut rule
$(\forall\omega)$, $(\exists\omega)$, 294	infinitary level quantifier rules
$(\forall)^b$, $(\exists)^b$, 297	bounded level quantifier rules
(N), $(\neg N)$, 305	rules for N

Chapter 11

LOG, 314	logical axioms
$OPER$, 314	operational axioms
$PERS_{ij}$, 314	persistence axioms
$CONS_i$, 314	level i consistency
FIX_i, 314	fixed point axioms for level i
INI_n, 315	initial axioms for level n
$(T^{\alpha+1})$ $(\neg T^{\alpha+1})$, 315	ramified successor rules for T of level n

(T-LIM$^{\alpha}$), (F-LIM$^{\alpha}$), 315	ramified limit rules for T of level n

Chapter 12

T-elem, T-out, T-univ, T-log, T-imp, T-rep, T-cons, 352	T-schemata
$WF(\prec)$.1, $WF(\prec)$.2, 356	axioms for the largest $\prec$-wellfounded part
T-Herb, 372	Herbrand's T-schema
I-CL-NIND, 372	internal class-N-induction
I-NIND, 372	internal N-induction
T^{+}-elem, T^{+}-elim, T^{+}-univ, T^{+}-log, 372	strengthened T-axioms and rules
T^{+}-imp, T^{+}-rep, T^{+}-cons, $T(T\rightarrow)$, T_{ax}-imp, T-univ$_{ax}$, 372	strengthened T-axioms and rules

Chapter 13

T-intro, T-elim, $\neg T$-intro, $\neg T$-elim, 388	T-rules
T-neg$T$, 388	$T\neg TA \rightarrow TA$
T^{+}-neg$T$, 389	$T(T\neg TA \rightarrow TA)$
T-Rcomp, T-S4comp, T-S5comp, 389, 396	Turner's schemata

Chapter 14

SE.1-2, 409	special axioms
LE.1-LE.3, 409	lazy evaluation axioms
NEV.1-NEV.2, 409	number evaluation axioms
PT.1-PT.4$_{ij}$, 411	axioms for propositions and truth

III. Other Symbols

Introduction

$E[x := t]$, FV(E), 9
$\emptyset$, ω, 9
$\mathcal{P}(X)$, c_Z, 10
$S \vdash A$, 10
$\mathcal{M} \models A$, 10

Chapter 1

N,14
K, S, 14
$\langle - , - \rangle$, 14
$(-)_1, (-)_2$, 14
$t+1$, 14
$\langle \ldots \rangle$, $(\ldots)_k$, 15
$\lambda x t$, 15
$\overline{n}$, 15
FP, 16
R_N, 19
$\forall n, \exists n$, 19
$\forall n<m$, $\exists n<m$, 20
CR, 23
RED, 23
$\geq_n$, 23
CTM, TM, 26
e_n, 28
$P\omega$, 28
RE, 28
$FUN(a)$, 28
$GRAPH(F)$, 28
$[\![t]\!]_\rho$, 30
D_M, 34
D_∞, 34

Chapter 2

$[A]$, 45
ID, NEG, TR, 45
ALL, AND, OR, 45
$\Gamma(S)$, 47
$FIX(\Gamma, \mathcal{M})$, 48
$\mathbb{O}(\mathcal{M})$, 48
$\mathbb{O}(\mathcal{M}, \alpha)$, 48
$Prop(x)$, 51
$A\hat{}$, 54
$a \Rightarrow b$, 54
$a \Leftrightarrow b$, 54
$\{x : A\}$, 55
η, $\overline{\eta}$, 55
$Cl(x)$, 55
CL, 55
$\forall x \eta y$, $\exists x \eta y$, 58
$\Sigma(a, f)$, 58
$\Pi(a, f)$, 58
$\cap$, $-$, 59
$\mathbb{N}$, 59
$a \equiv b$, 60
$a =_e b$, 60
$[a \rightarrow b]$, 61
V, 61
$a \oplus b$, 61
$IxyA(x, y)$, $I(A)$, 63
I_A, 66, 67
$Sat(x, y)$, 69

Chapter 3

$Clos_A(-)$, 87
$ENV(\mathcal{M}, S)$, 88
$HYP(\mathcal{M}, S)$, 89
$IND(\mathcal{M}, S)$, 89
$SEC(\mathcal{M}, S)$, 90
$|a|$, 92
$\pi a x$, 98

Chapter 4

$x \leq_z y$, $x <_z y$, 105
ξa, 105
$RD(x, y)$, 107

$CSP(x,y)$, 107
$SEP(x,y)$, 108
$ER(y,z)$, 109
b is extensional$_2$, 110
f is extensional$_1$, 110
$I(f)$, 112
$E(b)$, 113
$V(e)$, 114
$Cl\text{-}\mathfrak{T}$, 114
$ECL\text{–}OPEN$, 116
RS, 116
EFF, 118
ECL, 118

Chapter 5

$\prec_w$, 126
$\delta(a, \prec_w)$, 127
$Ext_2(f)$, 128
$2\text{-}Ext(f)$, 129
$AD(U)$, 132
$U\text{-}AT$, 132
$U\text{-}SET$, 132
$\equiv_U, \not\equiv_U$, 133
$x \in_U y$, $x \bar{\in}_U y$, 134
$\forall x \in y$, $\exists x \in y$, 138
$RC(f)$, 147

Chapter 6

$\Gamma \vdash^R a$, 153
$Adr(R)$, 153
$_R\supset$, 154
$_n\supset$, 157
$A \equiv B$, 162
$Funcl(f)$, 166
$\Rightarrow$, 166
$RT(r)$, 167
LR, 168
$\rightarrow_\alpha$, 169
$\Rightarrow_\alpha$, 169

Chapter 7

$COMP(\mathcal{M})$, 179
$CONS(\mathcal{M})$, 179
$FIX(\mathcal{M})$, 179
$FIX_{cs}(\mathcal{M})$, 180
$FIX_{cp}(\mathcal{M})$, 180
$\mathfrak{IT}$, 180
$\mathfrak{ST}$, 180
S^d, 181
$UP(S)$, $DOWN(S)$, 181
$\mathbb{I}(\mathcal{M})$, 181
$\sqcup \mathfrak{C}$, $\sqcap \mathfrak{C}$, 182
$INT(\mathcal{M})$, 186
$\mathbb{P}(\mathcal{M})$, 186
$MAX(\mathcal{M})$, 187
$\mathrm{Gr}(\mathcal{M})$, 204
$\mathrm{Paradox}(\mathcal{M})$, 204
$\mathrm{Intr}(\mathcal{M})$, 204

Chapter 8

$\mathfrak{L}_V$, 218
$i_0, i_1, \ldots$, 218
LT, 218
$\preceq$, $=_l$, 218
T_it , F_it, 218
$t\eta_i s$, $t\bar{\eta}_i s$, 218
$Cl_i(t)$, 218
Δ^+, 218
R, $R(i)$, 222
$x \Leftrightarrow_i y$, 222
$Univ(y)$, 226
$TI(\prec_w, b)$, 227
$TR(y, A, \prec_w, z)$, 227
L_α, 230
$\Sigma_1(L_\alpha)$, $\Delta_1(L_\alpha)$, 231
β^+, 232
τ_α, 232
ι , 232
IN, 232
$\mathfrak{V}(\delta)$, 235
$\mathfrak{C}_\iota$, 237

$\mathcal{L}_2$, 239

Δ_0, Σ^1_0, Π^1_0, 239

Σ^1_1, Π^1_1, Π^1_2, Σ^1_2, 239

T^P_e, 245

$W(X)$, 245

$\mathcal{W}(P)$, 245

Chapter 9

$f: CL \to CL$, 259

$\mathcal{R}_\alpha$, 262

ω_1^{ck}, 262

$GD(E)$, 264

$\mathcal{R}^p_\alpha(z)$, 267

$f: \Omega \to \Omega$, 269

E_X, 269

$fix(f)$, 269

$E_{fix(f)}$, f', 269

$\phi\beta\alpha$, 270

ϵ_0, 270

Γ_0, Γ_α, 270

ω^γ, 270

$f(t_1, s_1) * \ldots * f(t_n, s_n)$, 272

CN, 272

$\dot{<}$, 272

$t[n]$, 276

$Good(\mathcal{R}^p_\alpha(z))$, 278

$I^U(\beta)$, 278

Chapter 10

$\mathcal{L}^+_V$, 288

Δ^+, Σ, Π, Δ_0, 287

$Lc(A)$, 288

$rk(A)$, 288

T_i-$Clause(t)$, F_i-$Clause(t)$, 289

$NF(A)$, 293

$\alpha\#\beta$, 295 (see ch.11)

$\mathrm{STLR}^\infty \vdash^\alpha_k \Gamma$, 294

$\mathcal{L}^*_n$, 297

$A[m,n]$, $\Gamma[m,n]$, 298

$\mathcal{D} \vdash^\alpha_k \Gamma$, 299

$LevPar(\Gamma)$, 299

$|\Gamma|$, 299

$\mathcal{R}(\omega_k)$, 302

POS_n and NEG_n, 303

$K_n(A)$, 303

e_n, 306

Chapter 11

$\mathcal{L}_{n,r}$, 312

$T^\alpha_n(t)$, $F^\alpha_n(t)$, 313

$t\eta^\alpha s$, $t\overline{\eta}^\alpha s$, 313

$Lev(B)$, 313

$St_n(A)$, 313

$R_n(\mathrm{A})$, 313

A_α, 313

$\mathrm{RS}_n \vdash^\alpha_\rho \Gamma$, 316

$\alpha\#\beta$, 318

$A[\beta,\gamma]$, 324

$\mathrm{OP}^\infty \vdash^\alpha_\rho \Gamma$, 327

$TI(\alpha)$, 335

$TI(<\gamma)$, 335

$TI_{op}(<\alpha)$, 335

$\ulcorner E \urcorner$, 336

$Dim_k(d, \ulcorner A \urcorner, \mathcal{T})$, 336

$True_n(\ulcorner A \urcorner)$, 336

$[e](x)$, 338

$RF(f)$, $END(f)$, 339

$LAT(f)$, $AF(f)$, 339

$DEP(f)$, 339

$CODE$, 340

OT^*, 341
$Length(s)$, 341
$DER(f)$, 342
$LC(DER(f))$, 342
$f \vdash^{\alpha}_{\rho} \Gamma$, 344
$CF(f)$, 347

Chapter 12

$IMPLY$, 352
$WF(\prec)$, 354
$X \Vdash A$, 357
$SENT(\mathcal{M})$, 357
$FIX_0(\mathcal{M})$, 357
$At_{\infty}(a)$, $Eat_{\infty}(a)$, $Sent_{\infty}(a)$, 359
ON, 360
$Ord(x)$, 360
AD, 360
$TI(ON, B)$, 360
$\vdash^{\alpha}_{\rho} \Gamma \Rightarrow \Delta$, 361

Chapter 13

$J(X)$, 383
$Th_{\infty}(\mathcal{M})$, 383
$ko(A)$, 383
$Th(\mathcal{M})$, 383
$Diag(\mathcal{M})$, 386
$\mathcal{M}$-NOR, 386
$X \Vdash_2 A$, 387
$\Phi_2(X)$, 387
$N_{\infty}(\mathcal{M})$, 387
$liminf$, 390
$In(\Delta, X)$, $In(X)$, 390
$Out(\Delta, X)$, $Out(X)$, 390
$Stab(\Delta, X)$, 390
$Unstab(\Delta, X)$, 390
$Conf(X)$, 393
$Cycle(X)$, 393
$Init(X)$, 393

Chapter 14

OB, 403
$BOOL$, 403
$(\alpha \rightarrow \beta)$, $(\alpha_1 \ldots \alpha_n) \rightarrow \alpha$, 403
$\perp$, 403
$\mathfrak{L}_{TC}$, 403
$(xy)f(x, y)$, 404
LEV, 404
NEV, 404
$P_i(x)$, T, 404
Pa, $Decidenat$, 404
$Spread$, $Decide$, 404
$inl(x)$, $inr(x)$, Ind, 404
$E \equiv E' : \sigma$, 405
$\equiv_{\sigma}$, 406
$[E]_{\sigma}$, 406
M_{σ}, 406
$N(x)$, 409
$Nat(x)$, 410
$EqNat(x, y)$, 410

www.ingramcontent.com/pod-product-compliance
Ingram Content Group UK Ltd.
Pitfield, Milton Keynes, MK11 3LW, UK
UKHW020226290926
15470UKWH00001B/405

9 780444 823069